Barycentric Calculus
in Euclidean and Hyperbolic Geometry
A Comparative Introduction

SECOND EDITION

Barycentric Calculus
in Euclidean and Hyperbolic Geometry
A Comparative Introduction

SECOND EDITION

Abraham Albert Ungar
North Dakota State University, USA

World Scientific

NEW JERSEY • LONDON • SINGAPORE • BEIJING • SHANGHAI • HONG KONG • TAIPEI • CHENNAI • TOKYO

Published by

World Scientific Publishing Co. Pte. Ltd.
5 Toh Tuck Link, Singapore 596224
USA office: 27 Warren Street, Suite 401-402, Hackensack, NJ 07601
UK office: 57 Shelton Street, Covent Garden, London WC2H 9HE

Library of Congress Control Number: 2025040241

British Library Cataloguing-in-Publication Data
A catalogue record for this book is available from the British Library.

BARYCENTRIC CALCULUS IN EUCLIDEAN AND HYPERBOLIC GEOMETRY
A Comparative Introduction (Second Edition)

ISBN 978-981-98-2129-7 (hardcover)
ISBN 978-981-98-2130-3 (ebook for institutions)
ISBN 978-981-98-2131-0 (ebook for individuals)

For any available supplementary material, please visit
https://www.worldscientific.com/worldscibooks/10.1142/14528#t=suppl

Desk Editor: Kwong Lai Fun

Preface to the First Edition

Historically, Euclidean geometry became analytic with the appearance of Cartesian coordinates that followed the publication of René Descartes' (1596–1650) masterpiece in 1637, allowing algebra to be applied to Euclidean geometry. About 200 years later hyperbolic geometry was discovered following the publications of Nikolai Ivanovich Lobachevsky (1792–1856) in 1830 and János Bolyai (1802–1860) in 1832, and about 370 years later the hyperbolic geometry of Bolyai and Lobachevsky became analytic following the adoption of Cartesian coordinates for use in hyperbolic geometry in [1–4], allowing novel nonassociative algebra to be applied to hyperbolic geometry.

The history of Vector Algebra dates back to the end of the eighteenth century, considering complex numbers as the origin of vector algebra as we know today. Indeed, complex numbers are ordered pairs of real numbers with addition given by the parallelogram addition law. In the beginning of the nineteenth century there were attempts to extend this addition law into three dimensions leading Hamilton to the discovery of the quaternions in 1843. Quaternions, in turn, led to the notion of scalar multiplication in modern vector algebra. The key role in the creation of modern vector analysis as we know today, played by Willar Gibbs (1839–1903) and Oliver Heaviside (1850–1952), along with the contribution of Möbius' barycentric coordinates to vector analysis, is described in [5].

The success of the use of vector algebra along with Cartesian coordinates in Euclidean geometry led Varičak to admit in 1924 [6], for his chagrin, that the adaption of vector algebra for use in hyperbolic space was just not possible, as the renowned historian Scott Walter notes in [7, p. 121]. Fortunately however, along with the adaption of Cartesian coordinates for use in hyperbolic geometry, trigonometry and vector algebra have been adapted for use in hyperbolic geometry as well in [1–3], leading to the adaption in this book of Möbius barycentric coordinates for use in hyperbolic geometry. As a result, powerful tools that are commonly available in the study of Euclidean geometry became available in the study of hyperbolic geometry as well, enabling one to explore hyperbolic geometry in novel ways.

The notion of Euclidean barycentric coordinates dates back to Möbius, 1827, when he published his book *Der Barycentrische Calcul* (*The Barycentric Calculus*). The word *barycentric* is derived from the Greek word *barys* (heavy), and refers to center of gravity. Barycentric calculus is a method of treating geometry by considering a point as the center of gravity of certain other points to which weights are ascribed. Hence, in particular, barycentric calculus provides excellent insight into triangle and tetrahedron centers. This unique book provides a comparative introduction to the fascinating and beautiful subject of triangle and tetrahedron centers in hyperbolic geometry along with analogies they share with familiar triangle and tetrahedron centers in Euclidean geometry. As such, the book uncovers magnificent unifying notions that Euclidean and hyperbolic triangle and tetrahedron centers share.

The hunt for Euclidean triangle centers is an old tradition in Euclidean geometry, resulting in a repertoire of more than three thousands triangle centers that are determined by their barycentric coordinate representations with respect to the vertices of their reference triangles. Several triangle and tetrahedron centers are presented in the book as an illustration of the use of Euclidean barycentric calculus in the determination of Euclidean triangle centers, and in order to set the stage for analogous determination of triangle and tetrahedron centers in the hyperbolic geometry of Bolyai and Lobachevsky.

The adaption of Cartesian coordinates, barycentric coordinates, trigonometry and vector algebra for use in various models of hyperbolic geometry naturally leads to the birth of *comparative analytic geometry* in this book, in which triangles and tetrahedra in three models of geometry are studied comparatively along with their comparative advantages, comparative features and comparative patterns. Indeed, the term "comparative analytic geometry" affirms the idea that the three models of geometry that are studied in this book are to be compared. These three models of analytic geometry are:

(1) The standard Cartesian model of n-dimensional Euclidean geometry. It is regulated by the associative-commutative algebra of vector spaces, and it possesses the comparative advantage of being relatively simple and familiar.
(2) The Cartesian-Beltrami-Klein model of n-dimensional hyperbolic geometry. It is regulated by the gyroassociative-gyrocommutative algebra of Einstein gyrovector spaces, and it possesses the comparative advantage that its hyperbolic geodetic lines, called gyrolines, coincide with Euclidean line segments. As a result, points of concurrency of gyrolines in this model of hyperbolic geometry can be determined by familiar methods of linear algebra.
(3) The Cartesian-Poincaré model of n-dimensional hyperbolic geometry. It is regulated by the gyroassociative-gyrocommutative algebra of Möbius gyrovector spaces, and it possesses the comparative advantage of being conformal so that, in particular, its hyperbolic circles, called gyrocircles, coincide with Euclidean circles (noting, however, that the center and gyrocenter of a given circle/gyrocircle need not coincide).

The idea of comparative study of the three models of geometry is revealed with particular brilliance in comparative features, one of which emerges from the result that barycentric coordinates that are expressed trigonometrically in the three models are model invariant.

Following the adaption of barycentric coordinates for use in hyperbolic geometry, this book heralds the birth of comparative analytic geometry, and provides the starting-point for the hunt for novel centers of hyperbolic triangles and hyperbolic tetrahedra.

Abraham A. Ungar
2010

Preface to the Second Edition

In the fifteen years since the appearance of the first edition, barycentric calculus in both Euclidean and hyperbolic geometry is ever fresh. There are always important extensions, new applications and examples. After the first appearance of this book, remarkable patterns and analogies with classical results have emerged.

A vivid example is provided by the chain of equations (6.22), p. 280, the extreme right-hand side of which is the determinant of an elegant 4×4 matrix called the 4×4 *gamma matrix*. Unfortunately, this elegant determinant representation has gone unnoticed at the time the first edition was published, along with the pattern it shares with the 3×3 determinant in (2.195), p. 117, and the 2×2 determinant in (7.11), p. 313.

A major augmentation appears in the new three chapters 8, 9 and 10.

Chapter 8 presents novel results in hyperbolic geometry associated with hyperbolic barycentric coordinates, called gyrobarycentric coordinates, along with their descent to well known results in Euclidean geometry. Special attention is paid to the gyrodistance between gyropoints which are given by their gyrobarycentric coordinates. These descend to distance between points which are given by their barycentric coordinates.

Chapter 9 is devoted to the use of the trigonometry – gyrotrigonometry duality in the study of the well known Ptolemy's Theorem in Euclidean geometry and its novel counterpart in hyperbolic geometry.

Chapter 10 is motivated by the study in Chapter 9. Employing the trigonometry – gyrotrigonometry duality, it presents a novel study of the so called *cyclic antipodal segments* in both Euclidean and hyperbolic geometry. Thus, for instance, it reveals novel hyperbolic Pythagorean identities.

Abraham A. Ungar
2025

Contents

Chapter 1

Euclidean Barycentric Coordinates and the Classic Triangle Centers

In order to set the stage for the comparative introduction of barycentric calculus, we introduce in this Chapter Euclidean barycentric coordinates, employ them for the determination of several triangle centers, and exemplify their use for tetrahedron centers.

Unlike parallelograms and circles, triangles have many centers, four of which have already been known to the ancient Greeks. These four classic centers of the triangle are: the centroid, G, the orthocenter, H, the incenter, I, and the circumcenter O. Three of these, G, H, and O, are collinear, lying on the so called *Euler line*.

(1) The centroid, G, of a triangle is the point of concurrency of the triangle medians. The triangle centroid is also known as the triangle barycenter.
(2) The orthocenter, H, of a triangle is the point of concurrency of the triangle altitudes.
(3) The incenter, I, of a triangle is the point of concurrency of the triangle angle bisectors. Equivalently, it is the point on the interior of the triangle that is equidistant from the triangle three sides.
(4) The circumcenter, O, of a triangle is the point in the triangle plane equidistant from the three triangle vertices.

There are many other triangle centers. In fact, an on-line Encyclopedia of Triangle Centers that contains more that 3000 triangle centers is maintained by Clark Kimberling [8, 9].

1.1 Points, Lines, Distance and Isometries

In the Cartesian model $\mathbb{R}^n$ of the n-dimensional Euclidean geometry, where n is any positive integer, we introduce a Cartesian coordinate system relative to which points of $\mathbb{R}^n$ are given by n-tuples, like $X = (x_1, x_2, \ldots, x_n)$ or $Y = (y_1, y_2, \ldots, y_n)$, etc., of real numbers. The point $\mathbf{0} = (0, 0, \ldots) \in \mathbb{R}^n$ is called the *origin* of $\mathbb{R}^n$. The Cartesian model $\mathbb{R}^n$ of the n-dimensional Euclidean geometry is a real inner product space [10] with addition, subtraction, scalar multiplication and inner product given, respectively, by the equations

$$\begin{aligned}(x_1, x_2, \ldots, x_n) + (y_1, y_2, \ldots, y_n) &= (x_1 + y_1, x_2 + y_2, \ldots, x_n + y_n)\\ (x_1, x_2, \ldots, x_n) - (y_1, y_2, \ldots, y_n) &= (x_1 - y_1, x_2 - y_2, \ldots, x_n - y_n)\\ r(x_1, x_2, \ldots, x_n) &= (rx_1, rx_2, \ldots, rx_n)\\ (x_1, x_2, \ldots, x_n)\cdot(y_1, y_2, \ldots, y_n) &= x_1y_1 + x_2y_2 + \ldots + x_ny_n\end{aligned} \tag{1.1}$$

for any real number $r \in \mathbb{R}$ and any points $X, Y \in \mathbb{R}^n$. Unless it is otherwise specifically stated, we shall always adopt the convention that $n \geq 2$. In the study of spheres and tetrahedra it is assumed that $n \geq 3$.

In our Cartesian model $\mathbb{R}^n$ of Euclidean geometry, it is convenient to define a line by the set of its points. Let $A, B \in \mathbb{R}^n$ be any two distinct points. The unique line L_{AB} that passes through these points is the set of all points

$$L_{AB} = A + (-A + B)t \tag{1.2}$$

for all $t \in \mathbb{R}$, that is, for all $-\infty < t < \infty$. Equation (1.2) is said to be the line representation in terms of points A and B. Obviously, the same line can be represented by any two distinct points that lie on the line.

The *norm* $\|X\|$ of $X \in \mathbb{R}^n$ is given by

$$\|X\|^2 = X\cdot X \tag{1.3}$$

satisfying the Cauchy – Schwartz inequality

$$|X\cdot Y| \leq \|X\|\|Y\| \tag{1.4}$$

and the triangle inequality

$$\|X + Y\| \leq \|X\| + \|Y\| \tag{1.5}$$

for all $X, Y \in \mathbb{R}^n$.

The distance $d(X, Y)$ between points $X, Y \in \mathbb{R}^n$ is given by the distance function

$$d(X, Y) = \| - X + Y\| \tag{1.6}$$

that obeys the triangle inequality

$$\| -X+Y\| + \| -Y+Z\| \ge \| -X+Z\| \tag{1.7}$$

or, equivalently,

$$d(X,Y) + d(Y,Z) \ge d(X,Z) \tag{1.8}$$

for all $X, Y, Z \in \mathbb{R}^n$.

A map $f : \mathbb{R}^n \to \mathbb{R}^n$ is *isometric*, or an *isometry*, if it preserves distance, that is, if

$$d(f(X), f(Y)) = d(X,Y) \tag{1.9}$$

for all $X, Y \in \mathbb{R}^n$.

The set of all isometries of $\mathbb{R}^n$ forms a group that contains, as subgroups, the set of all translations of $\mathbb{R}^n$ and the set of all rotations of $\mathbb{R}^n$ about its origin. The group of all translations of $\mathbb{R}^n$ and all rotations of $\mathbb{R}^n$ about its origin, known as the Euclidean group of motions, plays an important role in Euclidean geometry. The formal definition of groups, therefore, follows.

Definition 1.1. (Groups). *A group is a pair* $(G, +)$ *of a nonempty set and a binary operation in the set, whose binary operation satisfies the following axioms. In* G *there is at least one element, 0, called a left identity, satisfying*

(G1) $\qquad 0+a=a$

for all $a \in G$. *There is an element* $0 \in G$ *satisfying Axiom* $(G1)$ *such that for each* $a \in G$ *there is an element* $-a \in G$, *called a left inverse of* a, *satisfying*

(G2) $\qquad -a+a=0$.

Moreover, the binary operation obeys the associative law

(G3) $\qquad (a+b)+c = a+(b+c)$

for all $a, b, c \in G$.

Definition 1.2. (Commutative Groups). *A group* $(G, +)$ *is commutative if its binary operation obeys the commutative law*

(G6) $\qquad a+b=b+a$

for all $a, b \in G$.

A natural extension of (commutative) groups into (gyrocommutative) gyrogroups [11], which is sensitive to the needs of exploring hyperbolic geometry, will be presented in Defs. 2.2 – 2.3, p. 71.

A translation $T_X A$ of a point A by a point X in $\mathbb{R}^n$, is given by

$$T_X A = X + A \tag{1.10}$$

for all $X, A \in \mathbb{R}^n$. Translation composition is given by point addition. Indeed,

$$T_X T_Y A = X + (Y + A) = (X + Y) + A = T_{X+Y} A \tag{1.11}$$

for all $X, Y, A \in \mathbb{R}^n$, thus obtaining the translation composition law

$$T_X T_Y = T_{X+Y} \tag{1.12}$$

for translations of $\mathbb{R}^n$. The set of all translations of $\mathbb{R}^n$, accordingly, forms a commutative group under translation composition.

Let SO(n) be the *special orthogonal group* of order n, that is, the group of all $n \times n$ orthogonal matrices with determinant 1. A rotation R of a point $A \in \mathbb{R}^n$, denoted RA, is given by the matrix product RA^t of a matrix $R \in$ SO(n) and the transpose A^t of $A \in \mathbb{R}^n$. A rotation of $\mathbb{R}^n$ is a linear map of $\mathbb{R}^n$, so that it leaves the origin of $\mathbb{R}^n$ invariant. Rotation composition is given by matrix multiplication, so that the set of all rotations of $\mathbb{R}^n$ about its origin forms a noncommutative group under rotation composition.

Translations of $\mathbb{R}^n$ and rotations of $\mathbb{R}^n$ about its origin are isometries. The set of all translations of $\mathbb{R}^n$ and all rotations of $\mathbb{R}^n$ about its origin forms a group under transformation composition, known as the Euclidean group of motions. In group theory, this group of motions turns out to be the so called *semidirect product* of the group of translations and the group of rotations.

Following Klein's 1872 *Erlangen Program* [12] [13, p. 253], the geometric objects of a geometry are the invariants of the group of motions of the geometry so that, conversely, objects that are invariant under the group of motions of a geometry possess geometric significance. Accordingly, for instance, the distance between two points of $\mathbb{R}^n$ is geometrically significant in Euclidean geometry since it is invariant under the group of motions, translations and rotations, of the Euclidean geometry of $\mathbb{R}^n$.

1.2 Vectors, Angles and Triangles

Definition 1.3. (Equivalence Relations and Classes). *A relation on a nonempty set S is a subset R of $S \times S$, $R \subset S \times S$, written as $a \sim b$ if $(a, b) \in R$. A relation $\sim$ on a set S is*

(1) *Reflexive if $a \sim a$ for all $a \in S$.*

(2) *Symmetric if* $a \sim b$ *implies* $b \sim a$ *for all* $a, b \in S$.
(3) *Transitive if* $a \sim b$ *and* $b \sim c$ *imply* $a \sim c$ *for all* $a, b, c \in S$.

A relation is an equivalence relation if it is reflexive, symmetric and transitive.

An equivalence relation $\sim$ *on a set* S *gives rise to equivalence classes. The equivalence class of* $a \in S$ *is the subset* $\{x \in S : x \sim a\}$ *of* S *of all the elements* $x \in S$ *that are related to* a *by the relation* $\sim$.

Two equivalence classes in a set S with an equivalence relation $\sim$ are either equal or disjoint, and the union of all the equivalence classes in S equals S. Accordingly, we say that the equivalence classes of a set S with an equivalence relation form a *partition* of S.

Points of $\mathbb{R}^n$, denoted by capital italic letters A, B, P, Q, etc., give rise to vectors in $\mathbb{R}^n$, denoted by bold roman lowercase letters $\mathbf{u}, \mathbf{v}$, etc. Any two ordered points $P, Q \in \mathbb{R}^n$ give rise to a unique rooted vector $\mathbf{v} \in \mathbb{R}^n$, rooted at the point P. It has a tail at the point P and a head at the point Q, and it has the value $-P + Q$,

$$\mathbf{v} = -P + Q \tag{1.13}$$

The length of the rooted vector $\mathbf{v} = -P + Q$ is the distance between its tail, P, and its head, Q, given by the equation

$$\|\mathbf{v}\| = \| - P + Q\| \tag{1.14}$$

Two rooted vectors $-P+Q$ and $-R+S$ are equivalent if they have the same value, $-P + Q = -R + S$, that is,

$$-P+Q \quad \sim \quad -R+S \qquad \text{if and only if} \qquad -P+Q = -R+S \tag{1.15}$$

The relation $\sim$ in (1.15) between rooted vectors is reflexive, symmetric and transitive. Hence, it is an equivalence relation that gives rise to equivalence classes of rooted vectors. To liberate rooted vectors from their roots we define a *vector* to be an equivalence class of rooted vectors. The vector $-P + Q$ is thus a representative of all rooted vectors with value $-P + Q$.

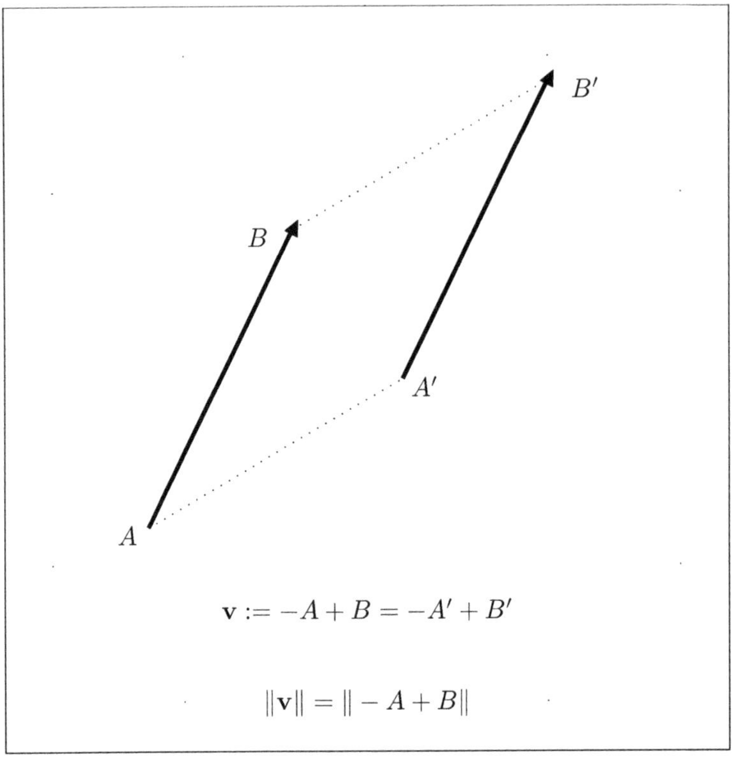

Fig. 1.1 The vectors $-A + B$ and $-A' + B'$ have equal values, that is, $-A + B = -A' + B'$, in a Euclidean space $\mathbb{R}^n$. As such, these two vectors are equivalent and, hence, indistinguishable in their vector space and its underlying Euclidean geometry. Two equivalent nonzero vectors in Euclidean geometry are parallel, and possess equal lengths, as shown here for $n = 2$. Vectors in hyperbolic geometry are called gyrovectors. For the hyperbolic geometric counterparts, see Fig. 2.2, p. 98, and Fig. 2.13, p. 138.

As an example, the two distinct rooted vectors $-A+B$ and $-A'+B'$ in Fig. 1.1 possess the same value so that, as vectors, they are indistinguishable.

Vectors add according to the parallelogram addition law. Hence, vectors in Euclidean geometry are equivalence classes of ordered pairs of points that add according to the parallelogram law.

A point $P \in \mathbb{R}^n$ is identified with the vector $-O + P$, O being the arbitrarily selected origin of the space $\mathbb{R}^n$. Hence, the algebra of vectors can be applied to points as well.

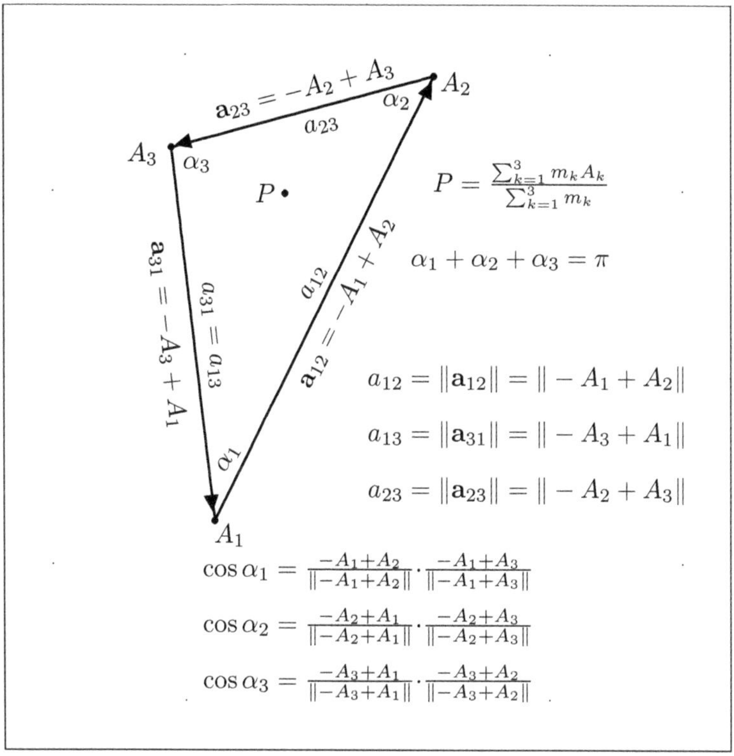

Fig. 1.2 A triangle $A_1A_2A_3$ in $\mathbb{R}^n$ is shown here for $n = 2$, along with its associated standard index notation. The triangle vertices, A_1, A_2 and A_3, are any non-collinear points of $\mathbb{R}^n$. Its sides are presented graphically as line segments that join the vertices. They form vectors, $\mathbf{a}_{ij}$, side-lengths, $a_{ij} = \|\mathbf{a}_{ij}\|$, $1 \le i, j \le 3$, and angles, α_k, $k = 1, 2, 3$. The triangle angle sum is π. The cosine function of the triangle angles is presented. The point P is a generic point in the triangle plane, with barycentric coordinates $(m_1 : m_1 : m_3)$ with respect to the triangle vertices.

Let $-A_1 + A_2$ and $-A_1 + A_3$ be two rooted vectors with a common tail A_1, Fig. 1.2. They include an angle $\alpha_1 = \angle A_2A_1A_3 = \angle A_3A_1A_2$, the measure of which is given by the equation

$$\cos\alpha_1 = \frac{-A_1 + A_2}{\| - A_1 + A_2\|} \cdot \frac{-A_1 + A_3}{\| - A_1 + A_3\|} \tag{1.16}$$

Accordingly, the angle α_1 in Fig. 1.2 has the radian measure

$$\alpha_1 = \cos^{-1} \frac{-A_1 + A_2}{\| - A_1 + A_2\|} \cdot \frac{-A_1 + A_3}{\| - A_1 + A_3\|} \tag{1.17}$$

The angle α_1 is invariant under translations. Indeed,

$$\begin{aligned}\cos\alpha_1' &= \frac{-(X+A_1)+(X+A_2)}{\|-(X+A_1)+(X+A_2)\|}\cdot\frac{-(X+A_1)+(X+A_3)}{\|-(X+A_1)+(X+A_3)\|}\\ &= \frac{-A_1+A_2}{\|-A_1+A_2\|}\cdot\frac{-A_1+A_3}{\|-A_1+A_3\|}\\ &= \cos\alpha_1\end{aligned} \tag{1.18}$$

for all $A_1, A_2, A_3, X \in \mathbb{R}^n$. Similarly, the angle α_1 is invariant under rotations of $\mathbb{R}^n$ about its origin. Indeed,

$$\begin{aligned}\cos\alpha_1'' &= \frac{-RA_1+RA_2}{\|-RA_1+RA_2\|}\cdot\frac{-RA_1+RA_3}{\|-RA_1+RA_3\|}\\ &= \frac{R(-A_1+A_2)}{\|R(-A_1+A_2)\|}\cdot\frac{R(-A_1+A_3)}{\|R(-A_1+A_3)\|}\\ &= \frac{-A_1+A_2}{\|-A_1+A_2\|}\cdot\frac{-A_1+A_3}{\|-A_1+A_3\|}\\ &= \cos\alpha_1\end{aligned} \tag{1.19}$$

for all $A_1, A_2, A_3 \in \mathbb{R}^n$ and $R \in \mathrm{SO(n)}$, since rotations $R \in \mathrm{SO(n)}$ are linear maps that preserve the inner product in $\mathbb{R}^n$.

Being invariant under the motions of $\mathbb{R}^n$, angles are geometric objects of the Euclidean geometry of $\mathbb{R}^n$. Triangle angle sum in Euclidean geometry is π. The standard index notation that we use with a triangle $A_1A_2A_3$ in $\mathbb{R}^n$, $n \geq 2$, is presented in Fig. 1.2 for $n = 2$. In our notation, triangle $A_1A_2A_3$, thus, has (i) three vertices, A_1, A_2 and A_3; (ii) three angles, α_1, α_2 and α_3; and (iii) three sides, which form the three vectors $\mathbf{a}_{12}$, $\mathbf{a}_{23}$ and $\mathbf{a}_{31}$; with respective (iv) three side-lengths a_{12}, a_{23} and a_{31}.

1.3 Euclidean Barycentric Coordinates

A barycenter in astronomy is the point between two objects where they balance each other. It is the center of gravity where two or more celestial bodies orbit each other. In 1827 Möbius published a book whose title, *Der Barycentrische Calcul*, translates as *The Barycentric Calculus*. The word *barycenter* means center of gravity, but the book is entirely geometrical and, hence, called by Jeremy Gray [14], *Möbius's Geometrical Mechanics*. The 1827 Möbius book is best remembered for introducing a new system of coordinates, the *barycentric coordinates.* The historical contribution of Möbius' barycentric coordinates to vector analysis is described in [5, pp. 48–50].

The Möbius idea, for a triangle as an illustrative example, is to attach masses, m_1, m_2, m_3, respectively, to three non-collinear points, A_1, A_2, A_3, in the Euclidean plane $\mathbb{R}^2$, and consider their center of mass, or momentum, P, called *barycenter*, given by the equation

$$P = \frac{m_1A_1 + m_2A_2 + m_3A_3}{m_1 + m_2 + m_3} \tag{1.20}$$

The barycentric coordinates of the point P in (1.20) in the plane of triangle $A_1A_2A_3$ relative to this triangle may be considered as weights, m_1, m_2, m_3, which if placed at vertices A_1, A_2, A_3, cause P to become the balance point for the plane. The point P turns out to be the center of mass when the points of $\mathbb{R}^2$ are viewed as position vectors, and the center of momentum when the points of $\mathbb{R}^2$ are viewed as relative velocity vectors.

Definition 1.4. (Euclidean Pointwise Independence – Hocking and Young [15, pp. 195–200]). *A set S of N points $S = \{A_1, \ldots, A_N\}$ in $\mathbb{R}^n$, $n \geq 2$, is pointwise independent if the $N-1$ vectors $-A_1+A_k$, $k = 2, \ldots, N$, are linearly independent.*

The notion of pointwise independence proves useful in the following definition of Euclidean barycentric coordinates.

Definition 1.5. (Euclidean Barycentric Coordinates). *Let $S=\{A_1, \ldots, A_N\}$ be a pointwise independent set of N points in $\mathbb{R}^n$. Then, the real numbers $m_1, \ldots, m_N$, satisfying*

$$\sum_{k=1}^{N} m_k \neq 0 \tag{1.21}$$

are barycentric coordinates of a point $P \in \mathbb{R}^n$ with respect to the set S if

$$P = \frac{\sum_{k=1}^{N} m_k A_k}{\sum_{k=1}^{N} m_k} \tag{1.22}$$

Equation (1.22) *is said to be a barycentric coordinate representation of P with respect to the set $S = \{A_1, \ldots, A_N\}$.*

Barycentric coordinates are homogeneous in the sense that the barycentric coordinates $(m_1, \ldots, m_N)$ of the point P in (1.22) *are equivalent to the barycentric coordinates $(\lambda m_1, \ldots, \lambda m_N)$ for any real nonzero number $\lambda \in \mathbb{R}$, $\lambda \neq 0$. Since in barycentric coordinates only ratios of coordinates* *are relevant, the barycentric coordinates* $(m_1, \ldots, m_N)$ *are also written as* $(m_1 : \ldots : m_N)$ *so that*

$$(m_1 : m_2 : \ \ldots \ : m_N) = (\lambda m_1 : \lambda m_2 : \ \ldots \ : \lambda m_N) \tag{1.23}$$

for any real $\lambda \neq 0$.

Barycentric coordinates that are normalized by the condition

$$\sum_{k=1}^{N} m_k = 1 \tag{1.24}$$

are called special barycentric coordinates.

The point P *in* (1.22) *is said to be a barycentric combination of the points of the set* S, *possessing the barycentric coordinate representation* (1.22).

The barycentric combination (1.22) *is positive if all the coefficients* m_k, $k = 1, \ldots, N$, *are positive. The set of all positive barycentric combinations of the points of the set* S *is called the convex span of* S.

The constant

$$m_0 = \sum_{k=1}^{N} m_k \tag{1.25}$$

is called the constant of the point P *with respect to the set* S.

The pointwise independence of the set S in Def. 1.5 insures that the barycentric coordinate representation of a point with respect to the set S is unique.

Definition 1.6. (Euclidean Simplex). *The convex span (see Def. 1.5) of the pointwise independent set* $S = \{A_1, \ldots, A_N\}$ *of* $N \geq 2$ *points in* $\mathbb{R}^n$ *is an* $(N-1)$*-dimensional simplex, called an* $(N-1)$*-simplex and denoted* $A_1 \ldots A_N$. *The points of* S *are the vertices of the simplex. The convex span of* $N-1$ *of the points of* S *is a face of the simplex, said to be the face opposite to the remaining vertex. The convex span of each two of the vertices is an edge of the simplex.*

Any two distinct points A, B of $\mathbb{R}^n$ are pointwise independent, and their convex span is the interior of the segment AB, which is a 1-simplex. Similarly, any three non-collinear points A, B, C of $\mathbb{R}^n$, $n \geq 2$, are pointwise independent, and their convex span is the interior of the triangle ABC, which is a 2-simplex, and the convex span of any four pointwise independent points A, B, C, D of $\mathbb{R}^n$, $n \geq 3$, is the interior of the tetrahedron $ABCD$, which is a 3-simplex.

1.4 Analogies with Classical Mechanics

Barycentric coordinate representations of points of the Euclidean space $\mathbb{R}^n$ with respect to the set $S = \{A_1, \ldots, A_N\}$ of vertices of a simplex in $\mathbb{R}^n$ admit a classical mechanical interpretation.

Guided by analogies with classical mechanics, the $(N-1)$-simplex of the N points of the pointwise independent set $S = \{A_1, A_2, \ldots, A_N\}$ along with barycentric coordinates $(m_1 : m_2 : \ldots : m_N)$ may be viewed as an isolated system $S = \{A_k, m_k, k = 1, \ldots, N\}$ of N noninteracting particles, where $m_k \in \mathbb{R}$ is the mass of the kth particle and $A_k \in \mathbb{R}^n$ is the velocity of the kth particle, $k = 1, \ldots, N$, relative to the arbitrarily selected origin $O = \mathbf{0} = (0, \ldots, 0)$ of the Newtonian velocity space $\mathbb{R}^n$. Each point of the Newtonian velocity space $\mathbb{R}^n$ represents a velocity of an inertial frame. In particular, the origin $O = \mathbf{0}$ of $\mathbb{R}^n$ represents the rest frame.

By analogy with classical mechanics, the point P in (1.22) is the velocity of the center of momentum (CM) frame of the particle system S relative to the rest frame. The CM frame of S, in turn, is an inertial reference frame relative to which the momentum, $\sum_{k=1}^{N} m_k A_k$, of the particle system S vanishes.

Finally, the constant m_0 in (1.25) of the point P with respect to the set S in (1.25) is viewed in the context of classical mechanics as the total mass of the particle system S.

Along these remarkable analogies between Euclidean geometry and classical mechanics, there is an important disanalogy. As opposed to classical mechanics, where masses are always positive, in Euclidean geometry the "masses" m_k, $k = 1, \ldots, N$, considered as barycentric coordinates of points, need not be positive.

The analogies with classical mechanics will help us in this book to form a bridge to hyperbolic geometry, where analogies with classical mechanics are replaced by corresponding analogies with relativistic mechanics. Thus, specifically, in our transition from Euclidean to hyperbolic geometry,

(1) the Euclidean space of Newtonian velocities is replaced by the Euclidean ball of Einsteinian velocities, that is, by the ball of all relativistically admissible velocities,
(2) the Newtonian velocity addition law, which is the ordinary vector addition in Euclidean space, is replaced by Einstein velocity addition law in the ball of relativistically admissible velocities, and
(3) the Newtonian mass is replaced by the relativistic mass, which is velocity dependent.

1.5 Barycentric Representations are Covariant

It is easy to see from (1.22) that barycentric coordinates are independent of the choice of the origin of their vector space, that is,

$$W + P = \frac{\sum_{k=1}^{N} m_k(W + A_k)}{\sum_{k=1}^{N} m_k} \tag{1.26}$$

for all $W \in \mathbb{R}^n$. The proof that (1.26) follows from (1.22) is immediate, owing to the result that scalar multiplication in vector spaces is distributive over vector addition.

It follows from (1.26) that the barycentric coordinate representation (1.22) of a point P is *covariant* with respect to translations of $\mathbb{R}^n$ since the point P and the points A_k, $k = 1, \ldots, N$, of its generating set $S = \{A_1, \ldots, A_N\}$ vary in (1.26) together under translations.

Let $R \in \mathrm{SO(n)}$ be an element of the *special orthogonal group* SO(n) of all $n \times n$ orthogonal matrices with determinant 1, which represent rotations of the space $\mathbb{R}^n$ about its origin. Since R is linear, it follows from (1.22) that

$$RP = \frac{\sum_{k=1}^{N} m_k R A_k}{\sum_{k=1}^{N} m_k} \tag{1.27}$$

for all $R \in \mathrm{SO(n)}$.

It follows from (1.27) that the *barycentric coordinate representation*, (1.22), of a point P is covariant with respect to rotations of $\mathbb{R}^n$ since the point P and the points A_k, $k = 1, \ldots, N$, of its generating set S vary in (1.27) together under rotations.

The group of all translations and all rotations of $\mathbb{R}^n$ forms the group of motions of $\mathbb{R}^n$, which is the group of all direct isometries of $\mathbb{R}^n$ (that is, isometries preserving orientation) for the Euclidean distance function (1.6).

The point P in (1.22) is determined by the points A_k, $k = 1, \ldots, N$, of its generating set S. It is said to be covariant since the point P and the points of its generating set S vary together in $\mathbb{R}^n$ under the motions of $\mathbb{R}^n$, as we see from (1.26) and (1.27).

The set of all points in $\mathbb{R}^n$ for which the barycentric coordinates with respect to S are all positive forms an open convex subset of $\mathbb{R}^n$, that is, the open *N-simplex* with the N vertices $A_1, \ldots, A_N$. The N-simplex with vertices $A_1, \ldots, A_N$, is denoted by $A_1 \ldots A_N$ so that, for instance, A_1A_2 is the open segment joining points A_1 with A_2 in $\mathbb{R}^n$, $n \geq 1$, and $A_1A_2A_3$ is the interior of the triangle with vertices A_1, A_2 and A_3 in $\mathbb{R}^n$, $n \geq 2$. If the positive number m_k is viewed as the mass of a massive object with

Newtonian velocity $A_k \in \mathbb{R}^n$, $1 \le k \le N$, the point P in (1.22) turns out to be the center of momentum of the N masses m_k, $1 \le k \le N$. If, furthermore, all the masses are equal, the center of momentum is the *centroid* of the N-simplex.

As an application of the covariance of barycentric coordinate representations in (1.26) and for later reference we present the following lemma:

Lemma 1.7. *Let $A_1A_2A_3$ be a triangle in a Euclidean space $\mathbb{R}^n$, and let*

$$P = \frac{m_1A_1 + m_2A_2 + m_3A_3}{m_1 + m_2 + m_3} \tag{1.28}$$

be the barycentric coordinate representation of a point $P \in \mathbb{R}^n$ with respect to the set $\{A_1, A_2, A_3\}$ of the triangle vertices. Then,

$$\begin{aligned}
\|-A_1 + P\|^2 &= \frac{m_2^2a_{12}^2 + m_3^2a_{13}^2 + m_2m_3(\ \ a_{12}^2 + a_{13}^2 - a_{23}^2)}{(m_1 + m_2 + m_3)^2} \\
\|-A_2 + P\|^2 &= \frac{m_1^2a_{12}^2 + m_3^2a_{23}^2 + m_1m_3(\ \ a_{12}^2 - a_{13}^2 + a_{23}^2)}{(m_1 + m_2 + m_3)^2} \\
\|-A_3 + P\|^2 &= \frac{m_1^2a_{13}^2 + m_2^2a_{23}^2 + m_1m_2(-a_{12}^2 + a_{13}^2 + a_{23}^2)}{(m_1 + m_2 + m_3)^2}
\end{aligned} \tag{1.29}$$

Proof. By the covariance property (1.26) of barycentric coordinate representations and the standard triangle index notation in Fig. 1.2, p. 7, it follows from (1.28) that

$$\begin{aligned}
-A_1 + P &= \frac{m_1(-A_1 + A_1) + m_2(-A_1 + A_2) + m_3(-A_1 + A_3)}{m_1 + m_2 + m_3} \\
&= \frac{m_2\mathbf{a}_{12} + m_3\mathbf{a}_{13}}{m_1 + m_2 + m_3}
\end{aligned} \tag{1.30}$$

so that

$$\|-A_1 + P\|^2 = \frac{m_2^2a_{12}^2 + m_3^2a_{13}^2 + 2m_2m_3a_{12}a_{13}\cos\alpha_1}{(m_1 + m_2 + m_3)^2} \tag{1.31}$$

Applying the law of cosines to triangle $A_1A_2A_3$ and its angle α_1 in Fig. 1.2, p. 7, we have

$$2a_{12}a_{13}\cos\alpha_1 = a_{12}^2 + a_{13}^2 - a_{23}^2 \tag{1.32}$$

Eliminating $\cos\alpha_1$ between (1.31) and (1.32) we obtain the first equation in (1.29). The second and the third equations in (1.29) are obtained from the first by cyclic permutations of the triangle vertices. □

1.6 Vector Barycentric Representation

Two points $P, P' \in \mathbb{R}^n$ define a vector $\mathbf{v} = -P' + P$ in $\mathbb{R}^n$ with a tail P' and a head P. In the following theorem we show that if each of the points P and P' possesses a barycentric representation with respect to a pointwise independent set $S = \{A_1, \ldots, A_N\}$ of N points in $\mathbb{R}^n$, then the vector $\mathbf{v} = -P' + P$ possesses an induced representation with respect to the vectors $\mathbf{a}_{ij} = -A_j + A_i$, $i, j = 1, \ldots, N$, $i < j$, called a *vector barycentric representation.*

Theorem 1.8. (The Vector Barycentric Representation). *Let*

$$P = \frac{\sum_{i=1}^{N} m_i A_i}{\sum_{i=1}^{N} m_i} \tag{1.33}$$

and

$$P' = \frac{\sum_{j=1}^{N} m'_j A_j}{\sum_{j=1}^{N} m'_j} \tag{1.34}$$

be barycentric representations of two points $P, P' \in \mathbb{R}^n$ with respect to a pointwise independent set $S = \{A_1, \ldots, A_N\}$ of N points of $\mathbb{R}^n$.

Then, the vector $\mathbf{v}$ formed by the point difference $\mathbf{v} = -P + P'$ possesses the vector barycentric representation

$$\begin{aligned} \mathbf{v} = -P + P' &= \frac{\sum_{\substack{i,j=1 \\ i<j}}^{N} (m_i m'_j - m'_i m_j)(-A_i + A_j)}{\sum_{k=1}^{N} m_i \sum_{j=1}^{N} m'_j} \\ &= \frac{\sum_{\substack{i,j=1 \\ i<j}}^{N} (m_i m'_j - m'_i m_j)\mathbf{a}_{ij}}{\sum_{k=1}^{N} m_i \sum_{j=1}^{N} m'_j} \end{aligned} \tag{1.35}$$

where $\mathbf{a}_{ij}$ is the vector $\mathbf{a}_{ij} = -A_i + A_j$.

Proof. The proof is given by the following chain of equations, which are numbered for subsequent explanation.

$$
\begin{aligned}
-P+P' &\overset{(1)}{=\!=} \frac{\sum_{j=1}^N m'_j(-P+A_j)}{\sum_{j=1}^N m'_j} = -\frac{\sum_{j=1}^N m'_j(-A_j+P)}{\sum_{j=1}^N m'_j} \\
&\overset{(2)}{=\!=} -\frac{\sum_{j=1}^N m'_j\left(-A_j+\frac{\sum_{i=1}^N m_iA_i}{\sum_{i=1}^N m_i}\right)}{\sum_j^N m'_j} \\
&\overset{(3)}{=\!=} -\frac{\sum_{j=1}^N m'_j\frac{\sum_{i=1}^N m_i(-A_j+A_i)}{\sum_{i=1}^N m_i}}{\sum_j^N m'_j} \\
&\overset{(4)}{=\!=} \frac{\sum_{j=1}^N m'_j\sum_{i=1}^N m_i(-A_i+A_j)}{\sum_{i=1}^N m_i\sum_{j=1}^N m'_j} \\
&\overset{(5)}{=\!=} \frac{\sum_{\substack{i,j=1\\i<j}}^N m_im'_j(-A_i+A_j)+\sum_{\substack{i,j=1\\i>j}}^N m_im'_j(-A_i+A_j)}{\sum_{i=1}^N m_i\sum_{j=1}^N m'_j} \\
&\overset{(6)}{=\!=} \frac{\sum_{\substack{i,j=1\\i<j}}^N m_im'_j(-A_i+A_j)-\sum_{\substack{i,j=1\\i<j}}^N m_jm'_i(-A_i+A_j)}{\sum_{i=1}^N m_i\sum_j^N m'_{j=1}} \\
&\overset{(7)}{=\!=} \frac{\sum_{\substack{i,j=1\\i<j}}^N (m_im'_j-m'_im_j)(-A_i+A_j)}{\sum_{i=1}^N m_i\sum_{j=1}^N m'_j} \\
&\overset{(8)}{=\!=} \frac{\sum_{\substack{i,j=1\\i<j}}^N (m_im'_j-m'_im_j)\mathbf{a}_{ij}}{\sum_{i=1}^N m_i\sum_{j=1}^N m'_j}
\end{aligned}
\tag{1.36}
$$

Derivation of the numbered equalities in (1.36) follows:

(1) Follows from the barycentric representation (1.34) of P' along with the covariance property (1.26) of barycentric representations.
(2) Follows from (1) by a substitution of the barycentric representation (1.33) of P.
(3) Follows from (2) by the covariance property (1.26) of barycentric representations.
(4) Follows from (3) immediately.

(5) Follows from (4) straightforwardly, noting that the contribution of pairs (i, j) vanishes when $i = j$.
(6) Follows from (5) by interchanging the labels i and j of the two summation indexes in the argument of the second Σ on the numerator of (5).
(7) Follows from (6) immediately.
(8) The passage from (7) to (8) is merely a matter of notation that we introduce here for its importance in the book. In this notation, the vector $-A_i+A_j$ with tail A_i and head A_j is denoted by $\mathbf{a}_{ij} = -A_i+A_j$, and its magnitude is denoted by $a_{ij} = \|-A_i + A_j\|$.

□

Example 1.9. (A Vector Barycentric Representation). Let

$$I = \frac{a_{23}A_1 + a_{13}A_2 + a_{12}A_3}{a_{12} + a_{13} + a_{23}} \tag{1.37}$$

and

$$P = \frac{a_{13}A_2 + a_{12}A_3}{a_{12} + a_{13}} \tag{1.38}$$

be barycentric representations of points $I, P \in \mathbb{R}^n$, where $a_{12}, a_{13}, a_{23} > 0$, and $S = \{A_1, A_2, A_3\}$ is a pointwise independent set in $\mathbb{R}^n$, $n \geq 2$.

Then, in the barycentric coordinate notation in Theorem 1.8,

$$I = \frac{m_1A_1 + m_2A_2 + m_3A_3}{m_1 + m_2 + m_3} \tag{1.39}$$

and

$$P = \frac{m'_1A_1 + m'_2A_2 + m'_3A_3}{m'_1 + m'_2 + m'_3} \tag{1.40}$$

where

$$\begin{aligned} m_1 &= a_{23} \\ m_2 &= a_{13} \\ m_3 &= a_{12} \end{aligned} \tag{1.41}$$

and

$$\begin{aligned} m'_1 &= 0 \\ m'_2 &= a_{13} \\ m'_3 &= a_{12} \end{aligned} \tag{1.42}$$

Hence, by Identity (1.35) of Theorem 1.8, we have the vector barycentric representation

$$-I+P = \frac{(m_1m_2' - m_1'm_2)\mathbf{a}_{12} + (m_1m_3' - m_1'm_3)\mathbf{a}_{13} + (m_2m_3' - m_2'm_3)\mathbf{a}_{23}}{(m_1+m_2+m_3)(m_1'+m_2'+m_3')}$$

$$= a_{23}\frac{a_{13}\mathbf{a}_{12} + a_{12}\mathbf{a}_{13}}{(a_{12}+a_{13})(a_{12}+a_{13}+a_{23})} \tag{1.43}$$

1.7 Triangle Centroid

The triangle centroid is located at the intersection of the triangle medians, Fig. 1.3.

Let $A_1A_2A_3$ be a triangle with vertices A_1, A_2 and A_3 in a Euclidean n-space $\mathbb{R}^n$, and let G be the triangle centroid, as shown in Fig. 1.3 for $n = 2$. Then, G is given in terms of its barycentric coordinates $(m_1 : m_2 : m_3)$ with respect to the set $\{A_1, A_2, A_3\}$ by the equation

$$G = \frac{m_1A_1 + m_2A_2 + m_3A_3}{m_1+m_2+m_3} \tag{1.44}$$

where the barycentric coordinates m_1, m_2 and m_3 of P_3 are to be determined in (1.50) below.

The midpoint of side A_1A_2 is given by

$$M_{A_1A_2} = \frac{A_1+A_2}{2} \tag{1.45}$$

so that an equation of the line L_{123} through the points $M_{A_1A_2}$ and A_3 is

$$L_{123}(t_1) = A_3 + \left(-A_3 + \frac{A_1+A_2}{2}\right)t_1 \tag{1.46}$$

with the line parameter $t_1 \in \mathbb{R}$.

The line $L_{123}(t_1)$ contains one of the three medians of triangle $A_1A_2A_3$. Invoking cyclicity, equations of the lines L_{123}, L_{231} and L_{312}, which contain, respectively, the three triangle medians are obtained from (1.46) by index cyclic permutations,

$$\begin{aligned} L_{123}(t_1) &= \frac{t_1}{2}A_1 + \frac{t_1}{2}A_2 + (1-t_1)A_3 \\ L_{231}(t_2) &= \frac{t_2}{2}A_2 + \frac{t_2}{2}A_3 + (1-t_2)A_1 \\ L_{312}(t_3) &= \frac{t_3}{2}A_3 + \frac{t_3}{2}A_1 + (1-t_3)A_2 \end{aligned} \tag{1.47}$$

$t_1, t_2, t_3 \in \mathbb{R}$.

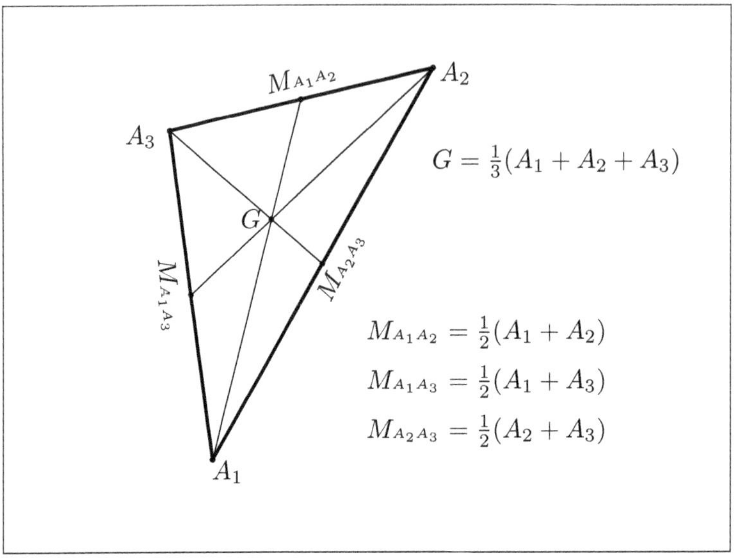

Fig. 1.3 The side midpoints M and the centroid G of triangle $A_1A_2A_3$ in a Euclidean plane $\mathbb{R}^2$.

The triangle centroid G, Fig. 1.3, is the point of concurrency of the three lines in (1.47). This point of concurrency is determined by solving the equation $L_{123}(t_1) = L_{231}(t_2) = L_{312}(t_3)$ for the unknowns $t_1, t_2, t_3 \in \mathbb{R}$, obtaining $t_1 = t_2 = t_3 = 2/3$. Hence, G is given by the equation

$$G = \frac{A_1 + A_2 + A_3}{3} \tag{1.48}$$

Comparing (1.48) with (1.44) we find that the special barycentric coordinates (m_1, m_2, m_3) of G with respect to the set $\{A_1, A_2, A_3\}$ are given by

$$m_1 = m_2 = m_3 = \frac{1}{3} \tag{1.49}$$

Hence, convenient barycentric coordinates $(m_1 : m_2 : m_3)$ of G may be given by

$$(m_1 : m_2 : m_3) = (1 : 1 : 1) \tag{1.50}$$

as it is well-known in the literature; see, for instance, [8, 9].

$$\begin{aligned} \mathbf{a}_{12} &= -A_1 + A_2, & a_{12} &= \|\mathbf{a}_{12}\| \\ \mathbf{a}_{13} &= -A_1 + A_3, & a_{13} &= \|\mathbf{a}_{13}\| \\ \mathbf{a}_{23} &= -A_2 + A_3, & a_{23} &= \|\mathbf{a}_{23}\| \\ \mathbf{h}_3 &= -A_3 + P_3, & h_3 &= \|\mathbf{h}_3\| \end{aligned}$$

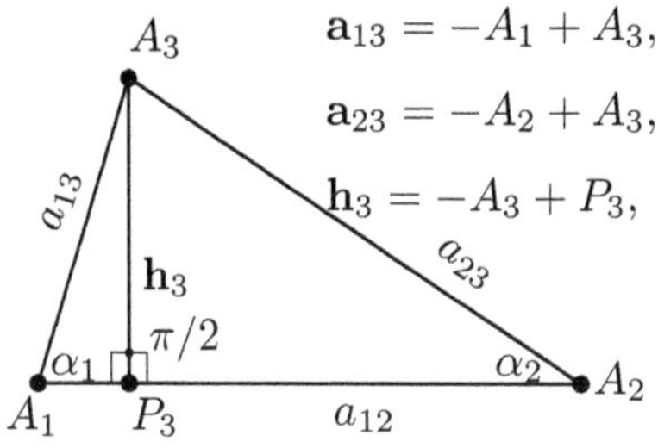

$$\begin{aligned} \mathbf{p}_1 &= -A_1 + P_3, & p_1 &= \|\mathbf{p}_1\| \\ \mathbf{p}_2 &= -A_2 + P_3, & p_2 &= \|\mathbf{p}_2\| \end{aligned}$$

Fig. 1.4 Orthogonal projection, P_3, of a point, A_3, onto a segment, A_1A_2, in a Euclidean n-space $\mathbb{R}^n$. The segment A_3P_3 is the altitude $\mathbf{h}_3$ of triangle $A_1A_2A_3$ dropped perpendicularly from vertex A_3 to its foot P_3 on its base, which is side A_1A_2 of the triangle. Barycentric coordinates $\{m_1 : m_2\}$ of the point P_3 with respect to the set of points $\{A_1, A_2\}$, satisfying (1.51), are determined in (1.61).

1.8 Triangle Altitude

Let $A_1A_2A_3$ be a triangle with vertices A_1, A_2, and A_3 in a Euclidean n-space $\mathbb{R}^n$, and let the point P_3 be the orthogonal projection of vertex A_3 onto its opposite side, A_1A_2 (or its extension), as shown in Fig. 1.4 for $n = 2$. Furthermore, let $(m_1 : m_2)$ be barycentric coordinates of P_3 with respect to the set $\{A_1, A_2\}$. Then, P_3 is given in terms of its barycentric coordinates $(m_1 : m_2)$ with respect to the set $\{A_1, A_2\}$ by the equation

$$P_3 = \frac{m_1A_1 + m_2A_2}{m_1 + m_2} \tag{1.51}$$

where the barycentric coordinates m_1 and m_2 of P_3 are to be determined in (1.61) below.

By the covariance (1.26) of barycentric coordinate representations with respect to translations we have, in particular, for $W = -A_1$ and $W = -A_2$,

$$\begin{aligned} \mathbf{p}_1 &= -A_1 + P_3 = \frac{m_2(-A_1 + A_2)}{m_1 + m_2} = \frac{m_2\mathbf{a}_{12}}{m_1 + m_2} \\ \mathbf{p}_2 &= -A_2 + P_3 = \frac{m_1(-A_2 + A_1)}{m_1 + m_2} = \frac{-m_1\mathbf{a}_{12}}{m_1 + m_2} \end{aligned} \tag{1.52}$$

As indicated in Fig. 1.2, p. 7, we use the notation

$$\mathbf{a}_{ij} = -A_i + A_j, \qquad a_{ij} = \|\mathbf{a}_{ij}\| \tag{1.53}$$

$i, j = 1, 2, 3$, $i \neq j$. Clearly, in general, $\mathbf{a}_{ij} \neq \mathbf{a}_{ji}$, but $a_{ij} = a_{ji}$.

We also use the notation

$$\begin{aligned} \mathbf{p}_1 &= -A_1 + P_3, \qquad & p_1 &= \|\mathbf{p}_1\| \\ \mathbf{p}_2 &= -A_2 + P_3, \qquad & p_2 &= \|\mathbf{p}_2\| \end{aligned} \tag{1.54}$$

and

$$\mathbf{h} = -A_3 + P_3, \qquad h = \|\mathbf{h}\| \tag{1.55}$$

In this notation, the vector equations (1.52) lead to the scalar equations

$$\begin{aligned} p_1 &= \frac{m_2 a_{12}}{m_1 + m_2} \\ p_2 &= \frac{m_1 a_{12}}{m_1 + m_2} \end{aligned} \tag{1.56}$$

The Pythagorean identity for the right-angled triangles $A_1P_3A_3$ and $A_2P_3A_3$ in Fig. 1.4 implies

$$h^2 = a_{13}^2 - p_1^2 = a_{23}^2 - p_2^2 \tag{1.57}$$

Hence, by (1.56)–(1.57),

$$a_{13}^2 - \frac{m_2^2 a_{12}^2}{(m_1 + m_2)^2} = a_{23}^2 - \frac{m_1^2 a_{12}^2}{(m_1 + m_2)^2} \tag{1.58}$$

Normalizing m_1 and m_2,

$$m_1 + m_2 = 1 \tag{1.59}$$

and solving (1.58)–(1.59) for m_1 and m_2, the special barycentric coordinates $\{m_1, m_2\}$ of the point P_3 with respect to the set $\{A_1, A_2\}$ are

$$\begin{aligned} m_1 &= \frac{a_{12}^2 - a_{13}^2 + a_{23}^2}{2a_{12}^2} \\ m_2 &= \frac{a_{12}^2 + a_{13}^2 - a_{23}^2}{2a_{12}^2} \end{aligned} \tag{1.60}$$

so that convenient barycentric coordinates $(m_1 : m_2)$ of P_3 with respect to the set $\{A_1, A_2\}$ may be given by

$$\begin{aligned} m_1 &= a_{12}^2 - a_{13}^2 + a_{23}^2 \\ m_2 &= a_{12}^2 + a_{13}^2 - a_{23}^2 \end{aligned} \tag{1.61}$$

Hence, by (1.51) and either (1.60) or (1.61) we have

$$P_3 = \frac{1}{2}\left(\frac{a_{12}^2 - a_{13}^2 + a_{23}^2}{a_{12}^2}\right) A_1 + \frac{1}{2}\left(\frac{a_{12}^2 + a_{13}^2 - a_{23}^2}{a_{12}^2}\right) A_2 \tag{1.62}$$

Following the law of sines,

$$\frac{a_{23}}{\sin\alpha_1} = \frac{a_{13}}{\sin\alpha_2} = \frac{a_{12}}{\sin\alpha_3} \tag{1.63}$$

for triangle $A_1A_2A_3$ in Fig. 1.5, (1.62) can be written in terms of the triangle angles as

$$P_3 = \frac{\sin^2\alpha_1 - \sin^2\alpha_2 + \sin^2\alpha_3}{2\sin^2\alpha_3} A_1 + \frac{-\sin^2\alpha_1 + \sin^2\alpha_2 + \sin^2\alpha_3}{2\sin^2\alpha_3} A_2 \tag{1.64}$$

Taking advantage of the triangle π condition

$$\alpha_1 + \alpha_2 + \alpha_3 = \pi \tag{1.65}$$

that triangle angles obey, we have the trigonometric elegant identities

$$\begin{aligned} -\sin^2\alpha_1 + \sin^2\alpha_2 + \sin^2\alpha_3 &= 2\cos\alpha_1 \sin\alpha_2 \sin\alpha_3 \\ \sin^2\alpha_1 - \sin^2\alpha_2 + \sin^2\alpha_3 &= 2\sin\alpha_1 \cos\alpha_2 \sin\alpha_3 \end{aligned} \tag{1.66}$$

where $\alpha_3 = \pi - \alpha_1 - \alpha_2$.

Substituting (1.66) into (1.64), we have

$$P_3 = \frac{\sin\alpha_1 \cos\alpha_2}{\sin(\alpha_1 + \alpha_2)} A_1 + \frac{\cos\alpha_1 \sin\alpha_2}{\sin(\alpha_1 + \alpha_2)} A_2 \tag{1.67}$$

so that the special trigonometric barycentric coordinates (m_1, m_2) of P_3 with respect to the set $\{A_1, A_2\}$ is

$$(m_1, m_2) = \left(\frac{\sin\alpha_1 \cos\alpha_2}{\sin(\alpha_1 + \alpha_2)}, \frac{\cos\alpha_1 \sin\alpha_2}{\sin(\alpha_1 + \alpha_2)}\right) \tag{1.68}$$

and, accordingly, convenient trigonometric barycentric coordinates $(m_1' : m_2')$ of P_3 with respect to the set $\{A_1, A_2\}$ are

$$\begin{aligned} (m_1' : m_2') &= (\sin\alpha_1 \cos\alpha_2 : \cos\alpha_1 \sin\alpha_2) \\ &= (\tan\alpha_1 : \tan\alpha_2) \end{aligned} \tag{1.69}$$

The altitude $\mathbf{h}_3$ of triangle $A_1A_2A_3$ in Fig. 1.4 is the vector

$$\begin{aligned} \mathbf{h}_3 &= -A_3 + P_3 \\ &= \frac{1}{2}\left(\frac{a_{12}^2 - a_{13}^2 + a_{23}^2}{a_{12}^2}\right)(-A_3 + A_1) + \frac{1}{2}\left(\frac{a_{12}^2 + a_{13}^2 - a_{23}^2}{a_{12}^2}\right)(-A_3 + A_2) \\ &= \frac{1}{2}\left(\frac{a_{12}^2 - a_{13}^2 + a_{23}^2}{a_{12}^2}\right)\mathbf{a}_{31} + \frac{1}{2}\left(\frac{a_{12}^2 + a_{13}^2 - a_{23}^2}{a_{12}^2}\right)\mathbf{a}_{32} \end{aligned} \tag{1.70}$$

as we see from (1.62) by employing the covariance property (1.26), p. 12, of barycentric coordinate representations. Note that $\mathbf{a}_{31} = -\mathbf{a}_{13}$, so that $a_{31} = \|\mathbf{a}_{31}\| = \|\mathbf{a}_{13}\| = a_{13}$, etc.

Noting the law of cosines,

$$a_{12}^2 = a_{13}^2 + a_{23}^2 - 2a_{13}a_{23}\cos\alpha_3 \tag{1.71}$$

in the notation of Fig. 1.4, we have

$$\begin{aligned}2\mathbf{a}_{31}\cdot\mathbf{a}_{32} &= 2(-A_3 + A_1)\cdot(-A_3 + A_2)\\ &= 2a_{13}a_{23}\cos\alpha_3 \\ &= -a_{12}^2 + a_{13}^2 + a_{23}^2\end{aligned} \tag{1.72}$$

Hence, by (1.70) and (1.72), the squared length h_3^2 of altitude $\mathbf{h}_3$, Fig. 1.4, is given by

$$\begin{aligned}h_3^2 &= \|\mathbf{h}_3^2\|\\ &= \frac{1}{4}\left\{\left(\frac{a_{12}^2 - a_{13}^2 + a_{23}^2}{a_{12}^2}\right)^2 a_{13}^2 + \left(\frac{a_{12}^2 + a_{13}^2 - a_{23}^2}{a_{12}^2}\right)^2 a_{23}^2\right.\\ &\quad \left. + \frac{a_{12}^2 - a_{13}^2 + a_{23}^2}{a_{12}^2}\frac{a_{12}^2 + a_{13}^2 - a_{23}^2}{a_{12}^2}(-a_{12}^2 + a_{13}^2 + a_{23}^2)\right\}\\ &= \frac{(a_{12}+a_{13}+a_{23})(-a_{12}+a_{13}+a_{23})(a_{12}-a_{13}+a_{23})(a_{12}+a_{13}-a_{23})}{4a_{12}^2}\\ &= \frac{1}{2}\frac{F_2(a_{12},a_{13},a_{23})}{F_1(a_{12})}\end{aligned} \tag{1.73}$$

Here $F_1(a_{12})$ has the determinant representation

$$F_1(a_{12}) = 2a_{12}^2 = \begin{vmatrix} 0 & a_{12}^2 & 1\\ a_{21}^2 & 0 & 1\\ 1 & 1 & 0\end{vmatrix} \tag{1.74}$$

and $F_2(a_{12}, a_{13}, a_{23})$ has the determinant representation

$$\begin{aligned}&F_2(a_{12},a_{13},a_{23})\\ &= (a_{12}+a_{13}+a_{23})(-a_{12}+a_{13}+a_{23})(a_{12}-a_{13}+a_{23})(a_{12}+a_{13}-a_{23})\\ &= -\begin{vmatrix} 0 & a_{12}^2 & a_{13}^2 & 1\\ a_{21}^2 & 0 & a_{23}^2 & 1\\ a_{31}^2 & a_{32}^2 & 0 & 1\\ 1 & 1 & 1 & 0\end{vmatrix}\end{aligned} \tag{1.75}$$

The determinants in (1.74) and (1.75) are known as *Cayley-Menger determinants* [16]. The pattern these determinants exhibit here is extended in (1.194)–(1.195), p. 60.

Equation (1.73) gives rise to Heron's formula of the triangle area in the following theorem:

Theorem 1.10. (Heron's Formula). *The area $|A_1A_2A_3|$ of triangle $A_1A_2A_3$ in a Euclidean space $\mathbb{R}^n$ is given by Heron's Formula*

$$
\begin{aligned}
|A_1A_2A_3| &= \tfrac{1}{2}a_{12}h_3 \qquad (1.76)\\
&= \frac{1}{4}\sqrt{a_{12}+a_{13}+a_{23}}\sqrt{-a_{12}+a_{13}+a_{23}}\sqrt{a_{12}-a_{13}+a_{23}}\sqrt{a_{12}+a_{13}-a_{23}}
\end{aligned}
$$

Owing to their importance, we elevate the results in (1.62) and (1.73) to the status of theorems:

Theorem 1.11. (Point to Line Perpendicular Projection). *Let A_1 and A_2 be any two distinct points of a Euclidean space $\mathbb{R}^n$, and let $L_{A_1A_2}$ be the line passing through these points. Furthermore, let A_3 be any point of the space that does not lie on $L_{A_1A_2}$, as shown in Fig. 1.4. Then, in the notation of Fig. 1.4, the perpendicular projection of the point A_3 on the line $L_{A_1A_2}$ is the point P_3 on the line given by,* (1.62), (1.69),

$$
\begin{aligned}
P_3 &= \frac{1}{2}\left(\frac{a_{12}^2-a_{13}^2+a_{23}^2}{a_{12}^2}\right)A_1+\frac{1}{2}\left(\frac{a_{12}^2+a_{13}^2-a_{23}^2}{a_{12}^2}\right)A_2\\
&= \frac{\tan\alpha_1 A_1+\tan\alpha_2 A_2}{\tan\alpha_1+\tan\alpha_2}
\end{aligned}
\qquad (1.77)
$$

Theorem 1.12. (Point to Line Distance). *Let A_1 and A_2 be any two distinct points of a Euclidean space $\mathbb{R}^n$, and let $L_{A_1A_2}$ be the line passing through these points. Furthermore, let A_3 be any point of the space that does not lie on $L_{A_1A_2}$, as shown in Fig. 1.4. Then, in the notation of Fig. 1.4, the distance $h_3 = \|-A_3+P_3\|$ between the point A_3 and the line $L_{A_1A_2}$ is given by the equation*

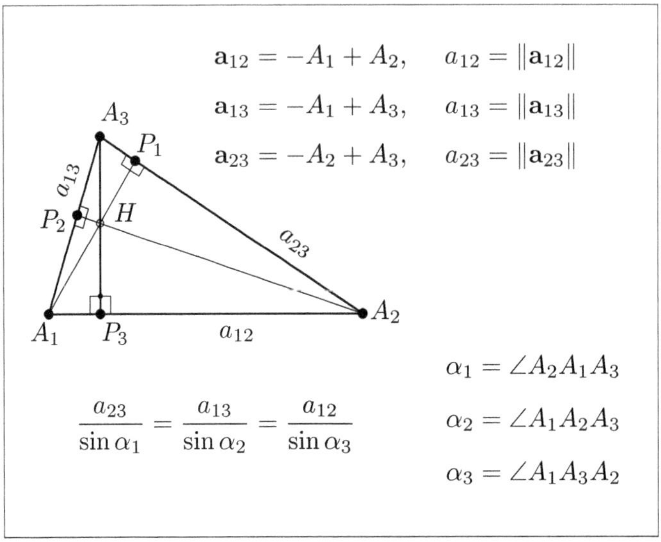

Fig. 1.5 The Triangle Orthocenter H. A triangle orthocenter is the point at which the three altitudes are concurrent. The standard triangle index notation along with its law of sines is presented.

$$
\begin{aligned}
h_3^2 &= \frac{F_2(a_{12}, a_{13}, a_{23})}{4a_{12}^2} \\
&:= \frac{(a_{12} + a_{13} + a_{23})(-a_{12} + a_{13} + a_{23})(a_{12} - a_{13} + a_{23})(a_{12} + a_{13} - a_{23})}{4a_{12}^2} \\
&= \frac{2(a_{12}^2 a_{13}^2 + a_{12}^2 a_{23}^2 + a_{13}^2 a_{23}^2) - (a_{12}^4 + a_{13}^4 + a_{23}^4)}{4a_{12}^2}
\end{aligned}
\tag{1.78}
$$

Following the result, (1.78), of Theorem 1.12 we have

$$
a_{12}^2 h_3^2 = \frac{2(a_{12}^2 a_{13}^2 + a_{12}^2 a_{23}^2 + a_{13}^2 a_{23}^2) - (a_{12}^4 + a_{13}^4 + a_{23}^4)}{4}
\tag{1.79}
$$

1.9 Triangle Orthocenter

The triangle orthocenter is located at the intersection of its altitudes, Fig. 1.5.

Let P_1, P_2 and P_3 be the feet of the three altitudes of a triangle $A_1A_2A_3$ in a Euclidean n-space $\mathbb{R}^n$, shown in Fig. 1.5 for $n = 2$. The barycentric coordinate representation of the altitude feet with respect to the set $\{A_1, A_2, A_3\}$ are

$$
\begin{aligned}
P_1 &= \frac{1}{2}\left(\frac{-a_{12}^2 + a_{13}^2 + a_{23}^2}{a_{23}^2}\right) A_2 + \frac{1}{2}\left(\frac{a_{12}^2 - a_{13}^2 + a_{23}^2}{a_{23}^2}\right) A_3 \\
P_2 &= \frac{1}{2}\left(\frac{-a_{12}^2 + a_{13}^2 + a_{23}^2}{a_{13}^2}\right) A_1 + \frac{1}{2}\left(\frac{a_{12}^2 + a_{13}^2 - a_{23}^2}{a_{13}^2}\right) A_3 \qquad (1.80) \\
P_3 &= \frac{1}{2}\left(\frac{a_{12}^2 - a_{13}^2 + a_{23}^2}{a_{12}^2}\right) A_1 + \frac{1}{2}\left(\frac{a_{12}^2 + a_{13}^2 - a_{23}^2}{a_{12}^2}\right) A_2
\end{aligned}
$$

The third equation in (1.80) is established in (1.62), and the first two equations in (1.80) are obtained from the third by vertex cyclic permutations.

The equations of the lines that contain the altitudes of triangle $A_1A_2A_3$, Fig. 1.5, are

$$
\begin{aligned}
L_{A_1P_1} &= A_1 + (-A_1 + P_1)t_1 \\
L_{A_2P_2} &= A_2 + (-A_2 + P_2)t_2 \qquad (1.81) \\
L_{A_3P_3} &= A_3 + (-A_3 + P_3)t_3
\end{aligned}
$$

for the three line parameters $-\infty < t_1, t_2, t_3 < \infty$, where the altitude feet P_1, P_2 and P_3 are given by (1.80).

In order to determine the point of concurrency H of the triangle altitudes, Fig. 1.5, if exists, we solve the vector equations

$$
A_1 + (-A_1 + P_1)t_1 = A_2 + (-A_2 + P_2)t_2 = A_3 + (-A_3 + P_3)t_3 \qquad (1.82)
$$

for the three scalar unknowns t_1, t_2 and t_3. The solution turns out to be

$$
\begin{aligned}
t_1 &= \frac{2(-a_{12} - a_{13} + a_{23})}{D} a_{23} \\
t_2 &= \frac{2(-a_{12} + a_{13} - a_{23})}{D} a_{13} \qquad (1.83) \\
t_3 &= \frac{2(\ a_{12} - a_{13} - a_{23})}{D} a_{12}
\end{aligned}
$$

where

$$
D = a_{12}^2 + a_{13}^2 + a_{23}^2 - 2(a_{12}a_{13} + a_{12}a_{23} + a_{13}a_{23}) \qquad (1.84)
$$

Substituting the solution for t_1 (respectively, for t_2, t_3) in the first (respectively, second, third) equation in (1.81) we determine the orthocenter H of triangle $A_1A_2A_3$ in terms of barycentric coordinates,

$$H = \frac{m_1A_1 + m_2A_2 + m_3A_3}{m_1 + m_2 + m_3} \tag{1.85}$$

where convenient barycentric coordinates are

$$\begin{aligned} m_1 &= a_{23}^4 - (a_{12}^2 - a_{13}^2)^2 \\ m_2 &= a_{13}^4 - (a_{12}^2 - a_{23}^2)^2 \\ m_3 &= a_{12}^4 - (a_{23}^2 - a_{13}^2)^2 \end{aligned} \tag{1.86}$$

Following the law of sines (1.63), the barycentric coordinates of H in (1.85) – (1.86) can be written in terms of the triangle angles as

$$\begin{aligned} m_1 &= \frac{1 - \cos 2\alpha_1 - \cos 2\alpha_2 + \cos 2\alpha_3}{1 + \cos 2\alpha_1 - \cos 2\alpha_2 - \cos 2\alpha_3} \\ m_2 &= \frac{1 - \cos 2\alpha_1 - \cos 2\alpha_2 + \cos 2\alpha_3}{1 - \cos 2\alpha_1 + \cos 2\alpha_2 - \cos 2\alpha_3} \\ m_3 &= 1 \end{aligned} \tag{1.87}$$

Taking advantage of the relationship (1.65) between triangle angles, and employing trigonometric identities, (1.87) can be simplified, obtaining the elegant barycentric coordinates of the orthocenter H of triangle $A_1A_2A_3$ in terms of its angles,

$$\begin{aligned} m_1 &= \frac{\tan\alpha_1}{\tan\alpha_3} \\ m_2 &= \frac{\tan\alpha_2}{\tan\alpha_3} \\ m_3 &= 1 \end{aligned} \tag{1.88}$$

or equivalently, owing to the homogeneity of the barycentric coordinates,

$$\begin{aligned} m_1 &= \tan\alpha_1 \\ m_2 &= \tan\alpha_2 \\ m_3 &= \tan\alpha_3 \end{aligned} \tag{1.89}$$

Following (1.85) and (1.89), the orthocenter H of a triangle $A_1A_2A_3$ with vertices A_1, A_2 and A_3, and with corresponding angles α_1, α_2 and α_3, Fig. 1.5, is given in terms of its barycentric coordinates with respect to the set $\{A_1, A_2, A_3\}$ by the equation

$$H = \frac{\tan\alpha_1 A_1 + \tan\alpha_2 A_2 + \tan\alpha_3 A_3}{\tan\alpha_1 + \tan\alpha_2 + \tan\alpha_3} \tag{1.90}$$

$$\mathbf{a}_{12} = -A_1 + A_2, \qquad a_{12} = \|\mathbf{a}_{12}\|$$
$$\mathbf{a}_{13} = -A_1 + A_3, \qquad a_{13} = \|\mathbf{a}_{13}\|$$
$$\mathbf{a}_{23} = -A_2 + A_3, \qquad a_{23} = \|\mathbf{a}_{23}\|$$
$$\mathbf{h} = -A_3 + P_3, \qquad h = \|\mathbf{h}\|$$

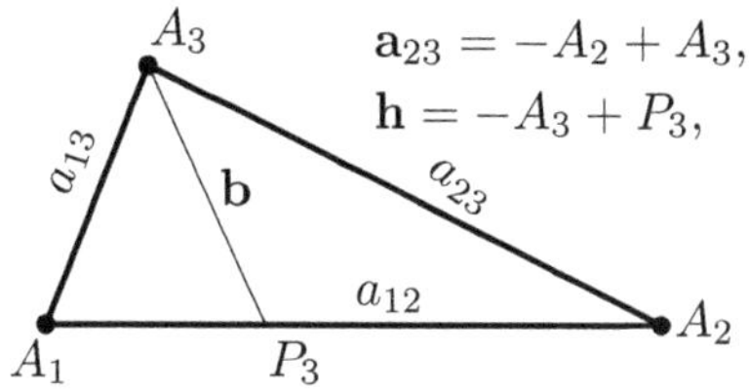

$$\mathbf{p}_1 = -A_1 + P_3, \qquad p_1 = \|\mathbf{p}_1\|$$
$$\mathbf{p}_2 = -A_2 + P_3, \qquad p_2 = \|\mathbf{p}_2\|$$

$$\alpha_1 = \angle A_2A_1A_3, \quad \alpha_2 = \angle A_1A_2A_3, \quad \alpha_3 = \angle A_1A_3A_2$$
$$\angle A_1A_3P_3 = \angle A_2A_3P_3$$

Fig. 1.6 Angle bisector, A_3P_3, of angle $\angle A_1A_3A_2$ in a Euclidean n-space $\mathbb{R}^n$, for $n = 2$. The segment A_3P_3 forms the angle bisector in triangle $A_1A_2A_3$, dropped from vertex A_3 to a point P_3 on its opposite side A_1A_2. Barycentric coordinates $\{m_1 : m_2\}$ of the point P_3 with respect to the set of points $\{A_1, A_2\}$, satisfying (1.91), are determined in (1.99) – (1.100).

1.10 Triangle Incenter

The incircle of a triangle is a circle lying inside the triangle, tangent to the triangle sides. The center, I, of the incircle is called the triangle incenter, shown in Fig. 1.8, p. 33. The triangle incenter is located at the intersection of the angle bisectors, shown in Fig. 1.7, p. 29.

Let P_3 be a point on side A_1A_2 of triangle $A_1A_2A_3$ in a Euclidean n space $\mathbb{R}^n$ such that A_3P_3 is an angle bisector of angle $\angle A_1A_3A_2$, as shown in Fig. 1.6. Then, P_3 is given in terms of its barycentric coordinates (m_1, m_2) with respect to the set $\{A_1, A_2\}$ by the equation

$$P_3 = \frac{m_1A_1 + m_2A_2}{m_1 + m_2} \tag{1.91}$$

where the barycentric coordinates m_1 and m_2 of P_3 are to be determined in (1.99) – (1.100) below.

As in (1.52), by the covariance (1.26) of barycentric coordinate representations with respect to translations we have, in particular for $W = -A_1$

and $W = -A_2$,

$$\begin{aligned}\mathbf{p}_1 &= -A_1 + P_3 = \frac{m_2(-A_1 + A_2)}{m_1 + m_2} = \frac{m_2\mathbf{a}_{12}}{m_1 + m_2} \\ \mathbf{p}_2 &= -A_2 + P_3 = \frac{m_1(-A_2 + A_1)}{m_1 + m_2} = \frac{-m_1\mathbf{a}_{12}}{m_1 + m_2}\end{aligned} \tag{1.92}$$

As indicated in Fig. 1.6, we use the notation

$$\begin{aligned}\mathbf{a}_{12} &= -A_1 + A_2, & a_{12} &= \|\mathbf{a}_{12}\| \\ \mathbf{a}_{13} &= -A_1 + A_3, & a_{13} &= \|\mathbf{a}_{13}\| \\ \mathbf{a}_{23} &= -A_2 + A_3, & a_{23} &= \|\mathbf{a}_{23}\|\end{aligned} \tag{1.93}$$

and

$$\begin{aligned}\mathbf{p}_1 &= -A_1 + P_3, & p_1 &= \|\mathbf{p}_1\| \\ \mathbf{p}_2 &= -A_2 + P_3, & p_2 &= \|\mathbf{p}_2\|\end{aligned} \tag{1.94}$$

so that, by (1.92) – (1.94),

$$\begin{aligned}p_1 &= \frac{m_2 a_{12}}{m_1 + m_2} \\ p_2 &= \frac{m_1 a_{12}}{m_1 + m_2}\end{aligned} \tag{1.95}$$

implying

$$\frac{p_1}{p_2} = \frac{m_2}{m_1} \tag{1.96}$$

By the *angle bisector theorem*, which follows immediately from the law of sines (1.63) and the equation $\sin\angle A_1P_3A_3 = \sin\angle A_2P_3A_3$ in Fig. 1.6, the angle bisector of an angle in a triangle divides the opposite side in the same ratio as the sides adjacent to the angle. Hence, in the notation of Fig. 1.6,

$$\frac{p_1}{p_2} = \frac{a_{13}}{a_{23}} \tag{1.97}$$

Hence, by (1.96) – (1.97), and by the law of sines (1.63),

$$\frac{m_2}{m_1} = \frac{a_{13}}{a_{23}} = \frac{\sin\alpha_2}{\sin\alpha_1} \tag{1.98}$$

so that barycentric coordinates of P_3 in (1.91) may be given by

$$\begin{aligned}m_1 &= a_{23} \\ m_2 &= a_{13}\end{aligned} \tag{1.99}$$

$$\mathbf{a}_{12} = -A_1 + A_2, \qquad a_{12} = \|\mathbf{a}_{12}\|$$

$$\mathbf{a}_{13} = -A_1 + A_3, \qquad a_{13} = \|\mathbf{a}_{13}\|$$

$$\mathbf{a}_{23} = -A_2 + A_3, \qquad a_{23} = \|\mathbf{a}_{23}\|$$

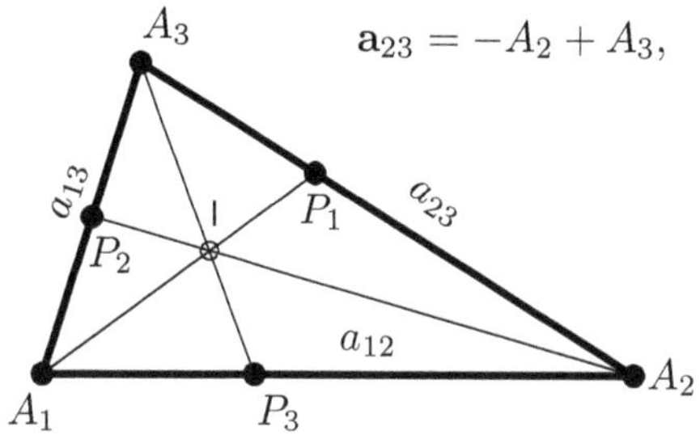

$$\angle A_2A_1P_1 = \angle A_3A_1P_1$$

$$\angle A_1A_2P_2 = \angle A_3A_2P_2$$

$$\angle A_1A_3P_3 = \angle A_2A_3P_3$$

$$\mathbf{p}_1 = -A_1 + P_1, \qquad p_1 = \|\mathbf{p}_1\|$$

$$\mathbf{p}_2 = -A_2 + P_2, \qquad p_2 = \|\mathbf{p}_2\|$$

$$\mathbf{p}_3 = -A_3 + P_3, \qquad p_3 = \|\mathbf{p}_3\|$$

$$\alpha_1 = \angle A_2A_1A_3, \quad \alpha_2 = \angle A_1A_2A_3, \quad \alpha_3 = \angle A_1A_3A_2$$

Fig. 1.7 The Triangle Incenter. The triangle angle bisectors are concurrent. The point of concurrency, I, is called the incenter of the triangle. Here $A_1A_2A_3$ is a triangle in a Euclidean n-space, $n = 2$, and the line A_kP_k is the angle bisector from vertex A_k to the intersection point P_k with the opposite side, $k = 1, 2, 3$.

or, equivalently, by

$$\begin{aligned} m_1 &= \sin\alpha_1 \\ m_2 &= \sin\alpha_2 \end{aligned} \tag{1.100}$$

Accordingly, P_3 in Fig. 1.7 is given in terms of its barycentric coordinates (m_1, m_2) with respect to the set $\{A_1, A_2\}$ by each of the two equations

$$\begin{aligned} P_3 &= \frac{a_{23}A_1 + a_{13}A_2}{a_{23} + a_{13}} \\ P_3 &= \frac{\sin\alpha_1 A_1 + \sin\alpha_2 A_2}{\sin\alpha_1 + \sin\alpha_2} \end{aligned} \tag{1.101}$$

The three bisector segments of triangle $A_1A_2A_3$ are A_1P_1, A_2P_2 and A_3P_3, as shown in Fig. 1.7. It follows from (1.101) by vertex cyclic permutations that barycentric coordinates of their feet, P_1, P_2 and P_3, with respect to the set of the triangle vertices $\{A_1, A_2, A_3\}$ are given by

$$\begin{aligned}
P_1 &= \frac{a_{13}A_2 + a_{12}A_3}{a_{13} + a_{12}} \\
P_2 &= \frac{a_{23}A_1 + a_{12}A_3}{a_{23} + a_{12}} \\
P_3 &= \frac{a_{23}A_1 + a_{13}A_2}{a_{23} + a_{13}}
\end{aligned} \tag{1.102}$$

or, equivalently, by

$$\begin{aligned}
P_1 &= \frac{\sin\alpha_2 A_2 + \sin\alpha_3 A_3}{\sin\alpha_2 + \sin\alpha_3} \\
P_2 &= \frac{\sin\alpha_1 A_1 + \sin\alpha_3 A_3}{\sin\alpha_1 + \sin\alpha_3} \\
P_3 &= \frac{\sin\alpha_1 A_1 + \sin\alpha_2 A_2}{\sin\alpha_1 + \sin\alpha_2}
\end{aligned} \tag{1.103}$$

The equations of the lines that contain the angle bisectors of triangle $A_1A_2A_3$, Fig. 1.7, are

$$\begin{aligned}
L_{A_1P_1} &= A_1 + (-A_1 + P_1)t_1 \\
L_{A_2P_2} &= A_2 + (-A_2 + P_2)t_2 \\
L_{A_3P_3} &= A_3 + (-A_3 + P_3)t_3
\end{aligned} \tag{1.104}$$

for the three line parameters $-\infty < t_1, t_2, t_3 < \infty$, where the angle bisector feet P_1, P_2 and P_3 are given by (1.102).

In order to determine the point of concurrency I of the triangle angle bisectors, Fig. 1.7, if exists, we solve the vector equations

$$A_1 + (-A_1 + P_1)t_1 = A_2 + (-A_2 + P_2)t_2 = A_3 + (-A_3 + P_3)t_3 \tag{1.105}$$

for the three scalar unknowns t_1, t_2 and t_3. The solution turn out to be

$$\begin{aligned}
t_1 &= \frac{a_{12} + a_{13}}{a_{12} + a_{13} + a_{23}} \\
t_2 &= \frac{a_{12} + a_{23}}{a_{12} + a_{13} + a_{23}} \\
t_3 &= \frac{a_{13} + a_{23}}{a_{12} + a_{13} + a_{23}}
\end{aligned} \tag{1.106}$$

Substituting the solution for t_1 (respectively, for t_2, t_3) in the first (respectively, second, third) equation in (1.104) we determine the incenter I of triangle $A_1A_2A_3$ in terms of barycentric coordinates,

$$I = \frac{m_1A_1 + m_2A_2 + m_3A_3}{m_1 + m_2 + m_3} \tag{1.107}$$

where the barycentric coordinates are

$$\begin{aligned} m_1 &= a_{23} \\ m_2 &= a_{13} \\ m_3 &= a_{12} \end{aligned} \tag{1.108}$$

or, equivalently by (1.98),

$$\begin{aligned} m_1 &= \sin\alpha_1 \\ m_2 &= \sin\alpha_2 \\ m_3 &= \sin\alpha_3 \end{aligned} \tag{1.109}$$

Following (1.107) and (1.108), the incenter I of a triangle $A_1A_2A_3$ with vertices A_1, A_2 and A_3, and with corresponding side lengths a_{23}, a_{13} and a_{12}, Fig. 1.7, is given in terms of its barycentric coordinates with respect to the set $\{A_1, A_2, A_3\}$ by the equation

$$I = \frac{a_{23}A_1 + a_{13}A_2 + a_{12}A_3}{a_{23} + a_{13} + a_{12}} \tag{1.110}$$

Following (1.107) and (1.109), the incenter I of a triangle $A_1A_2A_3$ with vertices A_1, A_2 and A_3, and with corresponding angles α_1, α_2 and α_3, Fig. 1.7, is given in terms of its trigonometric barycentric coordinates with respect to the set $\{A_1, A_2, A_3\}$ by the equation

$$I = \frac{\sin\alpha_1 A_1 + \sin\alpha_2 A_2 + \sin\alpha_3 A_3}{\sin\alpha_1 + \sin\alpha_2 + \sin\alpha_3} \tag{1.111}$$

The sine of any triangle angle is positive. Hence, by convexity considerations, the incenter I of a triangle lies on the interior of the triangle.

1.11 Triangle Inradius

Let $A_1A_2A_3$ be a triangle with incenter I in a Euclidean space $\mathbb{R}^n$. Following (1.110) and the covariance property (1.26), p. 12, of barycentric representations, we have, in the notation of Fig. 1.8,

$$\begin{aligned} -A_1 + I = -A_1 + \frac{a_{23}A_1 + a_{13}A_2 + a_{12}A_3}{a_{23} + a_{13} + a_{12}} \\ = \frac{a_{13}(-A_1 + A_2) + a_{12}(-A_1 + A_3)}{a_{23} + a_{13} + a_{12}} \\ = \frac{a_{13}\mathbf{a}_{12} + a_{12}\mathbf{a}_{13}}{a_{23} + a_{13} + a_{12}} \end{aligned} \tag{1.112}$$

Hence,

$$\bar{a}_{13}^2 := \| -A_1 + I\|^2 = \frac{2a_{12}^2 a_{13}^2 (1 + \cos\alpha_1)}{(a_{12} + a_{13} + a_{23})^2} \tag{1.113}$$

noting that

$$(-A_1 + A_2)\cdot(-A_1 + A_3) = a_{12}a_{13}\cos\alpha_1 \tag{1.114}$$

By the law of cosines (1.71) for triangle $A_1A_2A_3$, we have

$$2(1 + \cos\alpha_1) = \frac{(a_{12} + a_{13})^2 - a_{23}^2}{a_{12}a_{13}} \tag{1.115}$$

Hence, by (1.113)–(1.115),

$$\bar{a}_{13}^2 = \frac{a_{12}a_{13}}{(a_{12} + a_{13} + a_{23})^2}\{(a_{12} + a_{13})^2 - a_{23}^2\} \tag{1.116}$$

Similarly,

$$\bar{a}_{23}^2 := \| -A_2 + I\|^2 = \frac{2a_{12}^2 a_{23}^2 (1 + \cos\alpha_2)}{(a_{12} + a_{13} + a_{23})^2} \tag{1.117}$$

and hence

$$\bar{a}_{23}^2 = \frac{a_{12}a_{23}}{(a_{12} + a_{13} + a_{23})^2}\{(a_{12} + a_{23})^2 - a_{13}^2\} \tag{1.118}$$

The vectors $\bar{\mathbf{a}}_{13}$ and $\bar{\mathbf{a}}_{23}$ along with their magnitudes $\bar{a}_{13}$ and $\bar{a}_{23}$ are shown in Fig. 1.8

The tangency point T_3 where the incircle of triangle $A_1A_2A_3$ meets the triangle side A_1A_2 opposite to vertex A_3, Fig. 1.8, is the perpendicular projection of the incenter I on the line $L_{A_1A_2}$ that passes through the points A_1 and A_2. Hence, by the point to line perpendicular projection formula (1.77), p. 23,

$$T_3 = \frac{1}{2}\left(\frac{a_{12}^2 - \bar{a}_{13}^2 + \bar{a}_{23}^2}{a_{12}^2}\right)A_1 + \frac{1}{2}\left(\frac{a_{12}^2 + \bar{a}_{13}^2 - \bar{a}_{23}^2}{a_{12}^2}\right)A_2 \tag{1.119}$$

$$\mathbf{a}_{12} = -A_1 + A_2, \qquad a_{12} = \|\mathbf{a}_{12}\|$$

$$\mathbf{a}_{13} = -A_1 + A_3, \qquad a_{13} = \|\mathbf{a}_{13}\|$$

$$\mathbf{a}_{23} = -A_2 + A_3, \qquad a_{23} = \|\mathbf{a}_{23}\|$$

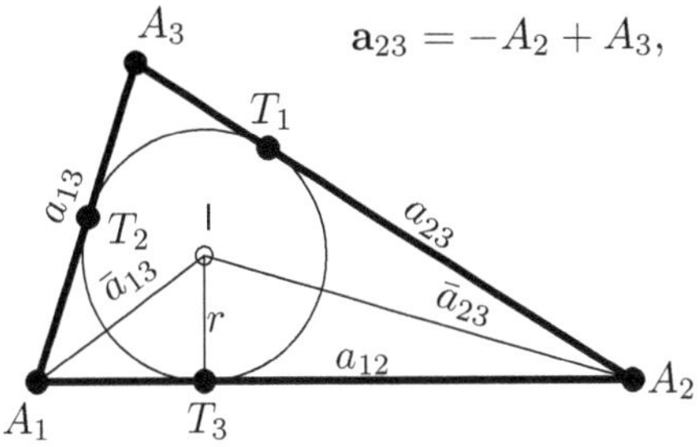

$$\alpha_1 = \angle A_3A_1A_2$$

$$\alpha_2 = \angle A_1A_2A_3$$

$$\alpha_3 = \angle A_2A_3A_1$$

$$\bar{\mathbf{a}}_{13} = -A_1 + I, \qquad \bar{a}_{13} = \| - A_1 + I\|$$

$$\bar{\mathbf{a}}_{23} = -A_2 + I, \qquad \bar{a}_{23} = \| - A_2 + I\|$$

Fig. 1.8 The Triangle Incircle, Incenter and Inradius. The triangle angle bisectors are concurrent. The point of concurrency, I, is called the incenter of the triangle. Here $A_1A_2A_3$ is a triangle in a Euclidean n-space, $\mathbb{R}^n$, and T_k is the point of tangency where the triangle incircle meets the triangle side opposite to vertex A_k, $k = 1, 2, 3$. The radius r of the triangle incircle, determined in (1.123), is called the triangle inradius.

Substituting (1.116) and (1.118) into (1.119), we obtain

$$T_3 = \frac{a_{12} - a_{13} + a_{23}}{2a_{12}} A_1 + \frac{a_{12} + a_{13} - a_{23}}{2a_{12}} A_2 \tag{1.120}$$

Equation (1.120) gives rise to the following theorem:

Theorem 1.13. (The Incircle Tangency Points). *Let $A_1A_2A_3$ be a triangle in a Euclidean space $\mathbb{R}^n$, and let T_k be the point of tangency where the triangle incircle meets the side opposite to vertex A_k, $k = 1, 2, 3$, Fig. 1.8. Then, in the standard triangle notation, Figs. 1.7–1.8,*

$$\begin{aligned}
T_1 &= \frac{-a_{12} + a_{13} + a_{23}}{2a_{23}} A_2 + \frac{a_{12} - a_{13} + a_{23}}{2a_{23}} A_3 \\
T_2 &= \frac{-a_{12} + a_{13} + a_{23}}{2a_{13}} A_1 + \frac{a_{12} + a_{13} - a_{23}}{2a_{13}} A_3 \\
T_3 &= \frac{a_{12} - a_{13} + a_{23}}{2a_{12}} A_1 + \frac{a_{12} + a_{13} - a_{23}}{2a_{12}} A_2
\end{aligned} \tag{1.121}$$

Proof. The third equation in (1.121) is established in (1.120). The first and the second equations in (1.121) are obtained from the first by vertex cyclic permutations. □

Applying the point to line distance formula (1.78), p. 24, to calculate the distance r between the point A_3 and the line $L_{A_1A_2}$ that contains the points A_1 and A_2, Fig. 1.8, we obtain the equation

$$r^2 = \frac{(a_{12} + \bar{a}_{13} + \bar{a}_{23})(-a_{12} + \bar{a}_{13} + \bar{a}_{23})(a_{12} - \bar{a}_{13} + \bar{a}_{23})(a_{12} + \bar{a}_{13} - \bar{a}_{23})}{4a_{12}^2} \tag{1.122}$$

Substituting $\bar{a}_{13}$ and $\bar{a}_{23}$ from (1.116) and (1.118) into (1.122) yields the following theorem.

Theorem 1.14. (The Triangle Inradius). *Let $A_1A_2A_3$ be a triangle in a Euclidean space $\mathbb{R}^n$. Then, in the standard triangle notation, Fig. 1.8, the triangle inradius r is given by the equation*

$$r = \sqrt{\frac{(p - a_{12})(p - a_{13})(p - a_{23})}{p}} \tag{1.123}$$

where p is the triangle semiperimeter,

$$p = \frac{a_{12} + a_{13} + a_{23}}{2} \tag{1.124}$$

Following Theorem 1.14 it is appropriate to present the well-known Heron's formula [17].

Theorem 1.15. (Heron's Formula). *Let $A_1A_2A_3$ be a triangle in a Euclidean space $\mathbb{R}^n$. Then, in the standard triangle notation, Fig. 1.2, the triangle area $|A_1A_2A_3|$ is given by Heron's formula,*

$$|A_1A_2A_3| = \sqrt{p(p - a_{12})(p - a_{13})(p - a_{23})} \tag{1.125}$$

or, equivalently,

$$|A_1A_2A_3|^2 = \frac{1}{16} F_2(a_{12}, a_{13}, a_{23}) \tag{1.126}$$

where $F_2(a_{12}, a_{13}, a_{23})$ is the 4×4 Cayley-Menger determinant (1.75), p. 22.

The determinant form (1.126) of Heron's formula possesses the comparative advantage of admitting a natural generalization to higher dimensions, as indicated in (1.195) – (1.196), p. 60.

Theorems 1.14 and 1.15 result in an elegant relationship between the triangle area $|A_1A_2A_3|$ and its inradius r,

$$r = \frac{|A_1A_2A_3|}{p} = \frac{2|A_1A_2A_3|}{a_{12} + a_{13} + a_{23}} \tag{1.127}$$

$$\mathbf{a}_{12} = -A_1 + A_2, \qquad a_{12} = \|\mathbf{a}_{12}\|$$

$$\mathbf{a}_{13} = -A_1 + A_3, \qquad a_{13} = \|\mathbf{a}_{13}\|$$

$$\mathbf{a}_{23} = -A_2 + A_3, \qquad a_{23} = \|\mathbf{a}_{23}\|$$

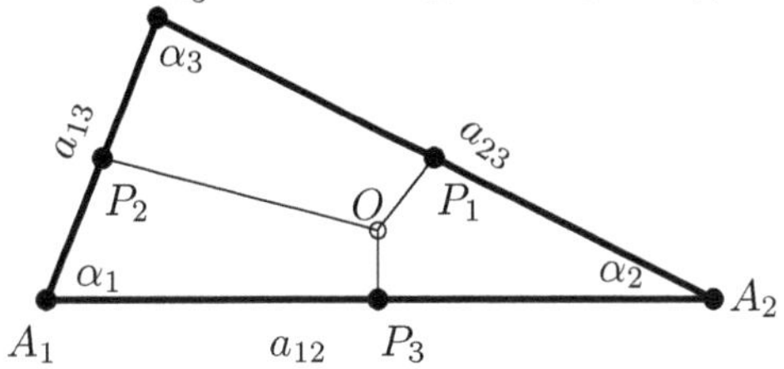

$$\mathbf{p}_1 = -A_1 + P_1, \qquad p_1 = \|\mathbf{p}_1\|$$

$$\mathbf{p}_2 = -A_2 + P_2, \qquad p_2 = \|\mathbf{p}_2\|$$

$$\mathbf{p}_3 = -A_3 + P_3, \qquad p_3 = \|\mathbf{p}_3\|$$

$$a_{23}^2 = a_{12}^2 + a_{13}^2 - 2a_{12}a_{13}\cos\alpha_1$$

$$a_{13}^2 = a_{12}^2 + a_{23}^2 - 2a_{12}a_{23}\cos\alpha_2$$

$$a_{12}^2 = a_{13}^2 + a_{23}^2 - 2a_{13}a_{23}\cos\alpha_3$$

Fig. 1.9 The Triangle Circumcenter is located at the intersection of its perpendicular bisectors. Accordingly, it is equidistant from the triangle vertices.

1.12 Triangle Circumcenter

The triangle circumcenter is located at the intersection of the perpendicular bisectors of its sides, Fig. 1.9. Accordingly, it is equidistant from the triangle vertices.

Let $A_1A_2A_3$ be a triangle with vertices A_1, A_2 and A_3 in a Euclidean n-space $\mathbb{R}^n$, and let O be the triangle circumcenter, as shown in Fig. 1.9. Then, O is given in terms of its barycentric coordinates $(m_1 : m_2 : m_3)$ with respect to the set $\{A_1, A_2, A_3\}$ by the equation

$$O = \frac{m_1A_1 + m_2A_2 + m_3A_3}{m_1 + m_2 + m_3} \tag{1.128}$$

where the barycentric coordinates m_1, m_2 and m_3 of O are to be determined in (1.137).

Applying Lemma 1.7, p. 13, to the point $P = O$ in (1.128) we obtain the equations

$$\begin{aligned}
\|-A_1+O\|^2 &= \frac{m_2^2a_{12}^2+m_3^2a_{13}^2+m_2m_3(\ \ a_{12}^2+a_{13}^2-a_{23}^2)}{(m_1+m_2+m_3)^2} \\
\|-A_2+O\|^2 &= \frac{m_1^2a_{12}^2+m_3^2a_{23}^2+m_1m_3(\ \ a_{12}^2-a_{13}^2+a_{23}^2)}{(m_1+m_2+m_3)^2} \\
\|-A_3+O\|^2 &= \frac{m_1^2a_{13}^2+m_2^2a_{23}^2+m_1m_2(-a_{12}^2+a_{13}^2+a_{23}^2)}{(m_1+m_2+m_3)^2}
\end{aligned} \tag{1.129}$$

Equations (1.129) along with the triangle circumcenter condition, Fig. 1.9,

$$\begin{aligned}
\|-A_1+O\|^2 &= \|-A_2+O\|^2 \\
\|-A_2+O\|^2 &= \|-A_3+O\|^2
\end{aligned} \tag{1.130}$$

and the normalization condition

$$m_1+m_2+m_3=1 \tag{1.131}$$

give the following system of three equations for the three unknowns m_1, m_2 and m_3:

$$\begin{aligned}
& m_2^2a_{12}^2+m_3^2a_{13}^2+m_2m_3(a_{12}^2+a_{13}^2-a_{23}^2) = \\
&\quad m_1^2a_{12}^2+m_3^2a_{23}^2+m_1m_3(\ \ a_{12}^2-a_{13}^2+a_{23}^2) \\
& m_1^2a_{12}^2+m_3^2a_{23}^2+m_1m_3(a_{12}^2-a_{13}^2+a_{23}^2) = \\
&\quad m_1^2a_{13}^2+m_2^2a_{23}^2+m_1m_2(-a_{12}^2+a_{13}^2+a_{23}^2) \\
& \qquad\qquad m_1+m_2+m_3=1
\end{aligned} \tag{1.132}$$

Substituting $m_3 = 1 - m_1 - m_2$ from the third equation in (1.132) into the first two equations in (1.132), and simplifying (the use of a computer system for algebra, like Mathematica or Maple, is recommended) we obtain two equations for the unknowns m_1 and m_2, each of which turns out to be linear in m_1 and quadratic in m_2. Eliminating m_2^2 between these two equations, we obtain the following single equation that relates m_1 to m_2 linearly:

$$a_{13}^2-a_{23}^2-m_1(a_{12}^2+a_{13}^2-a_{23}^2)+m_2(a_{12}^2-a_{13}^2+a_{23}^2)=0 \tag{1.133}$$

A vertex cyclic permutation in (1.133) gives a second linear connection between m_2 and m_3. A third linear connection, between m_1, m_2 and m_3, is provided by (1.131) thus obtaining the following system of three linear equations for the three unknowns m_1, m_2 and m_3:

$$
\begin{aligned}
a_{13}^2 - a_{23}^2 - m_1(a_{12}^2 + a_{13}^2 - a_{23}^2) + m_2(\ \ a_{12}^2 - a_{13}^2 + a_{23}^2) &= 0 \\
a_{12}^2 - a_{13}^2 - m_2(a_{12}^2 - a_{13}^2 + a_{23}^2) + m_3(-a_{12}^2 + a_{13}^2 + a_{23}^2) &= 0 \\
m_1 + m_2 + m_3 &= 1
\end{aligned}
\tag{1.134}
$$

The solution of the linear system (1.134) gives the special barycentric coordinates $\{m_1, m_2, m_3\}$ of the triangle circumcenter O:

$$
\begin{aligned}
m_1 &= \frac{a_{23}^2(\ \ a_{12}^2 + a_{13}^2 - a_{23}^2)}{D} \\
m_2 &= \frac{a_{13}^2(\ \ a_{12}^2 - a_{13}^2 + a_{23}^2)}{D} \\
m_3 &= \frac{a_{12}^2(-a_{12}^2 + a_{13}^2 + a_{23}^2)}{D}
\end{aligned}
\tag{1.135}
$$

in terms of its side lengths, where D is given by

$$
D = (a_{12}^2 + a_{13}^2 + a_{23}^2)(-a_{12}^2 + a_{13}^2 + a_{23}^2)(a_{12}^2 - a_{13}^2 + a_{23}^2)(a_{12}^2 + a_{13}^2 - a_{23}^2)
\tag{1.136}
$$

Finally, it follows from (1.135) that convenient barycentric coordinates $\{m_1 : m_2 : m_3\}$ of the triangle circumcenter O are given by

$$
\begin{aligned}
m_1 &= a_{23}^2(\ \ a_{12}^2 + a_{13}^2 - a_{23}^2) \\
m_2 &= a_{13}^2(\ \ a_{12}^2 - a_{13}^2 + a_{23}^2) \\
m_3 &= a_{12}^2(-a_{12}^2 + a_{13}^2 + a_{23}^2)
\end{aligned}
\tag{1.137}
$$

We now wish to find trigonometric barycentric coordinates for the triangle circumcenter, that is, barycentric coordinates that are expressed in terms of the triangle angles. Hence, we calculate m_1/m_3 and m_2/m_3 by means of (1.137) and the law of sines (1.63), p. 21, and employ the trigonometric identity $\sin^2\alpha = (1 - \cos 2\alpha)/2$, obtaining

$$
\begin{aligned}
\frac{m_1}{m_3} &= \frac{(1 + \cos 2\alpha_1 - \cos 2\alpha_2 - \cos 2\alpha_3)\sin^2\alpha_1}{(1 - \cos 2\alpha_1 - \cos 2\alpha_2 + \cos 2\alpha_3)\sin^2\alpha_3} \\
\frac{m_2}{m_3} &= \frac{(1 - \cos 2\alpha_1 + \cos 2\alpha_2 - \cos 2\alpha_3)\sin^2\alpha_2}{(1 - \cos 2\alpha_1 - \cos 2\alpha_2 + \cos 2\alpha_3)\sin^2\alpha_3}
\end{aligned}
\tag{1.138}
$$

Hence, convenient trigonometric barycentric coordinates $\{m_1 : m_2 : m_3\}$ for a triangle circumcenter are given by

$$\begin{aligned} m_1 &= (1 + \cos 2\alpha_1 - \cos 2\alpha_2 - \cos 2\alpha_3)\sin^2\alpha_1 \\ m_2 &= (1 - \cos 2\alpha_1 + \cos 2\alpha_2 - \cos 2\alpha_3)\sin^2\alpha_2 \\ m_3 &= (1 - \cos 2\alpha_1 - \cos 2\alpha_2 + \cos 2\alpha_3)\sin^2\alpha_3 \end{aligned} \tag{1.139}$$

Taking advantage of the relationship (1.65), p. 21, between triangle angles, and employing trigonometric identities, (1.139) can be simplified, obtaining the following elegant trigonometric barycentric coordinates of the circumcenter O of triangle $A_1A_2A_3$, Fig. 1.9, in terms of its angles,

$$\begin{aligned} m_1 &= \sin\alpha_1 \cos\alpha_1 \\ m_2 &= \sin\alpha_2 \cos\alpha_2 \\ m_3 &= \sin\alpha_3 \cos\alpha_3 \end{aligned} \tag{1.140}$$

For later reference we note that owing to the π-identity of triangles, (1.65), p. 21, the three equations in (1.140) are equivalent to the following three equations:

$$\begin{aligned} m_1 &= \sin\left(\frac{-\alpha_1 + \alpha_2 + \alpha_3}{2}\right)\sin\alpha_1 \\ m_2 &= \sin\left(\frac{\alpha_1 - \alpha_2 + \alpha_3}{2}\right)\sin\alpha_2 \\ m_3 &= \sin\left(\frac{\alpha_1 + \alpha_2 - \alpha_3}{2}\right)\sin\alpha_3 \end{aligned} \tag{1.141}$$

There is an important distinction between the elegant barycentric coordinates (1.141) and their simplified form (1.140). The former, (1.141), is free of the π-identity condition, while the latter, (1.140), embodies the π-identity. As a result, the validity of the latter is restricted to Euclidean geometry, where the π-identity holds. The former is also valid in Euclidean geometry but, unlike the latter, it survives unimpaired in hyperbolic geometry as well, where the π-identity does not hold.

Indeed, it will be found in (4.251), p. 237, that hyperbolic barycentric coordinates of the hyperbolic circumcenter of hyperbolic triangles in the Cartesian-Beltrami-Klein ball model of hyperbolic geometry are given by (1.141) as well.

1.13 Circumradius

The circumradius R of a triangle $A_1A_2A_3$ in a Euclidean space $\mathbb{R}^n$ is the radius of its circumcircle. Hence, in the notation of Fig. 1.10,

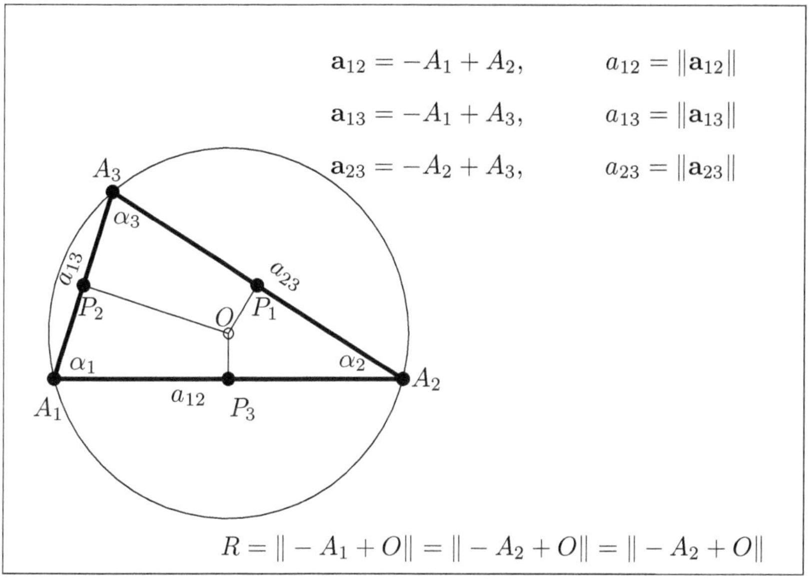

Fig. 1.10 The Circumcenter O and Circumradius R of a triangle $A_1A_2A_3$ in a Euclidean space $\mathbb{R}^n$.

$$\begin{aligned} R^2 &= \| - A_1 + O\|^2 \\ &= \| - A_2 + O\|^2 \\ &= \| - A_3 + O\|^2 \end{aligned} \tag{1.142}$$

where O is the triangle circumcenter.

The circumradius $R = \| - A_1 + O\|$ is determined by successively substituting $\| - A_1 + O\|^2$ from the first equation in (1.129), p. 36, and m_k, $k = 1, 2, 3$, from (1.137), p. 37, into (1.142), obtaining

$$\begin{aligned} R^2 &= \frac{a_{12}^2 a_{13}^2 a_{23}^2}{16p(p - a_{12})(p - a_{13})(p - a_{23})} \\ &= \frac{a_{12}^2 a_{13}^2 a_{23}^2}{(a_{12} + a_{13} + a_{23})(-a_{12} + a_{13} + a_{23})(a_{12} - a_{13} + a_{23})(a_{12} + a_{13} - a_{23})} \\ &= \frac{a_{12}^2 a_{13}^2 a_{23}^2}{16|A_1A_2A_3|^2} \end{aligned} \tag{1.143}$$

Hence, the triangle circumradius is given by

$$R = \frac{a_{12}a_{13}a_{23}}{4\sqrt{p(p-a_{12})(p-a_{13})(p-a_{23}}} = \frac{a_{12}a_{13}a_{23}}{4|A_1A_2A_3|} = \frac{a_{12}a_{13}a_{23}}{4rp} \tag{1.144}$$

where p and r are the triangle semiperimeter and inradius, and where $|A_1A_2A_3|$ is the triangle area given by Heron's formula (1.125), p. 34.

1.14 Triangle Incircle and Excircles

An incircle of a triangle is a circle lying inside the triangle, tangent to each of its sides, shown in Fig. 1.8, p. 33. The center and radius of the incircle of a triangle are called the triangle incenter and inradius. Similarly, an excircle of a triangle is a circle lying outside the triangle, tangent to one of its sides and tangent to the extensions of the other two. The centers and radii of the excircles of a triangle are called the triangle excenters and exradii. The incenter and excenters of a triangle, shown in Fig. 1.11, are equidistant from the triangles sides.

Let E be an incenter or an excenter of a triangle $A_1A_2A_3$, Fig. 1.11, in a Euclidean n space $\mathbb{R}^n$, and let

$$E = \frac{m_1A_1 + m_2A_2 + m_3A_3}{m_1 + m_2 + m_3} \tag{1.145}$$

be the barycentric coordinate representation of E with respect to the set $\{A_1, A_2, A_3\}$, where the barycentric coordinates m_k, $k = 1, 2, 3$, are to be determined in Theorem 1.16, p. 44.

Applying Lemma 1.7, p. 13, to the point $P = E$ in (1.145) we obtain the equations

$$\begin{aligned}
\|-A_1+E\|^2 &= \frac{m_2^2a_{12}^2 + m_3^2a_{13}^2 + m_2m_3(\ \ a_{12}^2 + a_{13}^2 - a_{23}^2)}{(m_1+m_2+m_3)^2} \\
\|-A_2+E\|^2 &= \frac{m_1^2a_{12}^2 + m_3^2a_{23}^2 + m_1m_3(\ \ a_{12}^2 - a_{13}^2 + a_{23}^2)}{(m_1+m_2+m_3)^2} \\
\|-A_3+E\|^2 &= \frac{m_1^2a_{13}^2 + m_2^2a_{23}^2 + m_1m_2(-a_{12}^2 + a_{13}^2 + a_{23}^2)}{(m_1+m_2+m_3)^2}
\end{aligned} \tag{1.146}$$

Imposing the normalization condition

$$m_1 + m_2 + m_3 = 1 \tag{1.147}$$

in (1.145) – (1.146) is clearly convenient.

Let E represent one of the incenter and excenters E_k, $k = 0, 1, 2, 3$, of a triangle $A_1A_2A_3$ in a Euclidean n-space $\mathbb{R}^n$, shown in Fig. 1.11 for $n = 2$.

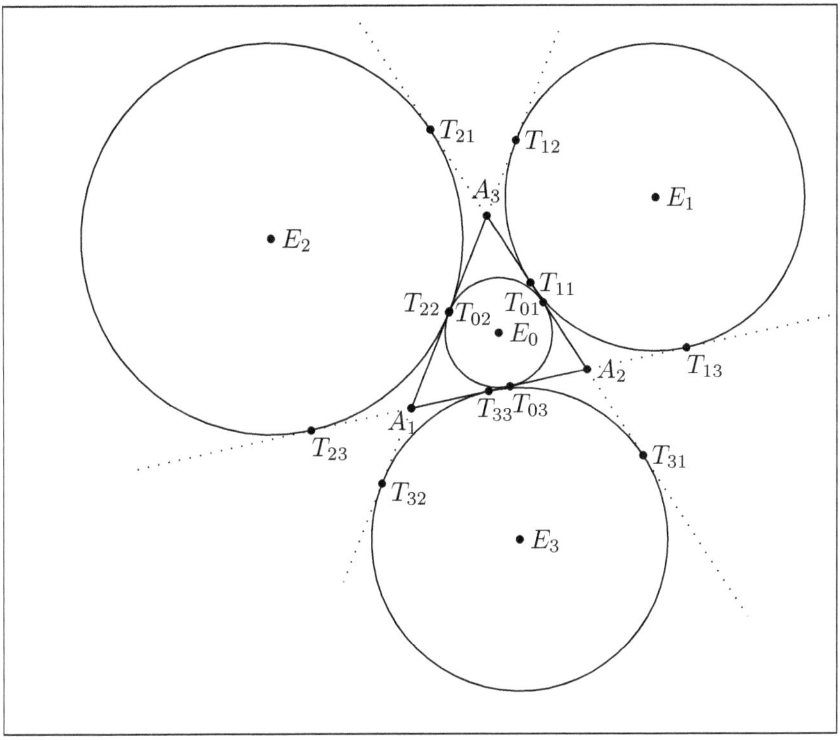

Fig. 1.11 The triangle incircle and excircles. Here T_{ij} is the tangency point where the in-excircle with center E_i, $i = 0, 1, 2, 3$, meets the triangle side, or its extension, opposite to vertex A_j, $j = 1, 2, 3$. The incircle points of tangency, T_{0j} are determined by Theorem 1.13, p. 33, and the excircle points of tangency, T_{ij}, $i = 1, 2, 3$, are determined by Theorem 1.17, p. 46. Trigonometric barycentric coordinate representations of the in-excircle tangency points T_{ij} are listed in (1.169), p. 48.

(1) The distance of E from the line $L_{A_1A_2}$ that passes through points A_1 and A_2, Fig. 1.11, is the altitude r_3 of triangle A_1A_2E drawn from base A_1A_2. Hence, by the point-line distance formula (1.78), p. 24, in Theorem 1.12, the distance r_3 between the point E and the line $L_{A_1A_2}$ is given by the equation

$$r_3^2 = \frac{(a_{12} + \bar{a}_{13} + \bar{a}_{23})(-a_{12} + \bar{a}_{13} + \bar{a}_{23})(a_{12} - \bar{a}_{13} + \bar{a}_{23})(a_{12} + \bar{a}_{13} - \bar{a}_{23})}{4a_{12}^2} \tag{1.148a}$$

where

$$\begin{aligned} \bar{a}_{13}^2 &:= \| -A_1 + E\|^2 \\ \bar{a}_{23}^2 &:= \| -A_2 + E\|^2 \end{aligned} \tag{1.148b}$$

Substituting successively, $\bar{a}_{13}$ and $\bar{a}_{23}$ from (1.148b) and $\| -A_1 + E\|$ and $\| -A_2 + E\|$ from (1.146) into (1.148a) we obtain an equation of the form

$$r_3 = f_3(a_{12}, a_{13}, a_{23}, m_1, m_2, m_3) \tag{1.148c}$$

where r_3 is expressed as a function of the sides of triangle $A_1A_2A_3$ and the unknown barycentric coordinates of the point E in (1.145).

(2) The distance of E, Fig. 1.11, from the line $L_{A_1A_3}$ that passes through points A_1 and A_3, Fig. 1.11, is the altitude r_2 of triangle A_1A_3E drawn from base A_1A_3. Hence, by the point-line distance formula (1.78), p. 24, the distance r_2 between the point E and the line $L_{A_1A_3}$ is given by the equation

$$r_2^2 = \frac{(\bar{a}_{12} + a_{13} + \bar{a}_{23})(-\bar{a}_{12} + a_{13} + \bar{a}_{23})(\bar{a}_{12} - a_{13} + \bar{a}_{23})(\bar{a}_{12} + a_{13} - \bar{a}_{23})}{4a_{13}^2} \tag{1.149a}$$

where

$$\begin{aligned} \bar{a}_{12}^2 &:= \| -A_1 + E\|^2 \\ \bar{a}_{23}^2 &:= \| -A_3 + E\|^2 \end{aligned} \tag{1.149b}$$

Substituting successively, $\bar{a}_{12}$ and $\bar{a}_{23}$ from (1.149b) and $\| -A_1 + E\|$ and $\| -A_3 + E\|$ from (1.146) into (1.149a) we obtain an equation of the form

$$r_2 = f_2(a_{12}, a_{13}, a_{23}, m_1, m_2, m_3) \tag{1.149c}$$

where r_2 is expressed as a function of the sides of triangle $A_1A_2A_3$ and the unknown barycentric coordinates of the point E in (1.145).

(3) The distance of E from the line $L_{A_2A_3}$ that passes through points A_2 and A_3, Fig. 1.11, is the altitude r_1 of triangle A_2A_3E drawn from base A_2A_3. Hence, by the point-line distance formula (1.78), p. 24, the distance r_1 between the point E and the line $L_{A_2A_3}$ is given by the equation

$$r_1^2 = \frac{(\bar{a}_{12} + \bar{a}_{13} + a_{23})(-\bar{a}_{12} + \bar{a}_{13} + a_{23})(\bar{a}_{12} - \bar{a}_{13} + a_{23})(\bar{a}_{12} + \bar{a}_{13} - a_{23})}{4a_{23}^2} \tag{1.150a}$$

where

$$\bar{a}_{12}^2 := \|-A_2 + E\|^2$$
$$\bar{a}_{13}^2 := \|-A_3 + E\|^2 \tag{1.150b}$$

Substituting successively, $\bar{a}_{12}$ and $\bar{a}_{13}$ from (1.150b) and $\|-A_2 + E\|$ and $\|-A_3 + E\|$ from (1.146) into (1.150a) we obtain an equation of the form

$$r_1 = f_1(a_{12}, a_{13}, a_{23}, m_1, m_2, m_3) \tag{1.150c}$$

where r_1 is expressed as a function of the sides of triangle $A_1A_2A_3$ and the unknown barycentric coordinates of the point E in (1.145).

The condition that the point E, which represents an appropriate one of the points E_k, $k = 0, 1, 2, 3$, Fig. 1.11, is equidistant from the sides of triangle $A_1A_2A_3$ is equivalent to the system of two equations

$$\begin{aligned} r_1 &= r_2 \\ r_1 &= r_3 \end{aligned} \tag{1.151}$$

These equations along with the normalization condition (1.147) form a system of three equations for the three unknowns m_1, m_2, m_3 in (1.145).

Solving the system (1.151) and, then, ignoring the normalization condition (1.147), we obtain

$$\begin{aligned} m_1^2 &= a_{23}^2 \\ m_2^2 &= a_{13}^2 \\ m_3^2 &= a_{12}^2 \end{aligned} \tag{1.152}$$

The equations in (1.152) present eight solutions for the triple (m_1, m_2, m_3). Owing to the homogeneity of barycentric coordinates, only four of the eight solutions are indistinguishable. These are:

$$\begin{aligned} (m_1 : m_2 : m_3) &= (\ \ a_{23} : \ \ a_{13} : \ \ a_{12}) \\ (m_1 : m_2 : m_3) &= (-a_{23} : \ \ a_{13} : \ \ a_{12}) \\ (m_1 : m_2 : m_3) &= (\ \ a_{23} : -a_{13} : \ \ a_{12}) \\ (m_1 : m_2 : m_3) &= (\ \ a_{23} : \ \ a_{13} : -a_{12}) \end{aligned} \tag{1.153}$$

Each of the four barycentric coordinate sets in (1.153) determines the barycentric coordinates of the point E in (1.145), which is equidistant from the sides of triangle $A_1A_2A_3$ in Fig. 1.11. Accordingly, the first equation in (1.153) gives the barycentric coordinates of the incenter, $E = E_0$, of triangle $A_1A_2A_3$, and the other equations in (1.153) give the barycentric coordinates of each of the excenters, $E = E_k$, $k = 1, 2, 3$, of the triangle.

By the law of sines (1.63), p. 21, and owing to their homogeneity, the barycentric coordinates in the first equation in (1.153) can be written as

$$\begin{aligned}(m_1 : m_2 : m_3) &= \left(\frac{a_{23}}{a_{12}} : \frac{a_{13}}{a_{12}} : 1\right) \\ &= \left(\frac{\sin\alpha_1}{\sin\alpha_3} : \frac{\sin\alpha_2}{\sin\alpha_3} : 1\right) \\ &= (\sin\alpha_1 : \sin\alpha_2 : \sin\alpha_3)\end{aligned} \tag{1.154}$$

Similarly, all the barycentric coordinates in (1.153) can be expressed trigonometrically in terms of the triangle angles.

Formalizing, we thus obtain the following theorem:

Theorem 1.16. (In-Excenters Barycentric Representations). *Let $A_1A_2A_3$ be a triangle with incenter E_0 and excenters E_k, $k = 1, 2, 3$, in a Euclidean space $\mathbb{R}^n$, Fig. 1.11. Then the barycentric coordinate representations of the triangle in-excenters E_k, $k = 0, 1, 2, 3$, are given by the equations*

$$\begin{aligned}E_0 &= \frac{a_{23}A_1 + a_{13}A_2 + a_{12}A_3}{a_{23} + a_{13} + a_{12}} \\ E_1 &= \frac{-a_{23}A_1 + a_{13}A_2 + a_{12}A_3}{-a_{23} + a_{13} + a_{12}} \\ E_2 &= \frac{a_{23}A_1 - a_{13}A_2 + a_{12}A_3}{a_{23} - a_{13} + a_{12}} \\ E_3 &= \frac{a_{23}A_1 + a_{13}A_2 - a_{12}A_3}{a_{23} + a_{13} - a_{12}}\end{aligned} \tag{1.155}$$

and their trigonometric barycentric coordinate representations are given by the equations

$$\begin{aligned}E_0 &= \frac{\sin\alpha_1 A_1 + \sin\alpha_2 A_2 + \sin\alpha_3 A_3}{\sin\alpha_1 + \sin\alpha_2 + \sin\alpha_3} \\ E_1 &= \frac{-\sin\alpha_1 A_1 + \sin\alpha_2 A_2 + \sin\alpha_3 A_3}{-\sin\alpha_1 + \sin\alpha_2 + \sin\alpha_3} \\ E_2 &= \frac{\sin\alpha_1 A_1 - \sin\alpha_2 A_2 + \sin\alpha_3 A_3}{\sin\alpha_1 - \sin\alpha_2 + \sin\alpha_3} \\ E_3 &= \frac{\sin\alpha_1 A_1 + \sin\alpha_2 A_2 - \sin\alpha_3 A_3}{\sin\alpha_1 + \sin\alpha_2 - \sin\alpha_3}\end{aligned} \tag{1.156}$$

1.15 Excircle Tangency Points

Let $A_1A_2A_3$ be a triangle in a Euclidean space $\mathbb{R}^n$, with excircles centered at the points E_k, $k = 1,2,3$, and let the tangency point where the A_3-excircle meets the triangle side A_1A_2 be T_{33}, as shown in Fig. 1.11, p. 41. The tangency point T_{33} is the perpendicular projection of the point E_3 on the line $L_{A_1A_2}$ that passes through the points A_1 and A_2, Fig. 1.11. Hence, by the point to line perpendicular projection formula (1.77), p. 23, the point T_{33} possesses the barycentric coordinate representation

$$T_{33} = \frac{1}{2}\left(\frac{a_{12}^2 - \bar{a}_{13}^2 + \bar{a}_{23}^2}{a_{12}^2}\right) A_1 + \frac{1}{2}\left(\frac{a_{12}^2 + \bar{a}_{13}^2 - \bar{a}_{23}^2}{a_{12}^2}\right) A_2 \qquad (1.157)$$

with respect to the set $S = \{A_1, A_2, A_3\}$ of the triangle vertices, where

$$\begin{aligned} \bar{a}_{13}^2 &:= \|-A_1 + E_3\|^2 \\ \bar{a}_{23}^2 &:= \|-A_2 + E_3\|^2 \end{aligned} \qquad (1.158)$$

By Lemma 1.7, p. 13, we have

$$\begin{aligned} \|-A_1 + E_3\|^2 &= \frac{m_2^2a_{12}^2 + m_3^2a_{13}^2 + m_2m_3(\ \ a_{12}^2 + a_{13}^2 - a_{23}^2)}{(m_1+m_2+m_3)^2} \\ \|-A_2 + E_3\|^2 &= \frac{m_1^2a_{12}^2 + m_3^2a_{23}^2 + m_1m_3(\ \ a_{12}^2 - a_{13}^2 + a_{23}^2)}{(m_1+m_2+m_3)^2} \\ \|-A_3 + E_3\|^2 &= \frac{m_1^2a_{13}^2 + m_2^2a_{23}^2 + m_1m_2(-a_{12}^2 + a_{13}^2 + a_{23}^2)}{(m_1+m_2+m_3)^2} \end{aligned} \qquad (1.159)$$

where m_k, $k = 1,2,3$, are the barycentric coordinates of E_3, given by

$$(m_1 : m_2 : m_3) = (a_{23} : a_{13} : -a_{12}) \qquad (1.160)$$

as indicated in the fourth equation in (1.155).

Substituting successively, $\bar{a}_{12}^2$ and $\bar{a}_{13}^2$ from (1.158), $\|-A_1 + E_3\|^2$ and $\|-A_2 + E_3\|^2$ from (1.159), and m_k, $k = 1,2,3$, from (1.160) into (1.157) we obtain

$$T_{33} = \frac{1}{2}\left(\frac{a_{12} + a_{13} - a_{23}}{a_{12}}\right) A_1 + \frac{1}{2}\left(\frac{a_{12} - a_{13} + a_{23}}{a_{12}}\right) A_2 \qquad (1.161)$$

Now let T_{32} be the tangency point where the A_3-excircle meets the extension of the triangle side A_1A_3, as shown in Fig. 1.11, p. 41. The tangency point T_{32} is the perpendicular projection of the point E_3 on the line $L_{A_1A_3}$ that passes through the points A_1 and A_3, Fig. 1.11. Hence, by the point to line perpendicular projection formula (1.77), p. 23, the point T_{32} possesses the barycentric coordinate representation

$$T_{32} = \frac{1}{2}\left(\frac{-\bar{a}_{12}^2 + a_{13}^2 + \bar{a}_{23}^2}{a_{13}^2}\right) A_1 + \frac{1}{2}\left(\frac{\bar{a}_{12}^2 + a_{13}^2 - \bar{a}_{23}^2}{a_{13}^2}\right) A_3 \tag{1.162}$$

with respect to the set $S = \{A_1, A_2, A_3\}$ of the triangle vertices, where

$$\begin{aligned} \bar{a}_{12}^2 &:= \|-A_1 + E_3\|^2 \\ \bar{a}_{23}^2 &:= \|-A_3 + E_3\|^2 \end{aligned} \tag{1.163}$$

Substituting successively, $\bar{a}_{12}^2$ and $\bar{a}_{23}^2$ from (1.163), $\|-A_1 + E_3\|^2$ and $\|-A_2 + E_3\|^2$ from (1.159), and m_k, $k = 1, 2, 3$, from (1.160) into (1.162) we obtain

$$T_{32} = \frac{1}{2}\left(\frac{a_{12} + a_{13} + a_{23}}{a_{13}}\right) A_1 + \frac{1}{2}\left(\frac{-a_{12} + a_{13} - a_{23}}{a_{13}}\right) A_3 \tag{1.164}$$

Finally, let T_{31} be the tangency point where the A_3-excircle meets the extension of the triangle side A_2A_3, as shown in Fig. 1.11, p. 41. Then,

$$T_{31} = \frac{1}{2}\left(\frac{a_{12} + a_{13} + a_{23}}{a_{23}}\right) A_2 + \frac{1}{2}\left(\frac{-a_{12} - a_{13} + a_{23}}{a_{23}}\right) A_3 \tag{1.165}$$

Here, (1.165) is obtained from (1.164) by interchanging the triangle vertices A_1 and A_2.

Equations (1.161) and (1.164)–(1.165) give rise to the following theorem:

Theorem 1.17. (Excircle Tangency Points). *Let $A_1A_2A_3$ be a triangle in a Euclidean space $\mathbb{R}^n$ and let T_{ij}, $i, j = 1, 2, 3$, be the points of tangency where the triangle A_i-excircle, $i = 1, 2, 3$, meets the side opposite to A_j, or its extension, of the triangle, Fig. 1.11, p. 41.*

Then, barycentric coordinate representations of the tangency points T_{ij} with respect to the pointwise independent set $S = \{A_1, A_2, A_3\}$ are given by the equations listed below.

$$\begin{aligned} T_{11} &= \frac{a_{12} - a_{13} + a_{23}}{2a_{23}} A_2 + \frac{-a_{12} + a_{13} + a_{23}}{2a_{23}} A_3 \\ T_{12} &= \frac{-a_{12} + a_{13} - a_{23}}{2a_{13}} A_1 + \frac{a_{12} + a_{13} + a_{23}}{2a_{13}} A_3 \\ T_{13} &= \frac{a_{12} - a_{13} - a_{23}}{2a_{12}} A_1 + \frac{a_{12} + a_{13} + a_{23}}{2a_{12}} A_2 \end{aligned} \tag{1.166a}$$

$$T_{21} = \frac{-a_{12} - a_{13} + a_{23}}{2a_{23}} A_2 + \frac{a_{12} + a_{13} + a_{23}}{2a_{23}} A_3$$

$$T_{22} = \frac{a_{12} + a_{13} - a_{23}}{2a_{13}} A_1 + \frac{-a_{12} + a_{13} + a_{23}}{2a_{13}} A_3 \qquad (1.166b)$$

$$T_{23} = \frac{a_{12} + a_{13} + a_{23}}{2a_{12}} A_1 + \frac{a_{12} - a_{13} - a_{23}}{2a_{12}} A_2$$

$$T_{31} = \frac{a_{12} + a_{13} + a_{23}}{2a_{23}} A_2 + \frac{-a_{12} - a_{13} + a_{23}}{2a_{23}} A_3$$

$$T_{32} = \frac{a_{12} + a_{13} + a_{23}}{2a_{13}} A_1 + \frac{-a_{12} + a_{13} - a_{23}}{2a_{13}} A_3 \qquad (1.166c)$$

$$T_{33} = \frac{a_{12} + a_{13} - a_{23}}{2a_{12}} A_1 + \frac{a_{12} - a_{13} + a_{23}}{2a_{12}} A_2$$

Proof. The proof of (1.166c) is established in (1.161) and (1.164) – (1.165). The proof of (1.166a) – (1.166b) follows from (1.166c) by invoking cyclicity, that is, by cyclic permutations of the triangle vertices. □

By the law of cosines for triangle $A_1A_2A_3$ in Theorem 1.17, with the triangle standard notation in Fig. 1.2, p. 7, and by the triangle π condition (1.65), p. 21, that triangle angles obey, we have the first equation in (1.167) below.

$$\begin{aligned}\frac{a_{12} - a_{13} + a_{23}}{2a_{23}} &= \frac{1}{2}\left(\frac{\sin\alpha_3}{\sin\alpha_1} - \frac{\sin\alpha_2}{\sin\alpha_1} + 1\right) = \frac{\tan\frac{\alpha_3}{2}}{\tan\frac{\alpha_3}{2} + \tan\frac{\alpha_2}{2}} \\ \frac{-a_{12} + a_{13} + a_{23}}{2a_{23}} &= \frac{1}{2}\left(\frac{\sin\alpha_3}{\sin\alpha_1} + \frac{\sin\alpha_2}{\sin\alpha_1} + 1\right) = \frac{\tan\frac{\alpha_2}{2}}{\tan\frac{\alpha_3}{2} + \tan\frac{\alpha_2}{2}}\end{aligned} \qquad (1.167)$$

The second equation in (1.167) is obtained in a similar way.

Hence, the barycentric coordinate representation of the tangency point T_{11} in (1.166a) can be written as the trigonometric barycentric coordinate representation

$$T_{11} = \frac{\tan\frac{\alpha_3}{2} A_2 + \tan\frac{\alpha_2}{2} A_3}{\tan\frac{\alpha_3}{2} + \tan\frac{\alpha_2}{2}} = \frac{\cot\frac{\alpha_2}{2} A_2 + \cot\frac{\alpha_2}{3} A_3}{\cot\frac{\alpha_2}{2} + \cot\frac{\alpha_2}{3}} \qquad (1.168)$$

As in (1.168), all the barycentric coordinate representations of the tangency points in Theorem 1.13, p. 33, and Theorem 1.17, p. 46, shown in Fig. 1.11, can be written as trigonometric barycentric coordinate representations, obtaining the following theorem:

Theorem 1.18. (In-Excircle Tangency Points). *Let $A_1A_2A_3$ be a triangle in a Euclidean space $\mathbb{R}^n$ and let T_{ij}, $i = 0, 1, 2, 3$, $j = 1, 2, 3$, be the points of tangency where the triangle in-excircle with center E_i meets the side opposite to A_j, or its extension, of the triangle, Fig. 1.11, p. 41.*

Then, trigonometric barycentric coordinate representations of the tangency points T_{ij} with respect to the pointwise independent set $S = \{A_1, A_2, A_3\}$, are given by the equations listed below.

$$\begin{aligned} T_{01} &= \frac{\tan\frac{\alpha_2}{2}A_2 + \tan\frac{\alpha_3}{2}A_3}{\tan\frac{\alpha_2}{2} + \tan\frac{\alpha_3}{2}} \\ T_{02} &= \frac{\tan\frac{\alpha_1}{2}A_1 + \tan\frac{\alpha_3}{2}A_3}{\tan\frac{\alpha_1}{2} + \tan\frac{\alpha_3}{2}} \\ T_{03} &= \frac{\tan\frac{\alpha_1}{2}A_1 + \tan\frac{\alpha_2}{2}A_2}{\tan\frac{\alpha_1}{2} + \tan\frac{\alpha_2}{2}} \end{aligned} \tag{1.169a}$$

$$\begin{aligned} T_{11} &= \frac{\cot\frac{\alpha_2}{2}A_2 + \cot\frac{\alpha_3}{2}A_3}{\cot\frac{\alpha_2}{2} + \cot\frac{\alpha_3}{2}} \\ T_{12} &= \frac{\tan\frac{\alpha_1}{2}A_1 - \cot\frac{\alpha_3}{2}A_3}{\tan\frac{\alpha_1}{2} - \cot\frac{\alpha_3}{2}} \\ T_{13} &= \frac{\tan\frac{\alpha_1}{2}A_1 - \cot\frac{\alpha_2}{2}A_2}{\tan\frac{\alpha_1}{2} - \cot\frac{\alpha_2}{2}} \end{aligned} \tag{1.169b}$$

$$\begin{aligned} T_{21} &= \frac{\tan\frac{\alpha_2}{2}A_2 - \cot\frac{\alpha_3}{2}A_3}{\tan\frac{\alpha_2}{2} - \cot\frac{\alpha_3}{2}} \\ T_{22} &= \frac{\cot\frac{\alpha_1}{2}A_1 + \cot\frac{\alpha_3}{2}A_3}{\cot\frac{\alpha_1}{2} + \cot\frac{\alpha_3}{2}} \\ T_{23} &= \frac{\cot\frac{\alpha_1}{2}A_1 - \tan\frac{\alpha_2}{2}A_2}{\cot\frac{\alpha_1}{2} - \tan\frac{\alpha_2}{2}} \end{aligned} \tag{1.169c}$$

$$\begin{aligned} T_{31} &= \frac{\cot\frac{\alpha_2}{2}A_2 - \tan\frac{\alpha_3}{2}A_3}{\cot\frac{\alpha_2}{2} - \tan\frac{\alpha_3}{2}} \\ T_{32} &= \frac{\cot\frac{\alpha_1}{2}A_1 - \tan\frac{\alpha_3}{2}A_3}{\cot\frac{\alpha_1}{2} - \tan\frac{\alpha_3}{2}} \\ T_{33} &= \frac{\cot\frac{\alpha_1}{2}A_1 + \cot\frac{\alpha_2}{2}A_2}{\cot\frac{\alpha_1}{2} + \cot\frac{\alpha_2}{2}} \end{aligned} \tag{1.169d}$$

Proof. The trigonometric barycentric coordinate representation of T_{11} in (1.169b) is established in (1.168). All the other trigonometric barycentric coordinate representations in the Theorem can be established in a similar way. □

Surprisingly, we will find in Theorem 5.2, p. 260, that the trigonometric barycentric coordinates in the trigonometric barycentric coordinate representations of tangency points in Theorem 1.18 survive unimpaired in the transition from Euclidean to hyperbolic geometry. This remarkable result indicates the comparative advantage of trigonometric barycentric coordinate representations in our comparative study of the transition from Euclidean to hyperbolic geometry.

1.16 From Triangle Tangency Points to Triangle Centers

The triangle tangency points T_{ij}, $i = 0, 1, 2, 3$, $j = 1, 2, 3$, in Theorem 1.18, shown in Fig. 1.11, p. 41, give rise to the following three triangle centers:

(1) *Gergonne Point* G_e. In the notation of Fig. 1.11, the lines A_kT_{0k}, $k = 1, 2, 3$, are concurrent, Fig. 1.12. The resulting point of concurrency, called the triangle Gergonne point, G_e, possesses the trigonometric barycentric coordinate representation (see Exercise 9, p. 61)

$$G_e = \frac{\tan\frac{\alpha_1}{2}A_1 + \tan\frac{\alpha_2}{2}A_2 + \tan\frac{\alpha_3}{2}A_3}{\tan\frac{\alpha_1}{2} + \tan\frac{\alpha_2}{2} + \tan\frac{\alpha_3}{2}} \tag{1.170}$$

with respect to the vertices of the reference triangle $A_1A_2A_3$.

(2) *Nagel Point* N_a. In the notation of Fig. 1.11, the lines A_kT_{kk}, $k = 1, 2, 3$, are concurrent, Fig. 1.13. The resulting point of concurrency, called the triangle Nagel point, N_a, possesses the trigonometric barycentric coordinate representation (see Exercise 10, p. 61)

$$N_a = \frac{\cot\frac{\alpha_1}{2}A_1 + \cot\frac{\alpha_2}{2}A_2 + \cot\frac{\alpha_3}{2}A_3}{\cot\frac{\alpha_1}{2} + \cot\frac{\alpha_2}{2} + \cot\frac{\alpha_3}{2}} \tag{1.171}$$

with respect to the vertices of the reference triangle $A_1A_2A_3$.

(3) *Point* P_u. In the notation of Fig. 1.11, the lines E_kT_{kk}, $k = 1, 2, 3$, are concurrent, Fig. 1.14. The resulting point of concurrency, called the triangle point P_u, possesses the trigonometric barycentric coordinate representation (see Exercise 11, p. 61)

$$P_u = \frac{m_1A_1 + m_2A_2 + m_3A_3}{m_1 + m_2 + m_3} \tag{1.172a}$$

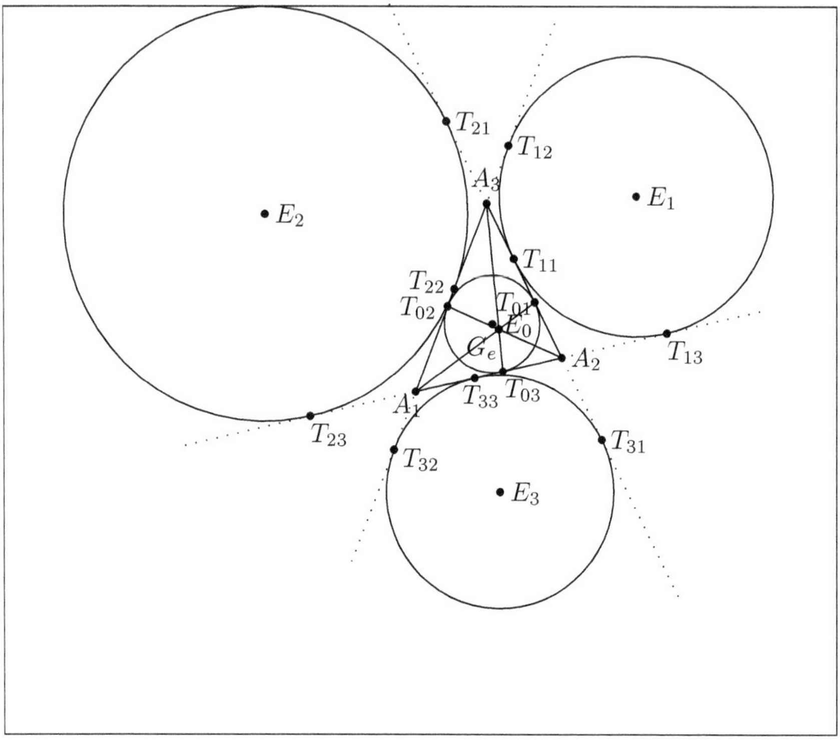

Fig. 1.12 Gergonne Point, G_e. In the notation of Fig. 1.11, p. 41, for the triangle in-excircle tangency points, the lines A_kT_{0k}, $k = 1, 2, 3$, are concurrent, and the resulting point of concurrency is the triangle Gergonne point G_e. Trigonometric barycentric coordinate representation of G_e is given by (1.170).

where gyrotrigonometric gyrobarycentric coordinates $(m_1 : m_2 : m_3)$ of P_u, Fig. 1.14, in (1.172a) are

$$\begin{aligned} m_1 &= \sin\alpha_1(1 + \cos\alpha_1 - \cos\alpha_2 - \cos\alpha_3) \\ m_2 &= \sin\alpha_2(1 - \cos\alpha_1 + \cos\alpha_2 - \cos\alpha_3) \\ m_3 &= \sin\alpha_3(1 - \cos\alpha_1 - \cos\alpha_2 + \cos\alpha_3) \end{aligned} \tag{1.172b}$$

with respect to the vertices of the reference triangle $A_1A_2A_3$.

1.17 Triangle In-Exradii

The A_3-exradius r_3 of a triangle $A_1A_2A_3$ in a Euclidean space $\mathbb{R}^n$ is the distance from excenter E_3 to side A_1A_2 of the triangle, as shown in Fig. 1.11,

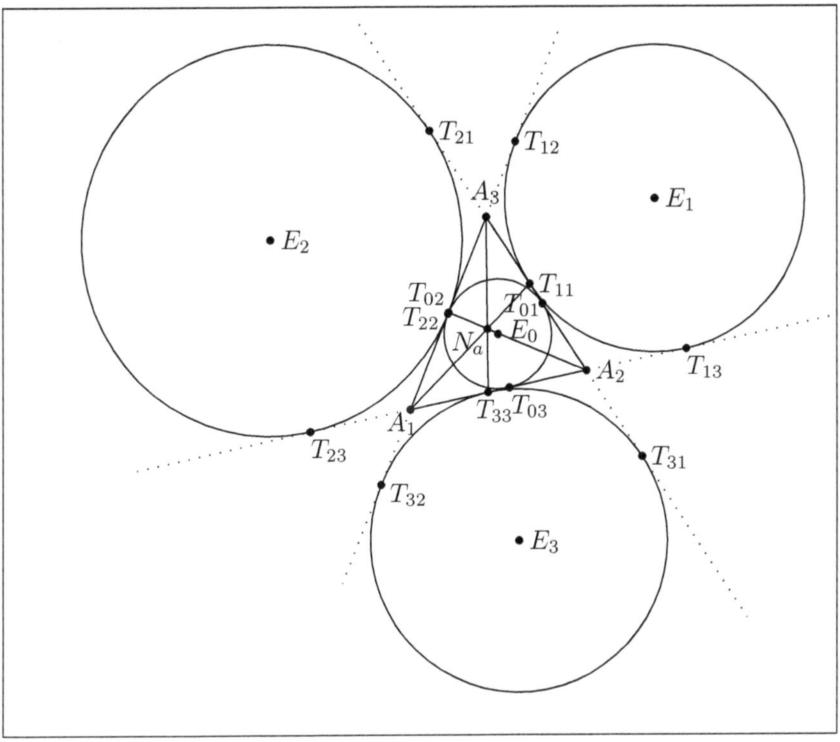

Fig. 1.13 Nagel Point, N_a. In the notation of Fig. 1.11, p. 41, for the triangle in-excircle tangency points, the lines A_kT_{kk}, $k = 1, 2, 3$, are concurrent, and the resulting point of concurrency is the triangle Nagel point. Trigonometric barycentric coordinate representation of N_a is given by (1.171).

p. 41. Applying the point to line distance formula (1.78), p. 24, to calculate the distance r_3 between the point E_3 and the line $L_{A_1A_2}$ that passes through the points A_1 and A_2, Figs. 1.11 – 1.14, we obtain the equation

$$r_3^2 = \frac{(a_{12} + \bar{a}_{13} + \bar{a}_{23})(-a_{12} + \bar{a}_{13} + \bar{a}_{23})(a_{12} - \bar{a}_{13} + \bar{a}_{23})(a_{12} + \bar{a}_{13} - \bar{a}_{23})}{4a_{12}^2} \tag{1.173}$$

where

$$\begin{aligned} \bar{a}_{13}^2 &:= \| - A_1 + E_3\|^2 \\ \bar{a}_{23}^2 &:= \| - A_2 + E_3\|^2 \end{aligned} \tag{1.174}$$

Substituting successively, $\bar{a}_{13}$ and $\bar{a}_{23}$ from (1.174) and $\|-A_1+E_3\|$ and $\|-A_2+E_3\|$ from (1.159), p. 45, along with the barycentric coordinates of

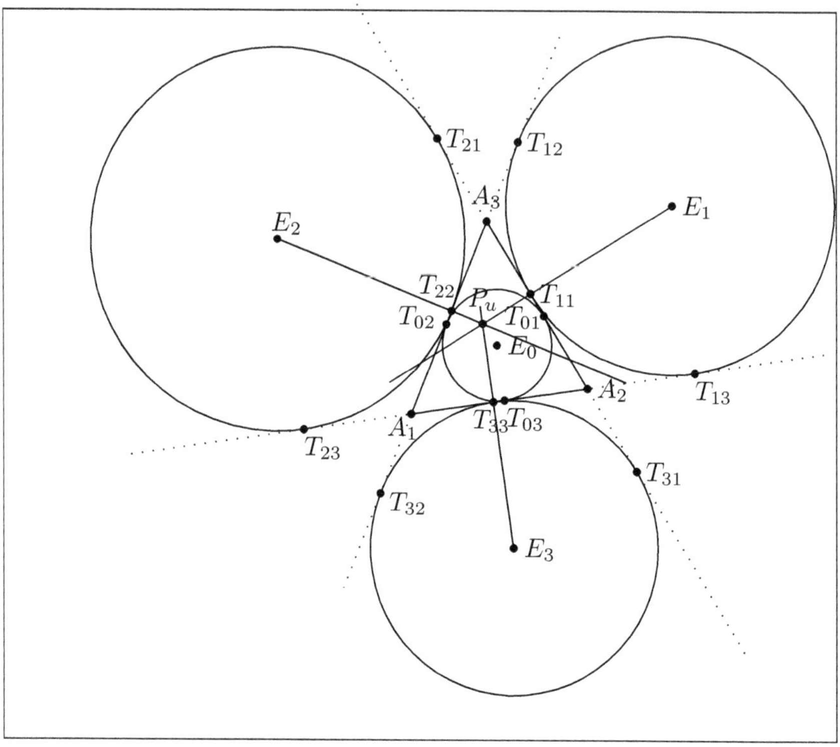

Fig. 1.14 The Triangle P_u Point. In the notation of Fig. 1.11, p. 41, for the triangle in-excircle tangency points, the lines E_kT_{kk}, $k = 1, 2, 3$, are concurrent, and the resulting point of concurrency is the triangle P_u point. Trigonometric barycentric coordinate representation of P_u is given by (1.172).

E_3 in (1.160) into (1.173) we obtain the equation

$$r_3^2 = \frac{p(p - a_{13})(p - a_{23})}{p - a_{12}} \tag{1.175}$$

where p is the triangle semiperimeter (1.124), p. 34.

Equation (1.175) gives rise to the following theorem:

Theorem 1.19. (Triangle In-Exradii). *Let $A_1A_2A_3$ be a triangle in a Euclidean space $\mathbb{R}^n$ with inradius r_0 and exradii r_k, $k = 1, 2, 3$, and with a* *perimeter p. Then*

$$
\begin{aligned}
r_0^2 &= \frac{(p-a_{12})(p-a_{13})(p-a_{23})}{p} \\
r_1^2 &= \frac{p(p-a_{12})(p-a_{13})}{p-a_{23}} \\
r_2^2 &= \frac{p(p-a_{12})(p-a_{23})}{p-a_{13}} \\
r_3^2 &= \frac{p(p-a_{13})(p-a_{23})}{p-a_{12}}
\end{aligned}
\tag{1.176}
$$

Proof. The first equation in (1.176) is the result of Theorem 1.14, p. 34. The fourth equation in (1.176) is established in (1.175). The second and the third equations in (1.176) follow from the fourth equation by cyclic permutations of the triangle vertices. □

By Heron's formula (1.15), p. 34, the equations in (1.176) for the in-exradii r_k, $k = 0, 1, 2, 3$, along with the equation in (1.144), p. 40, for the circumradius R of triangle $A_1A_2A_3$ can be written as

$$
\begin{aligned}
r_0 &= \frac{2|A_1A_2A_3|}{a_{12}+a_{13}+a_{23}} \\
r_1 &= \frac{2|A_1A_2A_3|}{a_{12}+a_{13}-a_{23}} \\
r_2 &= \frac{2|A_1A_2A_3|}{a_{12}-a_{13}+a_{23}} \\
r_3 &= \frac{2|A_1A_2A_3|}{-a_{12}+a_{13}+a_{23}} \\
R &= \frac{a_{12}a_{13}a_{23}}{4|A_1A_2A_3|}
\end{aligned}
\tag{1.177}
$$

implying

$$
\frac{1}{r_1}+\frac{1}{r_2}+\frac{1}{r_3}=\frac{1}{r_0}
\tag{1.178}
$$

and

$$
r_1+r_2+r_3=4R+r_0
\tag{1.179}
$$

It can be shown that the in-exradii r_k, $k = 0,1,2,3$, in (1.177) are related to the circumradius R of triangle $A_1A_2A_3$ by the equations

$$\begin{aligned} r_0 &= 4R\sin\frac{\alpha_1}{2}\sin\frac{\alpha_2}{2}\sin\frac{\alpha_3}{2} \\ r_1 &= 4R\sin\frac{\alpha_1}{2}\cos\frac{\alpha_2}{2}\cos\frac{\alpha_3}{2} \\ r_2 &= 4R\cos\frac{\alpha_1}{2}\sin\frac{\alpha_2}{2}\cos\frac{\alpha_3}{2} \\ r_3 &= 4R\cos\frac{\alpha_1}{2}\cos\frac{\alpha_2}{2}\sin\frac{\alpha_3}{2} \end{aligned} \tag{1.180}$$

or, equivalently, by the equations (compare with [18, Eq. 8.25] for Einstein gyrovector spaces)

$$\begin{aligned} r_0 &= \frac{1}{2}R\frac{\cos\frac{\alpha_1-\alpha_2-\alpha_3}{2}}{\cos\frac{\alpha_1}{2}}\frac{\cos\frac{-\alpha_1+\alpha_2-\alpha_3}{2}}{\cos\frac{\alpha_2}{2}}\frac{\cos\frac{-\alpha_1-\alpha_2+\alpha_3}{2}}{\cos\frac{\alpha_3}{2}} \\ r_1 &= \frac{1}{2}R\frac{\cos\frac{\alpha_1-\alpha_2-\alpha_3}{2}}{\cos\frac{\alpha_1}{2}}\frac{\cos\frac{-\alpha_1+\alpha_2-\alpha_3}{2}}{\sin\frac{\alpha_2}{2}}\frac{\cos\frac{-\alpha_1-\alpha_2+\alpha_3}{2}}{\sin\frac{\alpha_3}{2}} \\ r_2 &= \frac{1}{2}R\frac{\cos\frac{\alpha_1-\alpha_2-\alpha_3}{2}}{\sin\frac{\alpha_1}{2}}\frac{\cos\frac{-\alpha_1+\alpha_2-\alpha_3}{2}}{\cos\frac{\alpha_2}{2}}\frac{\cos\frac{-\alpha_1-\alpha_2+\alpha_3}{2}}{\sin\frac{\alpha_3}{2}} \\ r_3 &= \frac{1}{2}R\frac{\cos\frac{\alpha_1-\alpha_2-\alpha_3}{2}}{\sin\frac{\alpha_1}{2}}\frac{\cos\frac{-\alpha_1+\alpha_2-\alpha_3}{2}}{\sin\frac{\alpha_2}{2}}\frac{\cos\frac{-\alpha_1-\alpha_2+\alpha_3}{2}}{\cos\frac{\alpha_3}{2}} \end{aligned} \tag{1.181}$$

Remarkably, the equations in (1.180) fail in hyperbolic geometry while the equations in (1.181) remain valid in hyperbolic geometry as well. Accordingly, we say that the equations in (1.180) embody the π-identity condition of triangles, (1.65), p. 21, which fails in hyperbolic geometry, while the equations in (1.181) are free of the π-identity condition.

1.18 A Step Toward the Comparative Study

In this chapter we have introduced the notion of Möbius barycentric coordinates in the Cartesian model $\mathbb{R}^n$ of Euclidean geometry and used it for the determination of several barycentric coordinate representations, including those of the four classical triangle centers, with respect to the vertices of reference triangles. Using the standard notation for a triangle $A_1A_2A_3$, Fig. 1.2, p. 7, we expressed barycentric coordinates $\{m_1 : m_2 : m_3\}$ (i) in terms of triangle side-lengths, a_{12}, a_{13}, a_{23}, and (ii) in terms of triangle

Table 1.1 Trigonometric barycentric coordinates of the classical triangle centers and the triangle altitude foot.

Center	Symbol	Trigonometric Barycentric Coordinates
Centroid	G, (1.50), p. 18 Fig. 1.3, p. 18	$\begin{pmatrix} m_1 \\ m_2 \\ m_3 \end{pmatrix} = \begin{pmatrix} 1 \\ 1 \\ 1 \end{pmatrix}$
Orthocenter	H, (1.89), p. 26 Fig. 1.5, p. 24	$\begin{pmatrix} m_1 \\ m_2 \\ m_3 \end{pmatrix} = \begin{pmatrix} \tan\alpha_1 \\ \tan\alpha_2 \\ \tan\alpha_3 \end{pmatrix}$
Incenter	I, (1.109), p. 31 Fig. 1.7, p. 29	$\begin{pmatrix} m_1 \\ m_2 \\ m_3 \end{pmatrix} = \begin{pmatrix} \sin\alpha_1 \\ \sin\alpha_2 \\ \sin\alpha_3 \end{pmatrix}$
Circumcenter	O, (1.141), p. 38 Fig. 1.9, p. 35	$\begin{pmatrix} m_1 \\ m_2 \\ m_3 \end{pmatrix} = \begin{pmatrix} \sin\frac{-\alpha_1+\alpha_2+\alpha_3}{2}\sin\alpha_1 \\ \sin\frac{\alpha_1-\alpha_2+\alpha_3}{2}\sin\alpha_2 \\ \sin\frac{\alpha_1+\alpha_2-\alpha_3}{2}\sin\alpha_3 \end{pmatrix} = \begin{pmatrix} \sin\alpha_1\cos\alpha_1 \\ \sin\alpha_2\cos\alpha_2 \\ \sin\alpha_3\cos\alpha_3 \end{pmatrix}$
Altitude Foot	P_3, (1.77), p. 23 Fig. 1.4, p. 19	$\begin{pmatrix} m_1 \\ m_2 \\ m_3 \end{pmatrix} = \begin{pmatrix} \tan\alpha_1 \\ \tan\alpha_2 \\ 0 \end{pmatrix}$

angles $\alpha_1, \alpha_2, \alpha_3$. The resulting trigonometric barycentric coordinates, in which barycentric coordinates are expressed in terms of the angles of the reference triangle will prove useful in the extension of our study from Euclidean to hyperbolic geometry. Indeed, studying triangle centers comparatively, we will find in Tables 4.1 – 4.2, p. 243, that trigonometric barycentric coordinates that do not embody the π-identity condition of triangles, (1.65), p. 21, survive unimpaired the transition from Euclidean to hyperbolic geometry.

As a first step in the comparative study of triangle centers, a table of the trigonometric barycentric coordinates of the four classic triangle centers is presented in Table 1.1.

The comparative study of triangle centers, initiated here in Table 1.1, will be completed in Tables 4.1 – 4.2, p. 243.

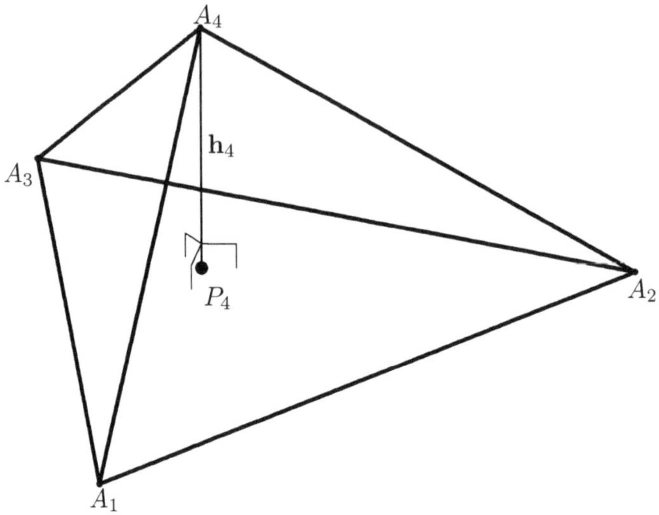

Fig. 1.15 The foot P_4 of the altitude $\mathbf{h}_4$ drawn from vertex A_4 of a tetrahedron $A_1A_2A_3A_4$ in a Euclidean 3-space $\mathbb{R}^3$. The special (that is, normalized) barycentric coordinates (m_1, m_2, m_3) of P_4 with respect to the set $\{A_1A_2A_3\}$ are determined in (1.189), p. 58.

1.19 Tetrahedron Altitude

Let $A_1A_2A_3A_4$ be a tetrahedron with vertices A_1, A_2, A_3 and A_4 in a Euclidean n-space $\mathbb{R}^n$, $n \geq 3$, and let the point P_4 be the orthogonal projection of vertex A_4 onto its opposite face, $A_1A_2A_3$ (or its extension), as shown in Fig. 1.15 for $n = 3$. Furthermore, let (m_1, m_2, m_3) be barycentric coordinates of P_4 with respect to the set $\{A_1, A_2, A_3\}$. Then, P_4 is given in terms of its barycentric coordinates (m_1, m_2, m_3) with respect to the set $\{A_1, A_2, A_3\}$ by the equation

$$P_4 = \frac{m_1A_1 + m_2A_2 + m_3A_3}{m_1 + m_2 + m_3} \tag{1.182}$$

where the barycentric coordinates m_1, m_2 and m_3 of P_4, Fig. 1.15, are to be determined in (1.189) below.

By the covariance (1.26), p. 12, of barycentric coordinate representations with respect to translations we have, in particular, for $W = -A_k$, $k =$

$1, 2, 3, 4$, the following equations:

$$\mathbf{p}_1 := -A_1 + P_4 = \frac{m_2(-A_1 + A_2) + m_3(-A_1 + A_3)}{m_1 + m_2 + m_3} = \frac{m_2\mathbf{a}_{12} + m_3\mathbf{a}_{13}}{m_1 + m_2 + m_3}$$

$$\mathbf{p}_2 := -A_2 + P_4 = \frac{m_1(-A_2 + A_1) + m_3(-A_2 + A_3)}{m_1 + m_2 + m_3} = \frac{m_1\mathbf{a}_{21} + m_3\mathbf{a}_{23}}{m_1 + m_2 + m_3}$$

$$\mathbf{p}_3 := -A_3 + P_4 = \frac{m_1(-A_3 + A_1) + m_2(-A_3 + A_2)}{m_1 + m_2 + m_3} = \frac{m_1\mathbf{a}_{31} + m_2\mathbf{a}_{32}}{m_1 + m_2 + m_3}$$

$$\begin{aligned}\mathbf{h}_4 := -A_4 + P_4 &= \frac{m_1(-A_4 + A_1) + m_2(-A_4 + A_2) + m_3(-A_4 + A_3)}{m_1 + m_2 + m_3} \\ &= \frac{m_1\mathbf{a}_{41} + m_2\mathbf{a}_{42} + m_3\mathbf{a}_{43}}{m_1 + m_2 + m_3}\end{aligned} \tag{1.183}$$

where we use the notation

$$\mathbf{a}_{ij} = -A_i + A_j\,, \qquad a_{ij} = \|\mathbf{a}_{ij}\| \tag{1.184}$$

for $i, j = 1, 2, 3$, noting that $a_{ij} = a_{ji}$.

Let α_{4k} be the angle with vertex A_k, $k = 1, 2, 3$, of triangle $A_1A_2A_3$ that, in turn, forms the face opposite to vertex A_4 of tetrahedron $A_1A_2A_3A_4$, Fig. 1.15.

Then, the squared norms $p_k^2 = \|\mathbf{p}_k\|^2$, $k = 1, 2, 3$ are obtained from (1.183), in (1.185) below, by the law of cosines (1.71), p. 22.

$$\begin{aligned}p_1^2 &= \frac{1}{m_0^2}(m_2^2a_{12}^2 + m_3^2a_{13}^2 + 2m_2m_3a_{12}a_{13}\cos\alpha_{41}) \\ &= \frac{1}{m_0^2}\{m_2^2a_{12}^2 + m_3^2a_{13}^2 + m_2m_3(a_{12}^2 + a_{13}^2 - a_{23}^2)\}\end{aligned} \tag{1.185a}$$

$$\begin{aligned}p_2^2 &= \frac{1}{m_0^2}(m_1^2a_{12}^2 + m_3^2a_{23}^2 + 2m_1m_3a_{12}a_{23}\cos\alpha_{42}) \\ &= \frac{1}{m_0^2}\{m_1^2a_{12}^2 + m_3^2a_{23}^2 + m_1m_3(a_{12}^2 - a_{13}^2 + a_{23}^2)\}\end{aligned} \tag{1.185b}$$

$$\begin{aligned}p_3^2 &= \frac{1}{m_0^2}(m_1^2a_{13}^2 + m_2^2a_{23}^2 + 2m_1m_2a_{13}a_{23}\cos\alpha_{43}) \\ &= \frac{1}{m_0^2}\{m_1^2a_{13}^2 + m_2^2a_{23}^2 + m_1m_2(-a_{12}^2 + a_{13}^2 + a_{23}^2)\}\end{aligned} \tag{1.185c}$$

where

$$m_0 = m_1 + m_2 + m_3 \tag{1.185d}$$

The condition that the tetrahedron altitude $\mathbf{h}_4$ is perpendicular to the vectors $\mathbf{p}_k$, $k = 1, 2, 3$, along with the Pythagorean theorem, imply

$$h_4^2 = a_{14}^2 - p_1^2 = a_{24}^2 - p_2^2 = a_{34}^2 - p_3^2 \tag{1.186}$$

thus obtaining the system of two equations

$$\begin{aligned} p_1^2 - a_{14}^2 &= p_2^2 - a_{24}^2 \\ p_1^2 - a_{14}^2 &= p_3^2 - a_{34}^2 \end{aligned} \tag{1.187}$$

for the three unknowns m_1, m_2 and m_3,

Substituting p_k^2, $k = 1, 2, 3$, from (1.185) into (1.187) along with the normalization condition

$$m_3 = 1 - m_1 - m_2 \tag{1.188}$$

the system (1.187) reduces to a system of two linear equations for the two unknowns m_1 and m_2. Solving the resulting linear system for m_1 and m_2, and calculating m_3 from (1.188) we obtain the unique special barycentric coordinates (m_1, m_2, m_3) of the point P_4 in (1.182), p. 56,

$$\begin{aligned} m_1 = \frac{1}{D}\{ & a_{24}^2(-a_{12}^2 + a_{13}^2 + a_{23}^2) \\ & + a_{34}^2(\ \ a_{12}^2 - a_{13}^2 + a_{23}^2) \\ & + a_{23}^2(\ \ a_{12}^2 + a_{13}^2 - a_{23}^2) - 2a_{14}^2 a_{23}^2\} \end{aligned} \tag{1.189a}$$

$$\begin{aligned} m_2 = \frac{1}{D}\{ & a_{14}^2(-a_{12}^2 + a_{13}^2 + a_{23}^2) \\ & + a_{13}^2(\ \ a_{12}^2 - a_{13}^2 + a_{23}^2) \\ & + a_{34}^2(\ \ a_{12}^2 + a_{13}^2 - a_{23}^2) - 2a_{13}^2 a_{24}^2\} \end{aligned} \tag{1.189b}$$

$$\begin{aligned} m_3 = \frac{1}{D}\{ & a_{12}^2(-a_{12}^2 + a_{13}^2 + a_{23}^2) \\ & + a_{14}^2(\ \ a_{12}^2 - a_{13}^2 + a_{23}^2) \\ & + a_{24}^2(\ \ a_{12}^2 + a_{13}^2 - a_{23}^2) - 2a_{12}^2 a_{34}^2\} \end{aligned} \tag{1.189c}$$

where D is given by the equation

$$\begin{aligned}D &= 2(a_{12}^2a_{13}^2 + a_{12}^2a_{23}^2 + a_{13}^2a_{23}^2) - (a_{12}^4 + a_{13}^4 + a_{23}^4)\\ &= (a_{12} + a_{13} + a_{23})(-a_{12} + a_{13} + a_{23})(a_{12} - a_{13} + a_{23})(a_{12} + a_{13} - a_{23})\\ &= 16|A_1A_2A_3|^2 \end{aligned} \qquad (1.189\text{d})$$

The third equation in (1.189d) follows from (1.76), p. 23, where, by Heron's formula, $|A_1A_2A_3|$ is the area of triangle $A_1A_2A_3$.

Convenient barycentric coordinates for the point P_4 can be obtained from (1.189) by omitting the nonzero common factor $1/D$.

Formalizing the main result of this section we obtain the following theorem:

Theorem 1.20. (Point to Plane Perpendicular Projection). *Let A_1, A_2 and A_3 be any three pointwise independent points of a Euclidean space $\mathbb{R}^n$, $n \geq 3$, and let $\pi_{A_1A_2A_3}$ be the plane passing through these points. Furthermore, let A_4 be any point of the space that does not lie on $\pi_{A_1A_2A_3}$, as shown in Fig. 1.15, p. 56. Then, the perpendicular projection P_4 of the point A_4 on the plane $\pi_{A_1A_2A_3}$ is given by,* (1.182),

$$P_4 = m_1A_1 + m_2A_2 + m_3A_3 \qquad (1.190)$$

where the special barycentric coordinates m_1, m_2 and m_3 are given by (1.189), *satisfying the normalization condition*

$$m_1 + m_2 + m_3 = 1 \qquad (1.191)$$

1.20 Tetrahedron Altitude Length

By the last equation in (1.183), p. 57, the squared length of the tetrahedron altitude $\mathbf{h}_4$ is given by the equation

$$\begin{aligned}h_4^2 &= (m_1\mathbf{a}_{41} + m_2\mathbf{a}_{42} + m_3\mathbf{a}_{43})^2\\ &= m_1^2a_{41}^2 + m_2^2a_{42}^2 + m_3^2a_{43}^2\\ &\quad + 2m_1m_2a_{14}a_{24}\cos\alpha_{34}\\ &\quad + 2m_1m_3a_{14}a_{34}\cos\alpha_{24}\\ &\quad + 2m_2m_3a_{24}a_{34}\cos\alpha_{14}\end{aligned} \qquad (1.192)$$

Hence, by the law of cosines,

$$\begin{aligned} h_4^2 &= m_1^2 a_{14}^2 + m_2^2 a_{24}^2 + m_3^2 a_{34}^2 \\ &\quad + m_1 m_2(-a_{12}^2 + a_{14}^2 + a_{24}^2) \\ &\quad + m_1 m_3(-a_{13}^2 + a_{14}^2 + a_{34}^2) \\ &\quad + m_2 m_3(-a_{23}^2 + a_{24}^2 + a_{34}^2) \end{aligned} \tag{1.193}$$

We now substitute the special barycentric coordinates m_1, m_2 and m_3 from (1.189), p. 58, into (1.193) obtaining

$$h_4^2 = \frac{1}{2} \frac{F_3(a_{12}, a_{13}, a_{14}, a_{23}, a_{24}, a_{34})}{F_2(a_{12}, a_{13}, a_{23})} \tag{1.194}$$

where F_2 is given by its 4×4 determinant representation in (1.75), p. 22, and where F_3 is given by the similar, 5×5 determinant representation

$$F_3(a_{12}, a_{13}, a_{14}, a_{23}, a_{24}, a_{34}) = \begin{vmatrix} 0 & a_{12}^2 & a_{13}^2 & a_{14}^2 & 1 \\ a_{21}^2 & 0 & a_{23}^2 & a_{24}^2 & 1 \\ a_{31}^2 & a_{32}^2 & 0 & a_{34}^2 & 1 \\ a_{41}^2 & a_{42}^2 & a_{43}^2 & 0 & 1 \\ 1 & 1 & 1 & 1 & 0 \end{vmatrix} \tag{1.195}$$

known as the 5×5 *Cayley-Menger determinant* [16]. The generalization of (1.194)–(1.195) to higher dimensions is found in [19, Eq. (10.492)].

We can now calculate the volume $|A_1A_2A_3A_4|$ of tetrahedron $A_1A_2A_3A_4$. By (1.126), p. 34, and (1.194) – (1.195),

$$|A_1A_2A_3A_4|^2 = \frac{1}{3^2}|A_1A_2A_3|^2 h_4^2 = \frac{1}{288} F_3(a_{12}, a_{13}, a_{14}, a_{23}, a_{24}, a_{34}) \,. \tag{1.196}$$

1.21 Exercises

(1) Show that the pointwise independence of the set S in Def. 1.5, p. 9, insures that the barycentric coordinate representation of a point with respect to the set S is unique.

(2) Prove the trigonometric identities in (1.66), p. 21.

(3) Employ the π-identity of triangles, (1.65), p. 21, to simplify the barycentric coordinates $(m_1 : m_2 : m_3)$ in (1.139), p. 38, into the barycentric coordinates $(m_1 : m_2 : m_3)$ in (1.140).

(4) Employ the π-identity of triangles, (1.65), p. 21, to simplify the barycentric coordinates $(m_1 : m_2 : m_3)$ in (1.141), p. 38, into the barycentric coordinates $(m_1 : m_2 : m_3)$ in (1.140).

(5) Solve the vector equations in (1.82), p. 25, for the three scalar unknowns t_1, t_2 and t_3. Substitute the solution in (1.81) and, hence, obtain the equation in (1.85) for the point of concurrency, H, of the three lines in (1.81).

(6) Derive the circumradius R of a triangle $A_1A_2A_3$ in (1.143), p. 39 by successively substituting $\|-A_1+O\|^2$ from the first equation in (1.129), p. 36, and m_k, $k=1,2,3$, from (1.137), p. 37, into (1.142).

(7) Derive the equations in (1.152), p. 43, from the equations in (1.151).

(8) Verify the trigonometric identities in (1.167), p. 47, under the triangle π condition (1.65), p. 21.

(9) Verify the trigonometric barycentric coordinate representation (1.170), p. 49, of the triangle Gergonne point G_e with respect to the vertices of its reference triangle in a Euclidean space $\mathbb{R}^n$. See also Exercise 2, p. 270.

(10) Verify the trigonometric barycentric coordinate representation (1.171), p. 49, of the triangle Nagel point N_a with respect to the vertices of its reference triangle in a Euclidean space $\mathbb{R}^n$. See also Exercise 3, p. 271.

(11) Verify the trigonometric barycentric coordinate representation (1.172), p. 49, of the triangle P_u Point with respect to the vertices of its reference triangle in a Euclidean space $\mathbb{R}^n$. See also Exercise 4, p. 271.

(12) Verify the equations in (1.180) and (1.181), p. 54.

(13) Substitute p_k^2, $k=1,2,3$, from (1.185), p. 57, into (1.187) along with the normalization condition (1.188) to obtain a linear system of two equations for m_1 and m_2. Then solve the resulting linear system and calculate m_3 from the normalization condition and, hence, obtain the solution (m_1, m_2, m_3) in (1.189).

(14) Substitute the special barycentric coordinates m_1, m_2 and m_3 from (1.189), p. 58, into (1.193), p. 60, to obtain h_4^2 in (1.194), p. 60.

Chapter 2

Gyrovector Spaces and Cartesian Models of Hyperbolic Geometry

In Chapter 1 we have determined the four classic triangle centers in terms of their barycentric coordinates with respect to the vertices of their reference triangle in the Cartesian model $\mathbb{R}^n$ of Euclidean geometry. The mission of this book is to extend the study in Chapter 1 of triangle centers in Euclidean geometry into the regime of the hyperbolic geometry of Bolyai and Lobachevsky.

The first task we face in the mission is to introduce Cartesian coordinates into various models of hyperbolic geometry. In this book we accomplish this task by introducing the

(1) Cartesian-Beltrami-Klein ball model $\mathbb{R}_s^n$ of hyperbolic geometry, regulated by the nonassociative algebra of Einstein addition in $\mathbb{R}_s^n$ of relativistically admissible velocities; and the

(2) Cartesian-Poincaré ball model $\mathbb{R}_s^n$ of hyperbolic geometry, regulated by the nonassociative algebra of Möbius addition in $\mathbb{R}_s^n$. A third interesting Cartesian model, which will not be considered in this book, is the

(3) Cartesian-PV space model $\mathbb{R}^n$ of hyperbolic geometry, regulated by the nonassociative algebra of Einstein addition in $\mathbb{R}^n$ of relativistically admissible proper velocities (PV), studied in [1–3, 20],

where $\mathbb{R}_s^n$ is the ball of the Euclidean n-space $\mathbb{R}^n$ with radius $s > 0$ centered at the origin of $\mathbb{R}^n$, $n \geq 2$.

The Cartesian-Beltrami-Klein ball model of hyperbolic geometry is appealing to us in this book since it is regulated by Einstein addition, $\oplus$, in $\mathbb{R}_s^n$ which, in turn, interplays harmoniously with the common vector addition, $+$, in $\mathbb{R}^n$, as explained in Chap. 3. It is owing to this interplay that the introduction of barycentric coordinates into hyperbolic geometry is straightforward along with their application for the determination of hyperbolic triangle centers.

The Cartesian-Poincaré ball model of hyperbolic geometry is regulated by Möbius addition, $\oplus_M$, in $\mathbb{R}_s^n$, which is a binary operation isomorphic in some sense to Einstein addition in $\mathbb{R}_s^n$. It is appealing to us in this book since it is *conformal to Euclidean geometry* in the sense that the measure of any hyperbolic angle between hyperbolic lines in this model equals the measure of a Euclidean angle between corresponding tangent lines, as illustrated in Fig. 2.14, p. 139.

Unlike the Cartesian-Beltrami-Klein ball model, into which the introduction of barycentric coordinates is straightforward owing to the harmonious interplay between $\oplus$ and $+$, the direct introduction of barycentric coordinates into the Cartesian-Poincaré ball model is not straightforward. We therefore determine, in this book, barycentric coordinates in the Cartesian-Poincaré model indirectly. We first determine these by solving a corresponding problem in the Cartesian-Beltrami-Klein model, and then transform the solution to the Cartesian-Poincaré model by means of the isomorphism between the two models studied in Sec. 2.29.

2.1 Einstein Addition

Relativistic mechanics and its Einstein addition law of Einsteinian, relativistically admissible velocities is woven into the fabric of hyperbolic geometry just as classical mechanics and its addition law of Newtonian velocities, which is the ordinary vector addition, is woven into the fabric of Euclidean geometry. We therefore start this chapter with the introduction of Einstein addition.

Let $\mathbb{R}^n$ be the Euclidean n-space equipped with Cartesian coordinates $(x_1, x_2, \ldots, x_n)$. These are all the n-tuples of real numbers satisfying

$$x_1^2 + x_2^2 + \ \ldots \ + x_n^2 < \infty \tag{2.1}$$

Similarly, let $\mathbb{R}_s^n$,

$$\mathbb{R}_s^n = \{X = (x_1, x_2, \ldots, x_n) \in \mathbb{R}^n : \|X\| < s\} \tag{2.2}$$

be the s-ball of $\mathbb{R}^n$ for any fixed $s > 0$. The ball $\mathbb{R}_s^n$ inherits from its ambient space $\mathbb{R}^n$ the Cartesian coordinates $(x_1, x_2, \ldots, x_n)$ which, contrasting (2.1), are all the n-tuples of real numbers satisfying

$$x_1^2 + x_2^2 + \ \ldots \ + x_n^2 < s^2 \tag{2.3}$$

Points of the resulting Cartesian model of the ball $\mathbb{R}^n_s$ are n-tuples like $X = (x_1, x_2, \ldots, x_n)$ or $Y = (y_1, y_2, \ldots, y_n)$, etc., of real numbers. The point $\mathbf{0} = (0, 0, \ldots, 0)$ (n zeros) is the *origin* of the ball $\mathbb{R}^n_s$.

Clearly, the point addition, $+$, in $\mathbb{R}^n$, given by (1.1), p. 2, is not closed in the ball. We, therefore, replace it by Einstein (velocity) addition, $\oplus$, in the ball.

Einstein addition, $\oplus$, of relativistically admissible velocities is a binary operation in the ball $\mathbb{R}^n_s$ of all relativistically admissible velocities, where, if $n = 3$, $s = c$ is the speed of light in empty space. It takes the form [21] [22, p. 55] [23, Eq. 2.9.2] [1],

$$X{\oplus}Y = \frac{1}{1 + \frac{X{\cdot}Y}{s^2}} \left\{ X + \frac{1}{\gamma_X} Y + \frac{1}{s^2} \frac{\gamma_X}{1 + \gamma_X} (X{\cdot}Y) X \right\} \tag{2.4}$$

$X, Y{\in}\mathbb{R}^n_s$, where γ_X is the gamma factor

$$\gamma_X = \frac{1}{\sqrt{1 - \frac{\|X\|^2}{s^2}}} \tag{2.5}$$

in $\mathbb{R}^n_s$, and where $\cdot$ and $\|{\cdot}\|$ are the inner product and norm that the ball $\mathbb{R}^n_s$ inherits from its ambient space $\mathbb{R}^n$.

Clearly, $X{\oplus}\mathbf{0} = \mathbf{0}{\oplus}X = X$. Naturally, Einstein subtraction, $\ominus$, is given by the equation

$$X{\ominus}Y = X{\oplus}(-Y) \tag{2.6}$$

so that, for instance, $X{\ominus}X = \mathbf{0}$, $\ominus X = \mathbf{0}{\ominus}X = -X$ and, in particular,

$$\begin{aligned} \ominus(\ominus X) &= X \\ \ominus(X{\oplus}Y) &= \ominus X{\ominus}Y \\ \ominus X{\oplus}(X{\oplus}Y) &= Y \end{aligned} \tag{2.7}$$

for all X, Y in the ball, in full analogy with vector addition and subtraction. The second and the third identities in (2.7) are called, respectively, the *automorphic inverse property* and the *left cancellation law* of Einstein addition.

We may note that Einstein addition does not obey the immediate right counterpart of the left cancellation law in (2.7) since, in general,

$$(X{\oplus}Y){\ominus}Y \neq X \tag{2.8}$$

However, this seemingly lack of a *right cancellation law* of Einstein addition will be remedied in (2.38), p. 73, by the introduction of a secondary binary operation.

The gamma factor (2.5), the hallmark of equations of special relativity, is linked to Einstein addition (2.4) by the *gamma identity*

$$\gamma_{X \oplus Y} = \gamma_X \gamma_Y \left(1 + \frac{X \cdot Y}{s^2}\right) \tag{2.9}$$

first studied by Sommerfeld [24] and Varičak [25]. It is this identity that revealed to Sommerfeld and Varičak that Einstein's special theory of relativity has natural interpretation in the hyperbolic geometry of Bolyai and Lobachevsky [26, Sec. 5].

Replacing X by $\ominus X$, the gamma identity (2.9) takes the form

$$\gamma_{\ominus X \oplus Y} = \gamma_X \gamma_Y \left(1 - \frac{X \cdot Y}{s^2}\right) \tag{2.10}$$

which will prove useful.

A frequently used identity that follows immediately from (2.5) is

$$\frac{\|X\|^2}{s^2} = \frac{\gamma_X^2 - 1}{\gamma_X^2} \tag{2.11}$$

In the special case when $X, Y \in \mathbb{R}_s^n \subset \mathbb{R}^n$ are nonzero, parallel vectors in $\mathbb{R}^n$, the general Einstein addition (2.4) of velocities that need not be parallel descends to the Einstein addition of parallel velocities,

$$X \oplus Y = \frac{X + Y}{1 + \frac{1}{s^2}\|X\|\|Y\|}, \qquad X \| Y \tag{2.12}$$

Einstein addition (2.4) of relativistically admissible velocities is used in this book as the addition of points of $\mathbb{R}_s^n$ in a way fully analogous to the point addition (1.1), p. 2, in $\mathbb{R}^n$, as demonstrated in Example 2.1 below.

Einstein (velocity) addition was introduced by Einstein for the special case when $n = 3$ in his 1905 paper [27] [28, p. 141] that founded the special theory of relativity. The Euclidean three-vector algebra was not so widely known in 1905 and, consequently, was not used by Einstein. Einstein calculated in [27] the behavior of the velocity components parallel and orthogonal to the relative velocity between inertial systems, which is as close as one can get without vectors to the vectorial version (2.4) of Einstein addition.

In the Newtonian limit, $s \to \infty$, the ball $\mathbb{R}_s^n$ of all relativistically admissible velocities expands to the whole of its ambient space $\mathbb{R}^n$, as we see from (2.2), and Einstein addition $\oplus$ in $\mathbb{R}_s^n$ descends to the ordinary vector addition $+$ in $\mathbb{R}^n$, as we see from (2.4) and (2.5). Obviously, in physical applications $n = 3$. In applications to geometry, however, n is any positive integer.

Example 2.1. (The Components of Einstein Addition in $\mathbb{R}_c^3$). Let $\mathbb{R}_c^3$ be the c-ball of the Euclidean 3-space, equipped with a Cartesian coordinate system Σ.

Accordingly, each point of the ball is represented by its coordinates $(x_1, x_2, x_3)^t$ (exponent t denotes transposition) relative to Σ, satisfying the condition

$$x_1^2 + x_2^2 + x_3^2 < c^2 \tag{2.13}$$

Each point $X = (x_1, x_2, x_3)^t$ of the ball is identified with the vector from the origin $(0, 0, 0)^t$ of Σ, which is the center of the ball, to the point $(x_1, x_2, x_3)^t$.

Furthermore, let $X, Y, Z \in \mathbb{R}_c^3$ be three points in $\mathbb{R}_c^3 \subset \mathbb{R}^3$ given by their components relative to Σ,

$$X = \begin{pmatrix} x_1 \\ x_2 \\ x_3 \end{pmatrix}, \qquad Y = \begin{pmatrix} y_1 \\ y_2 \\ y_3 \end{pmatrix}, \qquad Z = \begin{pmatrix} z_1 \\ z_2 \\ z_3 \end{pmatrix} \tag{2.14}$$

where

$$Z = X \oplus Y \tag{2.15}$$

The dot product of X and Y is given by the equation

$$X \cdot Y = x_1 y_1 + x_2 y_2 + x_3 y_3 \tag{2.16}$$

and the squared norm $\|X\|^2 = X \cdot X =: X^2$ of X is given by the equation

$$\|X\|^2 = x_1^2 + x_2^2 + x_3^2 \tag{2.17}$$

Hence, it follows from the coordinate independent vector representation (2.4) of Einstein addition that the coordinate dependent Einstein addition (2.15) relative to the Cartesian coordinate system Σ takes the form

$$\begin{aligned} \begin{pmatrix} z_1 \\ z_2 \\ z_3 \end{pmatrix} &= \frac{1}{1 + \dfrac{x_1 y_1 + x_2 y_2 + x_3 y_3}{c^2}} \\ &\times \left\{ [1 + \frac{1}{c^2} \frac{\gamma_X}{1 + \gamma_X} (x_1 y_1 + x_2 y_2 + x_3 y_3)] \begin{pmatrix} x_1 \\ x_2 \\ x_3 \end{pmatrix} + \frac{1}{\gamma_X} \begin{pmatrix} y_1 \\ y_2 \\ y_3 \end{pmatrix} \right\} \end{aligned} \tag{2.18}$$

where

$$\gamma_X = \frac{1}{\sqrt{1 - \dfrac{x_1^2 + x_2^2 + x_3^2}{c^2}}} \tag{2.19}$$

The three components of Einstein addition (2.15) are z_1, z_2 and z_3 in (2.18). For a two-dimensional illustration of Einstein addition (2.18) one may impose the condition $x_3 = y_3 = 0$, implying $z_3 = 0$.

In the Newtonian-Euclidean limit, $c \to \infty$, the ball $\mathbb{R}_c^3$ expands to the Euclidean 3-space $\mathbb{R}^3$, and Einstein addition (2.18) descends to the addition in (1.1), p. 2, of points of $\mathbb{R}^3$. Indeed, we will see in this book that Einstein addition plays in the Cartesian model of the Beltrami-Klein ball model of hyperbolic geometry the same role that vector addition plays in the Cartesian model of Euclidean geometry.

In this book vector equations and identities are represented by coordinate free expressions, like Einstein addition in (2.4). For numerical and graphical presentations, however, these must be converted into a coordinate dependent form relative to a Cartesian coordinate system that must be introduced. The latter, in turn, can be presented relative to Cartesian coordinates numerically and graphically, as we do in the generation of the Figures of this book.

2.2 Einstein Gyration

For any $X, Y \in \mathbb{R}_s^n$, let $\mathrm{gyr}[X, Y] : \mathbb{R}_s^n \to \mathbb{R}_s^n$ be the self-map of $\mathbb{R}_s^n$ given in terms of Einstein addition $\oplus$ by the equation [29, 30]

$$\mathrm{gyr}[X, Y]Z = \ominus(X \oplus Y) \oplus \{X \oplus (Y \oplus Z)\} \tag{2.20}$$

The self-map $\mathrm{gyr}[X, Y]$ of $\mathbb{R}_s^n$, which takes

$$Z \quad \to \quad \ominus(X \oplus Y) \oplus \{X \oplus (Y \oplus Z) \tag{2.21}$$

is called *Thomas gyration* (or, briefly, gyration) generated by X and Y. Thomas gyration is the mathematical abstraction of the relativistic effect known as Thomas precession [31], and it has an interpretation in hyperbolic geometry as a hyperbolic triangle defect [3, 32].

In the Euclidean limit, $s \to \infty$, Einstein addition $\oplus$ in $\mathbb{R}_s^n$, (2.4), descends to the common vector addition $+$ in $\mathbb{R}^n$, which is associative. Accordingly, in this limit the gyration $\mathrm{gyr}[X, Y]$ in (2.20) becomes *trivial*, that is, it descends to the identity map of $\mathbb{R}^n$. Hence, as expected, gyrations $\mathrm{gyr}[X, Y]$, $X, Y \in \mathbb{R}_s^n$, vanish (that is, they become *trivial*) in the Euclidean limit.

The gyration equation (2.20) can be manipulated (with the help of computer algebra, like Mathematica or Maple) into the equation

$$\mathrm{gyr}[X, Y]Z = Z + \frac{AX + BY}{D} \tag{2.22}$$

where

$$\begin{aligned}
A &= -\frac{1}{s^2}\frac{\gamma_X^2}{(\gamma_X+1)}(\gamma_Y-1)(X\cdot Z)+\frac{1}{s^2}\gamma_X\gamma_Y(Y\cdot Z)\\
&\quad+\frac{2}{s^4}\frac{\gamma_X^2\gamma_Y^2}{(\gamma_X+1)(\gamma_Y+1)}(X\cdot Y)(Y\cdot Z)\\
B &= -\frac{1}{s^2}\frac{\gamma_Y}{\gamma_Y+1}\{\gamma_X(\gamma_Y+1)(X\cdot Z)+(\gamma_X-1)\gamma_Y(Y\cdot Z)\}\\
D &= \gamma_X\gamma_Y(1+\frac{X\cdot Y}{s^2})+1=\gamma_{X\oplus Y}+1>1
\end{aligned} \tag{2.23}$$

for all $X, Y, Z \in \mathbb{R}_s^n$.

The domain of X and Y cannot be extended beyond $\mathbb{R}_s^n$ owing to the presence of γ_X and γ_Y. This is, however, not the case with Z. Rather than $Z \in \mathbb{R}_s^n$, allowing $Z \in \mathbb{R}^n \supset \mathbb{R}_s^n$ in (2.22) – (2.23), gyrations $\mathrm{gyr}[X, Y]$ are expendable to linear maps of $\mathbb{R}^n$ for all $X, Y \in \mathbb{R}_s^n$. The identity between the two expressions of D in (2.23) is taken from (2.9).

In each of the three special cases when (i) $X = \mathbf{0}$, or (ii) $Y = \mathbf{0}$, or (iii) X and Y are parallel in $\mathbb{R}^n$, $X \| Y$, we have $AX + BY = \mathbf{0}$ so that in these cases $\mathrm{gyr}[X, Y]$ is trivial,

$$\begin{aligned}
\mathrm{gyr}[\mathbf{0}, Y]Z &= Z\\
\mathrm{gyr}[X, \mathbf{0}]Z &= Z\\
\mathrm{gyr}[X, Y]Z &= Z, \qquad\qquad X\|Y
\end{aligned} \tag{2.24}$$

for all $X, Y \in \mathbb{R}_s^n$ and all $Z \in \mathbb{R}^n$.

It follows from (2.22) that

$$\mathrm{gyr}[Y, X](\mathrm{gyr}[X, Y]Z) = Z \tag{2.25}$$

for all $X, Y \in \mathbb{R}_s^n$, $Z \in \mathbb{R}^n$, so that gyrations are invertible linear maps of $\mathbb{R}^n$, the inverse of $\mathrm{gyr}[X, Y]$ being $\mathrm{gyr}[Y, X]$ for all $X, Y \in \mathbb{R}_s^n$.

Owing to the nonassociativity of Einstein addition $\oplus$, in general, a gyration is not the identity map. Interestingly, gyrations keep the inner product of elements of the ball $\mathbb{R}_s^n$ invariant, that is,

$$\mathrm{gyr}[X, Y]A\cdot\mathrm{gyr}[X, Y]B = A\cdot B \tag{2.26}$$

for all $A, B, X, Y \in \mathbb{R}_s^n$. As such, $\mathrm{gyr}[X, Y]$ is an *isometry* of $\mathbb{R}_s^n$, keeping the norm of elements of the ball $\mathbb{R}_s^n$ invariant,

$$\|\mathrm{gyr}[X, Y]Z\| = \|Z\| \tag{2.27}$$

Hence, $\text{gyr}[X,Y]$ represents a rotation of the ball $\mathbb{R}_s^n$ about its origin for any $X, Y \in \mathbb{R}_s^n$.

The invertible self-map $\text{gyr}[X,Y]$ of $\mathbb{R}_s^n$ respects Einstein addition in $\mathbb{R}_s^n$,

$$\text{gyr}[X,Y](A \oplus B) = \text{gyr}[X,Y]A \oplus \text{gyr}[X,Y]B \tag{2.28}$$

for all $A, B, X, Y \in \mathbb{R}_s^n$, so that $\text{gyr}[X,Y]$ is an automorphism of the Einstein groupoid $(\mathbb{R}_s^n, \oplus)$.

We recall that a groupoid is a nonempty set with a binary operation, and an automorphism of the groupoid $(\mathbb{R}_s^n, \oplus)$ is a bijective self-map of the groupoid $\mathbb{R}_s^n$ that respects its binary operation, that is, it satisfies (2.28).

Under bijection composition the automorphisms of a groupoid $(\mathbb{R}_s^n, \oplus)$ form a group known as the automorphism group, and denoted $\text{Aut}(\mathbb{R}_s^n, \oplus)$. Being special automorphisms, Thomas gyrations $\text{gyr}[X,Y]$, $X, Y \in \mathbb{R}_s^n$, are also called *gyroautomorphisms*, gyr being the gyroautomorphism generator called the *gyrator*.

The gyroautomorphisms $\text{gyr}[X,Y]$ regulate Einstein addition in the ball $\mathbb{R}_s^n$, giving rise to the following nonassociative algebraic laws that "repair" the breakdown of commutativity and associativity in Einstein addition:

$$\begin{aligned} &X \oplus Y = \text{gyr}[X,Y](Y \oplus X) && \text{Gyrocommutative Law} \\ &X \oplus (Y \oplus Z) = (X \oplus Y) \oplus \text{gyr}[X,Y]Z && \text{Left Gyroassociative Law} \\ &(X \oplus Y) \oplus Z = X \oplus (Y \oplus \text{gyr}[Y,X]Z) && \text{Right Gyroassociative Law} \end{aligned} \tag{2.29}$$

for all $X, Y, Z \in \mathbb{R}_s^n$.

It follows immediately from the gyrocommutative law and the norm invariance (2.27) under gyrations that while, in general, $X \oplus Y \neq Y \oplus X$, we have

$$\|X \oplus Y\| = \|Y \oplus X\| \tag{2.30}$$

An important property of Thomas gyration in Einstein gyrovector spaces is the *loop property* (left and right),

$$\begin{aligned} \text{gyr}[X \oplus Y, Y] &= \text{gyr}[X,Y] && \text{Left Loop Property} \\ \text{gyr}[X, Y \oplus X] &= \text{gyr}[X,Y] && \text{Right Loop Property} \end{aligned} \tag{2.31}$$

for all $X, Y \in \mathbb{R}_s^n$.

The grouplike groupoid $(\mathbb{R}_s^n, \oplus)$ that regulates Einstein addition, $\oplus$, in the ball $\mathbb{R}_s^n$ of the Euclidean 3-space $\mathbb{R}^n$ is a *gyrocommutative gyrogroup* called an *Einstein gyrogroup*. Einstein gyrogroups and gyrovector spaces, studied in [1–3, 33], will be introduced in the sequel.

2.3 From Einstein Velocity Addition to Gyrogroups

Taking the key features of the Einstein groupoid $(\mathbb{R}^n_s, \oplus)$ as axioms, and guided by analogies with groups in Def. 1.1, p. 3, we are led to the following formal gyrogroup definition.

Definition 2.2. (Gyrogroups). *A groupoid* $(G, \oplus)$ *is a gyrogroup if its binary operation satisfies the following axioms. In* G *there is at least one element,* 0*, called a left identity, satisfying*

(G1) $\qquad 0 \oplus a = a$

for all $a \in G$*. There is an element* $0 \in G$ *satisfying axiom* $(G1)$ *such that for each* $a \in G$ *there is an element* $\ominus a \in G$*, called a left inverse of* a*, satisfying*

(G2) $\qquad \ominus a \oplus a = 0$

Moreover, for any $a, b, c \in G$ *there exists a unique element* $\mathrm{gyr}[a, b]c \in G$ *such that the binary operation obeys the left gyroassociative law*

(G3) $\qquad a \oplus (b \oplus c) = (a \oplus b) \oplus \mathrm{gyr}[a, b]c$

The map $\mathrm{gyr}[a, b] : G \to G$ *given by* $c \mapsto \mathrm{gyr}[a, b]c$ *is an automorphism of the groupoid* $(G, \oplus)$*, that is,*

(G4) $\qquad \mathrm{gyr}[a, b] \in \mathrm{Aut}(G, \oplus)$

and the automorphism $\mathrm{gyr}[a, b]$ *of* G *is called the gyroautomorphism, or the gyration, of* G *generated by* $a, b \in G$*. The operator* $\mathrm{gyr} : G \times G \to \mathrm{Aut}(G, \oplus)$ *is called the gyrator of* G*. Finally, the gyroautomorphism* $\mathrm{gyr}[a, b]$ *generated by any* $a, b \in G$ *possesses the left loop property*

(G5) $\qquad \mathrm{gyr}[a, b] = \mathrm{gyr}[a \oplus b, b]\,.$

The gyrogroup axioms $(G1)-(G5)$ in Definition 2.2 split up into three classes:

(1) The first pair of axioms, $(G1)$ and $(G2)$, is a reminiscent of the group axioms.
(2) The last pair of axioms, $(G4)$ and $(G5)$, presents the gyrator axioms.
(3) The middle axiom, $(G3)$, is a hybrid axiom linking the two pairs of axioms in (1) and (2).

As in group theory, we use the notation $a \ominus b = a \oplus (\ominus b)$ in gyrogroup theory as well.

In full analogy with groups, gyrogroups split up into gyrocommutative and non-gyrocommutative gyrogroups.

Definition 2.3. (Gyrocommutative Gyrogroups). *A gyrogroup* $(G, \oplus)$ *is gyrocommutative if its binary operation obeys the gyrocommutative law*

(G6) $a \oplus b = \mathrm{gyr}[a, b](b \oplus a)$
for all $a, b \in G$.

Clearly, a (commutative) group is a degenerate (gyrocommutative) gyrogroup whose gyroautomorphisms are all trivial. The algebraic structure of gyrogroups is, accordingly, richer than that of groups. Thus, without losing the flavor of the group structure we have generalized it into the gyrogroup structure to suit the needs of Einstein addition in the ball. Fortunately, the gyrogroup structure is by no means restricted to Einstein addition in the ball. Rather, it abounds in group theory as demonstrated, for instance, in [34–36], where finite and infinite gyrogroups, both gyrocommutative and non-gyrocommutative, are studied. Some first gyrogroup theorems, some of which are analogous to group theorems, are presented in [37], [2, Chap. 2] and [3, Chap. 2], and in Sec. 2.4 below.

In order to capture analogies with groups, we introduce into the abstract gyrogroup $(G, \oplus)$ a second operation $\boxplus$ called the *cooperation*, or coaddition, which shares useful duality symmetries with its gyrogroup operation $\oplus$ [1,2].

Definition 2.4. (The Gyrogroup Cooperation (Coaddition)). *Let* $(G, \oplus)$ *be a gyrogroup. The gyrogroup cooperation (or, coaddition)* $\boxplus$ *is a second binary operation in* G *related to the gyrogroup operation (or, addition)* $\oplus$ *by the equation*

$$a \boxplus b = a \oplus \mathrm{gyr}[a, \ominus b]b \tag{2.32}$$

for all $a, b \in G$.

Naturally, we use the notation $a \boxminus b = a \boxplus (\ominus b)$ where $\ominus b = -b$, so that

$$a \boxminus b = a \ominus \mathrm{gyr}[a, b]b \tag{2.33}$$

The gyrogroup cooperation is commutative if and only if the gyrogroup operation is gyrocommutative [2, Theorem 3.4]. Hence, in particular, Einstein coaddition $\boxplus$ is commutative since Einstein addition $\oplus$ is gyrocommutative. The commutativity of Einstein coaddition proves useful in the hyperbolic parallelogram (gyroparallelogram) law of relativistic velocities, presented in [3, Sec. 10.8], [33] and [38].

The gyrogroup cooperation $\boxplus$ is expressed in (2.32) in terms of the gyrogroup operation $\oplus$ and the gyrator gyr. It can be shown that, similarly, the gyrogroup operation $\oplus$ can be expressed in terms of the gyrogroup cooperation $\boxplus$ and the gyrator gyr by the identity [2, Theorem 2.10],

$$a \oplus b = a \boxplus \mathrm{gyr}[a, b]b \tag{2.34}$$

for all a, b in a gyrogroup $(G, \oplus)$. Identities (2.32) and (2.34) exhibit one of the duality symmetries that the gyrogroup operation and cooperation share.

First gyrogroup theorems are presented in [1,2] and in Sec. 2.4 below. In particular, it is found that any gyrogroup possesses a unique identity (left and right) and each element of any gyrogroup possesses a unique inverse (left and right). Similarly, the left gyroassociative law $(G3)$ and the left loop property $(G5)$ have the following respective right counterparts:

$$(a\oplus b)\oplus c = a\oplus(b\oplus \mathrm{gyr}[b, a]c) \tag{2.35}$$

and

$$\mathrm{gyr}[a, b] = \mathrm{gyr}[a, b\oplus a] \tag{2.36}$$

Furthermore, any gyrogroup obeys the left cancellation law,

$$\ominus a\oplus(a\oplus b) = b \tag{2.37}$$

and the two right cancellation laws,

$$(b\oplus a) \boxminus a = b \tag{2.38}$$

and

$$(b \boxplus a)\ominus a = b \tag{2.39}$$

Like Identities (2.32) and (2.34), Identities (2.38) and (2.39) present a duality symmetry between the gyrogroup operation $\oplus$ and cooperation $\boxplus$.

Applying the left cancellation law (2.37) to the left gyroassociative law $(G3)$ of a gyrogroup we obtain the *gyrator identity*

$$\mathrm{gyr}[a, b]x = \ominus(a\oplus b)\oplus\{a\oplus(b\oplus x)\} \tag{2.40}$$

The gyrator identity demonstrates that the gyrations of a gyrogroup are uniquely determined by the gyrogroup operation.

Furthermore, it is clear from the (gyrocommutative law and the) gyroassociative law that gyrations measure the extent to which the gyrogroup operation deviates from (both commutativity and) associativity. A (commutative) group is accordingly a (gyrocommutative) gyrogroup whose gyrations are trivial. Hence, the gyrogroup structure is richer than the group structure and, in particular, the algebra of Einstein velocity addition in hyperbolic geometry is richer than that of Newtonian velocity addition in Euclidean geometry. Elements of gyrogroup theory are presented, for instance, in [3] and [33].

2.4 First Gyrogroup Theorems

While it is clear how to define a right identity and a right inverse in a gyrogroup, the existence of such elements is not presumed. Indeed, the existence of a unique identity and a unique inverse, both left and right, is a consequence of the gyrogroup axioms, as the following theorem shows, along with other immediate results.

Theorem 2.5. *Let $(G, +)$ be a gyrogroup. For any elements $a, b, c, x \in G$ we have:*

(1) *If $a + b = a + c$, then $b = c$ (general left cancellation law; see item (9) below).*
(2) $\text{gyr}[0, a] = I$ *for any left identity* 0 *in* G.
(3) $\text{gyr}[x, a] = I$ *for any left inverse* x *of* a *in* G.
(4) $\text{gyr}[a, a] = I$
(5) *There is a left identity which is a right identity.*
(6) *There is only one left identity.*
(7) *Every left inverse is a right inverse.*
(8) *There is only one left inverse,* $-a$, *of* a, *and* $-(-a) = a$.
(9) $-a + (a + b) = b$ *(Left Cancellation Law).*
(10) $\text{gyr}[a, b]x = -(a + b) + \{a + (b + x)\}$ *(The Gyrator Identity).*
(11) $\text{gyr}[a, b]0 = 0$.
(12) $\text{gyr}[a, b](-x) = -\text{gyr}[a, b]x$.
(13) $\text{gyr}[a, 0] = I$.

Proof. (1) Let x be a left inverse of a corresponding to a left identity, 0, in G. We have $x + (a + b) = x + (a + c)$. By left gyroassociativity, $(x + a) + \text{gyr}[x, a]b = (x + a) + \text{gyr}[x, a]c$. Since 0 is a left identity, $\text{gyr}[x, a]b = \text{gyr}[x, a]c$. Since automorphisms are bijective, $b = c$.

(2) By left gyroassociativity we have for any left identity 0 of G, $a + x = 0 + (a + x) = (0 + a) + \text{gyr}[0, a]x = a + \text{gyr}[0, a]x$. Hence, by (1) above we have $x = \text{gyr}[0, a]x$ for all $x \in G$ so that $\text{gyr}[0, a] = I$.

(3) By the left loop property and by (2) above we have $\text{gyr}[x, a] = \text{gyr}[x + a, a] = \text{gyr}[0, a] = I$.

(4) Follows from an application of the left loop property to (2) above.

(5) Let x be a left inverse of a corresponding to a left identity, 0, of G. Then by left gyroassociativity and (3) above, $x + (a + 0) = (x + a) + \text{gyr}[x, a]0 = 0 + 0 = 0 = x + a$. Hence, by (1), $a + 0 = a$ for all $a \in G$ so that 0 is a right identity.

(6) Suppose 0 and 0^* are two left identities, one of which, say 0, is also a right identity. Then $0 = 0^* + 0 = 0^*$.
(7) Let x be a left inverse of a. Then $x + (a + x) = (x + a) + \mathrm{gyr}[x, a]x = 0 + x = x = x + 0$, by left gyroassociativity, (G2) of Def. 2.2, (3), and (5) and (6) above. By (1) we have $a + x = 0$ so that x is a right inverse of a.
(8) Suppose x and y are left inverses of a. By (7) above, they are also right inverses, so $a + x = 0 = a + y$. By (1), $x = y$. Let $-a$ be the resulting unique inverse of a. Then $-a + a = 0$ so that the inverse $-(-a)$ of $-a$ is a.
(9) By left gyroassociativity and by (3) above we have $-a + (a + b) = (-a + a) + \mathrm{gyr}[-a, a]b = b$.
(10) By an application of the left cancellation law (9) to the left gyroassociative law (G3) in Def. 2.2 we obtain (10).
(11) We obtain (11) from (10) with $x = 0$.
(12) Since $\mathrm{gyr}[a, b]$ is an automorphism of $(G, +)$ we have from (11) $\mathrm{gyr}[a, b](-x) + \mathrm{gyr}[a, b]x = \mathrm{gyr}[a, b](-x + x) = \mathrm{gyr}[a, b]0 = 0$, and hence the result.
(13) We obtain (13) from (10) with $b = 0$, and a left cancellation, (9).

□

Theorem 2.6. (Gyrosum Inversion Law). *For any two elements a, b of a gyrogroup $(G, +)$ we have the gyrosum inversion law*

$$-(a + b) = \mathrm{gyr}[a, b](-b - a) \tag{2.41}$$

Proof. By the gyrator identity in Theorem 2.5(10) and a left cancellation, Theorem 2.5(9), we have

$$\begin{aligned} \mathrm{gyr}[a, b](-b - a) &= -(a + b) + (a + (b + (-b - a))) \\ &= -(a + b) + (a - a) \\ &= -(a + b) \end{aligned} \tag{2.42}$$

□

Theorem 2.7. *Any three elements a, b, c of a gyrogroup $(G, +)$ satisfy the nested gyroautomorphism identities*

$$\mathrm{gyr}[a, b + c]\mathrm{gyr}[b, c] = \mathrm{gyr}[a + b, \mathrm{gyr}[a, b]c]\mathrm{gyr}[a, b] \tag{2.43}$$

$$\mathrm{gyr}[a + b, -\mathrm{gyr}[a, b]b]\mathrm{gyr}[a, b] = I \tag{2.44}$$

$$\mathrm{gyr}[a, -\mathrm{gyr}[a, b]b]\mathrm{gyr}[a, b] = I \tag{2.45}$$

and the gyroautomorphism product identities

$$\mathrm{gyr}[-a, a+b]\mathrm{gyr}[a, b] = I \tag{2.46}$$

and

$$\mathrm{gyr}[b, a+b]\mathrm{gyr}[a, b] = I \tag{2.47}$$

Proof. By two successive applications of the left gyroassociative law in two different ways, we obtain the following two chains of equations for all $a, b, c, x \in G$,

$$\begin{aligned} a + (b + (c + x)) &= a + ((b + c) + \mathrm{gyr}[b, c]x) \\ &= (a + (b + c)) + \mathrm{gyr}[a, b + c]\mathrm{gyr}[b, c]x \end{aligned} \tag{2.48}$$

and

$$\begin{aligned} a + (b + (c + x)) &= (a + b) + \mathrm{gyr}[a, b](c + x) \\ &= (a + b) + (\mathrm{gyr}[a, b]c + \mathrm{gyr}[a, b]x) \\ &= ((a + b) + \mathrm{gyr}[a, b]c) + \mathrm{gyr}[a + b, \mathrm{gyr}[a, b]c]\mathrm{gyr}[a, b]x \\ &= (a + (b + c)) + \mathrm{gyr}[a + b, \mathrm{gyr}[a, b]c]\mathrm{gyr}[a, b]x \end{aligned} \tag{2.49}$$

By comparing the extreme right-hand sides of these two chains of equations, and by employing the left cancellation law, Theorem 2.5(1), we obtain the identity

$$\mathrm{gyr}[a, b + c]\mathrm{gyr}[b, c]x = \mathrm{gyr}[a + b, \mathrm{gyr}[a, b]c]\mathrm{gyr}[a, b]x \tag{2.50}$$

for all $x \in G$, thus verifying (2.43).

In the special case when $c = -b$, (2.43) descends to (2.44), noting that the left-hand side of (2.43) is trivial owing to items (3) and (13) of Theorem 2.5. We recall that a map is *trivial* if it is the identity map.

Identity (2.45) results from the following chain of equations, which are numbered for subsequent derivation:

$$\begin{aligned} I &\overbrace{=\!=}^{(1)} \mathrm{gyr}[a + b, \ -\mathrm{gyr}[a, b]b]\mathrm{gyr}[a, b] \\ &\overbrace{=\!=}^{(2)} \mathrm{gyr}[(a + b) - \mathrm{gyr}[a, b]b, -\mathrm{gyr}[a, b]b]\mathrm{gyr}[a, b] \\ &\overbrace{=\!=}^{(3)} \mathrm{gyr}[a + (b - b), -\mathrm{gyr}[a, b]b]\mathrm{gyr}[a, b] \\ &\overbrace{=\!=}^{(4)} \mathrm{gyr}[a, -\mathrm{gyr}[a, b]b]\mathrm{gyr}[a, b] \end{aligned} \tag{2.51}$$

Derivation of the numbered equalities in (2.51) follows.

(1) Follows from (2.44).
(2) Follows from (1) by the left loop property (G5) of gyrogroups in Def. 2.2, p. 71.
(3) Follows from (2) by the left gyroassociative law (G3) of gyrogroups in Def. 2.2, p. 71. Indeed, an application of the left gyroassociative law to the first entry of the left gyration in (3) gives the first entry of the left gyration in (2), that is, $a + (b - b) = (a + b) - \text{gyr}[a, b]b$.
(4) Follows from (3) immediately, $a + (b - b) = a + 0 = a$.

To verify (2.46) we consider the special case of (2.43) when $b = -a$, obtaining

$$\text{gyr}[a, -a + c]\text{gyr}[-a, c] = \text{gyr}[0, \text{gyr}[a, -a]c]\text{gyr}[a, -a] = I \tag{2.52}$$

where the second identity in (2.52) follows from items (2) and (3) of Theorem 2.5. Replacing a by $-a$ and c by b in (2.52) we obtain (2.46).

Finally, (2.47) is derived from (2.46) by an application of the left loop property to the first gyroautomorphism in (2.46) followed by a left cancellation, Theorem 2.5(9),

$$\begin{aligned} I &= \text{gyr}[-a, a + b]\text{gyr}[a, b] \\ &= \text{gyr}[-a + (a + b), a + b]\text{gyr}[a, b] \\ &= \text{gyr}[b, a + b]\text{gyr}[a, b] \end{aligned} \tag{2.53}$$

□

The nested gyroautomorphism identity (2.45) in Theorem 2.7 allows the equation that defines the coaddition $\boxplus$ to be dualized with its corresponding equation in which the roles of the binary operations $\boxplus$ and $\oplus$ are interchanged, as shown in the following theorem.

Theorem 2.8. *Let $(G, \oplus)$ be a gyrogroup with cooperation $\boxplus$ given in Def. 2.4, p. 72, by the equation*

$$a \boxplus b = a \oplus \text{gyr}[a, \ominus b]b \tag{2.54}$$

Then

$$a \oplus b = a \boxplus \text{gyr}[a, b]b \tag{2.55}$$

Proof. Let a and b be any two elements of G. By (2.54) and (2.45) we have

$$\begin{aligned} a \boxplus \text{gyr}[a, b]b &= a \oplus \text{gyr}[a, \ominus \text{gyr}[a, b]b]\text{gyr}[a, b]b \\ &= a \oplus b \end{aligned} \tag{2.56}$$

thus verifying (2.55). □

In view of the duality symmetry that Identities (2.54) and (2.55) share, the gyroautomorphisms $\mathrm{gyr}[a,b]$ and $\mathrm{gyr}[a,\ominus b]$ may be considered dual to each other.

We naturally use the notation

$$a \boxminus b = a \boxplus (\ominus b) \tag{2.57}$$

in a gyrogroup $(G,\oplus)$, so that, by (2.57), (2.54) and Theorem 2.5(12),

$$\begin{aligned} a \boxminus b &= a \boxplus (\ominus b) \\ &= a \oplus \mathrm{gyr}[a,b](\ominus b) \\ &= a \ominus \mathrm{gyr}[a,b]b \end{aligned} \tag{2.58}$$

and, hence,

$$a \boxminus a = a \ominus a = 0 \tag{2.59}$$

as it should. Identity (2.59), in turn, implies the equality between the inverses of $a \in G$ with respect to $\oplus$ and $\boxplus$,

$$\boxminus a = \ominus a \tag{2.60}$$

for all $a \in G$.

As an application of the left cancellation law in Theorem 2.5(9) we present in the next theorem the gyrogroup counterpart (2.61) of the simple, but important, group identity $(-a+b)+(-b+c) = -a+c$.

Theorem 2.9. *Let $(G,+)$ be a gyrogroup. Then*

$$(-a+b) + \mathrm{gyr}[-a,b](-b+c) = -a+c \tag{2.61}$$

for all $a,b,c \in G$.

Proof. By the left gyroassociative law and the left cancellation law, and using the notation $d = -b+c$, we have,

$$\begin{aligned} (-a+b) + \mathrm{gyr}[-a,b](-b+c) &= (-a+b) + \mathrm{gyr}[-a,b]d \\ &= -a + (b+d) \\ &= -a + (b + (-b+c)) \\ &= -a + c \end{aligned} \tag{2.62}$$

□

Theorem 2.10. (The Gyrotranslation Theorem). *Let $(G,+)$ be a gyrogroup. Then*

$$-(-a+b) + (-a+c) = \mathrm{gyr}[-a,b](-b+c) \tag{2.63}$$

for all $a,b,c \in G$.

Proof. Identity (2.63) is a rearrangement of Identity (2.61) obtained by a left cancellation. □

The importance of Identity (2.63) is indicated by the analogy it shares with its group counterpart, $-(-a+b)+(-a+c)=-b+c$.

The identity of Theorem 2.9 can readily be generalized to any number of terms, for instance,

$$(-a+b)+\mathrm{gyr}[-a,b]\{(-b+c)+\mathrm{gyr}[-b,c](-c+d)\}=-a+d \tag{2.64}$$

which is analogous to the obvious group identity $(-a+b)+(-b+c)+(-c+d)=-a+d$.

2.5 The Two Basic Equations of Gyrogroups

The two basic equations of gyrogroup theory are

$$a\oplus x=b \tag{2.65}$$

and

$$x\oplus a=b \tag{2.66}$$

$a,b,x\in G$, for the unknown x in a gyrogroup $(G,\oplus)$.

Let x be a solution of the first basic equation, (2.65). Then we have by (2.65) and the left cancellation law, Theorem 2.5(9),

$$\ominus a\oplus b=\ominus a\oplus(a\oplus x)=x \tag{2.67}$$

Hence, if a solution x of (2.65) exists then it must be given by $x=\ominus a\oplus b$, as we see from (2.67).

Conversely, $x=\ominus a\oplus b$ is, indeed, a solution of (2.65) as we see by substituting $x=\ominus a\oplus b$ into (2.65) and applying the left cancellation law in Theorem 2.5(9). Hence, the gyrogroup equation (2.65) possesses the unique solution $x=\ominus a\oplus b$.

The solution of the second basic gyrogroup equation, (2.66), is quiet different from that of the first, (2.65), owing to the noncommutativity of the gyrogroup operation. Let x be a solution of (2.66). Then we have the following chain of equations, which are numbered for subsequent derivation:

$$
\begin{aligned}
x &\overset{(1)}{=\!=\!=} x\oplus 0 \\
&\overset{(2)}{=\!=\!=} x\oplus(a\ominus a) \\
&\overset{(3)}{=\!=\!=} (x\oplus a)\oplus\mathrm{gyr}[x,a](\ominus a) \\
&\overset{(4)}{=\!=\!=} (x\oplus a)\ominus\mathrm{gyr}[x,a]a \\
&\overset{(5)}{=\!=\!=} (x\oplus a)\ominus\mathrm{gyr}[x\oplus a,a]a \\
&\overset{(6)}{=\!=\!=} b\ominus\mathrm{gyr}[b,a]a \\
&\overset{(7)}{=\!=\!=} b\boxminus a
\end{aligned}
\tag{2.68}
$$

Derivation of the numbered equalities in (2.68) follows.

(1) Follows from the existence of a unique identity element, 0, in the gyrogroup $(G,\oplus)$ by Theorem 2.5.
(2) Follows from (1) by the existence of a unique inverse element $\ominus a$ of a in the gyrogroup $(G,\oplus)$ by Theorem 2.5.
(3) Follows from (2) by the left gyroassociative law in Axiom (G3) of gyrogroups in Def. 2.2, p. 71.
(4) Follows from (3) by Theorem 2.5(12).
(5) Follows from (4) by the left loop property (G5) of gyrogroups in Def. 2.2.
(6) Follows from (5) by the assumption that x is a solution of (2.66).
(7) Follows from (6) by (2.58).

Hence, if a solution x of (2.66) exists, then it must be given by $x = b\boxminus a$, as we see from (2.68).

Conversely, $x = b\boxminus a$ is, indeed, a solution of (2.66), as we see from the following chain of equations:

$$
\begin{aligned}
x\oplus a &\overbrace{=\!=}^{(1)} (b\boxminus a)\oplus a \\
&\overbrace{=\!=}^{(2)} (b\ominus \mathrm{gyr}[b,a]a)\oplus a \\
&\overbrace{=\!=}^{(3)} (b\ominus \mathrm{gyr}[b,a]a)\oplus \mathrm{gyr}[b,\ominus \mathrm{gyr}[b,a]a]\mathrm{gyr}[b,a]a \\
&\overbrace{=\!=}^{(4)} b\oplus(\ominus \mathrm{gyr}[b,a]a\oplus \mathrm{gyr}[b,a]a) \\
&\overbrace{=\!=}^{(5)} b\oplus 0 \\
&\overbrace{=\!=}^{(6)} b
\end{aligned}
\tag{2.69}
$$

Derivation of the numbered equalities in (2.69) follows.

(1) Follows from the assumption that $x = b\boxminus a$.
(2) Follows from (1) by (2.58).
(3) Follows from (2) by Identity (2.45) of Theorem 2.7, according to which the gyration product applied to a in (3) is trivial.
(4) Follows from (3) by the left gyroassociative law. Indeed, an application of the left gyroassociative law to (4) results in (3).
(5) Follows from (4) since $\ominus\mathrm{gyr}[b,a]a$ is the unique inverse of $\mathrm{gyr}[b,a]a$.
(6) Follows from (5) since 0 is the unique identity element of the gyrogroup $(G,\oplus)$.

Formalizing the results of this section, we have the following theorem.

Theorem 2.11. (The Two Basic Equations Theorem). *Let $(G,\oplus)$ be a gyrogroup, and let $a,b\in G$. The unique solution of the equation*

$$a\oplus x = b \tag{2.70}$$

in G for the unknown x is

$$x = \ominus a\oplus b \tag{2.71}$$

and the unique solution of the equation

$$x\oplus a = b \tag{2.72}$$

in G for the unknown x is

$$x = b\boxminus a \tag{2.73}$$

Let $(G, \oplus)$ be a gyrogroup, and let $a \in G$. The maps λ_a and ρ_a of G, given by

$$\begin{aligned} \lambda_a : \quad G &\to G, \qquad\qquad \lambda_a : \quad g \mapsto a \oplus g \\ \rho_a : \quad G &\to G, \qquad\qquad \rho_a : \quad g \mapsto g \oplus a \end{aligned} \tag{2.74}$$

are called, respectively, a *left gyrotranslation* of G by a and a *right gyrotranslation* of G by a. Theorem 2.11 asserts that each of these transformations of G is bijective, that is, it maps G onto itself in a one-to-one manner.

Substituting the solution (2.71) into its equation (2.70) we obtain the left cancellation law

$$a \oplus (\ominus a \oplus b) = b \tag{2.75}$$

for all $a, b \in G$, already verified in Theorem 2.5(9).

Similarly, substituting the solution (2.73) into its equation (2.72) we obtain the first right cancellation law

$$(b \boxminus a) \oplus a = b \tag{2.76}$$

for all $a, b \in G$. The latter can be dualized, obtaining the second right cancellation law

$$(b \ominus a) \boxplus a = b \tag{2.77}$$

for all $a, b \in G$. Indeed, (2.77) results from the following chain of equations

$$\begin{aligned} b &= b \oplus 0 \\ &= b \oplus (\ominus a \oplus a) \\ &= (b \ominus a) \oplus \mathrm{gyr}[b, \ominus a]a \\ &= (b \ominus a) \oplus \mathrm{gyr}[b \ominus a, \ominus a]a \\ &= (b \ominus a) \boxplus a \end{aligned} \tag{2.78}$$

where we employ the left gyroassociative law, the left loop property, and the definition of the gyrogroup cooperation.

Identities (2.75) – (2.77) form the three basic cancellation laws of gyrogroup theory. Indeed, these cancellation laws are used frequently in the study of gyrogroups and gyrovector spaces.

2.6 Einstein Gyrovector Spaces

Einstein addition $\oplus$ in balls $\mathbb{R}_s^n$, $n > 1$, gives rise to Einstein gyrocommutative gyrogroups $(\mathbb{R}_s^n, \oplus)$. The rich structure of Einstein addition is not limited to its gyrocommutative gyrogroup structure. Einstein addition admits scalar multiplication, giving rise to Einstein gyrovector spaces as well [11]. The latter, in turn, form the setting for the Beltrami-Klein ball model of hyperbolic geometry just as vector spaces form the setting for the standard model of Euclidean geometry, as shown in [3] and as we will see in this book.

Let $X \in \mathbb{R}_s^n$ be a point of an Einstein gyrocommutative gyrogroup $(\mathbb{R}_s^n, \oplus)$. Einstein addition of k copies of X, $k \geq 1$, denoted $k \otimes X$, gives

$$k \otimes X = s \frac{\left(1 + \frac{\|X\|}{s}\right)^k - \left(1 - \frac{\|X\|}{s}\right)^k}{\left(1 + \frac{\|X\|}{s}\right)^k + \left(1 - \frac{\|X\|}{s}\right)^k} \frac{X}{\|X\|} \tag{2.79}$$

Identity (2.79) of scalar multiplication of $X \in \mathbb{R}_s^n$ by any positive integer k suggests the following definition of Einstein scalar multiplication, which requires analytically continuing k off the positive integers.

Definition 2.12. (Einstein Scalar Multiplication, Einstein Gyrovector Spaces). *An Einstein gyrovector space* $(\mathbb{R}_s^n, \oplus, \otimes)$ *is an Einstein gyrogroup* $(\mathbb{R}_s^n, \oplus)$ *with scalar multiplication* $\otimes$ *given by the equation*

$$\begin{aligned} r \otimes X &= s \frac{\left(1 + \frac{\|X\|}{s}\right)^r - \left(1 - \frac{\|X\|}{s}\right)^r}{\left(1 + \frac{\|X\|}{s}\right)^r + \left(1 - \frac{\|X\|}{s}\right)^r} \frac{X}{\|X\|} \\ &= s \tanh(r \tanh^{-1} \frac{\|X\|}{s}) \frac{X}{\|X\|} \end{aligned} \tag{2.80}$$

where r is any real number, $r \in \mathbb{R}$, $X \in \mathbb{R}_s^n$, $X \neq \mathbf{0}$, and $r \otimes \mathbf{0} = \mathbf{0}$, and with which we use the notation $X \otimes r = r \otimes X$.

Einstein gyrovector spaces are studied in [3, Sec. 6.18] [33]. Einstein scalar multiplication does not distribute over Einstein addition, but it possesses other properties of vector spaces. For any positive integer k, and for all real numbers $r, r_1, r_2 \in \mathbb{R}$ and all $X \in \mathbb{R}_s^n$, we have

$$k \otimes X = X \oplus \ldots \oplus X \qquad k \text{ terms}$$

$$(r_1 + r_2) \otimes X = r_1 \otimes X \oplus r_2 \otimes X \qquad \text{Scalar Distributive Law}$$

$$(r_1 r_2) \otimes X = r_1 \otimes (r_2 \otimes X) \qquad \text{Scalar Associative Law}$$

in any Einstein gyrovector space $(\mathbb{R}_s^n, \oplus, \otimes)$.

As an example, the *Einstein half* is given by

$$\tfrac{1}{2} \otimes X = \frac{\gamma_X}{1 + \gamma_X} X \tag{2.81}$$

for all X in an Einstein gyrovector space $(\mathbb{R}_s^n, \oplus, \otimes)$. Indeed, by (2.12), p. 66,

$$\tfrac{1}{2} \otimes X \oplus \tfrac{1}{2} \otimes X = \frac{\gamma_X}{1 + \gamma_X} X \oplus \frac{\gamma_X}{1 + \gamma_X} X = X \tag{2.82}$$

in agreement with the scalar distributive law, as one can readily check.

Any Einstein gyrovector space $(\mathbb{R}_s^n, \oplus, \otimes)$ inherits an inner product and a norm from its ambient vector space $\mathbb{R}^n$. These turn out to be invariant under gyrations, that is,

$$\begin{aligned} \text{gyr}[A, B]X \cdot \text{gyr}[A, B]Y &= X \cdot Y \\ \|\text{gyr}[A, B]Y\| &= \|Y\| \end{aligned} \tag{2.83}$$

for all $A, B, X, Y \in \mathbb{R}_s^n$.

Unlike vector spaces, Einstein gyrovector spaces $(\mathbb{R}_s^n, \oplus, \otimes)$ do not possess the distributive law since, in general,

$$r \otimes (X \oplus Y) \neq r \otimes X \oplus r \otimes Y \tag{2.84}$$

for $r \in \mathbb{R}$ and $X, Y \in \mathbb{R}_s^n$. One might suppose that there is a price to pay in mathematical regularity when replacing ordinary vector addition with Einstein addition, but this is not the case as demonstrated in [1–4, 18, 19, 33, 39] and in this book, and as noted by S. Walter in [40].

The gamma factor of $r \otimes X$ is expressible in terms of the gamma factor of X by the identity

$$\gamma_{r \otimes X} = \frac{1}{2} \gamma_X^r \left\{ \left(1 + \frac{\|X\|}{s}\right)^r + \left(1 - \frac{\|X\|}{s}\right)^r \right\} \tag{2.85}$$

and hence, by (2.80),

$$\gamma_{r \otimes X}(r \otimes X) = \frac{1}{2} s \gamma_X^r \left\{ \left(1 + \frac{\|X\|}{s}\right)^r - \left(1 - \frac{\|X\|}{s}\right)^r \right\} \frac{X}{\|X\|} \tag{2.86}$$

for $X \neq \mathbf{0}$. The special case of $r = 2$ is of particular interest,

$$\gamma_{2\otimes X}(2\otimes X) = 2\gamma_X^2 X \tag{2.87}$$

Noting (2.11), p. 66, we have from (2.80),

$$2\otimes X = \frac{2\gamma_X^2}{2\gamma_X^2 - 1} X \tag{2.88}$$

so that

$$\gamma_{2\otimes X} = 2\gamma_X^2 - 1 = \frac{1 + \|X\|^2/s^2}{1 - \|X\|^2/s^2} \tag{2.89}$$

2.7 Gyrovector Spaces

To set the stage for the definition of gyrovector spaces, we present the common definition of vector spaces.

Definition 2.13. (Real Inner Product Vector Spaces). *A real inner product vector space* $(\mathbb{V}, +, \cdot)$ *(vector space, in short) is a real vector space together with a map*

$$\mathbb{V}\times\mathbb{V} \to \mathbb{R}, \qquad (\mathbf{u}, \mathbf{v}) \mapsto \mathbf{u}\cdot\mathbf{v} \tag{2.90}$$

called a real inner product, satisfying the following properties for all $\mathbf{u}, \mathbf{v}, \mathbf{w}\in\mathbb{V}$ *and* $r\in\mathbb{R}$*:*

(1) $\mathbf{v}\cdot\mathbf{v} \geq 0$*, with equality if, and only if,* $\mathbf{v} = 0$.
(2) $\mathbf{u}\cdot\mathbf{v} = \mathbf{v}\cdot\mathbf{u}$
(3) $(\mathbf{u} + \mathbf{v})\cdot\mathbf{w} = \mathbf{u}\cdot\mathbf{w} + \mathbf{v}\cdot\mathbf{w}$
(4) $(r\mathbf{u})\cdot\mathbf{v} = r(\mathbf{u}\cdot\mathbf{v})$

The norm $\|\mathbf{v}\|$ *of* $\mathbf{v}\in\mathbb{V}$ *is given by the equation* $\|\mathbf{v}\|^2 = \mathbf{v}\cdot\mathbf{v}$.

Note that the properties of vector spaces imply (i) the Cauchy-Schwarz inequality

$$|\mathbf{u}\cdot\mathbf{v}| \leq \|\mathbf{u}\|\|\mathbf{v}\| \tag{2.91}$$

for all $\mathbf{u}, \mathbf{v}\in\mathbb{V}$; and (ii) the *positive definiteness* of the inner product, according to which $\mathbf{u}\cdot\mathbf{v} = 0$ for all $\mathbf{u}\in\mathbb{V}$ implies $\mathbf{v} = 0$ [10].

Guided by analogies with vector spaces, we take key features of Einstein scalar multiplication as axioms in the following formal definition of gyrovector spaces.

Definition 2.14. (Real Inner Product Gyrovector Spaces). *A real inner product gyrovector space* $(G, \oplus, \otimes)$ *(gyrovector space, in short) is a gyrocommutative gyrogroup* $(G, \oplus)$ *that obeys the following axioms:*

(1) G is a subset of a real inner product vector space $\mathbb{V}$ *called the carrier of* G, $G \subset \mathbb{V}$, *from which it inherits its inner product,* $\cdot$, *and norm,* $\|\cdot\|$, *which are invariant under gyroautomorphisms, that is,*

(V1) $\mathrm{gyr}[\mathbf{u},\mathbf{v}]\mathbf{a}\cdot\mathrm{gyr}[\mathbf{u},\mathbf{v}]\mathbf{b} = \mathbf{a}\cdot\mathbf{b}$ *Inner Product Gyroinvariance*

for all points $\mathbf{a},\mathbf{b},\mathbf{u},\mathbf{v}\in G$.

(2) G admits a scalar multiplication, $\otimes$, *possessing the following properties. For all real numbers* $r, r_1, r_2 \in \mathbb{R}$ *and all points* $\mathbf{a}\in G$:

(V2) $1\otimes\mathbf{a} = \mathbf{a}$ *Identity Scalar Multiplication*

(V3) $(r_1 + r_2)\otimes\mathbf{a} = r_1\otimes\mathbf{a}\oplus r_2\otimes\mathbf{a}$ *Scalar Distributive Law*

(V4) $(r_1 r_2)\otimes\mathbf{a} = r_1\otimes(r_2\otimes\mathbf{a})$ *Scalar Associative Law*

(V5) $\dfrac{|r|\otimes\mathbf{a}}{\|r\otimes\mathbf{a}\|} = \dfrac{\mathbf{a}}{\|\mathbf{a}\|}, \quad \mathbf{a}\neq\mathbf{0},\ r\neq 0$ *Scaling Property*

(V6) $\mathrm{gyr}[\mathbf{u},\mathbf{v}](r\otimes\mathbf{a}) = r\otimes\mathrm{gyr}[\mathbf{u},\mathbf{v}]\mathbf{a}$ *Gyroautomorphism Property*

(V7) $\mathrm{gyr}[r_1\otimes\mathbf{v}, r_2\otimes\mathbf{v}] = I$ *Identity Gyroautomorphism.*

(3) Real, one-dimensional vector space structure $(\|G\|,\oplus,\otimes)$ *for the set* $\|G\|$ *of one-dimensional "vectors":*

(V8) $\|G\| = \{\pm\|\mathbf{a}\| : \mathbf{a}\in G\} \subset \mathbb{R}$ *Vector Space*

with vector addition $\oplus$ *and scalar multiplication* $\otimes$, *such that for all* $r\in\mathbb{R}$ *and* $\mathbf{a},\mathbf{b}\in G$,

(V9) $\|r\otimes\mathbf{a}\| = |r|\otimes\|\mathbf{a}\|$ *Homogeneity Property*

(V10) $\|\mathbf{a}\oplus\mathbf{b}\| \le \|\mathbf{a}\|\oplus\|\mathbf{b}\|$ *Gyrotriangle Inequality.*

Remark 2.15. We use the notation $(r_1\otimes\mathbf{a})\oplus(r_2\otimes\mathbf{b}) = r_1\otimes\mathbf{a}\oplus r_2\otimes\mathbf{b}$, and $\mathbf{a}\otimes r = r\otimes\mathbf{a}$. Our ambiguous use of $\oplus$ and $\otimes$ in Def. 2.14 as interrelated operations in the gyrovector space $(G,\oplus,\otimes)$ and in its associated vector space $(\|G\|,\oplus,\otimes)$ should raise no confusion since the sets in which these operations operate are always clear from the context. These operations in the former (gyrovector space $(G,\oplus,\otimes)$) are nonassociative-nondistributive gyrovector space operations, and in the latter (vector space $(\|G\|,\oplus,\otimes)$) are associative-distributive vector space operations. Additionally, the gyroaddition $\oplus$ is gyrocommutative in the former and commutative in the latter. Note that in the vector space $(\|G\|,\oplus,\otimes)$ gyrations are trivial so that $\boxplus = \oplus$ in $\|G\|$.

While the operations $\oplus$ and $\otimes$ have distinct interpretations in the gyrovector space G and in the vector space $\|G\|$, they are related to one another by the gyrovector space axioms $(V9)$ and $(V10)$. The analogies that conventions about the ambiguous use of $\oplus$ and $\otimes$ in G and $\|G\|$ share with similar vector space conventions are obvious. Indeed, in vector spaces we use the same notation, $+$, for the addition operation between vectors and between their magnitudes, and same notation for the scalar multiplication between two scalars and between a scalar and a vector. In full analogy, in gyrovector spaces we use the same notation, $\oplus$, for the gyroaddition operation between gyrovectors and between their magnitudes, and the same notation, $\otimes$, for the scalar gyromultiplication between two scalars and between a scalar and a gyrovector.

Immediate consequences of the gyrovector space axioms are presented in the following theorem.

Theorem 2.16. *Let* $(G, \oplus, \otimes)$ *be a gyrovector space whose carrier vector space is* $\mathbb{V}$, *and let* 0, $\mathbf{0}$ *and* $\mathbf{0}_{\mathbb{V}}$ *be the neutral elements of* $(\mathbb{R}, +)$, $(G, \oplus)$ *and* $(\mathbb{V}, +)$, *respectively. Then, for all* $n \in \mathbb{N}$, $r \in \mathbb{R}$, *and* $\mathbf{a} \in G$,

(1) $0 \otimes \mathbf{a} = \mathbf{0}$
(2) $n \otimes \mathbf{a} = \mathbf{a} \oplus \ \ldots \ \oplus \mathbf{a}$ $\qquad$ *(n terms).*
(3) $(-r) \otimes \mathbf{a} = \ominus (r \otimes \mathbf{a}) =: \ominus r \otimes \mathbf{a}$
(4) $r \otimes (\ominus \mathbf{a}) = \ominus r \otimes \mathbf{a}$
(5) $r \otimes \mathbf{0} = \mathbf{0}$
(6) $\|\ominus \mathbf{a}\| = \|\mathbf{a}\|$
(7) $\mathbf{0} = \mathbf{0}_{\mathbb{V}}$
(8) $r \otimes \mathbf{a} = \mathbf{0} \iff (r = 0 \ \mathit{or} \ \mathbf{a} = \mathbf{0})$

Proof. (1) follows from the scalar distributive law $(V3)$,

$$r \otimes \mathbf{a} = (r + 0) \otimes \mathbf{a} = r \otimes \mathbf{a} \oplus 0 \otimes \mathbf{a} \tag{2.92}$$

so that, by a left cancellation, $0 \otimes \mathbf{a} = \ominus (r \otimes \mathbf{a}) \oplus (r \otimes \mathbf{a}) = \mathbf{0}$.

(2) follows from $(V2)$, and the scalar distributive law $(V3)$. Indeed, with "..." signifying "n terms", we have

$$\mathbf{a} \oplus \ldots \oplus \mathbf{a} = 1 \otimes \mathbf{a} \oplus \ldots \oplus 1 \otimes \mathbf{a} = (1 + \ldots + 1) \otimes \mathbf{a} = n \otimes \mathbf{a} \tag{2.93}$$

(3) results from (1) and the scalar distributive law $(V3)$,

$$\mathbf{0} = 0 \otimes \mathbf{a} = (r - r) \otimes \mathbf{a} = r \otimes \mathbf{a} \oplus (-r) \otimes \mathbf{a} \tag{2.94}$$

implying $\ominus (r \otimes \mathbf{a}) = (-r) \otimes \mathbf{a}$.

(4) results from (3) and the scalar associative law,

$$r\otimes(\ominus\mathbf{a}) = r\otimes((-1)\otimes\mathbf{a}) = (-r)\otimes\mathbf{a} = \ominus r\otimes\mathbf{a} \tag{2.95}$$

(5) follows from (1), $(V4)$, $(V3)$, (3),

$$\begin{aligned} r\otimes\mathbf{0} &= r\otimes(0\otimes\mathbf{a}) \\ &= r\otimes((1-1)\otimes\mathbf{a}) \\ (r(1-1))\otimes\mathbf{a} \\ &= (r-r)\otimes\mathbf{a} \\ &= r\otimes\mathbf{a}\oplus(-r)\otimes\mathbf{a} \\ &= r\otimes\mathbf{a}\oplus(\ominus(r\otimes\mathbf{a})) \\ &= r\otimes\mathbf{a}\ominus r\otimes\mathbf{a} \\ &= \mathbf{0} \end{aligned} \tag{2.96}$$

(6) follows from (3), the homogeneity property $(V9)$, and $(V2)$,

$$\|\ominus\mathbf{a}\| = \|(-1)\otimes\mathbf{a}\| = |-1|\otimes\|\mathbf{a}\| = 1\otimes\|\mathbf{a}\| = \|1\otimes\mathbf{a}\| = \|\mathbf{a}\| \tag{2.97}$$

(7) results from (5), $(V9)$, and $(V8)$ as follows.

$$\|\mathbf{0}\| = \|2\otimes\mathbf{0}\| = 2\otimes\|\mathbf{0}\| = \|\mathbf{0}\|\oplus\|\mathbf{0}\| \tag{2.98}$$

implying $\|\mathbf{0}\| = \|\mathbf{0}\|\ominus\|\mathbf{0}\| = 0$ in the vector space $(\|G\|, \oplus, \otimes)$. This equation, $\|\mathbf{0}\| = 0$, is valid in the vector space $\mathbb{V}$ as well, where it implies $\mathbf{0} = \mathbf{0}_{\mathbb{V}}$.

(8) results from the following considerations. Suppose $r\otimes\mathbf{a} = \mathbf{0}$, but $r \neq 0$. Then, by $(V1)$, $(V4)$ and (5) we have

$$\mathbf{a} = 1\otimes\mathbf{a} = (1/r)\otimes(r\otimes\mathbf{a}) = (1/r)\otimes\mathbf{0} = \mathbf{0} \tag{2.99}$$

□

Clearly, in the special case when all the gyrations of a gyrovector space are trivial, the gyrovector space descends to a vector space.

In general, gyroaddition does not distribute with scalar multiplication,

$$r\otimes(\mathbf{a}\oplus\mathbf{b}) \neq r\otimes\mathbf{a}\oplus r\otimes\mathbf{b} \tag{2.100}$$

However, gyrovector spaces possess a weak distributive law, called the monodistributive law, presented in the following theorem.

Theorem 2.17. (The Monodistributive Law). *A gyrovector space* $(G, \oplus, \otimes)$ *possesses the monodistributive law*

$$r\otimes(r_1\otimes\mathbf{a}\oplus r_2\otimes\mathbf{a}) = r\otimes(r_1\otimes\mathbf{a})\oplus r\otimes(r_2\otimes\mathbf{a}) \tag{2.101}$$

for all $r, r_1, r_2 \in \mathbb{R}$ *and* $\mathbf{a} \in G$.

Proof. The proof follows from $(V3)$ and $(V4)$,

$$\begin{aligned} r\otimes(r_1\otimes\mathbf{a}\oplus r_2\otimes\mathbf{a}) &= r\otimes\{(r_1+r_2)\otimes\mathbf{a}\} \\ &= (r(r_1+r_2))\otimes\mathbf{a} \\ &= (rr_1+rr_2)\otimes\mathbf{a} \\ &= (rr_1)\otimes\mathbf{a}\oplus(rr_2)\otimes\mathbf{a} \\ &= r\otimes(r_1\otimes\mathbf{a})\oplus r\otimes(r_2\otimes\mathbf{a}) \end{aligned} \tag{2.102}$$

□

Definition 2.18. (Gyrovector Space Automorphisms). *An automorphism* τ *of a gyrovector space* $(G, \oplus, \otimes)$, $\tau \in \mathrm{Aut}(G, \oplus, \otimes)$, *is a bijective self-map of* G

$$\tau : G \rightarrow G \tag{2.103}$$

which preserves its structure, that is, (i) binary operation, (ii) scalar multiplication, and (iii) inner product,

$$\begin{aligned} \tau(\mathbf{a}\oplus\mathbf{b}) &= \tau\mathbf{a}\oplus\tau\mathbf{b} \\ \tau(r\otimes\mathbf{a}) &= r\otimes\tau\mathbf{a} \\ \tau\mathbf{a}\cdot\tau\mathbf{b} &= \mathbf{a}\cdot\mathbf{b} \end{aligned} \tag{2.104}$$

The automorphisms of the gyrovector space $(G, \oplus, \otimes)$ *form a group denoted* $\mathrm{Aut}(G, \oplus, \otimes)$, *with group operation given by automorphism composition.*

Following Axiom $(G4)$ of gyrogroups and Axioms $(V6)$ and $(V1)$ of gyrovector spaces, gyroautomorphisms are special gyrovector space automorphisms.

Scalar multiplication in a gyrovector space does not distribute over the gyrovector space operation. Hence, the *Two-Sum Identity* in the following theorem proves useful.

Theorem 2.19. (The Two-Sum Identity). *Let $(G, \oplus, \otimes)$ be a gyrovector space. Then*

$$2\otimes(\mathbf{a}\oplus\mathbf{b}) = \mathbf{a}\oplus(2\otimes\mathbf{b}\oplus\mathbf{a}) \tag{2.105}$$

for any $\mathbf{a}, \mathbf{b}\in G$.

Proof. Employing the right gyroassociative law, the identity $\mathrm{gyr}[\mathbf{b}, \mathbf{b}] = I$, the left gyroassociative law, and the gyrocommutative law we have the following chain of equations that gives (2.105),

$$\begin{aligned}\mathbf{a}\oplus(2\otimes\mathbf{b}\oplus\mathbf{a}) &= \mathbf{a}\oplus((\mathbf{b}\oplus\mathbf{b})\oplus\mathbf{a}) \\ &= \mathbf{a}\oplus(\mathbf{b}\oplus(\mathbf{b}\oplus\mathrm{gyr}[\mathbf{b}, \mathbf{b}]\mathbf{a})) \\ &= \mathbf{a}\oplus(\mathbf{b}\oplus(\mathbf{b}\oplus\mathbf{a})) \\ &= (\mathbf{a}\oplus\mathbf{b})\oplus\mathrm{gyr}[\mathbf{a}, \mathbf{b}](\mathbf{b}\oplus\mathbf{a}) \\ &= (\mathbf{a}\oplus\mathbf{b})\oplus(\mathbf{a}\oplus\mathbf{b}) \\ &= 2\otimes(\mathbf{a}\oplus\mathbf{b})\end{aligned} \tag{2.106}$$

□

A gyrovector space is a gyrometric space with a gyrodistance function that obeys the gyrotriangle inequality.

Definition 2.20. (The Gyrodistance Function). *Let $G = (G, \oplus, \otimes)$ be a gyrovector space. Its gyrometric is given by the gyrodistance function* $d_\oplus(\mathbf{a}, \mathbf{b}) : G\times G \to \mathbb{R}^{\geq 0}$,

$$d_\oplus(\mathbf{a}, \mathbf{b}) = \|\ominus\mathbf{a}\oplus\mathbf{b}\| = \|\mathbf{b}\ominus\mathbf{a}\| \tag{2.107}$$

Here $d_\oplus(\mathbf{a}, \mathbf{b})$ *is the gyrodistance between* $\mathbf{a}$ *and* $\mathbf{b}$, *and* $\mathbb{R}^{\geq 0}$ *is the set of all nonnegative Numbers.*

By Def. 2.14 of gyrovector spaces, gyroautomorphisms preserve the inner product. Hence, they are isometries, that is, they preserve the norm as well. Accordingly, the identity $\|\ominus\mathbf{a}\oplus\mathbf{b}\| = \|\mathbf{b}\ominus\mathbf{a}\|$ in Def. 2.20 follows from the gyrocommutative law,

$$\|\ominus\mathbf{a}\oplus\mathbf{b}\| = \|\mathrm{gyr}[\ominus\mathbf{a}, \mathbf{b}](\mathbf{b}\ominus\mathbf{a})\| = \|\mathbf{b}\ominus\mathbf{a}\| \tag{2.108}$$

Theorem 2.21. (The Gyrotriangle Inequality). *The gyrometric of a gyrovector space $(G, \oplus, \otimes)$ satisfies the gyrotriangle inequality*

$$\|\ominus\mathbf{a}\oplus\mathbf{c}\| \leq \|\ominus\mathbf{a}\oplus\mathbf{b}\|\oplus\|\ominus\mathbf{b}\oplus\mathbf{c}\| \tag{2.109}$$

for all $\mathbf{a}, \mathbf{b}, \mathbf{c}\in G$.

Proof. By Theorem 2.9, p. 78, we have,

$$\ominus\mathbf{a}\oplus\mathbf{c} = (\ominus\mathbf{a}\oplus\mathbf{b})\oplus\mathrm{gyr}[\ominus\mathbf{a},\mathbf{b}](\ominus\mathbf{b}\oplus\mathbf{c}) \tag{2.110}$$

Hence, by the gyrotriangle inequality $(V10)$ in Def. 2.14 we have

$$\begin{aligned}\|\ominus\mathbf{a}\oplus\mathbf{c}\| &= \|(\ominus\mathbf{a}\oplus\mathbf{b})\oplus\mathrm{gyr}[\ominus\mathbf{a},\mathbf{b}](\ominus\mathbf{b}\oplus\mathbf{c})\| \\ &\le \|\ominus\mathbf{a}\oplus\mathbf{b}\|\oplus\|\mathrm{gyr}[\ominus\mathbf{a},\mathbf{b}](\ominus\mathbf{b}\oplus\mathbf{c})\| \\ &= \|\ominus\mathbf{a}\oplus\mathbf{b}\|\oplus\|\ominus\mathbf{b}\oplus\mathbf{c}\|\end{aligned} \tag{2.111}$$

□

The basic properties of the gyrodistance function $d_\oplus$ are

(*i*) $d_\oplus(\mathbf{a},\mathbf{b}) \ge 0$
(*ii*) $d_\oplus(\mathbf{a},\mathbf{b}) = 0$ if and only if $\mathbf{a} = \mathbf{b}$.
(*iii*) $d_\oplus(\mathbf{a},\mathbf{b}) = d_\oplus(\mathbf{b},\mathbf{a})$
(*iv*) $d_\oplus(\mathbf{a},\mathbf{c}) \le d_\oplus(\mathbf{a},\mathbf{b})\oplus d_\oplus(\mathbf{b},\mathbf{c})$ (gyrotriangle inequality),

for any $\mathbf{a},\mathbf{b},\mathbf{c}\in G$.

Curves on which the gyrotriangle inequality descends to an equality, called geodesics or *gyrolines*, will be studied in this book in Einstein gyrovector spaces and in Möbius gyrovector spaces. A detailed proof that a Möbius gyrogroup with the Möbius scalar multiplication is a gyrovector space is presented in [3, Theorem 6.84]. A proof that an Einstein gyrogroup with the Einstein scalar multiplication is a gyrovector space is similar. Alternatively, the proof follows from the fact that Einstein gyrovector spaces and Möbius gyrovector spaces of the same dimension are isomorphic [41], as we will see in (2.278), p. 142.

2.8 Gyropoints, Gyrolines and Gyrodistance

In the Cartesian model $\mathbb{R}_s^n$ of the n-dimensional Einstein gyrovector space $(\mathbb{R}_s^n,\oplus,\otimes)$, where n is any positive integer, we introduce a Cartesian coordinate system relative to which points of $\mathbb{R}_s^n$, also called *gyropoints*, are given by n-tuples of real numbers, like $X = (x_1,x_2,\ldots,x_n)$, $\|X\|^2 < s^2$, or $Y = (y_1,y_2,\ldots,y_n)$, $\|Y\|^2 < s^2$, etc. The point $\mathbf{0} = (0,0,\ldots)$ (n zeros) is called the *origin* of $\mathbb{R}_s^n$. The Cartesian model $\mathbb{R}_s^n$ is a model of the n-dimensional hyperbolic geometry, as we will see in Sec. 2.9. It is a real inner product gyrovector space with addition and subtraction given by Einstein addition (2.4) and its associated subtraction, with scalar multiplication given by (2.80), and with the inner product and norm that it inherits from its ambient Euclidean n-space $\mathbb{R}^n$.

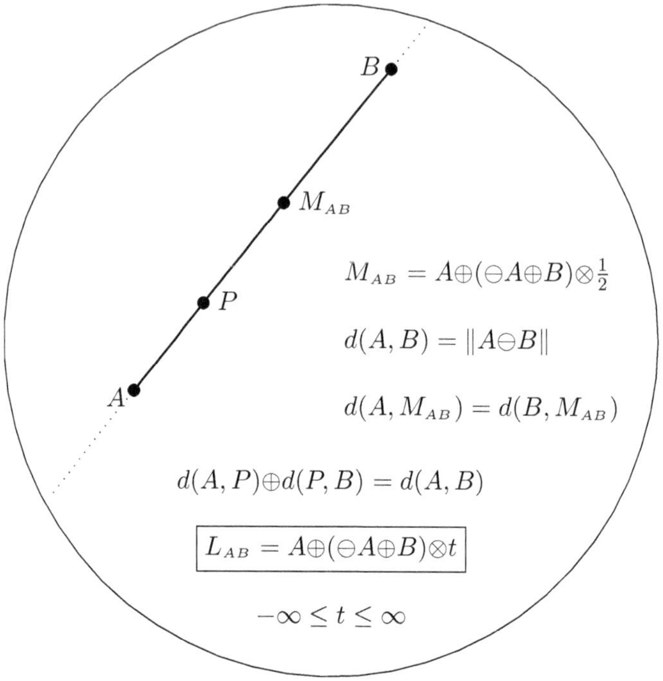

Fig. 2.1 The Einstein gyroline $L_{AB} = A\oplus(\ominus A\oplus B)\otimes t$, $t \in \mathbb{R}$, in an Einstein gyrovector space $(\mathbb{R}_s^n, \oplus, \otimes)$ is a geodesic line in the Beltrami-Klein ball model of hyperbolic geometry, fully analogous to the straight line $A + (-A + B)t$, $t \in \mathbb{R}$, in Euclidean geometry. The points A and B correspond to the gyroline parameter values $t = 0$ and $t = 1$, respectively. The point P with gyroline parameter $0 < t < 1$ is a generic point on the gyroline through the points A and B lying between these points. The Einstein sum, $\oplus$, of the Einstein gyrodistance $d(A,P)$ from A to P and $d(P,B)$ from P to B equals the Einstein gyrodistance $d(A,B)$ from A to B, that is, $d(A,P)\oplus d(P,B) = d(A,B)$, called a *gyrotriangle equality*. The point M_{AB} that corresponds to $t = 1/2$ is the hyperbolic midpoint, gyromidpoint, of the points A and B.

An illustrative example of the addition of points $X = (x_1, x_2, x_3)$ and $Y = (y_1, y_2, y_3)$ in the Cartesian model of $\mathbb{R}_c^3$ is presented in Example 2.1, p. 67.

In the Cartesian model $\mathbb{R}_s^n$ of the hyperbolic geometry of the n-dimensional Einstein gyrovector space $(\mathbb{R}_s^n, \oplus, \otimes)$, it is convenient to define a gyroline by the set of its points. Let $A, B\in\mathbb{R}_s^n$ be any two distinct points, which may also be called gyropoints. The unique gyroline L_{AB}, shown in Fig. 2.1, that passes through these points is the set of all points

$$L_{AB} = A\oplus(\ominus A\oplus B)\otimes t \tag{2.112}$$

for all $t\in\mathbb{R}$, $\mathbb{R}$ being the real line.

Equation (2.112) is said to be the gyroline representation in terms of points A and B with the parameter t. Obviously, the same gyroline can be represented by any two distinct points that it contains, as demonstrated in [3].

The Einstein gyrodistance function, $d(X, Y)$ in an Einstein gyrovector space $(\mathbb{R}_s^n, \oplus, \otimes)$ is given by the equation

$$d(X,Y) = \|\ominus X \oplus Y\| = \|X \ominus Y\| \tag{2.113}$$

$X, Y \in \mathbb{R}_s^n$. We call it a *gyrodistance function* in order to emphasize the analogies it shares with its Euclidean counterpart, the distance function $\|X - Y\|$ in $\mathbb{R}^n$. Among these analogies is the gyrotriangle inequality

$$\|X \oplus Y\| \le \|X\| \oplus \|Y\| \tag{2.114}$$

for all $X, Y \in \mathbb{R}_s^n$. For this and other analogies that distance and gyrodistance functions share see [2, 3, 33].

A left gyrotranslation $T_X A$ of a point A by a point X in an Einstein gyrovector space $(\mathbb{R}_s^n, \oplus, \otimes)$, is given by

$$T_X A = X \oplus A \tag{2.115}$$

for all $X, A \in \mathbb{R}_s^n$. Left gyrotranslation composition is given by point addition preceded by a gyration. Indeed, by the left gyroassociative law, which is Axiom $(G3)$ of gyrogroups in Def. 2.2, p. 71, a left gyrotranslation composition can be written as

$$T_X T_Y A = X \oplus (Y \oplus A) = (X \oplus Y) \oplus \mathrm{gyr}[X, Y]A = T_{X \oplus Y}\, \mathrm{gyr}[X, Y]A \tag{2.116}$$

for all $X, Y, A \in \mathbb{R}_s^n$, thus obtaining the left gyrotranslation composition law

$$T_X T_Y = T_{X \oplus Y}\, \mathrm{gyr}[X, Y] \tag{2.117}$$

of left gyrotranslations of an Einstein gyrovector space $(\mathbb{R}_s^n, \oplus, \otimes)$. According to (2.117), a left gyrotranslation by Y followed by a left gyrotranslation by X is equivalent to a single left gyrotranslation by $X \oplus Y$ preceded by a gyration $\mathrm{gyr}[X, Y]$.

Owing to the presence of a gyration in the composition law (2.117), the set of all left gyrotranslations of $\mathbb{R}_s^n$ does not form a group under left gyrotranslation composition. Rather, under left gyrotranslation composition it forms a gyrocommutative gyrogroup.

Each element of the special orthogonal group SO(n) of order n, that is, each element of the group of all $n \times n$ orthogonal matrices with determinant 1, represents a rotation R of points $A \in \mathbb{R}_s^n$ about the center of $\mathbb{R}_s^n$, denoted RA. It is given by the matrix product RA^t of a matrix $R \in \mathrm{SO(n)}$ and the transpose A^t of $A \in \mathbb{R}_s^n$. A rotation of $\mathbb{R}_s^n$ about its center is a linear map of $\mathbb{R}_s^n$ that keeps the inner product invariant. Hence, it leaves the origin of $\mathbb{R}^n$ invariant and respects Einstein addition, that is, $R(A \oplus B) = RA \oplus RB$ for all $A, B \in \mathbb{R}_s^n$.

Rotation composition is given by matrix multiplication, so that the set of all rotations of $\mathbb{R}_s^n$ about its origin forms a group under rotation composition.

Theorem 2.22. *Gyrodistance in any Einstein gyrovector space* $(\mathbb{R}_s^n, \oplus, \otimes)$ *is invariant under left gyrotranslations and rotations.*

Proof. Let $A, B, X \in \mathbb{R}_s^n$ be any three points, so that the left gyrotranslations of A and B by X are $X \oplus A$ and $X \oplus B$, respectively. Then, by (2.63) and (2.27) we have

$$\begin{aligned} d(X\oplus A, X\oplus B) &= \|\ominus(X\oplus A)\oplus(X\oplus B)\| \\ &= \|\mathrm{gyr}[X, A](\ominus A\oplus B)\| \\ &= \|\ominus A\oplus B\| \\ &= d(A, B) \end{aligned} \tag{2.118}$$

Similarly, let $R \in \mathrm{SO(n)}$. Being a rotation of the ball $\mathbb{R}_s^n$, R keeps the inner product in $\mathbb{R}_s^n$ invariant

$$RA \cdot RB = A \cdot B \tag{2.119}$$

for all $R \in \mathrm{SO(n)}$ and $A, B \in \mathbb{R}_s^n$. Hence, as we see from the definition of Einstein addition, (2.4), Rotations of the ball respect Einstein addition in the ball, that is,

$$R(A\oplus B) = RA \oplus RB \tag{2.120}$$

Identity (2.120), in turn, implies

$$d(RA, RB) = \|\ominus RA\oplus RB\| = \|R(\ominus A\oplus B)\| = \|\ominus A\oplus B\| = d(A, B) \tag{2.121}$$

as desired. □

We may remark, for later reference, that following (2.118) and the definition of the gamma factor in (2.5), we have the gyrogroup identity

$$\gamma_{\ominus(\ominus X\oplus A)\oplus(\ominus X\oplus B)} = \gamma_{\ominus A\oplus B} \tag{2.122}$$

for all $A, B, X \in (\mathbb{R}_s^n, \oplus, \otimes)$.

Left gyrotranslations of $\mathbb{R}_s^n$ and rotations of $\mathbb{R}_s^n$ about its origin are *gyroisometries* of $\mathbb{R}_s^n$ in the sense that they keep the Einstein gyrodistance function (2.113) invariant. The set of all left gyrotranslations of $\mathbb{R}_s^n$ and all rotations of $\mathbb{R}_s^n$ about its origin forms a group under transformation composition, called the hyperbolic group of motions. In gyrogroup theory, this group of motions turns out to be the so called *gyrosemidirect product* of the gyrogroup of left gyrotranslations and the group of rotations [3].

Following Klein's 1872 *Erlangen Program* [12] [13, p. 253], the geometric objects of a geometry are the invariants of the group of motions of the geometry so that, conversely, objects that are invariant under the group of motions of a geometry possess geometric significance. Accordingly, for instance, the Einstein gyrodistance, (2.113), between two points of $\mathbb{R}_s^n$ is geometrically significant in hyperbolic geometry since, by Theorem 2.22, it is invariant under the group of motions, left gyrotranslations and rotations, of the hyperbolic geometry of $\mathbb{R}_s^n$.

2.9 Linking Einstein Addition to Hyperbolic Geometry

On the one hand, it is known that geodesics of the Beltrami-Klein ball model of hyperbolic geometry are Euclidean chords of the ball [13] [42] and, on the other hand, Fig. 2.1 indicates that gyrolines in Einstein ball gyrovector spaces are Euclidean chords of the ball as well. This coincidence is not accidental. We will find in this section that the Einstein gyrodistance (2.113) leads to the Riemannian line element, in differential geometry, of the Beltrami-Klein ball model of hyperbolic geometry.

In a two dimensional Einstein gyrovector space $(\mathbb{R}_s^2, \oplus, \otimes)$ the squared gyrodistance between a point $X = (x_1, x_2) \in \mathbb{R}_s^2$ and an infinitesimally nearby point $X + dX \in \mathbb{R}_s^2$, where $dX = (dx_1, dx_2)$, is given by the equation [3, Sec. 7.5] [2, Sec. 7.5]

$$ds^2 = \|X \ominus (X + dX)\|^2 = E dx_1^2 + 2F dx_1 dx_2 + G dx_2^2 + \dots \qquad (2.123)$$

where, if we use the notation $r^2 = x_1^2 + x_2^2$, we have

$$\begin{aligned} E &= s^2 \frac{s^2 - x_2^2}{(s^2 - r^2)^2} \\ F &= s^2 \frac{x_1 x_2}{(s^2 - r^2)^2} \\ G &= s^2 \frac{s^2 - x_1^2}{(s^2 - r^2)^2} \end{aligned} \qquad (2.124)$$

The triple $(g_{11}, g_{12}, g_{22}) = (E, F, G)$ along with $g_{21} = g_{12}$ is known in differential geometry as the metric tensor g_{ij} [43]. It turns out to be the metric tensor of the Beltrami-Klein disc model of hyperbolic geometry [42, p. 220]. Hence, ds^2 in (2.123) – (2.124) is the Riemannian line element of the Beltrami-Klein disc model of hyperbolic geometry, linked to Einstein velocity addition (2.4) and to Einstein gyrodistance function (2.113) [44].

The Gaussian curvature K of an Einstein gyrovector plane, corresponding to the triple (E, F, G), turns out to be [42, p. 149] [3, Sec. 7.5] [2, Sec. 7.5]

$$K = -\frac{1}{s^2} \tag{2.125}$$

The link between Einstein gyrovector spaces and the Beltrami-Klein ball model of hyperbolic geometry, already noted by Fock [21, p. 39], has thus been established in (2.123) – (2.124) in $n = 2$ dimensions. The extension of the link to higher dimensions, $n \geq 2$, is presented in [1, Sec. 9, Chap. 3], [3, Sec. 7.5] [2, Sec. 7.5] and [44]. For a brief account of the history of linking Einstein's velocity addition law to hyperbolic geometry see [45, p. 943].

In full analogy with Euclidean geometry, the graph of the parametric expression (2.112) in an Einstein gyrovector space $(\mathbb{R}^n_s, \oplus, \otimes)$, for the parameter $t \in \mathbb{R}$, where $A, B \in \mathbb{R}^n_s$, describes a geodesic line in the Beltrami-Klein ball model of hyperbolic geometry. It is a chord of the ball, as shown in Fig. 2.1 for the disc. The geodesic (2.112) is the unique geodesic passing through the points A and B. It passes through the point A at "time" $t = 0$ and, owing to the left cancellation law, (2.37), p. 73, it passes through the point B at "time" $t = 1$. Furthermore, it passes through the gyromidpoint M_{AB} of gyrosegment AB at "time" $t = 1/2$. Hence, the gyrosegment that joins the points A and B in Fig. 2.1 is obtained from (2.112) by restricting its parameter t to the interval $0 \leq t \leq 1$.

We have thus seen the power and elegance of the Einstein gyrodistance function (2.113). It shares remarkable analogies with its Euclidean counterpart, the Euclidean distance function (1.6), p. 2, studied in [3] and [33]. In particular, it is shown in [3] and [33] that Einstein gyrodistance function gives rise to a gyrotriangle inequality in full analogy with the common triangle inequality in vector spaces. Furthermore, we have seen in this section that Einstein gyrodistance leads naturally to the Riemannian line element of the Beltrami-Klein disc model of hyperbolic geometry, which is well-known in differential geometry [42]. It is therefore important to note that Einstein gyrodistance function is in one-to-one correspondence with the commonly used distance function in the Beltrami-Klein ball model of hyperbolic geometry, as explained in [3, Sec. 6.19].

The Einstein gyrovector space $\mathbb{R}_s^n$ regulates algebraically the Beltrami-Klein model of n-dimensional hyperbolic geometry just as the vector space $\mathbb{R}^n$ regulates algebraically the standard model of n-dimensional Euclidean geometry. Euclidean geometry regulated by the vector space $\mathbb{R}^n$ is equipped with Cartesian coordinates and, hence, it is known as the standard Cartesian model of Euclidean geometry. In full analogy, the Beltrami-Klein model of n-dimensional hyperbolic geometry is regulated by the Einstein gyrovector space $\mathbb{R}_s^n$ which is equipped with Cartesian coordinates. Hence, the Beltrami-Klein model of hyperbolic geometry that is regulated by an Einstein gyrovector space is called the *Cartesian-Beltrami-Klein model of hyperbolic geometry.*

2.10 Einstein Gyrovectors, Gyroangles and Gyrotriangles

Points of an Einstein gyrovector space $(\mathbb{R}_s^n, \oplus, \otimes)$, denoted by capital italic letters A, B, etc., give rise to gyrovectors in $\mathbb{R}_s^n$, denoted by bold roman lowercase letters $\mathbf{u}, \mathbf{v}$, etc. Any two ordered points $P, Q \in \mathbb{R}_s^n$ give rise to a unique rooted gyrovector $\mathbf{v} \in \mathbb{R}_s^n$, rooted at the point P. It has a tail at the point P and a head at the point Q, and it has the value $\ominus P \oplus Q$,

$$\mathbf{v} = \ominus P \oplus Q \tag{2.126}$$

The gyrolength of the rooted gyrovector $\mathbf{v} = \ominus P \oplus Q$ is the gyrodistance between its tail, P, and its head, Q, given by the equation

$$\|\mathbf{v}\| = \|\ominus P \oplus Q\| \tag{2.127}$$

Two rooted gyrovectors $\ominus P \oplus Q$ and $\ominus R \oplus S$, Fig. 2.2, are equivalent if they have the same value, $\ominus P \oplus Q = \ominus R \oplus S$, that is,

$$\ominus P \oplus Q \quad \sim \quad \ominus R \oplus S \qquad \text{if and only if} \qquad \ominus P \oplus Q = \ominus R \oplus S \tag{2.128}$$

The relation $\sim$ in (2.128) between rooted gyrovectors in Einstein gyrovector spaces is reflexive, symmetric and transitive. As such, it is an equivalence relation that gives rise to equivalence classes of rooted gyrovectors. To liberate rooted gyrovectors from their roots we define a *gyrovector* to be an equivalence class of rooted gyrovectors. The gyrovector $\ominus P \oplus Q$ is thus a representative of all rooted gyrovectors with value $\ominus P \oplus Q$. Thus, for instance, the two distinct rooted gyrovectors $\ominus A \oplus B$ and $\ominus A' \oplus B'$ in Fig. 2.2 possess the same value so that, as gyrovectors in an Einstein gyrovector space, they are indistinguishable.

As it is well-known, vectors add according to the parallelogram addition law [33] so that vectors in Euclidean geometry are, in fact, equivalence

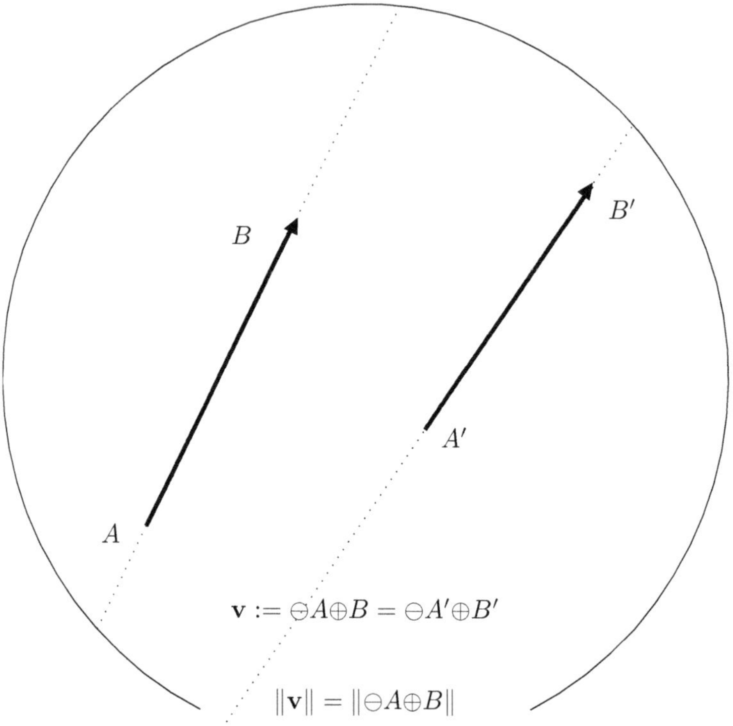

Fig. 2.2 The rooted gyrovectors $\ominus A \oplus B$ and $\ominus A' \oplus B'$ that are shown here in an Einstein gyrovector plane $(\mathbb{R}_s^2, \oplus, \otimes)$ have equal values, $\ominus A \oplus B = \ominus A' \oplus B'$. Hence, they are equivalent so that they represent the same gyrovector. Accordingly, as gyrovectors, $\ominus A \oplus B$ and $\ominus A' \oplus B'$ are indistinguishable in their gyrovector space and its underlying hyperbolic geometry. As an important disanalogy with Euclidean geometry, the gyroquadrilateral $AA'B'B$ is not a hyperbolic parallelogram. A hyperbolic parallelogram, called a gyroparallelogram, is a gyroquadrangle the two gyrodiagonals of which intersect at their gyromidpoints [3].

classes of ordered pairs of points that add according to the parallelogram law. In full analogy, gyrovectors in Einstein gyrovector spaces are equivalence classes of ordered pairs of points that add according to the gyroparallelogram law. This remarkable result about gyrovector addition, presented in [2, 3, 33], will not be used in this book.

A point $P \in \mathbb{R}_s^n$ is identified with the gyrovector $\ominus O \oplus P$, O being the arbitrarily selected origin of the space $\mathbb{R}_s^n$. Hence, the algebra of gyrovectors can be applied to the points of $\mathbb{R}_s^n$ as well.

Let $\ominus A_1 \oplus A_2$ and $\ominus A_1 \oplus A_3$ be two rooted gyrovectors with a common tail A_1, Fig. 2.3. They include a gyroangle $\alpha_1 = \angle A_2A_1A_3 = \angle A_3A_1A_2$, the measure of which is given by the equation

$$\cos\alpha_1 = \frac{\ominus A_1\oplus A_2}{\|\ominus A_1\oplus A_2\|}\cdot\frac{\ominus A_1\oplus A_3}{\|\ominus A_1\oplus A_3\|} \tag{2.129}$$

Accordingly, α_1 has the radian measure

$$\alpha_1 = \cos^{-1}\frac{\ominus A_1\oplus A_2}{\|\ominus A_1\oplus A_2\|}\cdot\frac{\ominus A_1\oplus A_3}{\|\ominus A_1\oplus A_3\|} \tag{2.130}$$

where cos and $\cos^{-1}$ = arccos are the standard cosine and arccosine functions of trigonometry. In the context of gyroangles, as in (2.129) – (2.130), we refer these functions of trigonometry to as the functions *gyrocosine* and *gyroarccosine* of gyrotrigonometry, for reasons explained in the chapter "gyrotrigonometry" of [33, Chap. 4].

The gyroangle α_1 is invariant under left gyrotranslations. Indeed, by (2.63), p. 78, and (2.83), p. 84, we have

$$\begin{aligned}\cos\alpha_1' &= \frac{\ominus(X\oplus A_1)\oplus(X\oplus A_2)}{\|\ominus(X\oplus A_1)\oplus(X\oplus A_2)\|}\cdot\frac{\ominus(X\oplus A_1)\oplus(X\oplus A_3)}{\|\ominus(X\oplus A_1)\oplus(X\oplus A_3)\|}\\ &= \frac{\mathrm{gyr}[X,A_1](\ominus A_1\oplus A_2)}{\|\mathrm{gyr}[X,A_1](\ominus A_1\oplus A_2)\|}\cdot\frac{\mathrm{gyr}[X,A_1](\ominus A_1\oplus A_3)}{\|\mathrm{gyr}[X,A_1](\ominus A_1\oplus A_3)\|}\\ &= \frac{\ominus A_1\oplus A_2}{\|\ominus A_1\oplus A_2\|}\cdot\frac{\ominus A_1\oplus A_3}{\|\ominus A_1\oplus A_3\|} = \cos\alpha_1\end{aligned} \tag{2.131}$$

for all $A_1, A_2, A_3, X\in\mathbb{R}_s^n$.

Remarkably, both trigonometry and gyrotrigonometry share the same elementary trigonometric/gyrotrigonometric functions, $\sin\alpha$, $\cos\alpha$, $\tan\alpha$, etc. This result, established in [33, Chap. 4] and in [46, 47], will be further enhanced in this book in the observation that a triangle center in Euclidean geometry and its counterpart in the Beltrami-Klein ball model of hyperbolic geometry share the same *trigonometric barycentric coordinates.*

Furthermore, the gyroangle α_1 is invariant under rotations of $\mathbb{R}_s^n$ about its origin. Indeed, by (2.120) and (2.119) we have

$$\begin{aligned}\cos\alpha_1'' &= \frac{\ominus RA_1\oplus RA_2}{\|\ominus RA_1\oplus RA_2\|}\cdot\frac{\ominus RA_1\oplus RA_3}{\|\ominus RA_1\oplus RA_3\|}\\ &= \frac{R(\ominus A_1\oplus A_2)}{\|R(\ominus A_1\oplus A_2)\|}\cdot\frac{R(\ominus A_1\oplus A_3)}{\|R(\ominus A_1\oplus A_3)\|}\\ &= \frac{\ominus A_1\oplus A_2}{\|\ominus A_1\oplus A_2\|}\cdot\frac{\ominus A_1\oplus A_3}{\|\ominus A_1\oplus A_3\|}\\ &= \cos\alpha_1\end{aligned} \tag{2.132}$$

for all $A_1, A_2, A_3 \in \mathbb{R}^n$ and $R \in \mathrm{SO(n)}$, since rotations $R \in \mathrm{SO(n)}$ preserve the inner product and the norm in $\mathbb{R}^n_s$.

Being invariant under the motions of $\mathbb{R}^n_s$, which are left gyrotranslations and rotations, gyroangles are geometric objects of the hyperbolic geometry of Einstein gyrovector spaces. Gyrotriangle gyroangle sum in hyperbolic geometry is less than π. The standard notation that we use with a gyrotriangle $A_1A_2A_3$ in $\mathbb{R}^n_s$, $n \geq 2$, is presented in Fig. 2.3 for $n = 2$.

In our notation, an Einstein gyrotriangle $A_1A_2A_3$, thus, has (i) three vertices, A_1, A_2 and A_3; (ii) three gyroangles, α_1, α_2 and α_3; and (iii) three gyrosides, which form the three gyrovectors $\mathbf{a}_{12}$, $\mathbf{a}_{23}$ and $\mathbf{a}_{31}$; with respective (iv) three gyroside-gyrolengths a_{12}, a_{23} and a_{31}, as shown in Fig. 2.3.

The standard notation that we use with gyrotriangles, presented in Fig. 2.3, thus include the following equations

$$
\begin{aligned}
\mathbf{a}_{12} &= \ominus A_1 \oplus A_2, & a_{12} &= \|\mathbf{a}_{12}\|, & \gamma_{21} &= \gamma_{12} = \gamma_{a_{12}} \\
\mathbf{a}_{13} &= \ominus A_1 \oplus A_3, & a_{13} &= \|\mathbf{a}_{13}\|, & \gamma_{31} &= \gamma_{13} = \gamma_{a_{13}} \\
\mathbf{a}_{23} &= \ominus A_2 \oplus A_3, & a_{23} &= \|\mathbf{a}_{23}\|, & \gamma_{32} &= \gamma_{23} = \gamma_{a_{23}}
\end{aligned}
\tag{2.133}
$$

2.11 The Law of Gyrocosines

Let $\ominus A \oplus B$ and $\ominus A \oplus C$ be two gyrovectors that form two gyrosides of gyrotriangle ABC and include the gyrotriangle gyroangle α in any Einstein gyrovector space $(\mathbb{R}^n_s, \oplus, \otimes)$, as shown in Fig. 2.4 for $n = 2$.

By the Gyrotranslation Theorem 2.10, p. 78,

$$\ominus(\ominus A \oplus B) \oplus (\ominus A \oplus C) = \mathrm{gyr}[\ominus A, B](\ominus B \oplus C) \tag{2.134}$$

Since gyrations preserve the norm, (2.27), p. 69,

$$\|\ominus(\ominus A \oplus B) \oplus (\ominus A \oplus C)\| = \|\mathrm{gyr}[\ominus A, B](\ominus B \oplus C)\| = \|\ominus B \oplus C\| \tag{2.135}$$

In the notation of Fig. 2.4 for gyrotriangle ABC, (2.135) is written as

$$\|\ominus \mathbf{c} \oplus \mathbf{b}\| = \|\mathbf{a}\| \tag{2.136}$$

implying

$$\gamma_{\ominus \mathbf{c} \oplus \mathbf{b}} = \gamma_{\mathbf{a}} = \gamma_a \tag{2.137}$$

By (2.10), p. 66, we have

$$\gamma_{\ominus \mathbf{c} \oplus \mathbf{b}} = \gamma_b \gamma_c \left(1 - \frac{\mathbf{b} \cdot \mathbf{c}}{s^2}\right) \tag{2.138}$$

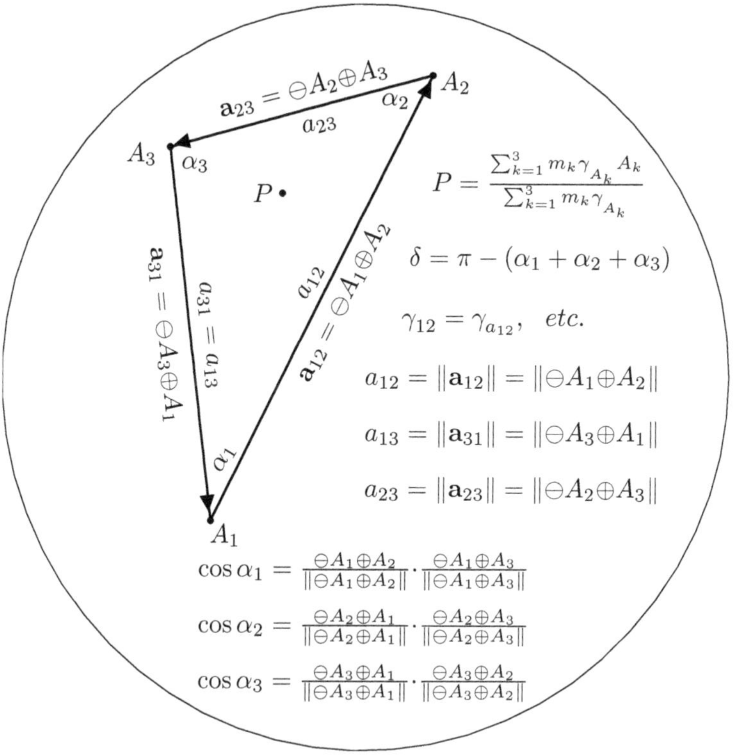

Fig. 2.3 The gyrotriangle $A_1A_2A_3$ in an Einstein gyrovector space $(\mathbb{R}_s^n, \oplus, \otimes)$ is shown for $n = 2$, along with the associated standard triangle index notation. The gyrotriangle vertices, A_1, A_2 and A_3, are any non-gyrocollinear points of $\mathbb{R}_s^n$. Its sides are presented graphically as gyrosegments that join the vertices. They form gyrovectors, $\mathbf{a}_{ij}$, side-gyrolengths, $a_{ij} = \|\mathbf{a}_{ij}\|$, $1 \le i, j \le 3$, $i \ne j$, and gyroangles, α_k, $k = 1, 2, 3$. The gyrotriangle gyroangle sum is less than π, the difference, $\delta = \pi - (\alpha_1 + \alpha_2 + \alpha_3)$, being the gyrotriangular defect. The gyrocosine function of the gyrotriangle gyroangles is presented. Remarkably, it assumes a form that is fully analogous to the cosine function of Euclidean trigonometry. The point P is a generic point on the interior of the gyrotriangle, with positive gyrobarycentric coordinates $(m_1 : m_1 : m_3)$ with respect to the gyrotriangle vertices, (4.2), p. 171. Both α_i and γ_{ij} are geometrically significant since they are invariant under the motions of $\mathbb{R}_s^n$.

The gyrocosine of gyroangle $\alpha = \angle BAC$ of gyrotriangle ABC in Fig. 2.4 is given by

$$\cos\alpha = \frac{\ominus A \oplus B}{\|\ominus A \oplus B\|} \cdot \frac{\ominus A \oplus C}{\|\ominus A \oplus C\|} = \frac{\mathbf{c}}{c} \cdot \frac{\mathbf{b}}{b} \tag{2.139}$$

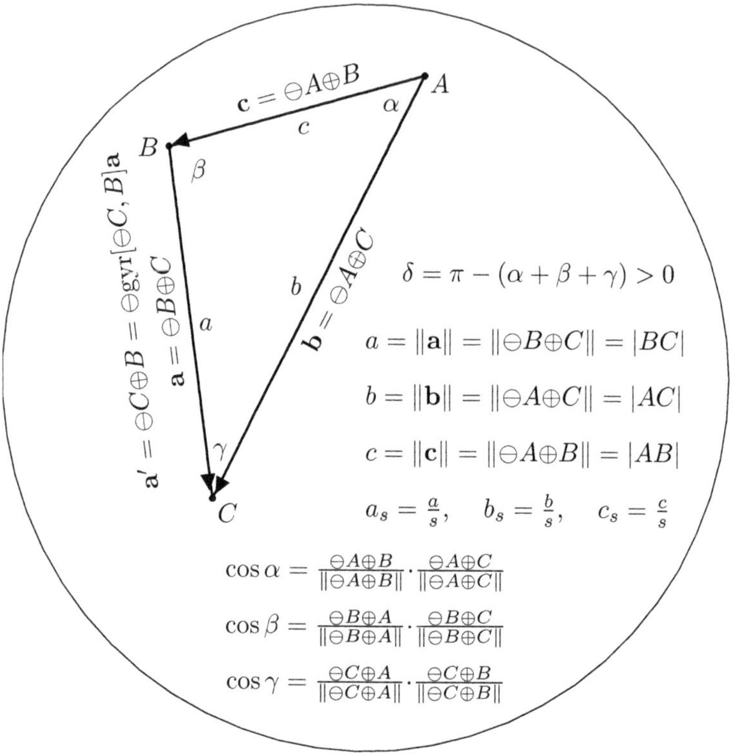

Fig. 2.4 The gyrotriangle, and its standard notation, in an Einstein gyrovector space. The notation that we use with a gyrotriangle ABC, its gyrovector sides, and its gyroangles in an Einstein gyrovector space $(\mathbb{R}_s^n, \oplus, \otimes)$ is shown here for the Einstein gyrovector plane $(\mathbb{R}_s^2, \oplus, \otimes)$. A gyrotriangle is *isosceles* if it has two sides congruent, and a gyrotriangle is *equilateral* if it has three sides congruent.

so that

$$\frac{\mathbf{b}\cdot\mathbf{c}}{s^2} = \frac{bc}{s^2}\cos\alpha = b_s c_s \cos\alpha \tag{2.140}$$

where $b_s = b/s$, etc.

Following (2.137) – (2.140) we have

$$\gamma_a = \gamma_b \gamma_c (1 - b_s c_s \cos\alpha) \tag{2.141}$$

where a, b, c are the gyroside-gyrolengths of gyrotriangle ABC in an Einstein gyrovector space $(\mathbb{R}_s^n, \oplus, \otimes)$, as shown in Fig. 2.4 for $n = 2$.

Identity (2.141) is the *law of gyrocosines* in the gyrotrigonometry of Einstein gyrovector spaces. As in trigonometry, it is useful for calculating one gyroside, a, of a gyrotriangle ABC, Fig. 2.4, when the gyroangle α opposite to gyroside a and the other two gyrosides (that is, their gyrolengths), b and c, are known.

Remarkably, in the Euclidean limit of large s, $s \to \infty$, gamma factors tend to 1 and the law of gyrocosines (2.141) descends to the trivial identity $1 = 1$. Hence, (2.141) has no immediate Euclidean counterpart, thus presenting a disanalogy between hyperbolic and Euclidean geometry. As a result, each of Theorems 2.23 and 2.26 below has no Euclidean counterpart as well.

2.12 The *SSS* to *AAA* Conversion Law

Let ABC be a gyrotriangle in an Einstein gyrovector space $(\mathbb{R}_s^n, \otimes, \oplus)$ with its standard notation in Fig. 2.4. According to (2.141) the gyrotriangle ABC possesses the following three identities, each of which represents its law of gyrocosines,

$$\begin{aligned}\gamma_a &= \gamma_b\gamma_c(1 - b_s c_s \cos\alpha)\\ \gamma_b &= \gamma_a\gamma_c(1 - a_s c_s \cos\beta)\\ \gamma_c &= \gamma_a\gamma_b(1 - a_s b_s \cos\gamma)\end{aligned} \tag{2.142}$$

Like Euclidean triangles, the gyroangles of a gyrotriangle are uniquely determined by its gyrosides. Solving the system (2.142) of three identities for the three unknowns $\cos\alpha$, $\cos\beta$ and $\cos\gamma$, and employing (2.11), p. 66, we obtain the following theorem.

Theorem 2.23. (The Law of Gyrocosines; The *SSS* to *AAA* Conversion Law). *Let ABC be a gyrotriangle in an Einstein gyrovector space $(\mathbb{R}_s^n, \oplus, \otimes)$. Then, in the gyrotriangle notation in Fig. 2.4,*

$$\begin{aligned}\cos\alpha &= \frac{-\gamma_a + \gamma_b\gamma_c}{\gamma_b\gamma_c b_s c_s} = \frac{-\gamma_a + \gamma_b\gamma_c}{\sqrt{\gamma_b^2 - 1}\sqrt{\gamma_c^2 - 1}}\\ \cos\beta &= \frac{-\gamma_b + \gamma_a\gamma_c}{\gamma_a\gamma_c a_s c_s} = \frac{-\gamma_b + \gamma_a\gamma_c}{\sqrt{\gamma_a^2 - 1}\sqrt{\gamma_c^2 - 1}}\\ \cos\gamma &= \frac{-\gamma_c + \gamma_a\gamma_b}{\gamma_a\gamma_b a_s b_s} = \frac{-\gamma_c + \gamma_a\gamma_b}{\sqrt{\gamma_a^2 - 1}\sqrt{\gamma_b^2 - 1}}\end{aligned} \tag{2.143}$$

The identities in (2.143) form the *SSS* (Side-Side-Side) to *AAA* (gyroAngle-gyroAngle-gyroAngle) conversion law in Einstein gyrovector spaces. This law is useful for calculating the gyroangles of a gyrotriangle in an Einstein gyrovector space when its gyrosides (that is, its gyroside-gyrolengths) are known.

In full analogy with the trigonometry of triangles, the *gyrosine* of a gyrotriangle gyroangle α is nonnegative, given by the equation

$$\sin\alpha = \sqrt{1-\cos^2\alpha} \geq 0 \tag{2.144}$$

Hence, it follows from Theorem 2.23 that the gyrosines of the gyrotriangle gyroangles in that Theorem are given by

$$\begin{aligned}
\sin\alpha &= \frac{\sqrt{1+2\gamma_a\gamma_b\gamma_c-\gamma_a^2-\gamma_b^2-\gamma_c^2}}{\sqrt{\gamma_b^2-1}\sqrt{\gamma_c^2-1}} \\
\sin\beta &= \frac{\sqrt{1+2\gamma_a\gamma_b\gamma_c-\gamma_a^2-\gamma_b^2-\gamma_c^2}}{\sqrt{\gamma_a^2-1}\sqrt{\gamma_c^2-1}} \\
\sin\gamma &= \frac{\sqrt{1+2\gamma_a\gamma_b\gamma_c-\gamma_a^2-\gamma_b^2-\gamma_c^2}}{\sqrt{\gamma_a^2-1}\sqrt{\gamma_b^2-1}}
\end{aligned} \tag{2.145}$$

Any gyrotriangle gyroangle α satisfies the inequality $0 < \alpha < \pi$, so that $\sin\alpha > 0$. Following (2.145) we have the inequality

$$1+2\gamma_a\gamma_b\gamma_c-\gamma_a^2-\gamma_b^2-\gamma_c^2 > 0 \tag{2.146}$$

for any gyrotriangle in an Einstein gyrovector space, in the notation of Theorem 2.23 and Fig. 2.4.

Identities (2.145) immediately give rise to the identities

$$\frac{\sin\alpha}{\sqrt{\gamma_a^2-1}} = \frac{\sin\beta}{\sqrt{\gamma_b^2-1}} = \frac{\sin\gamma}{\sqrt{\gamma_c^2-1}} \tag{2.147}$$

which form the *law of gyrosines* that we will study in Theorem 2.30, p. 110.

Noting that $\sqrt{\gamma_a^2-1} = \gamma_a a/s$, the law of gyrosines (2.147) can be written equivalently as

$$\frac{\sin\alpha}{\gamma_a a} = \frac{\sin\beta}{\gamma_b b} = \frac{\sin\gamma}{\gamma_c c} \tag{2.148}$$

A related *extended law of gyrosines* is presented in [18, Eq. (7.32)].

2.13 Inequalities for Gyrotriangles

Elegant inequalities for gyrotriangles in Einstein gyrovector spaces result immediately from (2.145), as we see in the following theorem.

Theorem 2.24. *Let ABC be a gyrotriangle in an Einstein gyrovector space, with the notation in Fig. 2.4, p. 102, for its gyroside-gyrolengths a, b, c. Then*

$$\gamma_a^2 + \gamma_b^2 + \gamma_c^2 - 1 < 2\gamma_a\gamma_b\gamma_c \le \gamma_a^2 + \gamma_b^2\gamma_c^2 \tag{2.149}$$

Proof. The left inequality in each of the three chains of inequalities in (2.149) follows immediately from (2.146).

The inequality $\sin^2\alpha \le 1$ implies, by means of (2.145), the inequality

$$\frac{1 + 2\gamma_a\gamma_b\gamma_c - \gamma_a^2 - \gamma_b^2 - \gamma_c^2}{(\gamma_b^2 - 1)(\gamma_c^2 - 1)} \le 1 \tag{2.150}$$

which, in turn, implies

$$2\gamma_a\gamma_b\gamma_c \le \gamma_a^2 + \gamma_b^2\gamma_c^2 \tag{2.151}$$

□

An interesting chain of inequalities that results from (2.149), for instance, is

$$0 < \gamma_c^2 - 1 < 2\gamma_a\gamma_b\gamma_c - \gamma_a^2 - \gamma_b^2 < \gamma_a^2 + \gamma_b^2\gamma_c^2 - \gamma_a^2 - \gamma_b^2 = \gamma_b^2(\gamma_c^2 - 1) \tag{2.152}$$

and, hence,

$$1 < \frac{2\gamma_a\gamma_b\gamma_c - \gamma_a^2 - \gamma_b^2}{\gamma_c^2 - 1} < \gamma_b^2 \tag{2.153}$$

Equality is attained in Inequality (2.151) when, and only when, $\alpha = \pi/2$. In this case gyrotriangle ABC is a right gyroangled gyrotriangle, satisfying the Einstein-Pythagoras identity $\gamma_a = \gamma_b\gamma_c$, (2.179), p. 113, that will be studied in Sec. 2.18.

Gyrotriangle gyroangles vary over the open interval $(0, \pi)$. Accordingly, we present the following definition about gyrotriangles and their gyroangles.

Definition 2.25. (Acute, Right, Obtuse Gyroangles and Gyrotriangles). *An acute gyroangle θ is a gyroangle measuring between 0 and $\pi/2$ radians, $0 < \theta < \pi/2$.*
A right gyroangle θ is a gyroangle measuring $\pi/2$ radians, $\theta = \pi/2$.
An obtuse gyroangle θ is a gyroangle measuring between $\pi/2$ and π radians, $\pi/2 < \theta < \pi$.
A gyrotriangle in which all three gyroangles are acute, $0 < \alpha, \beta, \gamma < \pi/2$, is acute.
A gyrotriangle in which one gyroangle is a right gyroangle, $\theta = \pi/2$, is right.
A gyrotriangle which has an obtuse gyroangle, $\pi/2 < \theta < \pi$, is obtuse.

It follows from (2.143), and from the result that the trigonometric cosine function, $\cos\alpha$, and the gyrotrigonometric gyrocosine function, ambiguously also denoted $\cos\alpha$, have the same behavior, that we have important equalities and inequalities for gyrotriangles in Einstein gyrovector spaces. These are:

$$\begin{cases} \gamma_{\mathbf{b}}\,\gamma_{\mathbf{c}} > \gamma_{\mathbf{a}} & \text{if and only if } \alpha < \frac{\pi}{2} \text{ (Acute);} \\ \gamma_{\mathbf{b}}\,\gamma_{\mathbf{c}} = \gamma_{\mathbf{a}} & \text{if and only if } \alpha = \frac{\pi}{2} \text{ (Right);} \\ \gamma_{\mathbf{b}}\,\gamma_{\mathbf{c}} < \gamma_{\mathbf{a}} & \text{if and only if } \alpha > \frac{\pi}{2} \text{ (Obtuse)} \end{cases} \tag{2.154}$$

and, similarly,

$$\begin{cases} \gamma_{\mathbf{a}}\,\gamma_{\mathbf{c}} > \gamma_{\mathbf{b}} & \text{if and only if } \beta < \frac{\pi}{2} \text{ (Acute);} \\ \gamma_{\mathbf{a}}\,\gamma_{\mathbf{c}} = \gamma_{\mathbf{b}} & \text{if and only if } \beta = \frac{\pi}{2} \text{ (Right);} \\ \gamma_{\mathbf{a}}\,\gamma_{\mathbf{c}} < \gamma_{\mathbf{b}} & \text{if and only if } \beta > \frac{\pi}{2} \text{ (Obtuse)} \end{cases} \tag{2.155}$$

and

$$\begin{cases} \gamma_{\mathbf{a}}\,\gamma_{\mathbf{b}} > \gamma_{\mathbf{c}} & \text{if and only if } \gamma < \frac{\pi}{2} \text{ (Acute);} \\ \gamma_{\mathbf{a}}\,\gamma_{\mathbf{b}} = \gamma_{\mathbf{c}} & \text{if and only if } \gamma = \frac{\pi}{2} \text{ (Right);} \\ \gamma_{\mathbf{a}}\,\gamma_{\mathbf{b}} < \gamma_{\mathbf{c}} & \text{if and only if } \gamma > \frac{\pi}{2} \text{ (Obtuse)} \end{cases} \tag{2.156}$$

where $\mathbf{a}, \mathbf{b}, \mathbf{c}$ are the gyrosides of any given gyrotriangle in an Einstein gyrovector space and α, β, γ are their respective opposing gyroangles, as shown in Fig. 2.4, p. 102. The equalities in (2.154) – (2.156) correspond to right gyroangles, giving rise to the Einstein-Pythagoras identity (2.179), p. 113, that will be studied in Sec. 2.18.

2.14 The *AAA* to *SSS* Conversion Law

Unlike Euclidean triangles, the gyroside-gyrolengths of a gyrotriangle are uniquely determined by its gyroangles, as the following theorem demonstrates.

Theorem 2.26. (The *AAA* to *SSS* Conversion Law I). *Let ABC be a gyrotriangle in an Einstein gyrovector space $(\mathbb{R}^n_s, \oplus, \otimes)$. Then, in the gyrotriangle notation in Fig. 2.4,*

$$\begin{aligned} \gamma_a &= \frac{\cos\alpha + \cos\beta\cos\gamma}{\sin\beta\sin\gamma} \\ \gamma_b &= \frac{\cos\beta + \cos\alpha\cos\gamma}{\sin\alpha\sin\gamma} \\ \gamma_c &= \frac{\cos\gamma + \cos\alpha\cos\beta}{\sin\alpha\sin\beta} \end{aligned} \tag{2.157}$$

where, following (2.144), the gyrosine of the gyrotriangle gyroangle α, $\sin\alpha$, *is the nonnegative value of* $\sqrt{1-\cos^2\alpha}$, *etc.*

Proof. Let ABC be a gyrotriangle in an Einstein gyrovector space $(\mathbb{R}_s^n, \otimes, \oplus)$ with its standard notation in Fig. 2.4. It follows straightforwardly from the SSS to AAA conversion law (2.143) that

$$\left(\frac{\cos\alpha + \cos\beta\cos\gamma}{\sin\beta\sin\gamma}\right)^2 = \frac{(\cos\alpha + \cos\beta\cos\gamma)^2}{(1-\cos^2\beta)(1-\cos^2\gamma)} = \gamma_a^2 \tag{2.158}$$

implying the first identity in (2.157). The remaining two identities in (2.157) are obtained from (2.143) in a similar way. □

The identities in (2.157) form the AAA to SSS conversion law. This law is useful for calculating the gyrosides (that is, the gyroside-gyrolengths) of a gyrotriangle in an Einstein gyrovector space when its gyroangles are known. Thus, for instance, γ_a is obtained from the first identity in (2.157), and a is obtained from γ_a by Identity (2.11), p. 66.

Solving the third identity in (2.157) for $\cos\gamma$ we have

$$\begin{aligned} \cos\gamma &= -\cos\alpha\cos\beta + \gamma_c \sin\alpha\sin\beta \\ &= -\cos(\alpha+\beta) + (\gamma_c - 1)\sin\alpha\sin\beta \end{aligned} \tag{2.159}$$

implying

$$\cos\gamma = \cos(\pi-\alpha-\beta) + (\gamma_c - 1)\sin\alpha\sin\beta \tag{2.160}$$

In the Euclidean limit of large s, $s\to\infty$, γ_c descends to 1, so that the gyrotrigonometric identity (2.160) descends to the trigonometric identity

$$\cos\gamma = \cos(\pi-\alpha-\beta) \qquad \text{(Euclidean Geometry)} \tag{2.161}$$

in Euclidean geometry. The latter, in turn, is equivalent to the familiar result,

$$\alpha+\beta+\gamma = \pi \qquad \text{(Euclidean Geometry)} \tag{2.162}$$

of Euclidean geometry, according to which the triangle angle sum is π.

As an immediate application of the SSS to AAA conversion law (2.143)–(2.145) and the AAA to SSS conversion law (2.157) we present the following two theorems.

Theorem 2.27. (The Isosceles Gyrotriangle Theorem). *A gyrotriangle is isosceles (that is, it has two gyrosides congruent) if and only if the two gyroangles opposing its two congruent gyrosides are congruent.*

Proof. Using the gyrotriangle notation in Fig. 2.4, p. 102, if gyrotriangle ABC has two gyrosides, say a and b, congruent, $a = b$, then $\gamma_a = \gamma_b$ so that, by (2.143) – (2.145), $\cos\alpha = \cos\beta$ and $\sin\alpha = \sin\beta$ implying $\alpha = \beta$.

Conversely, if gyrotriangle ABC has two gyroangles, say α and β, congruent, then by (2.157), $\gamma_a = \gamma_b$, implying $a = b$. □

Theorem 2.28. (The Equilateral Gyrotriangle Theorem). *A gyrotriangle is equilateral (that is, it has all three gyrosides congruent) if and only if the gyrotriangle is equigyroangular (that is, it has all three gyroangles congruent).*

Proof. Using the gyrotriangle notation in Fig. 2.4, p. 102, if gyrotriangle ABC has all three gyrosides congruent, $a = b = c$, then $\gamma_a = \gamma_b = \gamma_c$ so that, by (2.143) – (2.145), $\cos\alpha = \cos\beta = \cos\gamma$ and $\sin\alpha = \sin\beta = \sin\gamma$, implying $\alpha = \beta = \gamma$.

Conversely, if gyrotriangle ABC has all three gyroangles congruent, then $\alpha = \beta = \gamma$ implying, by (2.157), $\gamma_a = \gamma_b = \gamma_c$. The latter implies $a = b = c$ by means of (2.11), p. 66. □

Trigonometry deals with relationships between side-lengths and angles of triangles in Euclidean vector spaces $\mathbb{R}^n$, and with the trigonometric functions, $\sin\alpha$, $\cos\alpha$, etc., that describe these relationships.

In full analogy, Gyrotrigonometry deals with relationships between gyroside-gyrolengths and gyroangles of gyrotriangles in Einstein and Möbius gyrovector spaces, and with the gyrotrigonometric functions, also denoted $\sin\alpha$, $\cos\alpha$, etc., that describe these relationships.

A most important point of gyrotrigonometry is that its gyrotrigonometric functions $\sin\alpha$, $\cos\alpha$, etc., possess the same properties that their Euclidean counterparts possess. Hence, it is justified to use in gyrotrigonometry the same notation commonly used in trigonometry. As a result, one can use computer algebra for symbolic manipulation, like Mathematica and Maple, to manipulate gyrotrigonometric expressions in gyrotrigonometry. Surely, a computer software like Mathematica or Maple is designed to deal with trigonometry, but it can be used to deal with gyrotrigonometry as well owing to the above mentioned important point.

Interesting gyrotriangle identities that follow from (2.157) by gyrotrigonometric identities, and which can easily be obtained by the use of Mathematica or Maple in trigonometry, are in the notation of

Theorem 2.26,

$$
\begin{aligned}
\gamma_a^2 - 1 &= 4\frac{F(\alpha,\beta,\gamma)}{\sin^2\beta\,\sin^2\gamma} \\
\gamma_b^2 - 1 &= 4\frac{F(\alpha,\beta,\gamma)}{\sin^2\alpha\,\sin^2\gamma} \\
\gamma_c^2 - 1 &= 4\frac{F(\alpha,\beta,\gamma)}{\sin^2\alpha\,\sin^2\beta} \\
\gamma_a\gamma_b\gamma_c - 1 &= 4\frac{F(\alpha,\beta,\gamma)(1+\cos\alpha\,\cos\beta\,\cos\gamma)}{\sin^2\alpha\,\sin^2\beta\,\sin^2\gamma}
\end{aligned}
\tag{2.163}
$$

where $F(\alpha,\beta,\gamma)$ is a symmetric functions of α,β and γ, given by

$$
\begin{aligned}
F(\alpha,\beta,\gamma) &:= \cos\frac{\alpha+\beta+\gamma}{2}\cos\frac{\alpha-\beta-\gamma}{2}\cos\frac{-\alpha+\beta-\gamma}{2}\cos\frac{-\alpha-\beta+\gamma}{2} \\
&= \frac{1}{4}(2\cos\alpha\cos\beta\cos\gamma+\cos^2\alpha+\cos^2\beta+\cos^2\gamma-1) \\
&= \frac{1}{4}\frac{(1+2\gamma_a\gamma_b\gamma_c-\gamma_a^2-\gamma_b^2-\gamma_c^2)^2}{(\gamma_a^2-1)(\gamma_b^2-1)(\gamma_c^2-1)}
\end{aligned}
\tag{2.164}
$$

By (2.11), p. 66, (2.163) and (2.157),

$$
a_s = \frac{\sqrt{\gamma_a^2-1}}{\gamma_a} = 2\frac{\sqrt{F(\alpha,\beta,\gamma)}}{\cos\alpha+\cos\beta\cos\gamma}
\tag{2.165}
$$

thus obtaining the AAA to SSS conversion law in the following theorem.

Theorem 2.29. (The AAA to SSS Conversion Law II). *Let ABC be a gyrotriangle in an Einstein gyrovector space $(\mathbb{R}_s^n,\oplus,\otimes)$. Then, in the gyrotriangle notation in Fig. 2.4,*

$$
\begin{aligned}
a_s &= 2\frac{\sqrt{F(\alpha,\beta,\gamma)}}{\cos\alpha+\cos\beta\cos\gamma} \\
b_s &= 2\frac{\sqrt{F(\alpha,\beta,\gamma)}}{\cos\beta+\cos\alpha\cos\gamma} \\
c_s &= 2\frac{\sqrt{F(\alpha,\beta,\gamma)}}{\cos\gamma+\cos\alpha\cos\beta}
\end{aligned}
\tag{2.166}
$$

2.15 The Law of Gyrosines

Theorem 2.30. (The Law of Gyrosines). *Let ABC be a gyrotriangle in an Einstein gyrovector space $(\mathbb{R}^n_s, \oplus, \otimes)$. Then, in the gyrotriangle notation in Fig. 2.4, p. 102,*

$$\frac{\sin\alpha}{\gamma_a a} = \frac{\sin\beta}{\gamma_b b} = \frac{\sin\gamma}{\gamma_c c} = \frac{1}{s}\sqrt{\frac{1+2\gamma_a\gamma_b\gamma_c - \gamma_a^2 - \gamma_b^2 - \gamma_c^2}{(\gamma_a^2-1)(\gamma_b^2-1)(\gamma_c^2-1)}} \tag{2.167}$$

Proof. It follows from (2.143) – (2.144) and Identity (2.11), p. 66, that

$$\begin{aligned}
\left(\frac{\sin\alpha}{\gamma_a\, a}\right)^2 &= \frac{1}{s^2}\frac{1-\cos^2\alpha}{\gamma_a^2-1} = \frac{1}{s^2}\frac{1+2\gamma_a\gamma_b\gamma_c - \gamma_a^2 - \gamma_b^2 - \gamma_c^2}{(\gamma_a^2-1)(\gamma_b^2-1)(\gamma_c^2-1)} \\
\left(\frac{\sin\beta}{\gamma_b\, b}\right)^2 &= \frac{1}{s^2}\frac{1-\cos^2\beta}{\gamma_b^2-1} = \frac{1}{s^2}\frac{1+2\gamma_a\gamma_b\gamma_c - \gamma_a^2 - \gamma_b^2 - \gamma_c^2}{(\gamma_a^2-1)(\gamma_b^2-1)(\gamma_c^2-1)} \\
\left(\frac{\sin\gamma}{\gamma_c\, c}\right)^2 &= \frac{1}{s^2}\frac{1-\cos^2\gamma}{\gamma_c^2-1} = \frac{1}{s^2}\frac{1+2\gamma_a\gamma_b\gamma_c - \gamma_a^2 - \gamma_b^2 - \gamma_c^2}{(\gamma_a^2-1)(\gamma_b^2-1)(\gamma_c^2-1)}
\end{aligned} \tag{2.168}$$

The result (2.167) of the theorem follows immediately from (2.168). □

One should note that the extreme right-hand side of each equation in (2.168) is symmetric in a, b, c.

The law of gyrosines (2.167), excluding the extreme right-hand side in (2.167), is fully analogous to the law of sines, to which it descends in the Euclidean limit, $s \to \infty$, of large s, when gamma factors tend to 1, gyroside-gyrolengths tend to side-lengths, and gyroangles tend to angles.

2.16 The *ASA* to *SAS* Conversion Law

Theorem 2.31. (The *ASA* to *SAS* Conversion Law). *Let ABC be a gyrotriangle in an Einstein gyrovector space $(\mathbb{R}^n_s, \oplus, \otimes)$. Then, in the gyrotriangle notation in Fig. 2.4, p. 102,*

$$\begin{aligned}
a_s &= \frac{\gamma_c c_s \sin\alpha}{\cos\alpha\sin\beta + \gamma_c\sin\alpha\cos\beta} \\
b_s &= \frac{\gamma_c c_s \sin\beta}{\cos\beta\sin\alpha + \gamma_c\sin\beta\cos\alpha}
\end{aligned} \tag{2.169a}$$

and

$$\cos\gamma = \cos(\pi - \alpha - \beta) + (\gamma_c - 1)\sin\alpha\sin\beta \tag{2.169b}$$

Proof. Noting (2.11), p. 66, it follows from the third identity in (2.157) and from (2.143) that

$$\begin{aligned}\left(\frac{\gamma_c c_s \sin\alpha}{\cos\alpha\sin\beta+\gamma_c\sin\alpha\cos\beta}\right)^2 &= \left(\frac{\gamma_c c_s \sin\alpha\sin\beta}{\cos\alpha\sin^2\beta+\gamma_c\sin\alpha\sin\beta\cos\beta}\right)^2 \\ &= \frac{(\gamma_c^2-1)(1-\cos^2\alpha)(1-\cos^2\beta)}{(\cos\alpha(1-\cos^2\beta)+(\cos\gamma+\cos\alpha\cos\beta)\cos\beta)^2} \\ &= \frac{(\gamma_c^2-1)(1-\cos^2\alpha)(1-\cos^2\beta)}{(\cos\alpha+\cos\beta\cos\gamma)^2} \\ &= a_s^2\end{aligned} \tag{2.170}$$

thus obtaining the first identity in (2.169a). The second identity in (2.169a), for b_s, is obtained from the first identity, for a_s, by interchanging α and β. Finally, the identity in (2.169b) is established in (2.160), p. 107. □

Employing Identity (2.11), p. 66, for c_s^2, the first two identities in (2.169) imply

$$a_s b_s = \frac{(\gamma_c^2-1)\sin\alpha\sin\beta}{(\cos\alpha\sin\beta+\gamma_c\sin\alpha\cos\beta)(\cos\beta\sin\alpha+\gamma_c\sin\beta\cos\alpha)} \tag{2.171}$$

thus obtaining an identity that will prove useful in calculating the gyrotriangle defect in (2.172).

The system of identities (2.169) of Theorem 2.31 gives the ASA (gyroAngle-Side-gyroAngle) to SAS (Side-gyroAngle-Side) conversion law. It is useful for calculating two gyrosides and their included gyroangle of a gyrotriangle in an Einstein gyrovector space when the remaining two gyroangles and the gyroside included between them are known.

2.17 Gyrotriangle Defect

Some algebraic manipulations are too difficult to be performed by hand, but can easily be accomplished by computer algebra, that is, a computer software for symbolic manipulation, like Mathematica or Maple. It follows by straightforward substitution from (2.157) and (2.171), using computer algebra, that

$$\begin{aligned}\frac{\gamma_a\gamma_b a_s b_s \sin\gamma}{(1+\gamma_a)(1+\gamma_b)-\gamma_a\gamma_b a_s b_s\cos\gamma} &= \cot\frac{\alpha+\beta+\gamma}{2}\\ &= \tan\frac{\pi-(\alpha+\beta+\gamma)}{2}\\ &= \tan\tfrac{\delta}{2}\end{aligned} \tag{2.172}$$

where

$$\delta = \pi - (\alpha+\beta+\gamma) \tag{2.173}$$

is the gyrotriangular *defect* of the gyrotriangle ABC, presented in Fig. 2.4, p. 102.

Identity (2.172) is useful for calculating the defect of a gyrotriangle in an Einstein gyrovector space when two gyroside-gyrolengths and their included gyroangle of the gyrotriangle are known.

Let us now substitute $\cos\gamma$ from (2.143) and a_s^2 and b_s^2 from Identities like (2.11), p. 66, into $\tan^2(\delta/2)$ of (2.172), obtaining the identity

$$\begin{aligned}\tan^2\tfrac{\delta}{2} &= \frac{\gamma_a^2\gamma_b^2 a_s^2 b_s^2(1-\cos^2\gamma)}{((1+\gamma_a)(1+\gamma_b)-\gamma_a\gamma_b a_s b_s\cos\gamma)^2}\\ &= \frac{1+2\gamma_a\gamma_b\gamma_c-\gamma_a^2-\gamma_b^2-\gamma_c^2}{(1+\gamma_a+\gamma_b+\gamma_c)^2}\end{aligned} \tag{2.174}$$

which leads to the following theorem.

Theorem 2.32. (The Gyrotriangular Defect I). *Let ABC be a gyrotriangle in an Einstein gyrovector space $(\mathbb{R}_s^n, \oplus, \otimes)$ with the standard gyrotriangle notation in Fig. 2.4, p. 102. Then the gyrotriangular defect δ of ABC is given by the equation*

$$\tan\tfrac{\delta}{2} = \frac{\sqrt{1+2\gamma_a\gamma_b\gamma_c-\gamma_a^2-\gamma_b^2-\gamma_c^2}}{1+\gamma_a+\gamma_b+\gamma_c} \tag{2.175}$$

In the Newtonian-Euclidean limit of large s, $s\to\infty$, gamma factors tend to 1 so that (2.175) tends to $\tan(\delta/2)=0$ and, hence, δ tends to 0. Hence, the triangular defects of Euclidean triangles vanish, as expected.

Theorem 2.33. (The Gyrotriangular Defect II). *Let ABC be a gyrotriangle in an Einstein gyrovector space $(\mathbb{R}_s^n, \oplus, \otimes)$, with the standard gyrotriangle notation in Fig. 2.4, p. 102. Then the gyrotriangular defect δ of gyrotriangle ABC is given by the equation*

$$\tan\tfrac{\delta}{2} = \frac{p\sin\gamma}{1-p\cos\gamma} \tag{2.176}$$

where $\gamma, \delta, p > 0$, *and*

$$p^2 = \frac{\gamma_a - 1}{\gamma_a + 1}\frac{\gamma_b - 1}{\gamma_b + 1} \tag{2.177}$$

Proof. Employing (2.11), p. 66, the first equation in (2.174) can be written as (2.176) where $\gamma, \delta, p > 0$, and where p is given by (2.177). □

The gyrotriangular defect formula (2.175) of Theorem 2.32 is useful for calculating the defect of a gyrotriangle in an Einstein gyrovector space when the three gyrosides of the gyrotriangle are known.

The gyrotriangular defect formula (2.176) of Theorem 2.33 is useful for calculating the defect of a gyrotriangle in an Einstein gyrovector space when two gyrosides and their included gyroangle of the gyrotriangle are known.

2.18 Right Gyrotriangles

Let ABC be a right gyrotriangle in an Einstein gyrovector space $(\mathbb{R}^n_s, \oplus, \otimes)$ with the right gyroangle $\gamma = \pi/2$, as shown in Fig. 2.5 for $n = 2$. It follows from (2.157) with $\gamma = \pi/2$ that the gyrosides a, b and c of gyrotriangle ABC in Fig. 2.5 are related to the acute gyroangles α and β of the gyrotriangle by the equations

$$\begin{aligned}
\gamma_a &= \frac{\cos\alpha}{\sin\beta} \\
\gamma_b &= \frac{\cos\beta}{\sin\alpha} \\
\gamma_c &= \frac{\cos\alpha\cos\beta}{\sin\alpha\sin\beta}
\end{aligned} \tag{2.178}$$

The identities in (2.178) imply the Einstein-Pythagoras Identity

$$\gamma_a\,\gamma_b = \gamma_c \tag{2.179}$$

for a right gyrotriangle ABC with hypotenuse c and legs a and b in an Einstein gyrovector space, Fig. 2.5. It follows from (2.179) that $\gamma_a^2\gamma_b^2 = \gamma_c^2 = (1 - c_s^2)^{-1}$, implying that the gyrolength of the hypotenuse is given by the equation

$$c_s = \frac{\sqrt{\gamma_a^2\gamma_b^2 - 1}}{\gamma_a\gamma_b} \tag{2.180}$$

In terms of *rapidities*, (2.179) takes the standard form,

$$\cosh\phi_a \cosh\phi_b = \cosh\phi_c \tag{2.181}$$

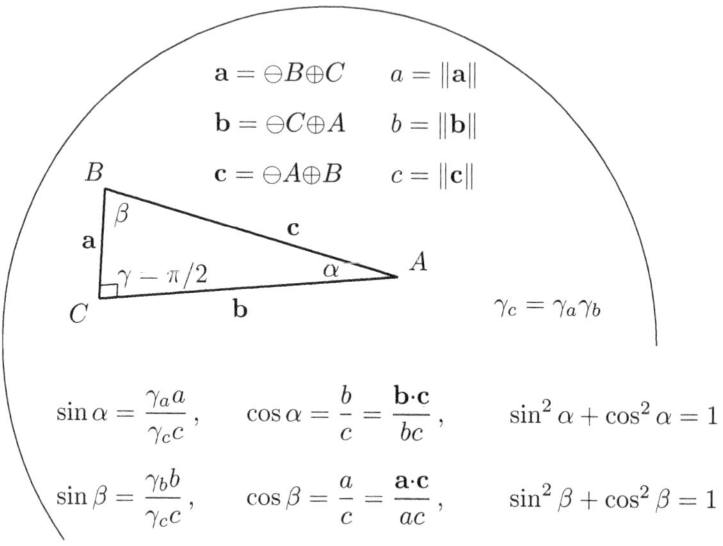

Fig. 2.5 Gyrotrigonometry in an Einstein gyrovector plane $(\mathbb{R}^2_s, \oplus, \otimes)$.

of the hyperbolic Pythagorean theorem [13, p. 334], where $\cosh\phi_a := \gamma_a$, etc.

In the special case of a right gyrotriangle in an Einstein gyrovector space, as in Fig. 2.5 where $\gamma = \pi/2$, it follows from Einstein-Pythagoras Identity (2.179) that the gyrotriangle defect identity (2.174) specializes to the right gyroangled gyrotriangle defect identity

$$\tan^2 \tfrac{\delta}{2} = \frac{\gamma_a - 1}{\gamma_a + 1}\frac{\gamma_b - 1}{\gamma_b + 1} \tag{2.182}$$

Identity (2.182) also follows from (2.176) with $\gamma = \pi/2$.

2.19 Einstein Gyrotrigonometry and Gyroarea

Let a, b and c be the respective gyrolengths of the two legs $\mathbf{a}$, $\mathbf{b}$ and the hypotenuse $\mathbf{c}$ of a right gyrotriangle ABC in an Einstein gyrovector space $(\mathbb{R}^n_s, \oplus, \otimes)$, Fig. 2.5. By (2.11), p. 66, and (2.178) we have

$$\begin{aligned} \left(\frac{a}{c}\right)^2 &= \frac{(\gamma_a^2 - 1)/\gamma_a^2}{(\gamma_c^2 - 1)/\gamma_c^2} = \cos^2\beta \\ \left(\frac{b}{c}\right)^2 &= \frac{(\gamma_b^2 - 1)/\gamma_b^2}{(\gamma_c^2 - 1)/\gamma_c^2} = \cos^2\alpha \end{aligned} \tag{2.183}$$

where γ_a, γ_b and γ_c are related by (2.179).

Similarly, by (2.11), p. 66, and (2.178) we also have

$$\begin{aligned} \left(\frac{\gamma_a a}{\gamma_c c}\right)^2 &= \frac{\gamma_a^2 - 1}{\gamma_c^2 - 1} = \sin^2 \alpha \\ \left(\frac{\gamma_b b}{\gamma_c c}\right)^2 &= \frac{\gamma_b^2 - 1}{\gamma_c^2 - 1} = \sin^2 \beta \end{aligned} \tag{2.184}$$

Identities (2.183) and (2.184) imply

$$\begin{aligned} \left(\frac{a}{c}\right)^2 + \left(\frac{\gamma_b b}{\gamma_c c}\right)^2 &= 1 \\ \left(\frac{\gamma_a a}{\gamma_c c}\right)^2 + \left(\frac{b}{c}\right)^2 &= 1 \end{aligned} \tag{2.185}$$

and, as shown in Fig. 2.5,

$$\begin{aligned} \cos \alpha &= \frac{b}{c} \\ \cos \beta &= \frac{a}{c} \end{aligned} \tag{2.186}$$

and

$$\begin{aligned} \sin \alpha &= \frac{\gamma_a a}{\gamma_c c} \\ \sin \beta &= \frac{\gamma_b b}{\gamma_c c} \end{aligned} \tag{2.187}$$

Interestingly, we see from (2.186) – (2.187) that the gyrocosine function of an acute gyroangle of a right gyrotriangle in an Einstein gyrovector space has the same form as its Euclidean counterpart, the cosine function. In contrast, it is only modulo gamma factors that the gyrosine function has the same form as its Euclidean counterpart.

Identities (2.185) give rise to the following two distinct Einsteinian-Pythagorean identities,

$$\begin{aligned} a^2 + \left(\frac{\gamma_b}{\gamma_c}\right)^2 b^2 &= c^2 \\ \left(\frac{\gamma_a}{\gamma_c}\right)^2 a^2 + b^2 &= c^2 \end{aligned} \tag{2.188}$$

for a right gyrotriangle with hypotenuse c and legs a and b in an Einstein gyrovector space.

The two distinct Einsteinian-Pythagorean identities in (2.188) that each Einsteinian right gyrotriangle obeys converge in the Newtonian-Euclidean limit of large s, $s \to \infty$, to the single Pythagorean identity

$$a^2 + b^2 = c^2 \tag{2.189}$$

that each Euclidean right-angled triangle obeys.

The identities in (2.188) and in (2.189) form the comparative study of the Pythagorean theorem in Euclidean and hyperbolic geometry that Einstein gyrovector spaces offer. A totally different comparative study of the Pythagorean theorem in Euclidean and hyperbolic geometry is offered by Möbius gyrovector spaces. Contrasting Einstein-Pythagoras theorem, Möbius-Pythagoras theorem and Pythagoras theorem share visual analogies, as shown in Fig. 7.1, p. 317.

As an application of the gyrotrigonometry formed by (2.186)–(2.187) in Einstein gyrovector spaces, we verify the following theorem.

Theorem 2.34. (The Base-Gyroaltitude Gyrotriangle Theorem). *Let ABC be a gyrotriangle with gyrosides a, b, c in an Einstein gyrovector space $(\mathbb{R}^n_s, \oplus, \otimes)$, Fig. 2.4, p. 102, and let h_a, h_b, h_c be the three gyroaltitudes of ABC drawn, respectively, from vertices A, B, C perpendicular to their opposite gyrosides a, b, c or their extension (For instance, $h = h_c$ in Fig. 2.8, p. 122). Then*

$$\gamma_a\, a\gamma_{h_a} h_a = \gamma_b\, b\gamma_{h_b} h_b = \gamma_c\, c\gamma_{h_c} h_c \tag{2.190}$$

Proof. By (2.187) with the notation of Fig. 2.8 we have $\gamma_{h_a} h_a = \gamma_c\, c \sin\beta$. Hence, by (2.11), p. 66, (2.143), p. 103, and Identity (2.174), p. 112, for $\tan(\delta/2)$, we have the following chain of equations that culminates in a most unexpected, elegant result,

$$\begin{aligned}
\gamma_a^2 a^2 \gamma_{h_a}^2 h_a^2 &= \gamma_a^2 a^2 \gamma_c^2 c^2 \sin^2\beta \\
&= s^4(\gamma_a^2 - 1)(\gamma_c^2 - 1)(1 - \cos^2\beta) \\
&= s^4(\gamma_a^2 - 1)(\gamma_c^2 - 1)\left(1 - \frac{(\gamma_a\gamma_c - \gamma_b)^2}{(\gamma_a^2 - 1)(\gamma_c^2 - 1)}\right) \\
&= s^4(1 + 2\gamma_a\gamma_b\gamma_c - \gamma_a^2 - \gamma_b^2 - \gamma_c^2) \\
&= s^4(1 + \gamma_a + \gamma_b + \gamma_c)^2 \tan^2\tfrac{\delta}{2}
\end{aligned} \tag{2.191}$$

thus verifying the first identity in (2.192) below.

$$\begin{aligned}\gamma_a a \gamma_{h_a} h_a &= s^2(1+\gamma_a+\gamma_b+\gamma_c)\tan(\delta/2)\\ \gamma_b b \gamma_{h_b} h_b &= s^2(1+\gamma_a+\gamma_b+\gamma_c)\tan(\delta/2)\\ \gamma_c c \gamma_{h_c} h_c &= s^2(1+\gamma_a+\gamma_b+\gamma_c)\tan(\delta/2)\end{aligned} \qquad (2.192)$$

The remaining two identities in (2.192) follow from the first by interchanging a and b, and by interchanging a and c, respectively, noting that $\tan(\delta/2)$ is symmetric in a, b, c by Theorem 2.32, p. 112.

Finally, the *gyrotriangle base-gyroaltitude identity* (2.190) of the theorem follows from the three identities in (2.192). □

Theorem 2.34 suggests the following definition.

Definition 2.35. (The Gyrotriangle Constant). *Let a, b, c be the three gyrosides of a gyrotriangle ABC with corresponding gyroaltitudes h_a, h_b, h_c in an Einstein gyrovector space $(\mathbb{R}^n_s, \oplus, \otimes)$, Fig. 2.8, p. 122.*

The number S_{ABC},

$$S_{ABC} = \gamma_a a \gamma_{h_a} h_a = \gamma_b b \gamma_{h_b} h_b = \gamma_c c \gamma_{h_c} h_c \qquad (2.193)$$

is called the gyrotriangle constant of gyrotriangle ABC.

It follows from Def. 2.35 and (2.191) that the gyrotriangle constant S_{ABC} of a gyrotriangle ABC with gyrosides a, b, c and defect δ in any Einstein gyrovector space satisfies each of the three identities

$$\begin{aligned}S_{ABC} &= s^2(1+\gamma_a+\gamma_b+\gamma_c)\tan(\delta/2)\\ S_{ABC} &= s^2\sqrt{1+2\gamma_a\gamma_b\gamma_c-\gamma_a^2-\gamma_b^2-\gamma_c^2}\end{aligned} \qquad (2.194)$$

and the 3×3 *gamma matrix* representation

$$S^2_{A_1A_2A_3} = s^4 \begin{vmatrix} 1 & \gamma_{12} & \gamma_{13} \\ \gamma_{12} & 1 & \gamma_{23} \\ \gamma_{13} & \gamma_{23} & 1 \end{vmatrix} \qquad (2.195)$$

For the sake of pattern recognition, we use in (2.195) the convenient notation in (6.4), p. 275. Interestingly, (2.195) shares its determinantal pattern with (6.22), p. 280.

Calling S_{ABC} in (2.194) a *gyrotriangle constant* of gyrotriangle ABC is justified since, as (2.194) demonstrates, S_{ABC} is invariant under any permutation of the gyrosides a, b, c of a gyrotriangle ABC, shown in Fig. 2.4, p. 102.

We may note that following the elementary gyrotrigonometric identities in Fig. 2.8, p. 122, the gyrotriangle constant (2.193) can be written in several forms. Thus, for instance,

$$S_{ABC} = \gamma_c c \gamma_{h_c} h_c = \gamma_c c \gamma_a a \sin\beta \tag{2.196}$$

for the gyrotriangle ABC in Fig. 2.4, p. 102.

The gyrotriangle constant proves useful in [3, p. 563], where it gives rise to the *gyrotriangle constant principle* which, in turn, allows a novel proof of the hyperbolic version of the theorems of Ceva and Menelaus.

Guided by analogies with Euclidean geometry, the equations in (2.192) suggest the following definition of the gyrotriangle gyroarea in Einstein gyrovector spaces.

Definition 2.36. (Gyrotriangle Gyroarea). *Let ABC be a gyrotriangle in an Einstein or a Möbius gyrovector space $(\mathbb{R}^n_s, \oplus, \otimes)$. The gyroarea $|ABC|$ of the gyrotriangle is given by the equation*

$$|ABC| = \frac{-2}{K}\tan\tfrac{\delta}{2} = \begin{cases} 2s^2\tan\tfrac{\delta}{2}\,, & \textit{Einstein} \\ \tfrac{1}{2}s^2\tan\tfrac{\delta}{2}\,, & \textit{Möbius} \end{cases} \tag{2.197}$$

where δ is the gyrotriangle defect, and where K is the Gaussian curvature of the gyrovector space. The Gaussian curvature is given by $K = -1/s^2$, (2.125), p. 96, for Einstein gyrovector planes, and by $K = -4/s^2$, (2.269), p. 136, for Möbius gyrovector planes

Following Def. 2.36 and (2.192), the gyroarea $|ABC|$ of a gyrotriangle ABC, in the notation of Fig. 2.8, p. 122, is given by

$$\begin{aligned} |ABC| &= \frac{2\gamma_a a \gamma_{h_a} h_a}{1+\gamma_a+\gamma_b+\gamma_c} \\ &= \frac{2\gamma_b b \gamma_{h_b} h_b}{1+\gamma_a+\gamma_b+\gamma_c} \\ &= \frac{2\gamma_c c \gamma_{h_c} h_c}{1+\gamma_a+\gamma_b+\gamma_c} \end{aligned} \tag{2.198}$$

In the Euclidean limit of large s, $s \to \infty$, gamma factors tend to 1 and hence the gyrotriangle gyroarea in (2.198) descends to the triangle area (1.76), p. 23, in Euclidean geometry. Similarly, the gyrotriangle constant, (2.193), descends in the Euclidean limit to twice the area of its corresponding triangle. An extension to the gyrotetrahedron constant emerges in (6.22), p. 280.

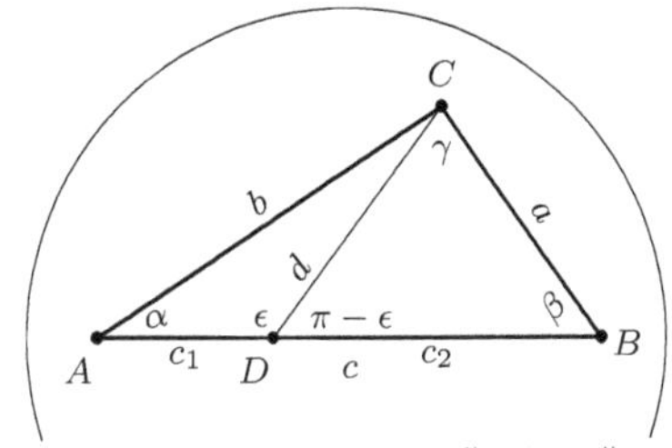

$$a = \|\ominus B \oplus C\|, \quad c_1 = \|\ominus A \oplus D\|$$
$$b = \|\ominus A \oplus C\|, \quad c_2 = \|\ominus B \oplus D\|$$
$$c = \|\ominus A \oplus B\|, \quad c_1 \oplus c_2 = c$$

Fig. 2.6 Gyrotriangle gyroarea addition in an Einstein gyrovector space. The gyroarea of a gyrotriangle ABC that is decomposed into two gyrotriangles, ADC and DBC, is shown. The point D is an arbitrary point between vertices A and B of gyrotriangle ABC.

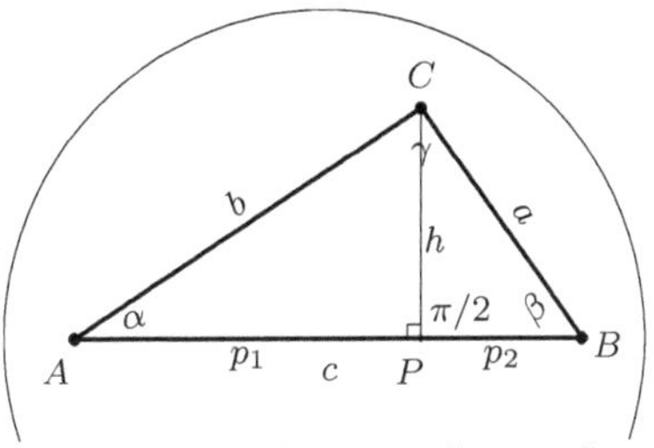

$$a = \|\ominus B \oplus C\|, \quad p_1 = \|\ominus A \oplus P\|$$
$$b = \|\ominus A \oplus C\|, \quad p_2 = \|\ominus B \oplus P\|$$
$$c = \|\ominus A \oplus B\|, \quad p_1 \oplus p_2 = c$$

Fig. 2.7 Gyrotriangle gyroarea addition in an Einstein gyrovector space. The gyroarea of a gyrotriangle ABC that is decomposed into two right gyrotriangles, APC and PBC, is shown. The point P is the orthogonal projection of vertex C on side AB of gyrotriangle ABC.

Following Def. 2.36 and (2.175), the gyrotriangle gyroarea is also given by the equation

$$\begin{aligned} |ABC| &= 2s^2 \tan \frac{\delta_{ABC}}{2} \\ &= 2s^2 \frac{\sqrt{1 + 2\gamma_a \gamma_b \gamma_c - \gamma_a^2 - \gamma_b^2 - \gamma_c^2}}{1 + \gamma_a + \gamma_b + \gamma_c} \end{aligned} \tag{2.199}$$

2.20 Gyrotriangle Gyroarea Addition Law

In Figs. 2.6 – 2.7, a gyrotriangle ABC in an Einstein gyrovector space is decomposed into two gyrotriangles, ADC and DBC by an arbitrarily selected point D between vertices A and B.

As a result, gyroangle $\gamma = \angle ACB$ in Fig. 2.6 is decomposed into the two gyroangles $\gamma_1 = \angle ACD$ and $\gamma_2 = \angle DCB$, so that $\gamma_1 + \gamma_2 = \gamma$.

Let the defect of a gyrotriangle ABC be denoted by δ_{ABC}. Then,

$$\begin{aligned} \delta_{ABC} &= \pi - (\alpha + \beta + \gamma) \\ \delta_{ADC} &= \pi - (\alpha + \epsilon + \gamma_1) \\ \delta_{DBC} &= \pi - (\pi - \epsilon + \beta + \gamma_2) \end{aligned} \tag{2.200}$$

so that

$$\delta_{ADC} + \delta_{DBC} = \delta_{ABC} \tag{2.201}$$

for the gyrotriangle defects in Fig. 2.6, thus demonstrating the additive property of the gyrotriangular defect.

The gyrotriangular defect identity (2.201) relates the gyroarea $|ABC|$ to its constituent gyroareas $|ADC|$ and $|DBC|$ in Fig. 2.6 by the addition rule of the trigonometric/gyrotrigonometric tangent function as follows.

$$\begin{aligned}
|ABC| &= 2s^2 \tan \frac{\delta_{ADC}}{2} \\
&= 2s^2 \tan \left(\frac{\delta_{ADC}}{2} + \frac{\delta_{DBC}}{2} \right) \\
&= 2s^2 \frac{\tan \frac{\delta_{ADC}}{2} + \tan \frac{\delta_{DBC}}{2}}{1 - \tan \frac{\delta_{ADC}}{2} \tan \frac{\delta_{DBC}}{2}} \\
&= \frac{2s^2 \tan \frac{\delta_{ADC}}{2} + 2s^2 \tan \frac{\delta_{DBC}}{2}}{1 - \frac{2s^2 \tan \frac{\delta_{ADC}}{2} 2s^2 \tan \frac{\delta_{DBC}}{2}}{4s^4}} \\
&= \frac{|ADC| + |DBC|}{1 - \frac{|ADC||DBC|}{4s^4}} \\
&=: |ADC| \oplus_{\mathrm{EA}} |DBC|
\end{aligned} \tag{2.202}$$

Here, in (2.202), gyrotriangle gyroarea addition in an Einstein gyrovector space $(\mathbb{R}_s^n, \oplus, \otimes)$ is denoted by the binary operation $\oplus_{\mathrm{EA}}$, and it is defined for gyrotriangle gyroareas S_1 and S_2 that obey the inequality $S_1 S_2 < 4s^4$.

Suggestively, the formal definition of gyroarea addition follows.

Definition 2.37. (Gyroarea Addition in Einstein Gyrovector Spaces). *Let $A_1B_1C_1$ and $A_2B_2C_2$ be two gyrotriangles in an Einstein gyrovector space $(\mathbb{R}_s^n, \oplus, \otimes)$, with gyroareas $S_1 = |A_1B_1C_1|$ and $S_2 = |A_2B_2C_2|$, satisfying the inequality*

$$S_1 S_2 < 4s^4 \tag{2.203}$$

Then, the gyroarea S_{12} of the union of the two gyrotriangles is given by the equation

$$S_{12} = S_1 \oplus_{EA} S_2 := \frac{S_1 + S_2}{1 - \frac{S_1 S_2}{4s^4}} \tag{2.204}$$

Example 2.38. (Gyrotriangle Decomposition Into Two Gyrotriangles). It follows from the gyrotriangle gyroarea identity (2.199), using the notation in Fig. 2.6, that

$$\begin{aligned}
|ABC| &= 2s^2 \tan\frac{\delta_{ABC}}{2} = 2s^2\frac{\sqrt{1+2\gamma_a\,\gamma_b\,\gamma_c\, -\gamma_a^2-\gamma_b^2-\gamma_c^2}}{1+\gamma_a\,+\gamma_b\,+\gamma_c} \\
|ADC| &= 2s^2 \tan\frac{\delta_{ADC}}{2} = 2s^2\frac{\sqrt{1+2\gamma_b\,\gamma_{c_1}\gamma_d\, -\gamma_b^2-\gamma_{c_1}^2-\gamma_d^2}}{1+\gamma_b\,+\gamma_{c_1}+\gamma_d} \\
|BDC| &= 2s^2 \tan\frac{\delta_{BDC}}{2} = 2s^2\frac{\sqrt{1+2\gamma_a\,\gamma_{c_2}\gamma_d\, -\gamma_a^2-\gamma_{c_2}^2-\gamma_d^2}}{1+\gamma_a\,+\gamma_{c_2}+\gamma_d}
\end{aligned} \tag{2.205}$$

where c, c_1, c_2 are related by the equation

$$c = c_1 \oplus c_2 \tag{2.206}$$

as shown in Fig. 2.6.

The gyrotriangles ADC and BDC form a partition of gyrotriangle ABC, where D is any point between A and B. Hence we have, by (2.202),

$$|ABC| = |ADC| \oplus_{\text{EA}} |DBC| \tag{2.207}$$

2.21 Gyrodistance Between a Point and a Gyroline

Let $A, B, C \in \mathbb{R}_s^n$ be any three non-gyrocollinear points of an Einstein gyrovector space $(\mathbb{R}_s^n, \oplus, \otimes)$, $n \geq 2$, and let L_{AB} be the gyroline passing through the points A and B, Fig. 2.8. The gyrodistance between the point C and the gyroline L_{AB} is the gyrolength $h_c = \|\mathbf{h}_c\|$ of the perpendicular $\mathbf{h}_c$ drawn from point C to gyroline L_{AB}. With the notation of Fig. 2.8, and with permutations of a, b, c, it follows from (2.191) that

$$\begin{aligned}
\gamma_a^2 a^2 \gamma_{h_a}^2 h_a^2 &= \gamma_b^2 b^2 \gamma_{h_b}^2 h_b^2 = \gamma_c^2 c^2 \gamma_{h_c}^2 h_c^2 \\
&= s^4(1+2\gamma_a\gamma_b\gamma_c - \gamma_a^2 - \gamma_b^2 - \gamma_c^2)
\end{aligned} \tag{2.208}$$

Hence, in particular, we have by the third equation in (2.208), and (2.11), p. 66,

$$\gamma_{h_c}^2 h_c^2 = s^2 \frac{1+2\gamma_a\gamma_b\gamma_c - \gamma_a^2 - \gamma_b^2 - \gamma_c^2}{\gamma_c^2 - 1} \tag{2.209}$$

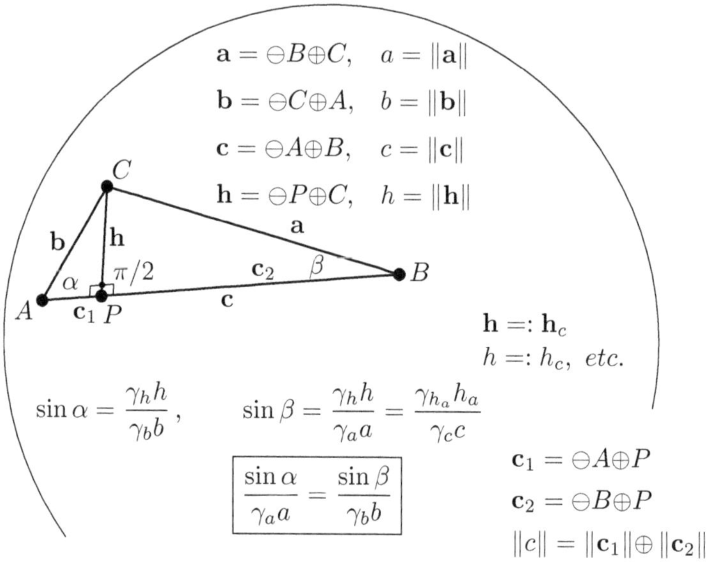

Fig. 2.8 The Law of Gyrosines. Drawing gyrotriangle gyroaltitudes in Einstein Gyrovector Spaces $(\mathbb{R}^n_s, \oplus, \otimes)$, and employing a basic gyrotrigonometric formula, shown in Fig. 2.5, we obtain the Law of Gyrosines, (2.167), in full analogy with the derivation of its Euclidean counterpart, the Law of Sines. The orthogonal projection of $\mathbf{a}$ on $\mathbf{c}' = \ominus B \oplus A$ is $\mathbf{c}_2$ and, similarly, the orthogonal projection of $\mathbf{b}' = \ominus A \oplus C$ on $\mathbf{c}$ is $\mathbf{c}_1$. Clearly, $\|\mathbf{b}'\| = \|\mathbf{b}\|$, and $\|\mathbf{c}'\| = \|\mathbf{c}\|$. It follows from the gyrotriangle equality that $\|\mathbf{c}\| = \|\mathbf{c}_1\| \oplus \|\mathbf{c}_2\|$.

By an obvious identity and by (2.209) we have

$$\begin{aligned} \frac{1}{s^2}h_c^2 &= 1 - \frac{1}{1 + \gamma_{h_c}^2 \frac{h_c^2}{s^2}} \\ &= \frac{1 + 2\gamma_a\gamma_b\gamma_c - \gamma_a^2 - \gamma_b^2 - \gamma_c^2}{2\gamma_a\gamma_b\gamma_c - \gamma_a^2 - \gamma_b^2} \end{aligned} \tag{2.210}$$

so that, by (2.209) and (2.210),

$$\gamma_{h_c}^2 = \frac{2\gamma_a\gamma_b\gamma_c - \gamma_a^2 - \gamma_b^2}{\gamma_c^2 - 1} \tag{2.211}$$

Formalizing the results in (2.208) – (2.211), we have the following theorem:

Theorem 2.39. (Gyrodistance Between a Gyropoint and a Gyroline). *Let A and B be any two distinct points of an Einstein gyrovector space $(\mathbb{R}^n_s, \oplus, \otimes)$, $n \geq 2$, and let L_{AB} be the gyroline passing through these* *points. Furthermore, let C be any point of the space that does not lie on L_{AB}, as shown in Fig. 2.8. Then, in the notation of Fig. 2.8, the gyrodistance $h_c = \|\mathbf{h}_c\| = \|\ominus C \oplus P\|$ between the point C and the gyroline L_{AB} is given by the equation*

$$h_c^2 = s^2 \frac{1 + 2\gamma_a\gamma_b\gamma_c - \gamma_a^2 - \gamma_b^2 - \gamma_c^2}{2\gamma_a\gamma_b\gamma_c - \gamma_a^2 - \gamma_b^2} = s^2 \left(1 - \frac{\gamma_c^2 - 1}{2\gamma_a\gamma_b\gamma_c - \gamma_a^2 - \gamma_b^2}\right) \tag{2.212}$$

satisfying

$$\gamma_{h_c}^2 = \frac{2\gamma_a\gamma_b\gamma_c - \gamma_a^2 - \gamma_b^2}{\gamma_c^2 - 1} \tag{2.213}$$

Furthermore, the product $\gamma_c c \gamma_{h_c} h_c$ is a symmetric function of a, b, c, given by the equation

$$\gamma_c^2 c^2 \gamma_{h_c}^2 h_c^2 = s^4 (1 + 2\gamma_a\gamma_b\gamma_c - \gamma_a^2 - \gamma_b^2 - \gamma_c^2) \tag{2.214}$$

As an immediate consequence of Theorem 2.39 we have the following Corollary.

Corollary 2.40. *Let A, B, C be any three non-gyrocollinear points of an Einstein gyrovector space, and let a, b, c be the gyrolengths of the gyrosides of the gyrotriangle ABC,*

$$\begin{aligned} a &= \|\ominus B \oplus C\| \\ b &= \|\ominus A \oplus C\| \\ c &= \|\ominus A \oplus B\| \end{aligned} \tag{2.215}$$

Then, the gamma factors of a, b, c satisfy the following inequalities.

$$\begin{aligned} 2\gamma_a\gamma_b\gamma_c &> \gamma_a^2 + \gamma_b^2 \\ 1 + 2\gamma_a\gamma_b\gamma_c &> \gamma_a^2 + \gamma_b^2 + \gamma_c^2 \end{aligned} \tag{2.216}$$

The second inequality in (2.216) descends to its corresponding equality

$$1 + 2\gamma_a\gamma_b\gamma_c = \gamma_a^2 + \gamma_b^2 + \gamma_c^2 \tag{2.217}$$

if and only if the points A, B, C are gyrocollinear.

Proof. If A, B, C are non-gyrocollinear then h_c is the gyrodistance between C and the line L_{AB} that passes through A and B, satisfying $h_c > 0$ and $\gamma_{h_c} > 1$. Hence, Inequalities (2.216) follow from (2.213) and from (2.214). Clearly, the second inequality in (2.216) descends to equality (2.217) if and only if $h_c = 0$, that is, if and only if A, B, C are gyrocollinear. The second inequality in (2.216) was also verified in (2.146), p. 104. □

The *orthogonal projection* of a point C onto a gyrosegment AB (or its extension, the gyroline L_{AB} that passes through the points A and B) is the foot P of the perpendicular CP from the point to the gyrosegment (or its extension), as shown in Fig. 2.8.

Accordingly, the orthogonal projection of gyroside $\mathbf{b}' = \ominus A \oplus C$ on gyroside $\mathbf{c} = \ominus A \oplus B$ of gyrotriangle ABC in Fig. 2.8 is

$$\mathbf{c}_1 = \ominus A \oplus P \tag{2.218}$$

the gyrolength of which is

$$c_1 = \|\mathbf{c}_1\| = \|\ominus A \oplus P\| = b \cos \alpha \tag{2.219}$$

as we see from Fig. 2.8 and from the relativistic gyrotrigonometry in Fig. 2.5, p. 114.

Hence, by (2.219), (2.11), p. 66, and (2.143), p. 103, we have

$$c_1^2 = b^2 \cos^2 \alpha = s^2 \frac{\gamma_b^2 - 1}{\gamma_b^2} \frac{(\gamma_b \gamma_c - \gamma_a)^2}{(\gamma_b^2 - 1)(\gamma_c^2 - 1)} = s^2 \frac{(\gamma_b \gamma_c - \gamma_a)^2}{\gamma_b^2 (\gamma_c^2 - 1)} \tag{2.220}$$

so that

$$\gamma_{c_1}^2 = \gamma_b^2 \frac{\gamma_c^2 - 1}{2\gamma_a \gamma_b \gamma_c - \gamma_a^2 - \gamma_b^2} \tag{2.221}$$

Furthermore, following (2.220) and (2.154), p. 106, we have

$$c_1 = \begin{cases} s \dfrac{\gamma_b \, \gamma_c \; - \gamma_a}{\gamma_b \, \sqrt{\gamma_c^2 - 1}}, & \text{if } 0 < \alpha \le \frac{\pi}{2}; \\ s \dfrac{\gamma_a \; - \gamma_b \, \gamma_c}{\gamma_b \, \sqrt{\gamma_c^2 - 1}}, & \text{if } \frac{\pi}{2} \le \alpha < \pi. \end{cases} \tag{2.222}$$

It follows from (2.213) and (2.221) that

$$\gamma_{h_c}^2 \gamma_{c_1}^2 = \gamma_b^2 \tag{2.223}$$

in agreement with Einstein-Pythagoras Identity (2.179) for the right gyrotriangle APC in Fig. 2.8.

Similarly, the orthogonal projection of gyroside $\mathbf{a} = \ominus B \oplus C$ on gyroside $\mathbf{c}' = \ominus B \oplus A$ of gyrotriangle ABC in Fig. 2.8 is

$$\mathbf{c}_2 = \ominus B \oplus P \tag{2.224}$$

the gyrolength of which is

$$c_2 = \|\mathbf{c}_2\| = \|\ominus B \oplus P\| = a \cos \beta \tag{2.225}$$

as we see from Fig. 2.8 and from the basic equations of relativistic gyrotrigonometry in Fig. 2.5, p. 114.

Hence, by (2.225), (2.11), p. 66, and (2.143), we have

$$c_2^2 = a^2 \cos^2 \beta = s^2 \frac{\gamma_a^2 - 1}{\gamma_a^2} \frac{(\gamma_a \gamma_c - \gamma_b)^2}{(\gamma_a^2 - 1)(\gamma_c^2 - 1)} = s^2 \frac{(\gamma_a \gamma_c - \gamma_b)^2}{\gamma_a^2 (\gamma_c^2 - 1)} \tag{2.226}$$

so that

$$\gamma_{c_2}^2 = \gamma_a^2 \frac{\gamma_c^2 - 1}{2\gamma_a \gamma_b \gamma_c - \gamma_a^2 - \gamma_b^2} \tag{2.227}$$

Furthermore, following (2.226) and (2.155), p. 106, we have

$$c_2 = \begin{cases} s\dfrac{\gamma_a\,\gamma_c\ - \gamma_b}{\gamma_a\,\sqrt{\gamma_c^2 - 1}}, & \text{if } 0 < \beta \le \frac{\pi}{2}; \\[2ex] s\dfrac{\gamma_b\ - \gamma_a\,\gamma_c}{\gamma_a\,\sqrt{\gamma_c^2 - 1}}, & \text{if } \frac{\pi}{2} \le \beta < \pi. \end{cases} \tag{2.228}$$

It follows from (2.213) and (2.227) that

$$\gamma_{h_c}^2 \gamma_{c_2}^2 = \gamma_a^2 \tag{2.229}$$

in agreement with Einstein-Pythagoras Identity (2.179) for the right gyrotriangle BCP in Fig. 2.8.

Interestingly, it follows from (2.221) and (2.227) or, equivalently, from (2.223) and (2.229), that gyrotriangle ABC in Fig. 2.8 satisfies the elegant hyperbolic geometric identity

$$\frac{\gamma_{c_1}}{\gamma_{c_2}} = \frac{\gamma_b}{\gamma_a} \tag{2.230}$$

In the Euclidean limit, $s \to \infty$, gamma factors tend to 1, so that (2.230) descends to the trivial identity $1 = 1$. Hence, the gyrotriangle identity (2.230) in hyperbolic geometry, shown in Fig. 2.8, has no analogue in Euclidean geometry.

Example 2.41. Let ABC be a gyrotriangle in an Einstein gyrovector space with acute gyroangles α and β, as shown in Fig. 2.9.

Following (2.220), and (2.154) for the acute gyroangle α in Fig. 2.9, we have

$$c_1 = s\frac{\gamma_b\,\gamma_c\ - \gamma_a}{\gamma_b\,\sqrt{\gamma_c^2 - 1}} \tag{2.231}$$

Similarly, following (2.226), and (2.155) for the acute gyroangle β in Fig. 2.9, we have

$$c_2 = s\frac{\gamma_a\,\gamma_c\ - \gamma_b}{\gamma_a\,\sqrt{\gamma_c^2 - 1}} \tag{2.232}$$

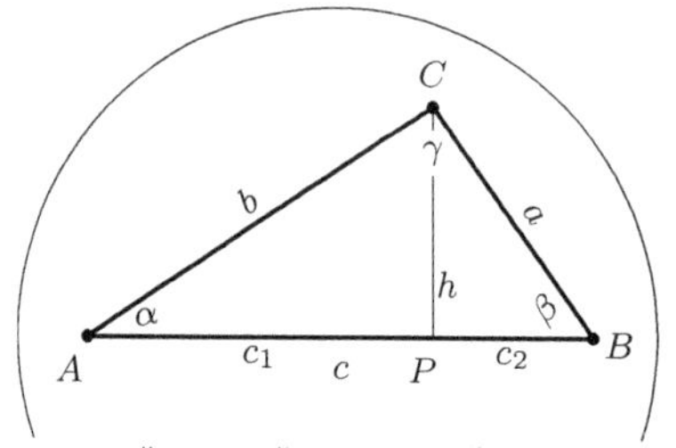

$$a = \|\ominus B \oplus C\|, \qquad c_1 = \|\ominus A \oplus P\|$$
$$b = \|\ominus A \oplus C\|, \qquad c_2 = \|\ominus B \oplus P\|$$
$$c = \|\ominus A \oplus B\|, \qquad c_1 \oplus c_2 = c$$

Fig. 2.9 Orthogonal Projection I. Illustrating Example 2.41: Gyroangles α and β of a gyrotriangle ABC in an Einstein gyrovector space $(\mathbb{V}_s, \oplus, \otimes)$ are acute. As a result, $c = c_1 \oplus c_2$.

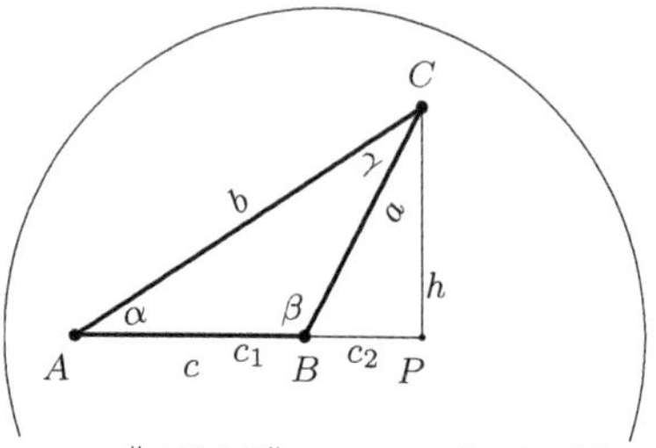

$$a = \|\ominus B \oplus C\|, \qquad c_1 = \|\ominus A \oplus P\|$$
$$b = \|\ominus A \oplus C\|, \qquad c_2 = \|\ominus B \oplus P\|$$
$$c = \|\ominus A \oplus B\|, \qquad c_1 \ominus c_2 = c$$

Fig. 2.10 Orthogonal Projection II. Illustrating Example 2.42: Gyroangle β of a gyrotriangle ABC in an Einstein gyrovector space $(\mathbb{V}_s, \oplus, \otimes)$ is obtuse. Contrasting Fig. 2.9, here $c = c_1 \ominus c_2$.

Hence, by (2.12), p. 66, by (2.231)–(2.232), and by (2.11), p. 66, we have

$$\frac{1}{s^2}(c_1 \oplus c_2)^2 = \frac{1}{s^2}\left(\frac{c_1 + c_2}{1 + \frac{c_1 c_2}{s^2}}\right)^2 = \frac{\gamma_c^2 - 1}{\gamma_c^2} = \frac{1}{s^2}c^2 \tag{2.233}$$

implying

$$c_1 \oplus c_2 = c \tag{2.234}$$

as expected in Fig. 2.9.

Example 2.42. Let ABC be a gyrotriangle in an Einstein gyrovector space with an acute gyroangle α and an obtuse gyroangle β, as shown in Fig. 2.10.

Following (2.220), and (2.154) for the acute gyroangle α in Fig. 2.10, we have

$$c_1 = s\frac{\gamma_b\,\gamma_c\ - \gamma_a}{\gamma_b\,\sqrt{\gamma_c^2 - 1}} \tag{2.235}$$

Similarly, following (2.226), and (2.155) for the obtuse gyroangle β in Fig. 2.10, we have

$$c_2 = -s\frac{\gamma_a\,\gamma_c\ - \gamma_b}{\gamma_a\,\sqrt{\gamma_c^2 - 1}} \tag{2.236}$$

Note that, unlike (2.232), the right-hand side of (2.236) is led by a negative sign.

Hence, by (2.12), p. 66, by (2.235) – (2.236), and by (2.11), p. 66, we have

$$\frac{1}{s^2}(c_1 \ominus c_2)^2 = \frac{1}{s^2}\left(\frac{c_1 - c_2}{1 - \frac{c_1 c_2}{s^2}}\right)^2 = \frac{\gamma_c^2 - 1}{\gamma_c^2} = \frac{1}{s^2}c^2 \tag{2.237}$$

implying

$$c_1 \ominus c_2 = c \tag{2.238}$$

as expected in Fig. 2.10.

Example 2.43. Let ABC be a gyrotriangle in an Einstein gyrovector space with acute gyroangles α and β, as shown in Fig. 2.9. For the product of the two orthogonal projections c_1 and c_2 on $\mathbf{c}$ we have, from (2.231) – (2.232),

$$c_1 c_2 = s^2 \frac{(\gamma_b \gamma_c - \gamma_a)(\gamma_a \gamma_c - \gamma_b)}{\gamma_a \gamma_b (\gamma_c^2 - 1)} \tag{2.239}$$

and from (2.227) and (2.221),

$$\gamma_{c_1} \gamma_{c_2} = \gamma_a \gamma_b \frac{\gamma_c^2 - 1}{2\gamma_a \gamma_b \gamma_c - \gamma_a^2 - \gamma_b^2} \tag{2.240}$$

Hence, by the gamma identity (2.10), p. 66, and by (2.239) and (2.240), we have

$$\gamma_{c_1 \oplus c_2} = \gamma_{c_1} \gamma_{c_2} (1 + \frac{c_1 c_2}{s^2}) = \gamma_c \tag{2.241}$$

Indeed, identity (2.241) is expected from (2.234).

Example 2.44. Let ABC be a gyrotriangle in an Einstein gyrovector space with an acute gyroangle α and an obtuse gyroangle β, as shown in Fig. 2.10. For the product of the two orthogonal projections c_1 and c_2 on $\mathbf{c}$ we have, from (2.235) – (2.236),

$$c_1 c_2 = -s^2 \frac{(\gamma_b \gamma_c - \gamma_a)(\gamma_a \gamma_c - \gamma_b)}{\gamma_a \gamma_b (\gamma_c^2 - 1)} \tag{2.242}$$

and from (2.227) and (2.221),

$$\gamma_{c_1} \gamma_{c_2} = \gamma_a \gamma_b \frac{\gamma_c^2 - 1}{2\gamma_a \gamma_b \gamma_c - \gamma_a^2 - \gamma_b^2} \tag{2.243}$$

Hence, by the gamma identity (2.10), p. 66, and by (2.242) and (2.243), we have

$$\gamma_{c_1 \ominus c_2} = \gamma_{c_1} \gamma_{c_2} (1 - \frac{c_1 c_2}{s^2}) = \gamma_c \tag{2.244}$$

Indeed, identity (2.244) is expected from (2.238).

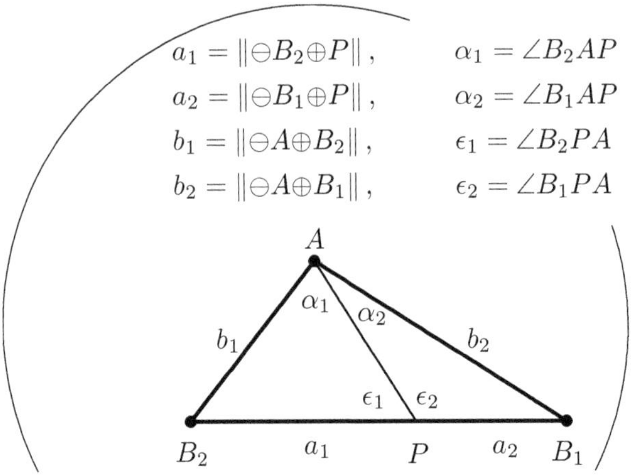

Fig. 2.11 The Generalized Gyroangle Bisector Theorem. Gyrotriangle AB_1B_2 in an Einstein gyrovector plane $(\mathbb{R}_s^2, \oplus, \otimes)$ illustrates the Generalized Gyroangle Bisector Theorem 2.46. In the special case when $\alpha_1 = \alpha_2$ it illustrates the Gyroangle Bisector Theorem 2.45.

2.22 The Gyroangle Bisector Theorem

Theorem 2.45. (The Gyrotriangle Bisector Theorem). *Let AB_1B_2 be a gyrotriangle in an Einstein gyrovector space $(\mathbb{R}_s^n, \oplus, \otimes)$, and let P be a point lying on gyroside B_1B_2 of the gyrotriangle such that AP is a bisector of gyroangle $\angle B_1AB_2$, as shown in Fig. 2.11, with $\alpha_1 = \alpha_2$, for an Einstein gyrovector plane $\mathbb{R}_s^n = \mathbb{R}_s^2$. Then, in the notation of Fig. 2.11,*

$$\frac{\gamma_{a_1} a_1}{\gamma_{a_2} a_2} = \frac{\gamma_{b_1} b_1}{\gamma_{b_2} b_2} \tag{2.245}$$

Theorem 2.45 above is a special case of theorem 2.46 below, corresponding to $\alpha_1 = \alpha_2$. The proof of Theorem 2.45 is therefore included in the proof of the following Theorem 2.46, for the special case when $\alpha_1 = \alpha_2$.

Theorem 2.46. (The Generalized Gyrotriangle Bisector Theorem). *Let AB_1B_2 be a gyrotriangle in an Einstein gyrovector space $(\mathbb{R}_s^n, \oplus, \otimes)$, and let P be a point lying on gyroside B_1B_2 of the gyrotriangle, as shown in Fig. 2.11 for an Einstein gyrovector plane $\mathbb{R}_s^n = \mathbb{R}_s^2$. Then, in the notation of Fig. 2.11,*

$$\frac{\gamma_{a_1} a_1}{\gamma_{a_2} a_2} = \frac{\gamma_{b_1} b_1 \sin \alpha_1}{\gamma_{b_2} b_2 \sin \alpha_2} \tag{2.246}$$

Proof. With the notation of Fig. 2.11 we have, by the law of gyrosines for each of the two gyrotriangles AB_2P and AB_1P,

$$\begin{aligned} \frac{\gamma_{a_1} a_1}{\sin\alpha_1} &= \frac{\gamma_{b_1} b_1}{\sin\epsilon_1} \\ \frac{\gamma_{a_2} a_2}{\sin\alpha_2} &= \frac{\gamma_{b_2} b_2}{\sin\epsilon_2} \end{aligned} \tag{2.247}$$

Since $\epsilon_1 + \epsilon_2 = \pi$, we have $\sin\epsilon_1 = \sin\epsilon_2$ so that (2.247) implies the desired result (2.246) of the theorem. □

2.23 Möbius Addition and Möbius Gyrogroups

We have seen that Einstein addition in the ball arises from Einstein's special theory of relativity, leading to gyrovector spaces that form the setting for our Cartesian-Beltrami-Klein ball model of hyperbolic geometry. Similarly, Möbius addition in the ball arises from Möbius transformations of the complex open unit disc [48], which is well-studied in complex analysis [49–51]. We will see that, like Einstein addition, Möbius addition in the ball leads to gyrovector spaces that form the setting for our Cartesian-Poincaré ball model of hyperbolic geometry. The definition of Möbius addition in the ball, thus, follows.

Möbius addition, $\oplus$, is a binary operation in the ball $\mathbb{R}_s^n$ [52], given by the equation

$$X \oplus Y = \frac{(1 + \frac{2}{s^2} X\cdot Y + \frac{1}{s^2}\|Y\|^2)X + (1 - \frac{1}{s^2}\|X\|^2)Y}{1 + \frac{2}{s^2} X\cdot Y + \frac{1}{s^4}\|X\|^2\|Y\|^2} \tag{2.248}$$

where $\cdot$ and $\|\cdot\|$ are the inner product and norm that the ball $\mathbb{R}_s^n$ inherits from its ambient space $\mathbb{R}^n$.

Without loss of generality, one may select $s = 1$ in (2.248). We, however, prefer to keep s as a free positive parameter in order to exhibit the result that in the limit as $s \to \infty$, the ball $\mathbb{R}_s^n$ expands to the whole of its Euclidean n-space $\mathbb{R}^n$, and Möbius addition, $\oplus$, descends to the vector addition, $+$, in $\mathbb{R}^n$. Like Einstein gyrocommutative gyrogroups $(\mathbb{R}_s^n, \oplus_{\text{E}})$, Möbius groupoids $(\mathbb{R}_s^n, \oplus_{\text{M}})$ are gyrocommutative gyrogroups. Interestingly, the right-hand side of (2.248) is known as a Möbius translation [53, p. 129]. Owing to the analogies it shares with vector addition we, however, call it Möbius addition.

Ambiguously, both Einstein addition and Möbius addition in the ball are, in general, denoted by $\oplus$ when no confusion may arise. However, when necessary, we use the notation $\oplus_{\text{E}}$ for Einstein addition, and $\oplus_{\text{M}}$ for Möbius addition.

2.24 Möbius Gyration

For any $X, Y \in \mathbb{R}_s^n$, let $\text{gyr}[X, Y] : \mathbb{R}_s^n \to \mathbb{R}_s^n$ be the self-map of $\mathbb{R}_s^n$ given in terms of Möbius addition $\oplus$ by the equation [29]

$$\text{gyr}[X, Y]Z = \ominus(X \oplus Y) \oplus \{X \oplus (Y \oplus Z)\} \tag{2.249}$$

where $\ominus Y = -Y$, for all $Z \in \mathbb{R}_s^n$. The self-map $\text{gyr}[X, Y]$ of $\mathbb{R}_s^n$, which takes $Z \in \mathbb{R}_s^n$ into $\ominus(X \oplus Y) \oplus \{X \oplus (Y \oplus Z) \in \mathbb{R}_s^n$, is called the *Thomas gyration* (or, briefly, gyration) generated by X and Y.

In the Euclidean limit, $s \to \infty$, Möbius addition $\oplus$ in $\mathbb{R}_s^n$ descends to the common vector addition $+$ in $\mathbb{R}^n$, which is associative. Accordingly, in this limit the gyration $\text{gyr}[X, Y]$ in (2.249) descends to the identity map of $\mathbb{R}^n$. Hence, as expected, gyrations $\text{gyr}[X, Y]$, $X, Y \in \mathbb{R}_s^n$, vanish (that is, they become *trivial*) in the Euclidean limit.

The gyration equation (2.249) can be manipulated (with the help of computer algebra) into the equation

$$\text{gyr}[X, Y]Z = Z + \frac{AX + BY}{D} \tag{2.250}$$

where

$$\begin{aligned} A &= -\frac{1}{s^4} X \cdot Z \|Y\|^2 + \frac{1}{s^2} Y \cdot Z + \frac{2}{s^4} (X \cdot Y)(Y \cdot Z) \\ B &= -\frac{1}{s^4} Y \cdot Z \|X\|^2 - \frac{1}{s^2} X \cdot Z \\ D &= 1 + \frac{2}{s^2} X \cdot Y + \frac{1}{s^4} \|X\|^2 \|Y\|^2 > \left(1 + \frac{X \cdot Y}{s^2}\right)^2 > 0 \end{aligned} \tag{2.251}$$

for all $X, Y, Z \in \mathbb{R}_s^n$. Owing to Cauchy-Schwarz inequality [10, p. 20], according to which $|X \cdot Y| \le \|X\| \|Y\|$, we have $D > 0$ in the ball $\mathbb{R}_s^n$. Allowing $Z \in \mathbb{R}^n \supset \mathbb{R}_s^n$ in (2.250) – (2.251), gyrations $\text{gyr}[X, Y]$ are expendable to linear maps of $\mathbb{R}^n$ for all $X, Y \in \mathbb{R}_s^n$.

In each of the three special cases when (i) $X = \mathbf{0}$, or (ii) $Y = \mathbf{0}$, or (iii) X and Y are parallel in $\mathbb{R}^n$, $X \| Y$, we have $AX + BY = \mathbf{0}$ so that $\text{gyr}[X, Y]$ is trivial,

$$\begin{aligned} \text{gyr}[\mathbf{0}, Y]Z &= Z \\ \text{gyr}[X, \mathbf{0}]Z &= Z \\ \text{gyr}[X, Y]Z &= Z, \qquad X \| Y \end{aligned} \tag{2.252}$$

for all $X, Y \in \mathbb{R}_s^n$ and all $Z \in \mathbb{R}^n$.

It follows from (2.250) that

$$\mathrm{gyr}[Y,X](\mathrm{gyr}[X,Y]Z) = Z \tag{2.253}$$

for all $X, Y \in \mathbb{R}_s^n$, $Z \in \mathbb{R}^n$, so that gyrations are invertible linear maps of $\mathbb{R}^n$, the inverse of $\mathrm{gyr}[X,Y]$ being $\mathrm{gyr}[Y,X]$ for all $X, Y \in \mathbb{R}_s^n$.

Owing to the nonassociativity of Möbius addition $\oplus$, in general, a gyration is not trivial. Interestingly, gyrations keep the inner product of elements of the ball $\mathbb{R}_s^n$ invariant, that is,

$$\mathrm{gyr}[X,Y]A{\cdot}\mathrm{gyr}[X,Y]B = A{\cdot}B \tag{2.254}$$

for all $A, B, X, Y \in \mathbb{R}_s^n$. As such, $\mathrm{gyr}[X,Y]$ is an *isometry* of $\mathbb{R}_s^n$, keeping the norm of elements of the ball $\mathbb{R}_s^n$ invariant,

$$\|\mathrm{gyr}[X,Y]Z\| = \|Z\| \tag{2.255}$$

Hence, $\mathrm{gyr}[X,Y]$ represents a rotation of the ball $\mathbb{R}_s^n$ about its origin for any $X, Y \in \mathbb{R}_s^n$.

The invertible self-map $\mathrm{gyr}[X,Y]$ of $\mathbb{R}_s^n$ respects Möbius addition in $\mathbb{R}_s^n$,

$$\mathrm{gyr}[X,Y](A \oplus B) = \mathrm{gyr}[X,Y]A \oplus \mathrm{gyr}[X,Y]B \tag{2.256}$$

for all $A, B, X, Y \in \mathbb{R}_s^n$, so that $\mathrm{gyr}[X,Y]$ is an automorphism of the Möbius groupoid $(\mathbb{R}_s^n, \oplus)$.

The gyroautomorphisms $\mathrm{gyr}[X,Y]$ regulate Möbius addition in the ball $\mathbb{R}_s^n$, giving rise to the following nonassociative algebraic laws that "repair" the breakdown of commutativity and associativity in Möbius addition:

$$\begin{aligned} X \oplus Y &= \mathrm{gyr}[X,Y](Y \oplus X) && \text{Gyrocommutative Law} \\ X \oplus (Y \oplus Z) &= (X \oplus Y) \oplus \mathrm{gyr}[X,Y]Z && \text{Left Gyroassociative Law} \\ (X \oplus Y) \oplus Z &= X \oplus (Y \oplus \mathrm{gyr}[Y,X]Z) && \text{Right Gyroassociative Law} \end{aligned} \tag{2.257}$$

for all $X, Y, Z \in \mathbb{R}_s^n$.

An important property of Thomas gyration in Möbius gyrovector spaces is the *loop property* (left and right),

$$\begin{aligned} \mathrm{gyr}[X \oplus Y, Y] &= \mathrm{gyr}[X,Y] && \text{Left Loop Property} \\ \mathrm{gyr}[X, Y \oplus X] &= \mathrm{gyr}[X,Y] && \text{Right Loop Property} \end{aligned} \tag{2.258}$$

for all $X, Y \in \mathbb{R}_s^n$.

The grouplike groupoid $(\mathbb{R}_s^n, \oplus)$ that regulates Möbius addition, $\oplus$, in the ball $\mathbb{R}_s^n$ of the Euclidean 3-space $\mathbb{R}^n$ is a *gyrocommutative gyrogroup* called a *Möbius gyrogroup*. Möbius gyrogroups and gyrovector spaces are studied in [2, 3, 33].

2.25 Möbius Gyrovector Spaces

Let $X \in \mathbb{R}_s^n$ be a point of a Möbius gyrocommutative gyrogroup $(\mathbb{R}^n, \oplus)$. Möbius addition of k copies of X, $k \geq 1$, denoted $k \otimes X$, gives

$$k \otimes X = \frac{\left(1 + \frac{\|X\|}{s}\right)^k - \left(1 - \frac{\|X\|}{s}\right)^k}{\left(1 + \frac{\|X\|}{s}\right)^k + \left(1 - \frac{\|X\|}{s}\right)^k} \frac{X}{\|X\|} \tag{2.259}$$

Identity (2.259) of scalar multiplication of any $X \in \mathbb{R}_s^n$ by any positive integer k is identical with its Einsteinian counterpart (2.79), suggesting the following definition of the Möbius scalar multiplication in the resulting Möbius gyrovector spaces.

Definition 2.47. (Möbius Scalar Multiplication, Möbius Gyrovector Spaces). *A Möbius gyrovector space* $(\mathbb{R}_s^n, \oplus, \otimes)$ *is a Möbius gyrogroup* $(\mathbb{R}_s^n, \oplus)$ *with scalar multiplication* $\otimes$ *given by the equation*

$$\begin{aligned} r \otimes X &= s \frac{\left(1 + \frac{\|X\|}{s}\right)^r - \left(1 - \frac{\|X\|}{s}\right)^r}{\left(1 + \frac{\|X\|}{s}\right)^r + \left(1 - \frac{\|X\|}{s}\right)^r} \frac{X}{\|X\|} \\ &= s \tanh\left(r \tanh^{-1} \frac{\|X\|}{s}\right) \frac{X}{\|X\|} \end{aligned} \tag{2.260}$$

where $r \in \mathbb{R}$, $X \in \mathbb{R}_s^n$, $X \neq \mathbf{0}$; *and* $r \otimes_M \mathbf{0} = \mathbf{0}$.

2.26 Möbius Points, Gyrolines and Gyrodistance

Möbius gyrolines are circular arcs in the ball $\mathbb{R}_s^n$ that approach the boundary of the ball orthogonally, as shown in Fig. 2.12 for $n = 2$.

The study of Möbius gyrovector spaces is similar to the study of Einstein gyrovector spaces. In the Cartesian model $\mathbb{R}_s^n$ of the n-dimensional Möbius gyrovector space $(\mathbb{R}_s^n, \oplus, \otimes)$, where n is any positive integer, we introduce a Cartesian coordinate system relative to which points of $\mathbb{R}_s^n$ are given by n-tuples of real numbers, like $X = (x_1, x_2, \ldots, x_n)$, $\|X\|^2 < s^2$, or $Y = (y_1, y_2, \ldots, y_n)$, $\|Y\|^2 < s^2$, etc. The point $\mathbf{0} = (0, 0, \ldots)$ (n zeros) is called the *origin* of $\mathbb{R}_s^n$. The Cartesian model $\mathbb{R}_s^n$ is a model of the n-dimensional hyperbolic geometry, as we will see in Sec. 2.27. It is a real inner product gyrovector space with addition and subtraction given by Möbius

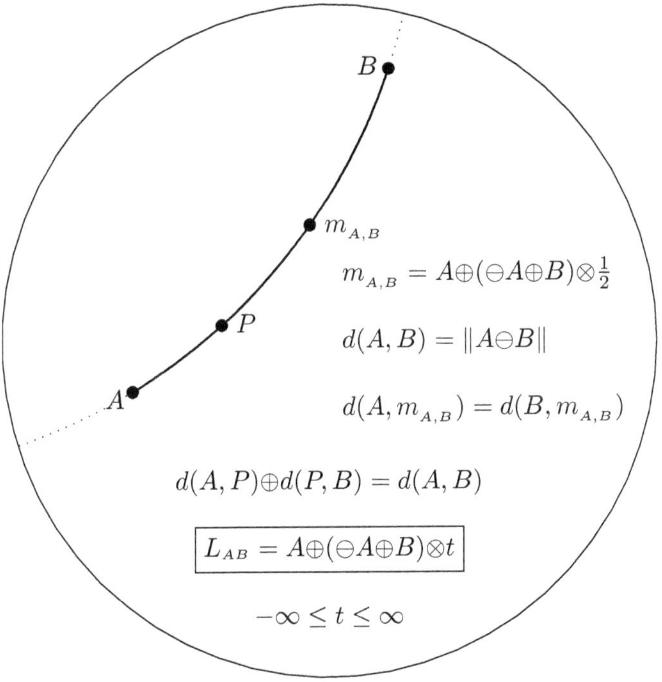

Fig. 2.12 The Möbius gyroline $L_{AB} = A\oplus(\ominus A\oplus B)\otimes t$, $t \in \mathbb{R}$, in a Möbius gyrovector space $(\mathbb{R}^n_s, \oplus, \otimes)$ is a geodesic line in the Poincaré ball model of hyperbolic geometry, fully analogous to the straight line $A + (-A + B)t$, $t \in \mathbb{R}$, in Euclidean geometry. The points A and B correspond to $t = 0$ and $t = 1$, respectively. The point P is a generic point on the gyroline through the points A and B lying between these points. The Möbius sum, $\oplus$, of the Möbius gyrodistance from A to P and from P to B equals the Möbius gyrodistance from A to B. The point $m_{A,B}$ that corresponds to $t = 1/2$ is the hyperbolic midpoint, gyromidpoint, of the points A and B.

addition (2.248) and its associated subtraction, with scalar multiplication given by (2.260), and with the inner product and norm that it inherits from its ambient Euclidean n-space $\mathbb{R}^n$.

In our Cartesian model $\mathbb{R}^n_s$ of the hyperbolic geometry of the n-dimensional Möbius gyrovector space $(\mathbb{R}^n_s, \oplus, \otimes)$, it is convenient to define a gyroline by the set of its points. Let $A, B\in\mathbb{R}^n_s$ be any two distinct points. The unique gyroline L_{AB} in $\mathbb{R}^n_s$ that passes through these points is the set of all points

$$L_{AB} = A\oplus(\ominus A\oplus B)\otimes t \tag{2.261}$$

for all $t\in\mathbb{R}$, that is, for all $-\infty < t < \infty$, shown in Fig. 2.12. Equation (2.261) is said to be the gyroline representation in terms of points A and

B. Obviously, the same gyroline can be represented by any two distinct points that lie on the gyroline, as demonstrated in [3].

The Möbius distance function, $d(X,Y)$ in a Möbius gyrovector space $(\mathbb{R}_s^n, \oplus, \otimes)$ is given by the equation

$$d(X,Y) = \|\ominus X \oplus Y\| \doteq \|X \ominus Y\| \tag{2.262}$$

$X, Y \in \mathbb{R}_s^n$. We call it a *gyrodistance function* in order to emphasize the analogies it shares with its Euclidean counterpart, the distance function $\|X - Y\|$ in $\mathbb{R}^n$. Among these analogies is the gyrotriangle inequality

$$\|X \oplus Y\| \leq \|X\| \oplus \|Y\| \tag{2.263}$$

for all $X, Y \in \mathbb{R}_s^n$. For this and other analogies that distance and gyrodistance functions share see [2, 3, 33].

A left gyrotranslation $T_X A$ of a point A by a point X in a Möbius gyrovector space $(\mathbb{R}_s^n, \oplus, \otimes)$, is given by

$$T_X A = X \oplus A \tag{2.264}$$

for all $X, A \in \mathbb{R}_s^n$. Left gyrotranslation composition is given by point addition preceded by a gyration. Indeed, by the left gyroassociative law, which is Axiom $(G3)$ of gyrogroups in Def. 2.2, p. 71, a left gyrotranslation composition can be written as

$$T_X T_Y A = X \oplus (Y \oplus A) = (X \oplus Y) \oplus \mathrm{gyr}[X,Y]A = T_{X \oplus Y} \mathrm{gyr}[X,Y]A \tag{2.265}$$

for all $X, Y, A \in \mathbb{R}_s^n$, thus obtaining the left gyrotranslation composition law

$$T_X T_Y = T_{X \oplus Y} \mathrm{gyr}[X,Y] \tag{2.266}$$

of left gyrotranslations of a Möbius gyrovector space $(\mathbb{R}_s^n, \oplus, \otimes)$. According to (2.266), a left gyrotranslation by Y followed by a left gyrotranslation by X is equivalent to a single left gyrotranslation by $X \oplus Y$ preceded by a gyration by $\mathrm{gyr}[X,Y]$, as in the case of Einstein gyrovector spaces in (2.116) – (2.117).

Owing to the presence of a gyration in the composition law (2.266), the set of all left gyrotranslations of $\mathbb{R}_s^n$ does not form a group under left gyrotranslation composition. Rather, under left gyrotranslation composition it forms a gyrocommutative gyrogroup.

As in the case of Einstein gyrovector spaces $\mathbb{R}_s^n$, each element of the special orthogonal group SO(n) of order n, that is, each element of the group of all $n \times n$ orthogonal matrices with determinant 1 represents a rotation R of points $A \in \mathbb{R}_s^n$ about the center of $\mathbb{R}_s^n$ in a Möbius gyrovector space $\mathbb{R}_s^n$, denoted RA. It is given by the matrix product RA^t of a matrix

$R \in SO(n)$ and the transpose A^t of $A \in \mathbb{R}_s^n$. A rotation of $\mathbb{R}_s^n$ is a linear map of $\mathbb{R}_s^n$ that keeps the inner product invariant. Hence, it leaves the origin of $\mathbb{R}^n$ invariant and respects Möbius addition, that is, $R(A \oplus B) = RA \oplus RB$ for all $A, B \in \mathbb{R}_s^n$.

Rotation composition is given by matrix multiplication, so that the set of all rotations of $\mathbb{R}_s^n$ about its origin forms a group under rotation composition.

Left gyrotranslations of $\mathbb{R}_s^n$ and rotations of $\mathbb{R}_s^n$ about its origin are *gyroisometries* of $\mathbb{R}_s^n$ in the sense that they keep the Möbius gyrodistance function (2.262) invariant. The set of all left gyrotranslations of $\mathbb{R}_s^n$ and all rotations of $\mathbb{R}_s^n$ about its origin forms a group under transformation composition, called the hyperbolic group of motions. In gyrogroup theory, this group of motions turns out to be the so called *gyrosemidirect product* of the gyrogroup of left gyrotranslations and the group of rotations [3].

Following Klein's 1872 *Erlangen Program* [12] [13, p. 253], the geometric objects of a geometry are the invariants of the group of motions of the geometry so that, conversely, objects that are invariant under the group of motions of a geometry possess geometric significance. Accordingly, for instance, the Möbius gyrodistance, (2.262), between two points of $\mathbb{R}_s^n$ is geometrically significant in hyperbolic geometry since it is invariant under the group of motions, left gyrotranslations and rotations, of the hyperbolic geometry of $\mathbb{R}_s^n$, as verified in [3].

2.27 Linking Möbius Addition to Hyperbolic Geometry

On the one hand, it is known that geodesics of the Poincaré ball model of hyperbolic geometry are Euclidean circular arcs in the ball that approach the boundary of the ball orthogonally [13] [42] and, on the other hand, Fig. 2.12 indicates that gyrolines in Möbius gyrovector planes are Euclidean circular arcs in the disc that approach the boundary of the disc orthogonally, as well. This coincidence is not accidental. We will find in this section that the Möbius gyrodistance (2.262) leads to the Riemannian line element, in differential geometry, of the Poincaré ball model of hyperbolic geometry.

In a two dimensional Möbius gyrovector space $(\mathbb{R}_s^2, \oplus, \otimes)$ the squared gyrodistance between a point $X \in \mathbb{R}_s^2$ and an infinitesimally nearby point $X + dX \in \mathbb{R}_s^2$, $dX = (dx_1, dx_2)$, is given by the equation [3, Sec. 7.3] [2, Sec. 7.3]

$$ds^2 = \|X \ominus (X + dX)\|^2 = E dx_1^2 + 2F dx_1 dx_2 + G dx_2^2 + \dots \qquad (2.267)$$

where, if we use the notation $r^2 = x_1^2 + x_2^2$, we have

$$
\begin{aligned}
E &= \frac{s^4}{(s^2 - r^2)^2} \\
F &= 0 \\
G &= \frac{s^4}{(s^2 - r^2)^2}
\end{aligned}
\tag{2.268}
$$

The triple $(g_{11}, g_{12}, g_{22}) = (E, F, G)$ along with $g_{21} = g_{12}$ is known in differential geometry as the metric tensor g_{ij} [43]. It turns out to be the metric tensor of the Poincaré disc model of hyperbolic geometry [42, p. 226]. Hence, ds^2 in (2.267) – (2.268) is the Riemannian line element of the Poincaré disc model of hyperbolic geometry, linked to Möbius addition (2.248) and to Möbius gyrodistance function (2.262) [44].

The Gaussian curvature K of a Möbius gyrovector plane, corresponding to the triple (E, F, G), turns out to be [42, p. 149] [3, Sec. 7.3] [2, Sec. 7.3]

$$
K = -\frac{4}{s^2}
\tag{2.269}
$$

The link between Möbius gyrovector spaces and the Poincaré ball model of hyperbolic geometry has thus been established in (2.267) – (2.268) in two dimensions. The extension of the link to higher dimensions is presented in [3, Sec. 7.3] [2, Sec. 7.3] and [44].

In full analogy with Euclidean geometry, the graph of the parametric expression (2.261) in a Möbius gyrovector space $(\mathbb{R}_s^n, \oplus, \otimes)$, for the parameter $t \in \mathbb{R}$, where $A, B \in \mathbb{R}_s^n$, describes a geodesic line in the Poincaré ball model of hyperbolic geometry. It is a circular arch in the ball that approaches the boundary of the ball orthogonally, as shown in Fig. 2.12 for the disc. The geodesic (2.261) is the unique geodesic passing through the points A and B. It passes through the point A at "time" $t = 0$ and, owing to the left cancellation law, (2.37), it passes through the point B at "time" $t = 1$. Hence, the geodesic segment that joins the points A and B in Fig. 2.12 is obtained from (2.261) with $0 \le t \le 1$.

The Möbius gyrovector space $\mathbb{R}_s^n$ regulates algebraically the Poincaré ball model of n-dimensional hyperbolic geometry, just as the vector space $\mathbb{R}^n$ regulates algebraically the standard model of n-dimensional Euclidean geometry. Euclidean geometry, regulated by the vector space $\mathbb{R}^n$, is equipped with Cartesian coordinates and, hence, it is known as the standard Cartesian model of Euclidean geometry. In full analogy, the Poincaré model of n-dimensional hyperbolic geometry is regulated by the Möbius gyrovector space $\mathbb{R}_s^n$ which is equipped with Cartesian coordinates. Hence, the Poincaré model of hyperbolic geometry that is regulated by a Möbius gyrovector space is called the *Cartesian-Poincaré model of hyperbolic geometry.*

2.28 Möbius Gyrovectors, Gyroangles and Gyrotriangles

The study of Möbius gyrovectors, gyroangles and gyrotriangles in Möbius gyrovector spaces is similar to the study in Sec. 2.10 of these geometric objects in Einstein gyrovector spaces.

Points of a Möbius gyrovector space $(\mathbb{R}_s^n, \oplus, \otimes)$, denoted by capital italic letters A, B, P, Q, etc., give rise to gyrovectors in $\mathbb{R}_s^n$, denoted by bold roman lowercase letters $\mathbf{u}, \mathbf{v}$, etc. Any two ordered points $A, B \in \mathbb{R}_s^n$ give rise to a unique rooted gyrovector $\mathbf{v} \in \mathbb{R}_s^n$, rooted at the point A. It has a tail at the point A and a head at the point B, and it has the value $\ominus A \oplus B$,

$$\mathbf{v} = \ominus A \oplus B \tag{2.270}$$

The gyrolength of the rooted gyrovector $\mathbf{v} = \ominus A \oplus B$ is the gyrodistance between its tail, A, and its head, B, given by the equation

$$\|\mathbf{v}\| = \|\ominus A \oplus B\| \tag{2.271}$$

Möbius gyrovectors in $\mathbb{R}_s^n$, $n = 2, 3$, are described graphically as directed gyroline gyrosegments with arrows, as shown in Fig. 2.13.

Two rooted gyrovectors $\ominus P \oplus Q$ and $\ominus R \oplus S$ are equivalent if they have the same value, $\ominus P \oplus Q = \ominus R \oplus S$, that is,

$$\ominus P \oplus Q \quad \sim \quad \ominus R \oplus S \qquad \text{if and only if} \qquad \ominus P \oplus Q = \ominus R \oplus S \tag{2.272}$$

The relation $\sim$ in (2.272) between rooted gyrovectors in Möbius gyrovector spaces is reflexive, symmetric and transitive. Hence, it is an equivalence relation that gives rise to equivalence classes of rooted gyrovectors. To liberate rooted gyrovectors from their roots we define a *gyrovector* to be an equivalence class of rooted gyrovectors. The gyrovector $\ominus P \oplus Q$ is thus a representative of all rooted gyrovectors with value $\ominus P \oplus Q$. Thus, for instance, the two distinct rooted gyrovectors $\ominus A \oplus B$ and $\ominus A' \oplus B'$ in Fig. 2.13 possess the same value so that, as gyrovectors in a Möbius gyrovector space, they are indistinguishable.

Vectors in Euclidean geometry are equivalence classes of ordered pairs of points that add according to the parallelogram law. In full analogy, gyrovectors in Möbius gyrovector spaces are equivalence classes of ordered

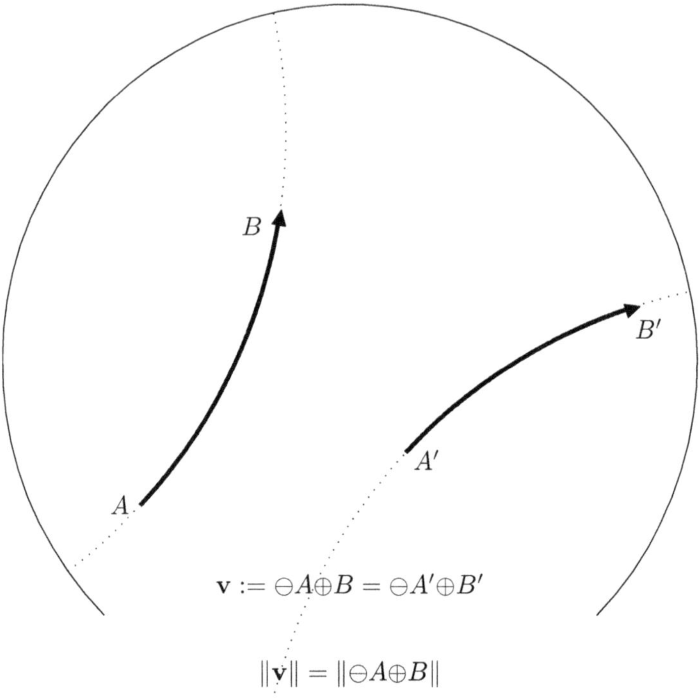

Fig. 2.13 The rooted gyrovectors $\ominus A \oplus B$ and $\ominus A' \oplus B'$ that are shown here in a Möbius gyrovector plane $(\mathbb{R}^2_s, \oplus, \otimes)$ have equal values, $\ominus A \oplus B = \ominus A' \oplus B'$. As such, these two gyrovectors are equivalent and, hence, indistinguishable in their gyrovector space and its underlying hyperbolic geometry. This figure, compared with Fig. 2.2, p. 98, and Fig. 1.1, p. 6, illustrates the comparative study of Euclidean and hyperbolic geometry that gyrogeometry offers.

pairs of points that add according to the gyroparallelogram law. This remarkable result about gyrovector addition, presented in [2, 3, 33], will not be used in this book.

A point $P \in \mathbb{R}^n_s$ is identified with the gyrovector $\ominus O \oplus P$, O being the arbitrarily selected origin of the space $\mathbb{R}^n_s$. Hence, the algebra of gyrovectors can be applied to the points of $\mathbb{R}^n_s$ as well. Accordingly, points of $\mathbb{R}^n_s$ are also known as *gyropoints*.

Let $\ominus A_1 \oplus A_2$ and $\ominus A_1 \oplus A_3$ be two rooted gyrovectors with a common tail A_1, Fig. 2.15. They include a gyroangle $\alpha_1 = \angle A_2 A_1 A_3 = \angle A_3 A_1 A_2$, the measure of which is given by the equation, Figs. 2.14 – 2.15,

$$\cos \alpha_1 = \frac{\ominus A_1 \oplus A_2}{\|\ominus A_1 \oplus A_2\|} \cdot \frac{\ominus A_1 \oplus A_3}{\|\ominus A_1 \oplus A_3\|} \tag{2.273}$$

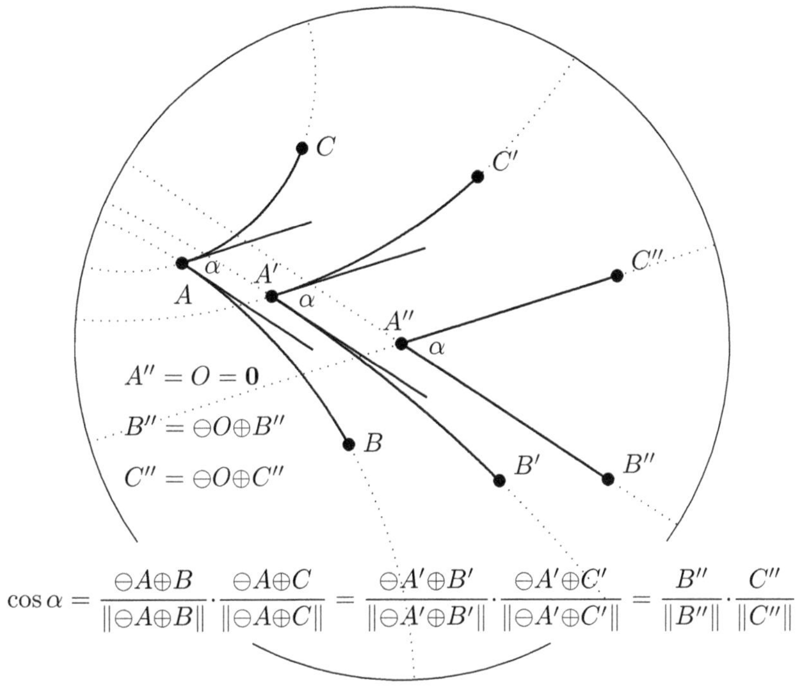

Fig. 2.14 A Möbius gyroangle α formed by two intersecting gyrolines in a Möbius gyrovector space $(\mathbb{R}_s^n, \oplus, \otimes)$. Its measure equals the measure of the Euclidean angle formed by corresponding intersecting tangent lines. As such, the Möbius gyroangle is *additive*, that is, if a Möbius gyroangle is split into two gyroangles then its measure equals the sum of the measures of these gyroangles. Gyroangles are invariant under left gyrotranslations, (2.275). Shown are two successive left gyrotranslations of a gyroangle $\alpha = \angle BAC$ into $\alpha = \angle B'A'C'$, and the latter into $\alpha = \angle B''A''C'' = \angle B''OC''$, so that $\cos\alpha = (B''/\|B''\|)\cdot(C''/\|C''\|)$, as in Euclidean geometry.

or, equivalently, by the equation

$$\alpha_1 = \cos^{-1} \frac{\ominus A_1 \oplus A_2}{\|\ominus A_1 \oplus A_2\|} \cdot \frac{\ominus A_1 \oplus A_3}{\|\ominus A_1 \oplus A_3\|} \tag{2.274}$$

where cos and $\cos^{-1} = \arccos$ are the standard cosine and arccosine functions of trigonometry. In the context of gyroangles, as in (2.273) – (2.274), we refer these functions of trigonometry to as the functions *gyrocosine* and *gyroarccosine* of gyrotrigonometry.

Thus, the elementary gyrotrigonometric functions $\sin\alpha$, $\cos\alpha$, $\tan\alpha$, etc., called respectively, gyrosine, gyrocosine, gyrotangent, etc., are identical with their counterparts in trigonometry with one exception:

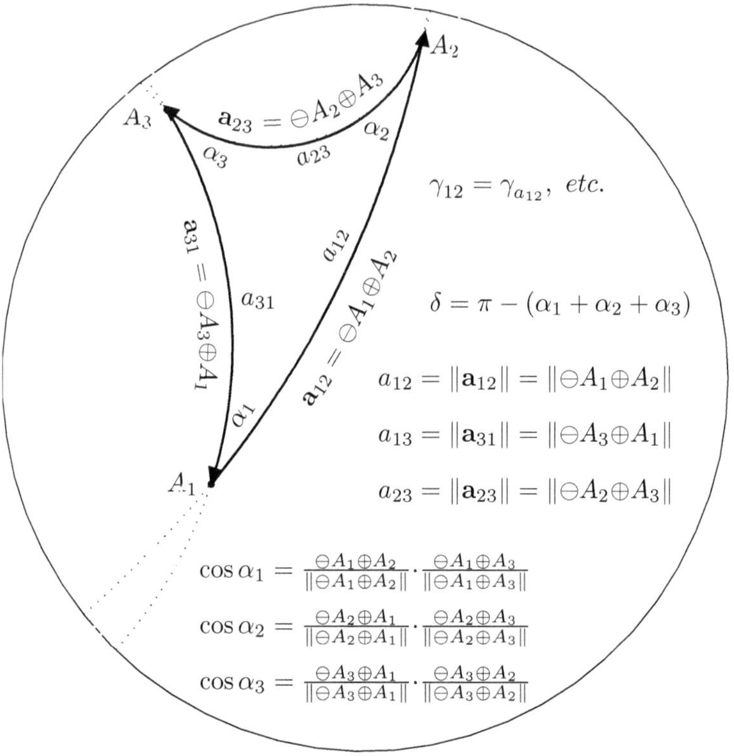

Fig. 2.15 A gyrotriangle $A_1A_2A_3$ in a Möbius gyrovector space $(\mathbb{R}^n_s, \oplus, \otimes)$ is presented for $n = 2$, along with its associated standard index notation. This figure, compared with Fig. 2.3, p. 101, and Fig. 1.2, p. 7, illustrates the comparative study of Euclidean and hyperbolic geometry that gyrogeometry offers.

gyrotrigonometric functions are applied to gyroangles while trigonometric functions are applied to angles.

The gyroangle α_1 is invariant under left gyrotranslations. Indeed,

$$\begin{aligned}
\cos \alpha_1' &= \frac{\ominus(X \oplus A_1) \oplus (X \oplus A_2)}{\|\ominus(X \oplus A_1) \oplus (X \oplus A_2)\|} \cdot \frac{\ominus(X \oplus A_1) \oplus (X \oplus A_3)}{\|\ominus(X \oplus A_1) \oplus (X \oplus A_3)\|} \\
&= \frac{\ominus A_1 \oplus A_2}{\|\ominus A_1 \oplus A_2\|} \cdot \frac{\ominus A_1 \oplus A_3}{\|\ominus A_1 \oplus A_3\|} \qquad (2.275) \\
&= \cos \alpha_1
\end{aligned}$$

for all $A_1, A_2, A_3, X \in \mathbb{R}^n_s$, as in (2.131) for Einstein gyrovector spaces.

Remarkably, both trigonometry and gyrotrigonometry share the same elementary trigonometric/gyrotrigonometric functions, $\sin\alpha$, $\cos\alpha$, $\tan\alpha$, etc. This result, established in [33, Chap. 4] and in [46, 47], will be further enhanced in this book in the observation that triangle centers in Euclidean geometry and their counterparts in the Beltrami-Klein and the Poincaré ball model of hyperbolic geometry may share the same trigonometric/gyrotrigonometric barycentric coordinates.

Similarly, as in (2.132) for Einstein gyrovector spaces, the gyroangle α_1 in any Möbius gyrovector space $(\mathbb{R}_s^n, \oplus, \otimes)$ is invariant under rotations of $\mathbb{R}_s^n$ about its origin. Indeed,

$$\begin{aligned}\cos\alpha_1'' &= \frac{\ominus RA_1 \oplus RA_2}{\|\ominus RA_1 \oplus RA_2\|} \cdot \frac{\ominus RA_1 \oplus RA_3}{\|\ominus RA_1 \oplus RA_3\|} \\ &= \frac{R(\ominus A_1 \oplus A_2)}{\|R(\ominus A_1 \oplus A_2)\|} \cdot \frac{R(\ominus A_1 \oplus A_3)}{\|R(\ominus A_1 \oplus A_3)\|} \\ &= \frac{\ominus A_1 \oplus A_2}{\|\ominus A_1 \oplus A_2\|} \cdot \frac{\ominus A_1 \oplus A_3}{\|\ominus A_1 \oplus A_3\|} \\ &= \cos\alpha_1\end{aligned} \tag{2.276}$$

for all $A_1, A_2, A_3 \in \mathbb{R}^n$ and $R \in \mathrm{SO(n)}$, since rotations $R \in \mathrm{SO(n)}$ preserve the inner product and the norm in $\mathbb{R}_s^n$, (2.83).

Being invariant under the motions of $\mathbb{R}_s^n$, which are left gyrotranslations and rotations, gyroangles are geometric objects of the hyperbolic geometry of Möbius gyrovector spaces. Gyrotriangle gyroangle sum in hyperbolic geometry is less than π. The standard notation that we use with a gyrotriangle $A_1A_2A_3$ in $\mathbb{R}_s^n$, $n \geq 2$, is presented in Fig. 2.15 for $n = 2$. In our notation, a Möbius gyrotriangle $A_1A_2A_3$, thus, has (i) three vertices, A_1, A_2 and A_3; (ii) three gyroangles, α_1, α_2 and α_3; and (iii) three gyrosides, which form the three gyrovectors $\mathbf{a}_{12}$, $\mathbf{a}_{23}$ and $\mathbf{a}_{31}$; with respective (iv) three gyroside-gyrolengths a_{12}, a_{23} and a_{31}, as shown in Fig. 2.15.

2.29 Gyrovector Space Isomorphism

Einstein and Möbius gyrovector spaces are isomorphic and, accordingly, they form the algebraic setting of two models of the same hyperbolic geometry, along with spectacular gain in clarity and simplicity. Isomorphisms between gyrovector spaces are studied in [3, Sec. 6.21]. In particular, it is shown there that an Einstein gyrovector space in the ball $\mathbb{R}_s^n$,

$\mathbb{R}_s^n = (\mathbb{R}_s^n, \oplus = \oplus_{\mathrm{E}}, \otimes = \otimes_{\mathrm{E}})$, and the Möbius gyrovector space in the same ball $\mathbb{R}_s^n$, $\mathbb{R}_s^n = (\mathbb{R}_s^n, \oplus = \oplus_{\mathrm{M}}, \otimes = \otimes_{\mathrm{M}})$, are isomorphic. The resulting isomorphism between Einstein and Möbius Gyrovector Spaces, $(\mathbb{R}_s^n, \oplus_{\mathrm{E}}, \otimes_{\mathrm{E}})$ and $(\mathbb{R}_s^n, \oplus_{\mathrm{M}}, \otimes_{\mathrm{M}})$, is given by the pair of identities

$$\begin{aligned} \tfrac{1}{2}\otimes_{\mathrm{E}}(X\oplus_{\mathrm{E}}Y) &= \tfrac{1}{2}\otimes_{\mathrm{M}}X\oplus_{\mathrm{M}}\tfrac{1}{2}\otimes_{\mathrm{M}}Y, && X,Y\in(\mathbb{R}_s^n.\oplus_{\mathrm{E}},\otimes_{\mathrm{E}}) \\ 2\otimes_{\mathrm{M}}(X\oplus_{\mathrm{M}}Y) &= 2\otimes_{\mathrm{E}}X\oplus_{\mathrm{E}}2\otimes_{\mathrm{E}}Y, && X,Y\in(\mathbb{R}_s^n.\oplus_{\mathrm{M}},\otimes_{\mathrm{M}}) \end{aligned} \tag{2.277}$$

for all $X, Y\in\mathbb{R}_s^n$.

The operations $\otimes_{\mathrm{E}}$ and $\otimes_{\mathrm{M}}$, which represent scalar gyromultiplication in Einstein and Möbius gyrovector spaces respectively, are identical to each other, $\otimes_{\mathrm{E}} = \otimes_{\mathrm{M}} =: \otimes$. Hence, Identities (2.277) can be written equivalently as

$$\begin{aligned} X\oplus_{\mathrm{E}}Y &= 2\otimes(\tfrac{1}{2}\otimes X\oplus_{\mathrm{M}}\tfrac{1}{2}\otimes Y), && X,Y\in(\mathbb{R}_s^n.\oplus_{\mathrm{E}},\otimes_{\mathrm{E}}) \\ X\oplus_{\mathrm{M}}Y &= \tfrac{1}{2}\otimes(2\otimes X\oplus_{\mathrm{E}}2\otimes Y), && X,Y\in(\mathbb{R}_s^n.\oplus_{\mathrm{M}},\otimes_{\mathrm{M}}) \end{aligned} \tag{2.278}$$

for all $X, Y\in\mathbb{R}_s^n$. The isomorphism in (2.278) is not trivial owing to the result that scalar gyromultiplication, $\otimes$, is non-distributive, that is, it does not distribute over gyrovector space addition, $\oplus$.

As examples of the use of the isomorphism in (2.277) – (2.278), and for later reference, let $A_e\in(\mathbb{R}_s^n, \oplus_{\mathrm{E}}, \otimes)$ and $A_m\in(\mathbb{R}_s^n, \oplus_{\mathrm{M}}, \otimes)$ be points of an Einstein and a Möbius gyrovector space that are isomorphic to each other under the isomorphism in (2.277) –(2.278). Then,

$$\begin{aligned} A_e &= 2\otimes A_m \\ A_m &= \tfrac{1}{2}\otimes A_e \end{aligned} \tag{2.279}$$

It follows from (2.279) that

$$\begin{aligned} \gamma_{A_e} &= \gamma_{2\otimes A_m} = 2\gamma_{A_m}^2 - 1 \\ \gamma_{A_e}A_e &= \gamma_{2\otimes A_m}(2\otimes A_m) = 2\gamma_{A_m}^2 A_m \end{aligned} \tag{2.280}$$

as shown in (2.89) and in (2.87), p. 85.

Furthermore, it follows from the first equation in (2.280) for $A_{i,e}, A_{j,e}\in(\mathbb{R}_s^n, \oplus_{\mathrm{E}}, \otimes)$, and their isomorphic image $A_{i,m}, A_{j,m}\in(\mathbb{R}_s^n, \oplus_{\mathrm{M}}, \otimes)$, that

$$\gamma_{ij,e} := \gamma_{\ominus_{\mathrm{E}}A_{i,e}\oplus_{\mathrm{E}}A_{j,e}} = 2\gamma^2_{\ominus_{\mathrm{M}}A_{i,m}\oplus_{\mathrm{M}}A_{j,m}} - 1 =: 2\gamma_{ij,m}^2 - 1 \tag{2.281}$$

as we see from the following chain of equations, in which equalities are numbered for subsequent derivation.

$$\begin{aligned}
\gamma_{12,e} &\overbrace{:=}^{(1)} \gamma_{\ominus_E A_{1,e} \oplus_E A_{2,e}} \\
&\overbrace{=}^{(2)} \gamma_{\ominus_E (2\otimes_E A_{1,m}) \oplus_E (2\otimes_E A_{2,m})} \\
&\overbrace{=}^{(3)} \gamma_{2\otimes_M (\ominus_M A_{1,m} \oplus_M A_{2,m})} \\
&\overbrace{=}^{(4)} 2\gamma^2_{\ominus_M A_{1,m} \oplus_M A_{2,m}} - 1 \\
&\overbrace{=:}^{(5)} 2\gamma^2_{12,m} - 1
\end{aligned} \tag{2.282}$$

Derivation of the numbered equalities in (2.282) follows.

(1) This is our standard index notation into which the subscript "e" is introduced to emphasize that the equality in (1) is considered in an Einstein gyrovector space.
(2) Follows from (1) by the gyrogroup isomorphism (2.279).
(3) Follows from (2) and the isomorphism between $\oplus_E$ and $\oplus_M$ in (2.278).
(4) Follows from (3) and the identity $\gamma_{2\otimes a} = 2\gamma_a^2 - 1$ in (2.89), p. 85.
(5) As in (1) above, this is our standard index notation into which the subscript "m" is introduced to emphasize that the equation under (5) is considered in a Möbius gyrovector space.

As an example illustrating the transformation in (2.281) of an expression from an Einstein gyrovector space to a corresponding Möbius gyrovector space, and for later reference, we note that following (2.281) we have

$$\sqrt{\gamma^2_{ij,e} - 1} = 2\gamma_{ij,m}\sqrt{\gamma^2_{ij,m} - 1} \tag{2.283}$$

In each of Einstein and Möbius gyrovector spaces, $(\mathbb{R}^n_s, \oplus_E, \otimes)$ and $(\mathbb{R}^n_s, \oplus_M, \otimes)$, the zero element is the origin,

$$O_e = O_m = \mathbf{0} \tag{2.284}$$

and the inverses are related by (2.279), that is

$$\begin{aligned}
\ominus_E A_e &= 2\otimes(\ominus_M A_m) = \ominus_M 2\otimes A_m \\
\ominus_M A_m &= \tfrac{1}{2}\otimes(\ominus_E A_e) = \ominus_E \tfrac{1}{2}\otimes A_e
\end{aligned} \tag{2.285}$$

Theorem 2.48. *The isomorphism* (2.279) *between Einstein and Möbius gyrovector spaces,* $(\mathbb{R}^n_s, \oplus_E, \otimes)$ *and* $(\mathbb{R}^n_s, \oplus_M, \otimes)$, *preserves the value of unit gyrovectors and the measure of gyroangles.*

Proof. Let A_e, B_e, C_e, be three distinct points of an Einstein gyrovector space $(\mathbb{R}_s^n, \oplus_E, \otimes)$, and let A_m, B_m, C_m, be their image under the isomorphism (2.279) in the corresponding Möbius gyrovector space $(\mathbb{R}_s^n, \oplus_M, \otimes)$.

The points A_e, B_e determine the unit gyrovector

$$\mathbf{v}_e = \frac{\ominus_E A_e \oplus_E B_e}{\|\ominus_E A_e \oplus_E B_e\|} \tag{2.286}$$

in the Einstein gyrovector space, the image of which in the corresponding Möbius gyrovector space is

$$\mathbf{v}_m = \frac{\ominus_M A_m \oplus_M B_m}{\|\ominus_M A_m \oplus_M B_m\|} \tag{2.287}$$

Then $\mathbf{v}_m = \mathbf{v}_e$, as we see from the following chain of equations:

$$\begin{aligned}
\mathbf{v}_m &\overset{(1)}{=\!=} \frac{\ominus_M A_m \oplus_M B_m}{\|\ominus_M A_m \oplus_M B_m\|} \\
&\overset{(2)}{=\!=} \frac{\tfrac{1}{2}\otimes(\ominus_E A_e) \oplus_M \tfrac{1}{2}\otimes B_e}{\|\tfrac{1}{2}\otimes(\ominus_E A_e) \oplus_M \tfrac{1}{2}\otimes B_e\|} \\
&\overset{(3)}{=\!=} \frac{\tfrac{1}{2}\otimes\{\ominus_E A_e \oplus_E B_e\}}{\|\tfrac{1}{2}\otimes\{\ominus_E A_e \oplus_E B_e\}\|} \\
&\overset{(4)}{=\!=} \frac{\ominus_E A_e \oplus_E B_e}{\|\ominus_E A_e \oplus_E B_e\|} \\
&\overset{(5)}{=\!=} \mathbf{v}_e
\end{aligned} \tag{2.288}$$

Derivation of the numbered equalities in (2.288) follows.

(1) Follows from the definition of $\mathbf{v}_m$ in (2.287).
(2) Follows from (1) by (2.279) and (2.285).
(3) Follows from (2) by (2.278).
(4) Follows from (3) by the scaling property $(V5)$ of gyrovector spaces in Def. 2.14, p. 85.
(5) Follows from (4) by the definition of $\mathbf{v}_e$ in (2.286).

Let

$$\alpha_e = \angle B_e A_e C_e \tag{2.289}$$

be the gyroangle included between the two gyrovectors $\ominus_{\mathrm{E}} A_e \oplus_{\mathrm{E}} B_e$ and $\ominus_{\mathrm{E}} A_e \oplus_{\mathrm{E}} C_e$ in the Einstein gyrovector space, and let

$$\alpha_m = \angle B_m A_m C_m \tag{2.290}$$

be the corresponding gyroangle included between the two corresponding gyrovectors $\ominus_{\mathrm{M}} A_m \oplus_{\mathrm{M}} B_m$ and $\ominus_{\mathrm{M}} A_m \oplus_{\mathrm{M}} C_m$ in the Möbius gyrovector space. Then

$$\alpha_e = \alpha_m \tag{2.291}$$

or, equivalently, $\cos\alpha_e = \cos\alpha_m$, as shown in the following chain of equations, which are numbered for subsequent derivation:

$$\begin{aligned}
\cos\alpha_m &\overbrace{=\!=\!=}^{(1)} \frac{\ominus_{\mathrm{M}} A_m \oplus_{\mathrm{M}} B_m}{\|\ominus_{\mathrm{M}} A_m \oplus_{\mathrm{M}} B_m\|} \cdot \frac{\ominus_{\mathrm{M}} A_m \oplus_{\mathrm{M}} C_m}{\|\ominus_{\mathrm{M}} A_m \oplus_{\mathrm{M}} C_m\|} \\
&\overbrace{=\!=\!=}^{(2)} \frac{\ominus_{\mathrm{E}} A_e \oplus_{\mathrm{E}} B_e}{\|\ominus_{\mathrm{E}} A_e \oplus_{\mathrm{E}} B_e\|} \cdot \frac{\ominus_{\mathrm{E}} A_e \oplus_{\mathrm{E}} C_e}{\|\ominus_{\mathrm{E}} A_e \oplus_{\mathrm{E}} C_e\|} \\
&\overbrace{=\!=\!=}^{(3)} \cos\alpha_e
\end{aligned} \tag{2.292}$$

Derivation of the numbered equalities in (2.292) follows.

(1) Follows from the gyroangle definition in Möbius gyrovector spaces.
(2) Follows from (1) by (2.288)
(3) Follows from the gyroangle definition in Einstein gyrovector spaces.

□

The isomorphism between Einstein and Möbius gyrovector spaces is an isomorphism between two distinct models of hyperbolic geometry. This isomorphism is useful, enabling us to perform calculations in one model and transform the results into the other model. Indeed, we will find in this book that it is simpler to determine gyrotriangle gyrocenters in Einstein gyrovector spaces since we can employ methods of linear algebra to calculate points of intersection of straight lines. In contrast, it is instructive to visualize the results in Möbius gyrovector spaces since their associated Poincaré ball model of hyperbolic geometry is conformal to Euclidean geometry. Hence, rather than determining gyrotriangle gyrocenters in the Poincaré ball model directly, we will first determine in this book gyrotriangle gyrocenters in the Cartesian-Beltrami-Klein ball model and, then, transform the results into their isomorphic image in the Cartesian-Poincaré ball model.

2.30 Möbius Gyrotrigonometry

Möbius gyrotrigonometry is presented in detail in [3]. In this book we do not employ gyrotrigonometry in Möbius gyrovector spaces directly. Rather, we employ directly gyrotrigonometry in Einstein gyrovector spaces and transform gyrotrigonometric results from Einstein to Möbius gyrovector spaces. The transformation of gyroangle measures from Einstein to Möbius gyrovector spaces is trivial, as demonstrated in (2.292).

IIowever, in order to demonstrate the elegance of gyrotrigonometry in Möbius gyrovector spaces we present in this section the AAA to SSS conversion law and its converse, the SSS to AAA conversion law, in the following two theorems:

Theorem 2.49. (The AAA to SSS Conversion Law, Möbius). *Let $A_1A_2A_3$ be a gyrotriangle in a Möbius gyrovector space $(\mathbb{R}^n_s, \oplus, \otimes)$. Then, in the standard gyrotriangle index notation in Fig. 2.15, p. 140,*

$$\begin{aligned}
\frac{a_{23}^2}{s^2} &= \frac{\cos\alpha_1 + \cos(\alpha_2+\alpha_3)}{\cos\alpha_1 + \cos(\alpha_2-\alpha_3)} \\
\frac{a_{13}^2}{s^2} &= \frac{\cos\alpha_2 + \cos(\alpha_1+\alpha_3)}{\cos\alpha_2 + \cos(\alpha_1-\alpha_3)} \\
\frac{a_{12}^2}{s^2} &= \frac{\cos\alpha_3 + \cos(\alpha_1+\alpha_2)}{\cos\alpha_3 + \cos(\alpha_1-\alpha_2)}
\end{aligned} \tag{2.293}$$

Theorem 2.50. (The SSS to AAA Conversion Law, Möbius). *Let $A_1A_2A_3$ be a gyrotriangle in a Möbius gyrovector space $(\mathbb{R}^n_s, \oplus, \otimes)$. Then, in the standard gyrotriangle index notation in Fig. 2.15, p. 140,*

$$\begin{aligned}
\cos\alpha_1 &= \frac{-a_{23}^2 + a_{13}^2 + a_{12}^2 - a_{23}^2a_{13}^2a_{12}^2/s^4}{2a_{13}a_{12}}\gamma_{23}^2 \\
\cos\alpha_2 &= \frac{a_{23}^2 - a_{13}^2 + a_{12}^2 - a_{23}^2a_{13}^2a_{12}^2/s^4}{2a_{23}a_{12}}\gamma_{13}^2 \\
\cos\alpha_3 &= \frac{a_{23}^2 + a_{13}^2 - a_{12}^2 - a_{23}^2a_{13}^2a_{12}^2/s^4}{2a_{23}a_{13}}\gamma_{12}^2
\end{aligned} \tag{2.294}$$

The proof of Theorems 2.49 and 2.50 is found in [3, Sec. 8.10].

In the Euclidean limit, $s \to \infty$, the identities of Theorem 2.49 descend to

$$\begin{aligned}
0 &= \cos\alpha_1 + \cos(\alpha_2+\alpha_3) \\
0 &= \cos\alpha_2 + \cos(\alpha_1+\alpha_3) \\
0 &= \cos\alpha_3 + \cos(\alpha_1+\alpha_2)
\end{aligned} \tag{2.295}$$

The first equation in (2.295) can be written as

$$\cos\alpha_1 = -\cos(\alpha_2+\alpha_3) = \cos(\pi-\alpha_2-\alpha_3) \tag{2.296}$$

implying the Euclidean triangle condition

$$\alpha_1+\alpha_2+\alpha_3 = \pi \tag{2.297}$$

Similarly, the other two equations in (2.295) are equivalent to (2.297) as well.

In the Euclidean limit, $s\to\infty$, the identities of Theorem 2.50 reduce to

$$\begin{aligned} a_{23}^2 &= a_{13}^2 + a_{12}^2 - 2a_{13}a_{12}\cos\alpha_1 \\ a_{13}^2 &= a_{23}^2 + a_{12}^2 - 2a_{23}a_{12}\cos\alpha_2 \\ a_{12}^2 &= a_{23}^2 + a_{13}^2 - 2a_{23}a_{13}\cos\alpha_3 \end{aligned} \tag{2.298}$$

Each of the equations in (2.298) represents the law of cosines in Euclidean geometry.

It follows from Theorem 2.50 that, as in Euclidean geometry, the gyrotriangle gyroside gyrolengths determine the gyrotriangle gyroangles uniquely. Similarly, it follows from Theorem 2.49 that, unlike Euclidean geometry, the gyrotriangle gyroangles determine the gyrotriangle gyroside gyrolengths uniquely.

2.31 Exercises

(1) Verify Identity (2.82), p. 84.

(2) Show that (2.143), p. 103, forms the unique solution of (2.142) for the unknowns $\cos\alpha$, $\cos\beta$ and $\cos\gamma$.

(3) Show that (2.143) – (2.144) imply (2.145), p. 104.

(4) Show that (2.158), p. 107, follows from (2.143).

(5) Prove the identities in (2.280) –(2.283), p. 142.

(6) Let, in the notation of Fig. 2.4, p. 102, $A_eB_eC_e$ be a gyrotriangle in an Einstein gyrovector space $(\mathbb{R}_s^n, \oplus_{\rm E}, \otimes)$, and let $\gamma_{a,e}, \gamma_{b,e}$ and $\gamma_{c,e}$ be the gamma factors of its gyrosides. Then, by (2.146), p. 104, we have the inequality

$$1 + 2\gamma_{a,e}\gamma_{b,e}\gamma_{c,e} - \gamma_{a,e}^2 - \gamma_{b,e}^2 - \gamma_{c,e}^2 > 0 \tag{2.299}$$

Now, in a similar gyrotriangle notation, let $A_mB_mC_m$ be a gyrotriangle in a Möbius gyrovector space $(\mathbb{R}_s^n, \oplus_{\rm M}, \otimes)$, with vertices A_m, B_m and C_m isomorphic, respectively, to the vertices A_e, B_e and C_e by isomorphism (2.279), p. 142. Furthermore, let $\gamma_{a,m}, \gamma_{b,m}$ and $\gamma_{c,m}$ be the gamma factors of its gyrosides.

Show that inequality (2.299) in Einstein gyrovector spaces is valid in Möbius gyrovector spaces as well, that is, show that

$$1 + 2\gamma_{a,m}\gamma_{b,m}\gamma_{c,m} - \gamma_{a,m}^2 - \gamma_{b,m}^2 - \gamma_{c,m}^2 > 0 \tag{2.300}$$

Solution: Let us transform the expression

$$E_e := 1 + 2\gamma_{a,e}\gamma_{b,e}\gamma_{c,e} - \gamma_{a,e}^2 - \gamma_{b,e}^2 - \gamma_{c,e}^2 \tag{2.301}$$

in (2.299) from an Einstein gyrovector space $(\mathbb{R}_s^n, \oplus_{\rm E}, \otimes)$ into a corresponding Möbius gyrovector space $(\mathbb{R}_s^n, \oplus_{\rm M}, \otimes)$ by means of the transformation formula (2.281), p. 142, that is,

$$\gamma_{a,e} = 2\gamma_{a,m}^2 - 1 \tag{2.302}$$

Then the expression E_e becomes E_m given by the equation

$$\begin{aligned} E_m &= (1 + 2\gamma_{a,m}\gamma_{b,m}\gamma_{c,m} - \gamma_{a,m}^2 - \gamma_{b,m}^2 - \gamma_{c,m}^2) \\ &\quad \times (-1 + 2\gamma_{a,m}\gamma_{b,m}\gamma_{c,m} + \gamma_{a,m}^2 + \gamma_{b,m}^2 + \gamma_{c,m}^2) \end{aligned} \tag{2.303}$$

The expression E_e is positive owing to its gyrotrigonometric interpretation in (2.145), p. 104. Since gyroangles are preserved by gyrovector space isomorphisms, the positivity of E_e implies that E_m is positive as well. The second factor of E_m in (2.303) is clearly positive since gamma factors are greater than 1. Hence, the first factor of E_m in (2.303) is positive, as desired.

Chapter 3

The Interplay of Einstein Addition and Vector Addition

The linearity of vector addition in Euclidean n-dimensional vector spaces $\mathbb{R}^n$ plays a crucially important role in the development and the use of Euclidean barycentric coordinates in Sec. 1.3. Contrasting vector addition, $+$, in vector spaces $\mathbb{R}^n$, Einstein addition, $\oplus$, in Einstein gyrovector spaces $\mathbb{R}^n_s$ is nonlinear. However, in order to adapt barycentric coordinates for use in Einstein gyrovector spaces $\mathbb{R}^n_s$, some linearity considerations are needed. These, indeed, are provided by the interplay between Einstein addition $\oplus$ and the common vector addition $+$.

Fortunately, there is an important interplay of $\oplus$ and $+$ that comes to the rescue. In order to uncover the interplay, we extend the n-dimensional Einstein addition in $\mathbb{R}^n_s$ to the $n+1$-dimensional *boost*, which turns out to be linear. This linearity will prove useful in the study of the interplay of $\oplus$ and $+$ that we need for the introduction of hyperbolic barycentric coordinates, which we naturally call gyrobarycentric coordinates.

Accordingly, this chapter is devoted to the study of the interplay of $\oplus$ and $+$, which will be used in the following chapter for the introduction of gyrobarycentric coordinates.

3.1 Extension of $\mathbb{R}^n_s$ into $\mathbb{T}^{n+1}_s$

In order to reveal the interplay of Einstein addition, $\oplus$, in $\mathbb{R}^n_s$ and the common vector addition, $+$, in $\mathbb{R}^n$ we extend the n-dimensional ball $\mathbb{R}^n_s$ to an $(n+1)$-dimensional object. Accordingly, let $\mathbb{T}^{n+1}_s$ be the set of all pairs $(t, tA)^t$, where $t{\in}\mathbb{R}^+$, $\mathbb{R}^+$ being the set of all positive numbers, where $A{\in}(\mathbb{R}^n_s, \oplus)$ is a point of the n-dimensional Einstein gyrogroup, and where exponent t denotes transposition. The condition $t > 0$ for $t{\in}\mathbb{R}$ will later be relaxed into the weaker condition $t \neq 0$.

The representation of elements $(t, tA)^t$ of $\mathbb{T}_s^{n+1}$ in terms of t and A is unique in the sense that

$$\begin{pmatrix} p \\ pA \end{pmatrix} = \begin{pmatrix} q \\ qB \end{pmatrix} \tag{3.1}$$

if and only if

$$\begin{aligned} p &= q \\ A &= B \end{aligned} \tag{3.2}$$

It follows from (3.1) – (3.2) that the pairs $(t, tA)^t$ are in one-to-one correspondence with the pairs $(t, A)^t$. The reason for using the former rather than the latter lies in the resulting Theorem 3.1 below, which necessarily involves the former rather than the latter.

Scalar multiplication in $\mathbb{T}_s^{n+1}$ is defined by the equation

$$m \begin{pmatrix} t \\ tA \end{pmatrix} = \begin{pmatrix} mt \\ mtA \end{pmatrix} \tag{3.3}$$

for all $m \in \mathbb{R}^+$ and $(t, tA)^t \in \mathbb{T}_s^{n+1}$.

Addition in $\mathbb{T}_s^{n+1}$ is defined by the equation

$$\begin{aligned} m_1 \begin{pmatrix} t_1 \\ t_1 A_1 \end{pmatrix} + m_2 \begin{pmatrix} t_2 \\ t_2 A_2 \end{pmatrix} &= \begin{pmatrix} m_1 t_1 + m_2 t_2 \\ m_1 t_1 A_1 + m_2 t_2 A_2 \end{pmatrix} \\ &= \begin{pmatrix} m_1 t_1 + m_2 t_2 \\ (m_1 t_1 + m_2 t_2) \frac{m_1 A_1 t_1 + m_2 A_2 t_2}{m_1 t_1 + m_2 t_2} \end{pmatrix} \\ &=: \begin{pmatrix} t_{12} \\ t_{12} A_{12} \end{pmatrix} \end{aligned} \tag{3.4}$$

for all $m_1, m_2 \in \mathbb{R}^+$ and $(t_1, A_1 t_1)^t, (t_2, A_2 t_2)^t \in \mathbb{T}_s^{n+1}$. Clearly, as anticipated in (3.4),

$$t_{12} = m_1 t_1 + m_2 t_2 \in \mathbb{R}^+ \tag{3.5}$$

and, since the ball $\mathbb{R}_s^n$ is a convex subset of its space $\mathbb{R}^n$,

$$A_{12} = \frac{m_1 A_1 t_1 + m_2 A_2 t_2}{m_1 t_1 + m_2 t_2} \in \mathbb{R}_s^n \tag{3.6}$$

Thus, addition, (3.4), in $\mathbb{T}_s^{n+1}$ is the common vector addition, for which we use the common sigma-notation as, for instance, in the equation

$$m_1 \begin{pmatrix} t_1 \\ t_1 A_1 \end{pmatrix} + m_2 \begin{pmatrix} t_2 \\ t_2 A_2 \end{pmatrix} = \sum_{k=1}^{2} m_k \begin{pmatrix} t_k \\ t_k A_k \end{pmatrix} \tag{3.7}$$

For any point $A \in \mathbb{R}_s^n$ of the Einstein gyrogroup $\mathbb{R}_s^n = (\mathbb{R}_s^n, \oplus)$, the *boost* $L(A)$ is a map of $\mathbb{T}_s^{n+1}$ into itself, given by the equation

$$L(A) \begin{pmatrix} t \\ tB \end{pmatrix} = \begin{pmatrix} \frac{\gamma_{A\oplus B}}{\gamma_B} t \\ \frac{\gamma_{A\oplus B}}{\gamma_B} t(A \oplus B) \end{pmatrix} \tag{3.8}$$

The application of $L(A)$ in (3.8) is expressed in terms of Einstein addition, $\oplus$. To express it directly, without Einstein addition, we note that Einstein addition (2.4) is given by

$$A \oplus B = \frac{1}{1 + \frac{A \cdot B}{s^2}} \left\{ A + \frac{1}{\gamma_A} B + \frac{1}{s^2} \frac{\gamma_A}{1 + \gamma_A} (A \cdot B) A \right\} \tag{3.9}$$

$A, B \in \mathbb{R}_s^n$, where γ_A is the gamma factor, (2.5),

$$\gamma_A = \frac{1}{\sqrt{1 - \frac{A^2}{s^2}}} \tag{3.10}$$

in $\mathbb{R}_s^n$, where we use the notation $A^2 = A \cdot A = \|A\|^2$. Clearly, $\gamma_A > 1$ for all $A \in \mathbb{R}_s^n$.

Einstein addition and the gamma factor are related by the *gamma identity*, (2.9),

$$\gamma_{A\oplus B} = \gamma_A \gamma_B \left(1 + \frac{A \cdot B}{s^2}\right) \tag{3.11}$$

that can be written, equivalently, as

$$\gamma_{\ominus A\oplus B} = \gamma_A \gamma_B \left(1 - \frac{A \cdot B}{s^2}\right) \tag{3.12}$$

for all $A, B \in \mathbb{R}_s^n$. Here, (3.12) is obtained from (3.11) by replacing A by $\ominus A = -A$ in (3.11).

An important identity that follows immediately from (3.10) is

$$\frac{A^2}{s^2} = \frac{\gamma_A^2 - 1}{\gamma_A^2} \tag{3.13}$$

and, similarly, an important identity that follows immediately from (3.12) is

$$\frac{A\cdot B}{s^2} = 1 - \frac{\gamma_{\ominus A\oplus B}}{\gamma_A \gamma_B} \tag{3.14}$$

It follows from (3.9) – (3.11) that

$$(A\oplus B)\frac{\gamma_{A\oplus B}}{\gamma_B}t = \gamma_A At + Bt + \frac{1}{s^2}\frac{\gamma_A^2}{1+\gamma_A}(A\cdot Bt)A \tag{3.15}$$

Substituting (3.11) and (3.15) into (3.8) yields

$$L(A)\begin{pmatrix} t \\ tB \end{pmatrix} = \begin{pmatrix} \gamma_A (1+\frac{A\cdot B}{s^2})t \\ \gamma_A At + Bt + \frac{1}{s^2}\frac{\gamma_A^2}{1+\gamma_A}(A\cdot Bt)A \end{pmatrix} \tag{3.16}$$

The boost $L(A)$ in (3.16) turns out to be linear, as stated in the following theorem:

Theorem 3.1. *Boosts $L(A)$, $A\in\mathbb{R}_s^n$, of $\mathbb{T}_s^{n+1}$ are linear, that is, for any positive integer N,*

$$L(A)\sum_{k=1}^{N} m_k \begin{pmatrix} t_k \\ t_k B_k \end{pmatrix} = \sum_{k=1}^{N} m_k L(A)\begin{pmatrix} t_k \\ t_k B_k \end{pmatrix} \tag{3.17}$$

for $m_k\in\mathbb{R}^+$, and $(t_k, B_k t_k)^t\in\mathbb{T}_s^{n+1}$, $k=1,\ldots,N$.

Proof. To prove the theorem it is enough to prove the following two properties, called homogeneity and additivity, of the boost:

$$L(A)\left\{m\begin{pmatrix} t \\ tB \end{pmatrix}\right\} = mL(A)\begin{pmatrix} t \\ tB \end{pmatrix} \tag{3.18}$$

and

$$L(A)\left\{\begin{pmatrix} t_1 \\ t_1B_1 \end{pmatrix} + \begin{pmatrix} t_2 \\ t_2B_2 \end{pmatrix}\right\} = L(A)\begin{pmatrix} t_1 \\ t_1B_1 \end{pmatrix} + L(A)\begin{pmatrix} t_2 \\ t_2B_2 \end{pmatrix} \tag{3.19}$$

for all $m\in\mathbb{R}^+$, $A\in\mathbb{R}_s^n$, and $(t, tB)^t, (t_k, t_kB_k)^t\in\mathbb{T}_s^{n+1}$, $k=1,2$.

Following (3.3) and (3.16) we have

$$\begin{aligned} L(A)\left\{m\begin{pmatrix} t \\ tB \end{pmatrix}\right\} &= L(A)\begin{pmatrix} mt \\ mtB \end{pmatrix} \\ &= \begin{pmatrix} \gamma_A (1+\frac{A\cdot B}{s^2})mt \\ \gamma_A mtA + mtB + \frac{1}{s^2}\frac{\gamma_A^2}{1+\gamma_A}(A\cdot mtB)A \end{pmatrix} \\ &= m\begin{pmatrix} \gamma_A (1+\frac{A\cdot B}{s^2})t \\ \gamma_A tA + tB + \frac{1}{s^2}\frac{\gamma_A^2}{1+\gamma_A}(A\cdot tB)A \end{pmatrix} = mL(A)\begin{pmatrix} t \\ tB \end{pmatrix} \end{aligned} \tag{3.20}$$

for all $m\in\mathbb{R}^+$, $A\in\mathbb{R}_s^n$, and $(t, tB)^t\in\mathbb{T}_s^{n+1}$, thus verifying (3.18).

Following (3.4) and (3.16) we have

$$
\begin{aligned}
L(A)\left\{\begin{pmatrix} t_1 \\ t_1B_1 \end{pmatrix} + \begin{pmatrix} t_2 \\ t_2B_2 \end{pmatrix}\right\} &= L(A)\begin{pmatrix} t_1+t_2 \\ (t_1+t_2)\frac{B_1t_1+B_2t_2}{t_1+t_2} \end{pmatrix} \\
&= \begin{pmatrix} \gamma_{A}\,(1+\frac{1}{s^2}A\cdot\frac{B_1t_1+B_2t_2}{t_1+t_2})(t_1+t_2) \\ \gamma_{A}\,(t_1+t_2)A+(t_1+t_2)\frac{B_1t_1+B_2t_2}{t_1+t_2}+\frac{1}{s^2}\frac{\gamma_A^2}{1+\gamma_{A}}(A\cdot(t_1+t_2)\frac{B_1t_1+B_2t_2}{t_1+t_2})A \end{pmatrix} \\
&= \begin{pmatrix} \gamma_{A}\,(t_1+t_2+\frac{1}{s^2}A\cdot(t_1B_1+t_2B_2)) \\ \gamma_{A}\,(t_1+t_2)A+t_1B_1+t_2B_2+\frac{1}{s^2}\frac{\gamma_A^2}{1+\gamma_{A}}A\cdot(t_1B_1+t_2B_2)A \end{pmatrix} \\
&= \begin{pmatrix} \gamma_{A}\,(1+\frac{A\cdot B_1}{s^2})t_1 \\ \gamma_{A}\,t_1A+t_1B_1+\frac{1}{s^2}\frac{\gamma_A^2}{1+\gamma_{A}}(A\cdot t_1B_1)A \end{pmatrix} \\
&+ \begin{pmatrix} \gamma_{A}\,(1+\frac{A\cdot B_2}{s^2})t_2 \\ \gamma_{A}\,t_2A+t_2B_2+\frac{1}{s^2}\frac{\gamma_A^2}{1+\gamma_{A}}(A\cdot t_2B_2)A \end{pmatrix} \\
&= L(A)\begin{pmatrix} t_1 \\ t_1B_1 \end{pmatrix} + L(A)\begin{pmatrix} t_2 \\ t_2B_2 \end{pmatrix}
\end{aligned}
\tag{3.21}
$$

for all $A\in\mathbb{R}_s^n$, $(t_k, t_kB_k)^t\in\mathbb{T}_s^{n+1}$, $k = 1, 2$, thus verifying (3.19), and the proof is complete. □

3.2 Scalar Multiplication and Addition in $\mathbb{T}_s^{n+1}$

The uniqueness of elements of $\mathbb{T}_s^{n+1}$, expressed in (3.1) – (3.2), is lost when $p = q = 0$. Hence, for the sake of simplicity, the upper entry, t, of elements $(t, tA)^t\in\mathbb{T}_s^{n+1}$ has been temporarily restricted in Sec. 3.1 to $t > 0$. We will now allow the upper entry of elements of $\mathbb{T}_s^{n+1}$ to be negative as well. Accordingly, $\mathbb{T}_s^{n+1}$ is now given by the equation

$$
\mathbb{T}_s^{n+1} = \left\{\begin{pmatrix} t \\ tA \end{pmatrix} : t\in\mathbb{R}-\{0\},\ A\in\mathbb{R}_s^n\right\}
\tag{3.22}
$$

that is, $\mathbb{T}_s^{n+1}$ is the set of all pairs $(t, tA)^t$ such that the upper entry is any real, nonzero number t, and the lower entry is any element A of $\mathbb{R}_s^n$ multiplied by t.

Definition 3.2. Let N be any positive integer, and let $m_k \in \mathbb{R} - \{0\}$ and $A_k \in \mathbb{R}_s^n$, $k = 1, \ldots, N$, be N nonzero numbers and N points of an Einstein gyrovector space $\mathbb{R}_s^n = (\mathbb{R}_s^n, \oplus, \otimes)$, such that

$$\sum_{k=1}^{N} m_k t_k \neq 0 \tag{3.23}$$

Then, scalar multiplication and addition in $\mathbb{T}_s^{n+1}$ are given by the equation

$$\sum_{k=1}^{N} m_k \begin{pmatrix} t_k \\ t_k A_k \end{pmatrix} = \begin{pmatrix} \sum_{k=1}^{N} m_k t_k \\ \sum_{k=1}^{N} m_k t_k A_k \end{pmatrix} = \begin{pmatrix} \sum_{k=1}^{N} m_k t_k \\ (\sum_{k=1}^{N} m_k t_k) A_0 \end{pmatrix} \tag{3.24}$$

where

$$A_0 = \frac{\sum_{k=1}^{N} m_k t_k A_k}{\sum_{k=1}^{N} m_k t_k} \tag{3.25}$$

3.3 Inner Product and Norm in $\mathbb{T}_s^{n+1}$

Definition 3.3. The inner product of any two elements $(p, pA)^t$ and $(q, qB)^t$ of $\mathbb{T}_s^{n+1}$ is given by the equation

$$\left\langle \begin{pmatrix} p \\ pA \end{pmatrix}, \begin{pmatrix} q \\ qB \end{pmatrix} \right\rangle = \frac{\gamma_{\ominus A \oplus B}}{\gamma_A \gamma_B} pq \tag{3.26}$$

Accordingly, the norm $\|(t, tA)^t\| > 0$ of any element $(t, tA)^t \in \mathbb{R}_s^n$ is given by the equation

$$\left\| \begin{pmatrix} t \\ tA \end{pmatrix} \right\|^2 = \left\langle \begin{pmatrix} t \\ tA \end{pmatrix}, \begin{pmatrix} t \\ tA \end{pmatrix} \right\rangle = \frac{\gamma_{\ominus A \oplus A}}{\gamma_A^2} t^2 = \frac{t^2}{\gamma_A^2} \tag{3.27}$$

noting that $\gamma_{\ominus A \oplus A} = \gamma_{\mathbf{0}} = 1$.

Theorem 3.4. *The inner product* (3.26) *in* $\mathbb{T}_s^{n+1}$ *is given by the equation*

$$\left\langle \begin{pmatrix} p \\ pA \end{pmatrix}, \begin{pmatrix} q \\ qB \end{pmatrix} \right\rangle = pq - \frac{1}{s^2}(pA)\cdot(qB) \tag{3.28a}$$

where we use the notation

$$(pA)\cdot(qB) := (A{\cdot}B)pq \tag{3.28b}$$

and the norm (3.27) *in* $\mathbb{T}_s^{n+1}$ *is given by the equation*

$$\left\| \begin{pmatrix} t \\ tA \end{pmatrix} \right\|^2 = t^2 - \frac{1}{s^2}(tA)^2 \tag{3.29a}$$

where we use the notation

$$(tA)^2 := (tA){\cdot}(tA) = (A{\cdot}A)t^2 = t^2A^2 \tag{3.29b}$$

Proof. By (3.12) we have

$$pq - \frac{1}{s^2}(pA){\cdot}(qB) = \left(1 - \frac{A{\cdot}B}{s^2}\right)pq = \frac{\gamma_{\ominus A\oplus B}}{\gamma_A\,\gamma_B}pq \tag{3.30}$$

Hence the right-hand sides of (3.28a) and (3.26) are equal, thus verifying (3.28a).

Similarly, by (3.10) we have

$$t^2 - \frac{1}{s^2}(tA)^2 = t^2\left(1 - \frac{A^2}{s^2}\right) = \frac{t^2}{\gamma_A^2} \tag{3.31}$$

Hence the right-hand sides of (3.29a) and (3.27) are equal, thus verifying (3.29a). ☐

Theorem 3.5. *Boosts of $\mathbb{T}_s^{n+1}$ preserve the inner product in $\mathbb{T}_s^{n+1}$ and, hence in particular, boosts of $\mathbb{T}_s^{n+1}$ preserve the norm in $\mathbb{T}_s^{n+1}$, that is,*

$$\left\langle L(W)\begin{pmatrix} p \\ pA \end{pmatrix}, L(W)\begin{pmatrix} q \\ qB \end{pmatrix}\right\rangle = \left\langle \begin{pmatrix} p \\ pA \end{pmatrix}, \begin{pmatrix} q \\ qB \end{pmatrix}\right\rangle \tag{3.32}$$

and

$$\left\| L(W)\begin{pmatrix} t \\ tA \end{pmatrix}\right\| = \left\| \begin{pmatrix} t \\ tA \end{pmatrix}\right\| \tag{3.33}$$

for all $p.q.t\in\mathbb{R}^+$ and $A, B, W\in\mathbb{R}_s^n$.

Proof. The proof is given by the following chain of equations, which are numbered for subsequent explanation.

$$\begin{aligned}
\left\langle L(W)\begin{pmatrix} p \\ pA \end{pmatrix}, L(W)\begin{pmatrix} q \\ qB \end{pmatrix}\right\rangle &\overset{(1)}{=\!=\!=} \left\langle \begin{pmatrix} \frac{\gamma_{W\oplus A}}{\gamma_A} p \\ \frac{\gamma_{W\oplus A}}{\gamma_A} p(W\oplus A) \end{pmatrix}, \begin{pmatrix} \frac{\gamma_{W\oplus B}}{\gamma_B} q \\ \frac{\gamma_{W\oplus B}}{\gamma_B} q(W\oplus B) \end{pmatrix}\right\rangle \\
&\overset{(2)}{=\!=\!=} \frac{\gamma_{\ominus(W\oplus A)\oplus(W\oplus B)}}{\gamma_{W\oplus A}\gamma_{W\oplus B}} \frac{\gamma_{W\oplus A}}{\gamma_A} \frac{\gamma_{W\oplus B}}{\gamma_B} pq \\
&\overset{(3)}{=\!=\!=} \frac{\gamma_{\ominus(W\oplus A)\oplus(W\oplus B)}}{\gamma_A\gamma_B} pq \\
&\overset{(4)}{=\!=\!=} \frac{\gamma_{\ominus A\oplus B}}{\gamma_A\gamma_B} pq \\
&\overset{(5)}{=\!=\!=} \left\langle \begin{pmatrix} p \\ pA \end{pmatrix}, \begin{pmatrix} q \\ qB \end{pmatrix}\right\rangle
\end{aligned} \tag{3.34}$$

Derivation of the numbered equalities in (3.34) follows:

(1) Follows from (3.8).
(2) Follows from (1) by (3.26).
(3) Follows from (2) by obvious cancellations.
(4) Follows from (3) by (2.122), p. 94.
(5) Follows from (4) by (3.26).

□

3.4 Unit Elements of $\mathbb{T}_s^{n+1}$

The Unit elements of $\mathbb{T}_s^{n+1}$, that is, the elements of $\mathbb{T}_s^{n+1}$ with norm 1, are of particular interest for the construction of hyperbolic barycentric coordinates in Einstein gyrovector spaces. These are the elements $(\gamma_A, \gamma_A A)^t \in \mathbb{T}_s^{n+1}$ for all $A \in \mathbb{R}_s^n$. Indeed, by (3.27),

$$\left\| \begin{pmatrix} \gamma_A \\ \gamma_A A \end{pmatrix} \right\| = \frac{\gamma_A^2}{\gamma_A^2} = 1 \tag{3.35}$$

Boost application to unit elements of $\mathbb{T}_s^{n+1}$ is particularly powerful and elegant. It is given by

$$L(A) \begin{pmatrix} \gamma_B \\ \gamma_B B \end{pmatrix} = \begin{pmatrix} \gamma_{A \oplus B} \\ \gamma_{A \oplus B}(A \oplus B) \end{pmatrix} \tag{3.36}$$

as we see from (3.8).

The linearity of $L(A)$ that Theorem 3.1 insures forms a powerful tool that enables the interplay of Einstein addition in $\mathbb{R}_s^n$ and the common vector addition in $\mathbb{R}^n$ to be revealed in (3.37) below. It follows from the result (3.17) of Theorem 3.1 and from (3.24), by the boost application to unit elements of $\mathbb{T}_s^{n+1}$, (3.36), that

$$\begin{aligned} L(W) \sum_{k=1}^{N} m_k \begin{pmatrix} \gamma_{A_k} \\ \gamma_{A_k} A_k \end{pmatrix} &= \sum_{k=1}^{N} m_k L(W) \begin{pmatrix} \gamma_{A_k} \\ \gamma_{A_k} A_k \end{pmatrix} \\ &= \sum_{k=1}^{N} m_k \begin{pmatrix} \gamma_{W \oplus A_k} \\ \gamma_{W \oplus A_k}(W \oplus A_k) \end{pmatrix} \\ &= \begin{pmatrix} \sum_{k=1}^{N} m_k \gamma_{W \oplus A_k} \\ \sum_{k=1}^{N} m_k \gamma_{W \oplus A_k}(W \oplus A_k) \end{pmatrix} \end{aligned} \tag{3.37}$$

The chain of equations (3.37) reveals the interplay of Einstein addition, $\oplus$, in $\mathbb{R}_s^n$ and vector addition, $+$, in $\mathbb{R}^n$ that appears implicitly in the sigma-notation for addition. In order to see clearly the interplay of $\oplus$ and $+$, let us consider (3.37) with $N = 2$, written in the following form:

$$L(W)\begin{pmatrix} m_1\gamma_{A_1} + m_2\gamma_{A_2} \\ m_1\gamma_{A_1}A_1 + m_2\gamma_{A_2}A_2 \end{pmatrix}$$

$$= \begin{pmatrix} m_1\gamma_{W\oplus A_1} + m_2\gamma_{W\oplus A_2} \\ m_1\gamma_{W\oplus A_1}(W\oplus A_1) + m_2\gamma_{W\oplus A_2}(W\oplus A_2) \end{pmatrix} \tag{3.38}$$

Remarkably, the common scalar and vector addition, $+$, on the left-hand side of (3.38) is balanced in equation (3.38) by an elegant interplay of $\oplus$ and $+$ on the right-hand side of (3.38).

The interplay of $\oplus$ and $+$ along with Lemma 3.6 below will prove useful in the adaptation of barycentric coordinates for use in Einstein gyrovector spaces and, hence, for use in Cartesian models of the hyperbolic geometry of Bolyai and Lobachevsky.

Lemma 3.6. *Let N be any positive integer, and let $m_k \in \mathbb{R}$ and $A_k \in \mathbb{R}_s^n$, $k = 1, \ldots, N$, be N real numbers and N points of an Einstein gyrovector space $\mathbb{R}_s^n = (\mathbb{R}_s^n, \oplus, \otimes)$. Furthermore, let m be a real or an imaginary number given by the equation*

$$m^2 := \left(\sum_{k=1}^{N} m_k\right)^2 + 2\sum_{\substack{j,k=1 \\ j<k}}^{N} m_j m_k (\gamma_{\ominus A_j \oplus A_k} - 1) \tag{3.39}$$

Then

$$\left(\sum_{k=1}^{N} m_k \gamma_{A_k} \frac{A_k}{s}\right)^2 = \left(\sum_{k=1}^{N} m_k \gamma_{A_k}\right)^2 - m^2 \tag{3.40}$$

We may note that in the Euclidean limit, when $s \to \infty$ and gamma factors tend to 1, Equations (3.39)–(3.40) descend to the simple equation

$$m = \sum_{k=1}^{N} m_k \qquad (\textit{Euclidean Geometry}) \tag{3.41}$$

so that the Euclidean counterpart of m in the Lemma is the total mass of a system of N particles of masses m_k.

Proof. The proof of the Lemma is given by the following chain of equations, which are numbered for subsequent explanation.

$$
\begin{aligned}
\left(\sum_{k=1}^{N} m_k\gamma_{A_k}\frac{A_k}{s}\right)^2 &\overset{(1)}{=\!=} \sum_{k=1}^{N} m_k^2\gamma_{A_k}^2\frac{A_k^2}{s^2} + 2\sum_{\substack{j,k=1\\ j<k}}^{N} m_j m_k\gamma_{A_j}\gamma_{A_k}\frac{A_j\cdot A_k}{s^2} \\
&\overset{(2)}{=\!=} \sum_{k=1}^{N} m_k^2(\gamma_{A_k}^2-1) + 2\sum_{\substack{j,k=1\\ j<k}}^{N} m_j m_k(\gamma_{A_j}\gamma_{A_k}-\gamma_{\ominus A_j\oplus A_k}) \\
&\overset{(3)}{=\!=} \sum_{k=1}^{N} m_k^2\gamma_{A_k}^2 - \sum_{k=1}^{N} m_k^2 + 2\sum_{\substack{j,k=1\\ j<k}}^{N} m_j m_k\gamma_{A_j}\gamma_{A_k} - 2\sum_{\substack{j,k=1\\ j<k}}^{N} m_j m_k\gamma_{\ominus A_j\oplus A_k} \\
&\overset{(4)}{=\!=} \left(\sum_{k=1}^{N} m_k\gamma_{A_k}\right)^2 - 2\sum_{\substack{j,k=1\\ j<k}}^{N} m_j m_k\gamma_{A_j}\gamma_{A_k} - \sum_{k=1}^{N} m_k^2 \\
&\qquad + 2\sum_{\substack{j,k=1\\ j<k}}^{N} m_j m_k\gamma_{A_j}\gamma_{A_k} - 2\sum_{\substack{j,k=1\\ j<k}}^{N} m_j m_k\gamma_{\ominus A_j\oplus A_k} \qquad (3.42) \\
&\overset{(5)}{=\!=} \left(\sum_{k=1}^{N} m_k\gamma_{A_k}\right)^2 - \left\{\sum_{k=1}^{N} m_k^2 + 2\sum_{\substack{j,k=1\\ j<k}}^{N} m_j m_k\gamma_{\ominus A_j\oplus A_k}\right\} \\
&\overset{(6)}{=\!=} \left(\sum_{k=1}^{N} m_k\gamma_{A_k}\right)^2 - \left\{\left(\sum_{k=1}^{N} m_k\right)^2 + 2\sum_{\substack{j,k=1\\ j<k}}^{N} m_j m_k(\gamma_{\ominus A_j\oplus A_k}-1)\right\} \\
&\overset{(7)}{=\!=} \left(\sum_{k=1}^{N} m_k\gamma_{A_k}\right)^2 - m^2
\end{aligned}
$$

The assumption $A_k\in\mathbb{R}_s^n$ implies, by (3.10), p. 151, that all gamma factors in (3.39) – (3.42) are real and greater than 1. Derivation of the numbered equalities in (3.42) follows:

(1) This equation is obtained by an expansion of the square of a sum of vectors in $\mathbb{R}^n$.
(2) Follows from (1) by (3.13) – (3.14).
(3) Follows from (2) by obvious expansions.
(4) Follows from (3) by an expansion of the square of a sum of real numbers.
(5) Follows from (4) by an obvious cancellation.
(6) Follows from (5) by an expansion of the square of a sum of real numbers.
(7) Follows from (6) by the definition of m^2 in (3.39)

□

We are now in a position to solve in the following theorem an equation for unit elements of $\mathbb{T}_s^{n+1}$ that will prove useful in the introduction of hyperbolic barycentric coordinates into Einstein gyrovector spaces.

Theorem 3.7. *Let N be any positive integer, and let $m_k \in \mathbb{R}$ and $A_k \in \mathbb{R}_s^n$, $k = 1, \ldots, N$, be N real numbers and N points of an Einstein gyrovector space $\mathbb{R}_s^n = (\mathbb{R}_s^n, \oplus, \otimes)$ satisfying*

$$\sum_{k=1}^{N} m_k \gamma_{A_k} \neq 0 \tag{3.43}$$

so that without loss of generality (otherwise, we replace m_k by $-m_k$),

$$\sum_{k=1}^{N} m_k \gamma_{A_k} > 0 \tag{3.44}$$

Furthermore, let

$$\sum_{k=1}^{N} m_k \begin{pmatrix} \gamma_{A_k} \\ \gamma_{A_k} A_k \end{pmatrix} = m_0 \begin{pmatrix} \gamma_{A_0} \\ \gamma_{A_0} A_0 \end{pmatrix} \tag{3.45}$$

be an equation in $\mathbb{T}_s^{n+1}$ for the two unknowns $m_0 \in \mathbb{R}$ and $A_0 \in \mathbb{R}^n$.

Then, (3.45) possesses a unique solution (m_0, A_0), where $m_0 > 0$ and $A_0 \in \mathbb{R}_s^n$ are given by the equations

$$m_0 = \sqrt{\left(\sum_{k=1}^{N} m_k\right)^2 + 2 \sum_{\substack{j,k=1 \\ j<k}}^{N} m_j m_k (\gamma_{\ominus A_j \oplus A_k} - 1)} \tag{3.46}$$

and

$$A_0 = \frac{\sum_{k=1}^{N} m_k \gamma_{A_k} A_k}{\sum_{k=1}^{N} m_k \gamma_{A_k}} \tag{3.47}$$

Proof. Clearly, under condition (3.44), (m_0, A_0) is a solution of (3.45) if and only if $m_0 \neq 0$ and

$$\gamma_{A_0} = \frac{\sum_{k=1}^N m_k \gamma_{A_k}}{m_0} \tag{3.48a}$$

$$\gamma_{A_0} A_0 = \frac{\sum_{k=1}^N m_k \gamma_{A_k} A_k}{m_0} \tag{3.48b}$$

Part I: The proof consists of two parts. In the first part we show that if equation (3.45) possesses a solution for the real unknown m_0 and the vector unknown A_0 in $\mathbb{R}^n$, then the solution must be given uniquely by (3.46) – (3.47), satisfying $m_0 > 0$, $A_0 \in \mathbb{R}_s^n$, and (3.48a) – (3.48b).

If m_0 and A_0 that satisfy (3.45) exist, then the norms of the two sides of (3.45) are equal while, by (3.35), the norm of the right-hand side of (3.45) is m_0. Hence, the norm of the left-hand side of (3.45) equals m_0 as well, obtaining the following chain of equations, which are numbered for subsequent explanation:

$$\begin{aligned}
m_0^2 &\overset{(1)}{=\!=} \left\| \sum_{k=1}^N m_k \begin{pmatrix} \gamma_{A_k} \\ \gamma_{A_k} A_k \end{pmatrix} \right\|^2 \\
&\overset{(2)}{=\!=} \left\| \begin{pmatrix} \sum_{k=1}^N m_k \gamma_{A_k} \\ \sum_{k=1}^N m_k \gamma_{A_k} A_k \end{pmatrix} \right\|^2 \\
&\overset{(3)}{=\!=} \left(\sum_{k=1}^N m_k \gamma_{A_k} \right)^2 - \left(\sum_{k=1}^N m_k \gamma_{A_k} \frac{A_k}{s} \right)^2 \\
&\overset{(4)}{=\!=} \left(\sum_{k=1}^N m_k \right)^2 + 2 \sum_{\substack{j,k=1 \\ j<k}}^N m_j m_k (\gamma_{\ominus A_j \oplus A_k} - 1)
\end{aligned} \tag{3.49}$$

Derivation of the numbered equalities in (3.49) follows:

(1) This equation follows from the result that the norm of the left-hand side of (3.45) equals the norm of the right-hand side of (3.45), the latter being m_0 by (3.35).
(2) Follows from (1) by (3.24).
(3) Follows from (2) by (3.29).
(4) Follows from (3) by Identity (3.40) of Lemma 3.6.

It follows from (3.44) and the upper entry of (3.45) that

$$m_0 > 0 \tag{3.50}$$

We thus obtain in (3.49) the desired equation, (3.46), for m_0.

Hence, if m_0 and A_0 that satisfy (3.45) exist, m_0 is positive and must be given by (3.46).

By assumption, A_0 satisfies (3.45). Equation (3.45) is equivalent to (3.48a) – (3.48b) formed by the upper entry and by the lower entry of (3.45). Dividing (3.48b) by (3.48a), noting that $m_0 \neq 0$ by (3.50), we obtain (3.47). Owing to the reality of m_0 in (3.45), γ_{A_0} is real implying $A_0 \in \mathbb{R}_s^n$.

Hence, if m_0 and A_0 that satisfy (3.45) exist, then $m_0 > 0$, $A_0 \in \mathbb{R}_s^n$, and they must be given by (3.46) – (3.47) and satisfy (3.48a) – (3.48b). The converse is also true as shown in Part II of the proof.

Part II: We now show that the pair consisting of $m_0 > 0$ and $A_0 \in \mathbb{R}_s^n$, given by (3.46) – (3.47), is indeed a solution of (3.45).

It follows from Identity (3.40) of Lemma 3.6, along with m_0 of (3.46) that

$$\left(\sum_{k=1}^{N} m_k \gamma_{A_k} \frac{A_k}{s} \right)^2 = \left(\sum_{k=1}^{N} m_k \gamma_{A_k} \right)^2 - m_0^2 \tag{3.51}$$

Hence, by (3.47) and (3.51), we have the following chain of equations, which are numbered for subsequent explanation.

$$\begin{aligned} \frac{A_0^2}{s^2} &\overset{(1)}{=\!=} \frac{(\sum_{k=1}^{N} m_k \gamma_{A_k} \frac{A_k}{s})^2}{(\sum_{k=1}^{N} m_k \gamma_{A_k})^2} \\ &\overset{(2)}{=\!=} \frac{(\sum_{k=1}^{N} m_k \gamma_{A_k})^2 - m_0^2}{(\sum_{k=1}^{N} m_k \gamma_{A_k})^2} \\ &\overset{(3)}{=\!=} 1 - \frac{m_0^2}{(\sum_{k=1}^{N} m_k \gamma_{A_k})^2} \end{aligned} \tag{3.52}$$

Derivation of the numbered equalities in (3.52) follows:

(1) This equation is valid by assumption, (3.47).
(2) Follows from (1) by (3.51).
(3) Follows from (2) by an obvious cancellation.

It follows from (3.52) that

$$\gamma_{A_0} = \frac{1}{\sqrt{1 - \frac{A_0^2}{s^2}}} = \frac{\sum_{k=1}^{N} m_k \gamma_{A_k}}{m_0} \tag{3.53}$$

thus verifying (3.48a).

Following (3.48a) and (3.47) we have

$$\gamma_{A_0} A_0 = \frac{\sum_{k=1}^{N} m_k \gamma_{A_k} A_k}{m_0} \tag{3.54}$$

thus verifying (3.48b).

Hence, the pair consisting of m_0 and A_0 satisfies (3.48a) – (3.48b), so that it forms a solution of (3.45), and the proof is complete. □

Employing Theorem 3.1, p. 152, the following theorem extends the results of Theorem 3.7, and reveals the interplay, emphasized in (3.38), p. 158, of Einstein addition, $\oplus$, in $\mathbb{R}_s^n$ and the common vector addition, $+$, in $\mathbb{R}^n$.

Theorem 3.8. *Let N be any positive integer, and let $m_k \in \mathbb{R}$ and $A_k \in \mathbb{R}_s^n$, $k = 1, \ldots, N$, be N real numbers and N points of an Einstein gyrovector space $\mathbb{R}_s^n = (\mathbb{R}_s^n, \oplus, \otimes)$ satisfying*

$$\sum_{k=1}^{N} m_k \gamma_{A_k} \neq 0 \tag{3.55}$$

so that without loss of generality (otherwise, we replace m_k by $-m_k$),

$$\sum_{k=1}^{N} m_k \gamma_{A_k} > 0 \tag{3.56}$$

Furthermore, let $m_0 > 0$ and $A_0 \in \mathbb{R}_s^n$ be the unique solution (determined by Theorem 3.7) of the equation

$$\sum_{k=1}^{N} m_k \begin{pmatrix} \gamma_{A_k} \\ \gamma_{A_k} A_k \end{pmatrix} = m_0 \begin{pmatrix} \gamma_{A_0} \\ \gamma_{A_0} A_0 \end{pmatrix} \tag{3.57}$$

Then, m_0 and A_0 satisfy the identities

$$m_0 = \sqrt{\left(\sum_{k=1}^{N} m_k\right)^2 + 2 \sum_{\substack{j,k=1 \\ j<k}}^{N} m_j m_k (\gamma_{\ominus(W \oplus A_j) \oplus (W \oplus A_k))} - 1)} \tag{3.58}$$

$$m_0 = \sqrt{\left(\sum_{k=1}^{N} m_k\right)^2 + 2\sum_{\substack{j,k=1\\j<k}}^{N} m_j m_k (\gamma_{\ominus A_j \oplus A_k} - 1)} \tag{3.59}$$

$$W \oplus A_0 = \frac{\sum_{k=1}^{N} m_k \gamma_{W\oplus A_k}(W\oplus A_k)}{\sum_{k=1}^{N} m_k \gamma_{W\oplus A_k}} \tag{3.60}$$

$$\gamma_{W\oplus A_0} = \frac{\sum_{k=1}^{N} m_k \gamma_{W\oplus A_k}}{m_0} \tag{3.61}$$

and

$$\gamma_{W\oplus A_0}(W\oplus A_0) = \frac{\sum_{k=1}^{N} m_k \gamma_{W\oplus A_k}(W\oplus A_k)}{m_0} \tag{3.62}$$

for all $W \in \mathbb{R}_s^n$.

Proof. The results of Theorem 3.8 in (3.58) and (3.60) – (3.62) descend to corresponding results of Theorem 3.7 in the special case when $W = \mathbf{0}$.

Applying the boost $L(W)$ to the right-hand side of (3.57) we have, by (3.37) with $N = 1$,

$$\begin{aligned} L(W)\left\{ m_0 \begin{pmatrix} \gamma_{A_0} \\ \gamma_{A_0} A_0 \end{pmatrix} \right\} &= m_0 \begin{pmatrix} \gamma_{W\oplus A_0} \\ \gamma_{W\oplus A_0}(W\oplus A_0) \end{pmatrix} \\ &= \begin{pmatrix} m_0 \gamma_{W\oplus A_0} \\ m_0 \gamma_{W\oplus A_0}(W\oplus A_0) \end{pmatrix} \end{aligned} \tag{3.63}$$

Similarly, applying the boost $L(W)$ to the left-hand side of (3.57) we have, by (3.37),

$$\begin{aligned} L(W)\left\{ \sum_{k=1}^{N} m_k \begin{pmatrix} \gamma_{A_k} \\ \gamma_{A_k} A_k \end{pmatrix} \right\} &= \sum_{k=1}^{N} m_k \begin{pmatrix} \gamma_{W\oplus A_k} \\ \gamma_{W\oplus A_k}(W\oplus A_k) \end{pmatrix} \\ &= \begin{pmatrix} \sum_{k=1}^{N} m_k \gamma_{W\oplus A_k} \\ \sum_{k=1}^{N} m_k \gamma_{W\oplus A_k}(W\oplus A_k) \end{pmatrix} \end{aligned} \tag{3.64}$$

Hence, following (3.63) – (3.64) and (3.57) we have

$$\sum_{k=1}^{N} m_k \begin{pmatrix} \gamma_{W\oplus A_k} \\ \gamma_{W\oplus A_k}(W\oplus A_k) \end{pmatrix} = m_0 \begin{pmatrix} \gamma_{W\oplus A_0} \\ \gamma_{W\oplus A_0}(W\oplus A_0) \end{pmatrix} \tag{3.65}$$

Equation (3.65) is identical with equation (3.45) of Theorem 3.7 with one exception. The points $A_k \in \mathbb{R}_s^n$ in (3.45) are replaced by the points $W \oplus A_k \in \mathbb{R}_s^n$, respectively, in (3.65), $k = 0, 1, \ldots, N$. Hence, by Theorem 3.7, the unique solution of (3.65) for the unknowns m_0 and $W \oplus A_0$ is given by (3.58) – (3.60), satisfying (3.61) – (3.62).

Remarkably, owing to Identity (2.122), p. 94, the constant m_0 in the solution of (3.65), which is given by (3.58), equals the constant m_0 in the solution of (3.45), which is given by (3.46) and by (3.59). □

The results (3.58) – (3.62) of Theorem 3.8 involve both vector addition, $+$, in $\mathbb{R}^n$, which is implicit in the sigma-notation as indicated in (3.7), p. 151, and Einstein addition, $\oplus$, in $\mathbb{R}_s^n$. As such, they exhibit a remarkable interplay of $\oplus$ and $+$ that will prove useful in the introduction of hyperbolic barycentric coordinates into Einstein gyrovector spaces.

3.5 From $\mathbb{T}_s^{n+1}$ back to $\mathbb{R}_s^n$

Our temporary trip from $\mathbb{R}_s^n$ to $\mathbb{T}_s^{n+1}$ in Secs. 3.1 – 3.4 was needed for the discovery of the linearity of boost application in Theorem 3.1, p. 152, and subsequently, for the discovery of the interplay of $\oplus$ and $+$ in (3.37) – (3.38). It is now appropriate to return from $\mathbb{T}_s^{n+1}$ to $\mathbb{R}_s^n$, obtaining the following theorem:

Theorem 3.9. *Let N be any positive integer, let $m_k \in \mathbb{R}$ and $A_k \in \mathbb{R}_s^n$, $k = 1, \ldots, N$, be N real numbers and N points of an Einstein gyrovector space $\mathbb{R}_s^n = (\mathbb{R}_s^n, \oplus, \otimes)$, satisfying*

$$\sum_{k=1}^{N} m_k \gamma_{A_k} \neq 0 \tag{3.66}$$

so that without loss of generality (otherwise, we replace m_k by $-m_k$),

$$\sum_{k=1}^{N} m_k \gamma_{A_k} > 0 \tag{3.67}$$

and let

$$P = \frac{\sum_{k=1}^{N} m_k \gamma_{A_k} A_k}{\sum_{k=1}^{N} m_k \gamma_{A_k}} \in \mathbb{R}_s^n \tag{3.68}$$

Then

$$\gamma_P = \frac{\sum_{k=1}^{N} m_k \gamma_{A_k}}{m_0} \tag{3.69}$$

and

$$\gamma_{_P} P = \frac{\sum_{k=1}^{N} m_k \gamma_{_{A_k}} A_k}{m_0} \tag{3.70}$$

where $m_0 > 0$ is given by

$$m_0 = \sqrt{\left(\sum_{k=1}^{N} m_k\right)^2 + 2\sum_{\substack{j,k=1\\ j<k}}^{N} m_j m_k (\gamma_{\ominus A_j \oplus A_k} - 1)} \tag{3.71}$$

Proof. Identity (3.40) of Lemma 3.6, and (3.71), imply

$$\left(\sum_{k=1}^{N} m_k \gamma_{_{A_k}} \frac{A_k}{s}\right)^2 = \left(\sum_{k=1}^{N} m_k \gamma_{_{A_k}}\right)^2 - m_0^2 \tag{3.72}$$

so that by (3.68) and (3.72) we have, as in (3.52),

$$\frac{P^2}{s^2} = \frac{(\sum_{k=1}^{N} m_k \gamma_{_{A_k}} \frac{A_k}{s})^2}{(\sum_{k=1}^{N} m_k \gamma_{_{A_k}})^2} = 1 - \frac{m_0^2}{(\sum_{k=1}^{N} m_k \gamma_{_{A_k}})^2} \tag{3.73}$$

Hence,

$$\gamma_{_P} = \frac{1}{\sqrt{1 - \frac{P^2}{s^2}}} = \frac{\sum_{k=1}^{N} m_k \gamma_{_{A_k}}}{m_0} \tag{3.74}$$

thus verifying (3.69), where $\gamma_{_P}$ is real by the assumption $P \in \mathbb{R}_s^n$ in (3.68), so that by (3.67) and (3.74) m_0 is positive.

Finally, (3.68) and (3.69) imply (3.70), and the proof is complete. □

In the following theorem we generalize Theorem 3.9 by introducing left gyrotranslations.

Theorem 3.10. *Let N be any positive integer, let $m_k \in \mathbb{R}$ and $A_k \in \mathbb{R}_s^n$, $k = 1, \ldots, N$, be N real numbers and N points of an Einstein gyrovector space $\mathbb{R}_s^n = (\mathbb{R}_s^n, \oplus, \otimes)$, satisfying*

$$\sum_{k=1}^{N} m_k \gamma_{_{A_k}} \neq 0 \tag{3.75}$$

so that without loss of generality (otherwise, we replace m_k by $-m_k$),

$$\sum_{k=1}^{N} m_k \gamma_{_{A_k}} > 0 \tag{3.76}$$

and let

$$P = \frac{\sum_{k=1}^{N} m_k \gamma_{A_k} A_k}{\sum_{k=1}^{N} m_k \gamma_{A_k}} \in \mathbb{R}_s^n \tag{3.77}$$

Then

$$W \oplus P = \frac{\sum_{k=1}^{N} m_k \gamma_{W \oplus A_k} (W \oplus A_k)}{\sum_{k=1}^{N} m_k \gamma_{W \oplus A_k}} \tag{3.78}$$

$$\gamma_{W \oplus P} = \frac{\sum_{k=1}^{N} m_k \gamma_{W \oplus A_k}}{m_0} \tag{3.79}$$

and

$$\gamma_{W \oplus P} (W \oplus P) = \frac{\sum_{k=1}^{N} m_k \gamma_{W \oplus A_k} (W \oplus A_k)}{m_0} \tag{3.80}$$

where $m_0 > 0$ *is given by*

$$m_0 = \sqrt{\left(\sum_{k=1}^{N} m_k \right)^2 + 2 \sum_{\substack{j,k=1 \\ j<k}}^{N} m_j m_k (\gamma_{\ominus(W \oplus A_j) \oplus (W \oplus A_k))} - 1)} \tag{3.81}$$

or, equivalently,

$$m_0 = \sqrt{\left(\sum_{k=1}^{N} m_k \right)^2 + 2 \sum_{\substack{j,k=1 \\ j<k}}^{N} m_j m_k (\gamma_{\ominus A_j \oplus A_k)} - 1)} \tag{3.82}$$

for all $W \in \mathbb{R}_s^n$.

Proof. Note that results of Theorem 3.10 descend to corresponding results of Theorem 3.9 in the special case when $W = \mathbf{0}$.

It is anticipated that m_0 in (3.81) and in (3.82) are equal. Following Identity (2.122), p. 94, this is indeed the case.

By Theorem 3.9, the point $P \in \mathbb{R}_s^n$, given by (3.77), satisfies the two equations

$$\gamma_P = \frac{\sum_{k=1}^{N} m_k \gamma_{A_k}}{m_0} \tag{3.83}$$

and

$$\gamma_P P = \frac{\sum_{k=1}^{N} m_k \gamma_{A_k} A_k}{m_0} \tag{3.84}$$

where m_0 is given by (3.82).

The two equations (3.83) and (3.84) are equivalent to the single equation with an upper and a lower entries,

$$\sum_{k=1}^{N} m_k \begin{pmatrix} \gamma_{A_k} \\ \gamma_{A_k} A_k \end{pmatrix} = m_0 \begin{pmatrix} \gamma_P \\ \gamma_P P \end{pmatrix} \tag{3.85}$$

where the upper entry of (3.85) is (3.83) and the lower entry of (3.85) is (3.84).

Applying the boost $L(W)$ to the right-hand side of (3.85) we have, by (3.37), p. 157, with $N = 1$,

$$L(W) \left\{ m_0 \begin{pmatrix} \gamma_P \\ \gamma_P P \end{pmatrix} \right\} = m_0 \begin{pmatrix} \gamma_{W \oplus P} \\ \gamma_{W \oplus P} (W \oplus P) \end{pmatrix} \tag{3.86}$$

Similarly, applying the boost $L(W)$ to the left-hand side of (3.85) we have, by (3.37),

$$L(W) \left\{ \sum_{k=1}^{N} m_k \begin{pmatrix} \gamma_{A_k} \\ \gamma_{A_k} A_k \end{pmatrix} \right\} = \sum_{k=1}^{N} m_k \begin{pmatrix} \gamma_{W \oplus A_k} \\ \gamma_{W \oplus A_k} (W \oplus A_k) \end{pmatrix} \tag{3.87}$$

Hence, it follows from (3.86) – (3.87) and (3.85) that

$$\sum_{k=1}^{N} m_k \begin{pmatrix} \gamma_{W \oplus A_k} \\ \gamma_{W \oplus A_k} (W \oplus A_k) \end{pmatrix} = m_0 \begin{pmatrix} \gamma_{W \oplus P} \\ \gamma_{W \oplus P} (W \oplus P) \end{pmatrix} \tag{3.88}$$

Dividing the lower entry of (3.88) by its upper entry, we obtain the equation

$$W \oplus P = \frac{\sum_{k=1}^{N} m_k \gamma_{W \oplus A_k} (W \oplus A_k)}{\sum_{k=1}^{N} m_k \gamma_{W \oplus A_k}} \in \mathbb{R}_s^n \tag{3.89}$$

thus verifying (3.78). Note that $W \oplus P \in \mathbb{R}_s^n$ since $W, P \in \mathbb{R}_s^n$.

Applying Theorem 3.9 to $W \oplus P$ in (3.89) instead of P in (3.68), we obtain the results in (3.79) – (3.81), as desired. □

In order to appreciate the power and elegance of Theorem 3.10 in the hyperbolic geometry of Einstein gyrovector spaces $\mathbb{R}_s^n$, we note that in the Euclidean limit, when $s \to \infty$, this theorem descends to the following immediate, but important, theorem in the Euclidean geometry of vector spaces $\mathbb{R}^n$:

Theorem 3.11. *Let N be any positive integer, let $m_k \in \mathbb{R}$ and $A_k \in \mathbb{R}^n$, $k = 1, \ldots, N$, be N real numbers and N points of a Euclidean space $\mathbb{R}^n$, such that*

$$\sum_{k=1}^{N} m_k \neq 0 \tag{3.90}$$

and let P be a point of $\mathbb{R}^n$ given by the equation

$$P = \frac{\sum_{k=1}^{N} m_k A_k}{\sum_{k=1}^{N} m_k} \tag{3.91}$$

Then,

$$W + P = \frac{\sum_{k=1}^{N} m_k (W + A_k)}{m_0} \tag{3.92}$$

where

$$m_0 = \sum_{k=1}^{N} m_k \tag{3.93}$$

Proof. A direct proof of the theorem is immediate. Instructively, however, our point is to prove the theorem as a special case of Theorem 3.10. Indeed, in the limit as $s \to \infty$, gamma factors tend to 1, as we see from (3.10), p. 151, and Einstein addition, $\oplus$, in $\mathbb{R}_s^n$ tends to ordinary vector addition, $+$, in $\mathbb{R}^n$, as we see from (3.9), p. 151. Accordingly, the statement and results of Theorem 3.10 descend to the statement and results of Theorem 3.11 in the limit as $s \to \infty$. □

Coincidentally, (3.90) – (3.93) appear in Chapter 1, where (3.91) is the Euclidean barycentric coordinate representation of a point $P \in \mathbb{R}^n$ in (1.22), p. 9, and (3.92) is the immediate, but important, covariance property (1.26), p. 12, of the barycentric coordinate representation. It is this property that insures that the barycentric coordinate representation of a point in a Euclidean space is independent of the choice of the origin of the space. Moreover, it is this property that allows us to determine in Chapter 1 triangle centers.

Being a most natural hyperbolic counterpart of the Euclidean barycentric coordinate representation (3.91) of a point $P \in \mathbb{R}^n$, the representation of $P \in \mathbb{R}_s^n$ in (3.77) is, suggestively, the corresponding hyperbolic barycentric coordinate representation of a point $P \in \mathbb{R}_s^n$ in an Einstein gyrovector space. We are thus naturally led to the definition of hyperbolic barycentric coordinates in Einstein gyrovector spaces. The formal definition of the resulting *gyrobarycentric coordinates* will be presented in Def, 4.2 of Chapter 4.

Chapter 4

Hyperbolic Barycentric Coordinates and Hyperbolic Triangle Centers

In gyrolanguage, hyperbolic barycentric coordinates are called *gyrobarycentric coordinates* [3, Sec. 11.3] [54, 55]. The definition of gyrobarycentric coordinates in Einstein gyrovector spaces is motivated by analogies that the Euclidean Theorem 3.11, p. 169, shares with its hyperbolic counterpart, Theorem 3.10, p. 166.

4.1 Gyrobarycentric Coordinates in Einstein Gyrovector Spaces

Definition 4.1. (Hyperbolic Pointwise Independence). *A set S of N points, $S = \{A_1, \dots, A_N\}$, $N \geq 2$, in a gyrovector space $(\mathbb{R}^n_s, \oplus, \otimes)$, $n \geq 2$, is pointwise independent if the $N-1$ gyrovectors $\ominus A_1 \oplus A_k$, $k = 2, \dots, N$, in $\mathbb{R}^n_s \subset \mathbb{R}^n$, considered as vectors in $\mathbb{R}^n$, are linearly independent in $\mathbb{R}^n$.*

Definition 4.2. (Gyrobarycentric Coordinates in Einstein Gyrovector Spaces). *Let $S = \{A_1, \dots, A_N\}$ be a pointwise independent set of $N \geq 2$ points in an Einstein gyrovector space $(\mathbb{R}^n_s, \oplus, \otimes)$. The N real numbers $m_1, \dots, m_N$ are gyrobarycentric coordinates of a point $P \in \mathbb{R}^n_s$ with respect to S if*

$$\sum_{k=1}^{N} m_k \gamma_{A_k} \neq 0 \tag{4.1}$$

and

$$P = \frac{\sum_{k=1}^{N} m_k \gamma_{A_k} A_k}{\sum_{k=1}^{N} m_k \gamma_{A_k}} \tag{4.2}$$

Gyrobarycentric coordinates are homogeneous in the sense that the gyrobarycentric coordinates $(m_1, \ldots, m_N)$ *of the point* P *in* (4.2) *are equivalent to the gyrobarycentric coordinates* $(\lambda m_1, \ldots, \lambda m_N)$ *for any* $\lambda \neq 0$. *Since in gyrobarycentric coordinates only ratios of coordinates are relevant, the gyrobarycentric coordinates* $(m_1, \ldots, m_N)$ *are also written as* $(m_1 :, \ldots, : m_N)$ *so that*

$$(m_1 : m_2 : \ \ldots \ : m_N) = (\lambda m_1 : \lambda m_2 : \ \ldots \ : \lambda m_N) \tag{4.3}$$

for any real $\lambda \neq 0$.

The point P *given by* (4.2) *is said to be a gyrobarycentric combination of the points of the set* S, *possessing the gyrobarycentric coordinate representation* (4.2).

A Gyrobarycentric combination (4.2) *is positive if all the coefficients* m_k, $k = 1, \ldots, N$, *are positive. The set of all positive gyrobarycentric combinations of the points of the set* S *is called the gyroconvex span of* S.

The constant

$$m_0 = \pm \sqrt{\left(\sum_{k=1}^{N} m_k\right)^2 + 2 \sum_{\substack{j,k=1 \\ j<k}}^{N} m_j m_k (\gamma_{\ominus A_j \oplus A_k} - 1)} \tag{4.4}$$

is called the constant of the gyrobarycentric coordinate representation of the point P *with respect to the set* S. *The ambiguous sign of* m_0 *in* (4.4) *is positive (negative) if the nonzero sum* $\sum_{k=1}^{N} m_k \gamma_{A_k}$ *in* (4.1) *is positive (negative).*

Finally, the gyrobarycentric coordinate representation (4.2) *of* P *is special if the gyrobarycentric coordinates are normalized by the condition*

$$\sum_{k=1}^{N} m_k = 1 \tag{4.5}$$

The pointwise independence of the set S in Def. 4.2 insures that a gyrobarycentric coordinate representation of a point with respect to the set S is unique.

Owing to the fact that gyrosegments in Einstein gyrovector spaces are Euclidean segments, the definition of the Euclidean simplex in Def. 1.6, p. 10, leads to the following definition of the Einsteinian gyrosimplex.

Definition 4.3. (Einsteinian Gyrosimplex). *The convex span of the pointwise independent set* $S = \{A_1, \ldots, A_N\}$ *of* $N \geq 2$ *points in* $\mathbb{R}^n_s$ *is an* $(N-1)$*-dimensional gyrosimplex, called an* $(N-1)$*-gyrosimplex and denoted*

$A_1 \dots A_N$. *The points of* S *are the vertices of the gyrosimplex. The convex span of* $N-1$ *of the points of* S *is a gyroface of the gyrosimplex, said to be the gyroface opposite to the remaining vertex. The convex span of each two of the vertices is an edge of the gyrosimplex.*

Any two distinct points A, B of an Einstein gyrovector space $\mathbb{R}_s^n$ are pointwise independent, and their convex span is the interior of the gyrosegment AB, which is a 1-gyrosimplex. Similarly, any three non-gyrocollinear points (that is, points that do not lie on the same gyroline; see [3, Remark 6.23] for this terminology) A, B, C of $\mathbb{R}_s^n$, $n \geq 2$, are pointwise independent, and their convex span is the interior of the gyrotriangle ABC, which is a 2-gyrosimplex.

Theorem 4.4. *Let* $S = \{A_1, \dots, A_N\}$ *be a pointwise independent set of* $N \geq 2$ *points in an Einstein gyrovector space* $(\mathbb{R}_s^n, \oplus, \otimes)$, *and let* P,

$$P = \frac{\sum_{k=1}^{N} m_k \gamma_{A_k} A_k}{\sum_{k=1}^{N} m_k \gamma_{A_k}} \in \mathbb{R}_s^n \tag{4.6}$$

$$\sum_{k=1}^{N} m_k \gamma_{A_k} \neq 0 \tag{4.7}$$

be a point in $\mathbb{R}_s^n$ *given by its gyrobarycentric coordinates* $(m_1 :, \dots, : m_N)$ *with respect to* S.

Then,

$$\gamma_P = \frac{\sum_{k=1}^{N} m_k \gamma_{A_k}}{m_0} \tag{4.8}$$

and

$$\gamma_P P = \frac{\sum_{k=1}^{N} m_k \gamma_{A_k} A_k}{m_0} \tag{4.9}$$

where $m_0 \neq 0$ *is given by the equation*

$$m_0 = \pm \sqrt{\left(\sum_{k=1}^{N} m_k \right)^2 + 2 \sum_{\substack{j,k=1 \\ j<k}}^{N} m_j m_k (\gamma_{\ominus A_j \oplus A_k} - 1)} \tag{4.10}$$

Moreover, m_0 *is invariant under left gyrotranslations, and* P, γ_P, *and* $\gamma_P P$ *are gyrocovariant under left gyrotranslations, that is, for all* $X \in \mathbb{R}_s^n$ *we have*

$$X \oplus P = \frac{\sum_{k=1}^{N} m_k \gamma_{X \oplus A_k} (X \oplus A_k)}{\sum_{k=1}^{N} m_k \gamma_{X \oplus A_k}}$$

$$\gamma_{X \oplus P} = \frac{\sum_{k=1}^{N} m_k \gamma_{X \oplus A_k}}{m_0} \tag{4.11}$$

$$\gamma_{X \oplus P}(X \oplus P) = \frac{\sum_{k=0}^{h} m_k \gamma_{X \oplus A_k} (X \oplus A_k)}{m_0}$$

where

$$m_0 = \pm \sqrt{\left(\sum_{k=1}^{N} m_k\right)^2 + 2 \sum_{\substack{j,k=1 \\ j<k}}^{N} m_j m_k (\gamma_{\ominus(X \oplus A_j) \oplus (X \oplus A_k))} - 1)} \tag{4.12}$$

and where the ambiguous sign of m_0 *in* (4.10) *and* (4.12) *is positive (negative) if the nonzero sum* $\sum_{k=1}^{N} m_k \gamma_{A_k}$ *in* (4.7) *is positive (negative).*

Proof. Theorem 4.4, stated in terms of gyrobarycentric coordinates of a point, is identical with Theorem 3.10, p. 166. □

Remark 4.5. *The scalar* m_0 *is given in Theorem 4.4 by each of the seemingly different equations* (4.10) *and* (4.12). *Indeed, these equations are identical to each other by* (2.122), *p. 94.*

It is assumed in Theorem 4.4 that the point P, (4.6), *lies inside the ball* $\mathbb{R}_s^n$, *implying that* $m_0^2 > 0$ *and that the gamma factor* γ_P *of* P *is a real number. If the coefficients* m_k, $k = 1, \ldots, N$, *in* (4.6) *are all positive or all negative, then the point* P *lies in the convex span of the points of the set* S, *that is,* P *lies inside the* $(N-1)$*-gyrosimplex* $A_1 \ldots A_N$. *This gyrosimplex, in turn, lies inside the ball* $\mathbb{R}_s^n$.

(1) *The point* P *lies inside (or, on the interior of) the* $(N-1)$*-gyrosimplex* $A_1 \ldots A_N$ *if and only if the coefficients* m_k, $k = 1, \ldots, N$, *of its gyrobarycentric coordinate representation* (4.6) *are all positive or all negative. Clearly, in this case* $m_0^2 > 0$.

Otherwise, when all the coefficients m_k *are nonzero but do not have the same sign, the location of the point* P *in* (4.6) *has the following three possibilities:*

(2) *The point P does not lie inside the $(N-1)$-gyrosimplex $A_1 \ldots A_N$, but it lies inside the ball $\mathbb{R}_s^n$. In this case the gamma factor γ_P of P is a real number and, hence, $m_0^2 > 0$.*
(3) *The point P lies on the boundary of the ball $\mathbb{R}_s^n$ if and only if the gamma factor γ_P of P is undefined, $\gamma_P = \infty$, so that $m_0^2 = 0$.*
(4) *The point $P \in \mathbb{R}^n$ does not lie in the ball $\mathbb{R}_s^n$ or on its boundary if and only if the gamma factor γ_P of P is purely imaginary, so that $m_0^2 < 0$.*

4.2 Analogies with Relativistic Mechanics

Guided by analogies with relativistic mechanics, the $(N-1)$-gyrosimplex of the $N \geq 2$ points of the pointwise independent set $S = \{A_1, \ldots, A_N\}$ along with gyrobarycentric coordinates $(m_1 :, \ldots, : m_N)$ may be viewed as an isolated system $S = \{A_k, m_k, k = 1, \ldots, N\}$ of N noninteracting particles. The constant $m_k \in \mathbb{R}$ is the invariant (that is, velocity independent, or, Newtonian) mass of the kth particle and $A_k \in \mathbb{R}_s^n$ is the relativistically admissible velocity of the kth particle, $k = 1, \ldots, N$, relative to the arbitrarily selected origin $O = \mathbf{0}$ of the Einsteinian velocity space $\mathbb{R}_s^n$, which represents the rest frame. Each point of the Einsteinian velocity space $\mathbb{R}_s^n$ represents a relativistically admissible velocity of an inertial frame relative to the rest frame. Accordingly, the relativistic (velocity dependent) mass of the k-th particle is $m_k \gamma_{A_k}$.

By analogies with relativistic mechanics, the point P in (4.2) represents the velocity of the center of momentum (CM) frame of the particle system S relative to the rest frame. The CM frame of S, in turn, is an inertial reference frame relative to which the relativistic momentum, $\sum_{k=1}^N m_k \gamma_{A_k} A_k$, of the particle system S vanishes (see Exercise 2, p. 245).

Finally, the constant

$$m_0 = \sqrt{\left(\sum_{k=1}^N m_k\right)^2 + 2 \sum_{\substack{j,k=1 \\ j<k}}^N m_j m_k (\gamma_{\ominus A_j \oplus A_k} - 1)} \qquad (4.13)$$

of the point P with respect to the set S in (4.4) turns out in the context of relativistic mechanics to be the total invariant, CM-velocity independent mass of the particle system S [56]. Accordingly, (4.8) implies that the relativistic mass $m_0 \gamma_P$ of the particle system S equals the sum of the relativistic masses of its constituent particles [57].

Along remarkable analogies between hyperbolic geometry and relativistic mechanics, there is an important disanalogy. Unlike relativistic mechanics, in hyperbolic geometry the "masses" m_k, $k = 1, \ldots, N$, considered as gyrobarycentric coordinates of points, need not be positive.

4.3 Gyrobarycentric Coordinates in Möbius Gyrovector Spaces

Owing to the isomorphism between Einstein and Möbius gyrovector spaces, studied in Sec. 2.29, p. 141, results in an Einstein gyrovector space can be transformed into corresponding results in a corresponding Möbius gyrovector space.

Employing the interplay of Einstein addition and vector addition, studied in Chap. 3, we have discovered the Einstein gyrobarycentric coordinate representation (3.77), p. 167, of a point P in an Einstein gyrovector space $(\mathbb{R}_s^n, \oplus_{\text{E}}, \otimes)$, along with its useful properties. This discovery, in turn, led to the formal definition of Einstein gyrobarycentric coordinates in Def. 4.2, p. 171. We now wish to transform the gyrobarycentric coordinate representation of a point, from points of Einstein gyrovector spaces $(\mathbb{R}_s^n, \oplus_{\text{E}}, \otimes)$ to points of Möbius gyrovector spaces $(\mathbb{R}_s^n, \oplus_{\text{M}}, \otimes)$.

In this section we use a notation that allows us to distinguish between Einstein and Möbius addition, as in Sec. 2.29. Einstein addition $\oplus = \oplus_{\text{E}}$ is given by (2.4), p. 65, and Möbius addition $\oplus = \oplus_{\text{M}}$ is given by (2.248), p. 129. Note that Einstein addition $\oplus_{\text{E}}$ and Möbius addition $\oplus_{\text{M}}$ admit the same scalar multiplication, so that $\otimes_{\text{E}} = \otimes_{\text{M}} = \otimes$, as we see from (2.80), p. 83, and (2.260), p. 132.

Emphasizing that P and A_k, $k = 1, \ldots, N$, in (4.2) are points of an Einstein gyrovector space $(\mathbb{R}_s^n, \oplus_{\text{E}}, \otimes)$, we rewrite the gyrobarycentric coordinate representation (4.2) of a point P in $(\mathbb{R}_s^n, \oplus_{\text{E}}, \otimes)$ as

$$P_e = \frac{\sum_{k=1}^{N} m_k \gamma_{A_{k,e}} A_{k,e}}{\sum_{k=1}^{N} m_k \gamma_{A_{k,e}}} \tag{4.14}$$

where the subscript e is attached to points of the Einstein gyrovector space $(\mathbb{R}_s^n, \oplus_{\text{E}}, \otimes)$.

Owing to the isomorphism between Einstein gyrovector spaces $(\mathbb{R}_s^n, \oplus_{\text{E}}, \otimes)$ and corresponding Möbius gyrovector spaces $(\mathbb{R}_s^n, \oplus_{\text{M}}, \otimes)$, the gyrobarycentric coordinate representation (4.14) of a point

$$P_e \in (\mathbb{R}_s^n, \oplus_{\text{E}}, \otimes) \tag{4.15}$$

with respect to a pointwise independent set $S = \{A_{1,e}, \ldots, A_{N,e}\}$ becomes the gyrobarycentric coordinate representation

$$P_m = \tfrac{1}{2}\otimes\frac{\sum_{k=1}^{N} m_k\gamma^2_{A_{k,m}} A_{k,m}}{\sum_{k=1}^{N} m_k(\gamma^2_{A_{k,m}} - \frac{1}{2})} \tag{4.16}$$

of the corresponding point

$$P_m \in (\mathbb{R}^n_s, \oplus_{\mathrm{M}}, \otimes) \tag{4.17}$$

with respect to the corresponding set $S = \{A_{1,m}, \ldots, A_{N,m}\}$, as we see from Theorem 4.6 below.

In accordance with (4.14), a subscript m is attached in (4.16) – (4.17) to points of the Möbius gyrovector space $(\mathbb{R}^n_s, \oplus_{\mathrm{M}}, \otimes)$.

Theorem 4.6. *Let*

$$P_e = \frac{\sum_{k=1}^{N} m_k\gamma_{A_{k,e}} A_{k,e}}{\sum_{k=1}^{N} m_k\gamma_{A_{k,e}}} \tag{4.18}$$

be the gyrobarycentric coordinate representation, (4.14)*, of a point* P_e *with respect to a pointwise independent set of points* $S_e = \{A_{1,e}, \ldots, A_{N,e}\}$ *in an Einstein gyrovector space* $(\mathbb{R}^n_s, \oplus_E, \otimes)$*, where each gyrobarycentric coordinate* $m_k = m_k(A_{1,e}, \ldots, A_{N,e})$ *is a function of the points of* S_e*,* $k = 1, \ldots, N$*.*

The isomorphic image P_m *of* P_e *in a corresponding Möbius gyrovector space* $(\mathbb{R}^n_s, \oplus_M, \otimes)$*, under the isomorphism* (2.277)*, p. 142, is the gyrobarycentric coordinate representation of the point* P_m *with respect to the pointwise independent set of points* $S_m = \{A_{1,m}, \ldots, A_{N,m}\}$ *in the Möbius gyrovector space* $(\mathbb{R}^n_s, \oplus_M, \otimes)$*, given by the equation*

$$P_m = \tfrac{1}{2}\otimes\frac{\sum_{k=1}^{N} m_k\gamma^2_{A_{k,m}} A_{k,m}}{\sum_{k=1}^{N} m_k(\gamma^2_{A_{k,m}} - \frac{1}{2})} \tag{4.19}$$

where now each gyrobarycentric coordinate $m_k = m_k(A_{1,m}, \ldots, A_{N,m})$ *is a function of the points of* S_m*,* $k = 1, \ldots, N$*.*

Proof. We transform the points P_e, $A_{k,e}$, $k = 1, \ldots, N$, of an Einstein gyrovector space $(\mathbb{R}^n_s, \oplus_{\mathrm{E}}, \otimes)$, which are involved in (4.18), into corresponding points P_m, $A_{k,m}$ of a corresponding Möbius gyrovector space, $(\mathbb{R}^n_s, \oplus_{\mathrm{M}}, \otimes)$, by means of isomorphism (2.277), and its resulting identities (2.279) – (2.280), p. 142. This transformation results in the gyrobarycentric coordinate representation (4.19) as shown in the following chain of equations, which are numbered for subsequent explanation.

$$
\begin{aligned}
2\otimes P_m &\overset{(1)}{=\!=} \frac{\sum_{k=1}^{N} m_k \gamma_{2\otimes A_{k,m}} (2\otimes A_{k,m})}{\sum_{k=1}^{N} m_k \gamma_{2\otimes A_{k,m}}} \\
&\overset{(2)}{=\!=} \frac{\sum_{k=1}^{N} m_k (2\gamma_{A_{k,m}}^2 A_{k,m})}{\sum_{k=1}^{N} m_k (2\gamma_{A_{k,m}}^2 - 1)} \\
&\overset{(3)}{=\!=} \frac{\sum_{k=1}^{N} m_k \gamma_{A_{k,m}}^2 A_{k,m}}{\sum_{k=1}^{N} m_k (\gamma_{A_{k,m}}^2 - \frac{1}{2})}
\end{aligned}
\tag{4.20}
$$

Derivation of the numbered equalities in (4.20) follows:

(1) This equation is the gyrobarycentric coordinate representation (4.18) in which points of an Einstein gyrovector space $(\mathbb{R}_s^n, \oplus_E, \otimes)$ are replaced by their image in a corresponding Möbius gyrovector space $(\mathbb{R}_s^n, \oplus_M, \otimes)$, under isomorphism (2.277), p. 142, as follows:

   (*i*) $P_e = 2\otimes P_m$ and $A_{k,e} = 2\otimes A_{k,m}$, according to (2.279), p. 142.

(2) Follows from (1) by employing algebraic identities as follows:

   (*ii*) $\gamma_{A_{k,e}} = \gamma_{2\otimes A_{k,m}} = 2\gamma_{A_{k,m}}^2 - 1$, according to (2.280), p. 142.

   (*iii*) $\gamma_{A_{k,e}} A_{k,e} = \gamma_{2\otimes A_{k,m}} (2\otimes A_{k,m}) = 2\gamma_{A_{k,m}}^2 A_{k,m}$, according to (2.280), p. 142.

(3) Follows from (2) immediately.

Finally, noting the scalar associative law of gyrovector spaces in Def. 2.14, p. 85, a scalar multiplication of both extreme sides of (4.20) by $\frac{1}{2}$ yields the desired result (4.19). □

It follows from Theorem 4.6 that the isomorphic image of the gyrobarycentric coordinate representation (4.18) of points in an Einstein gyrovector space $(\mathbb{R}_s^n, \oplus_E, \otimes)$ is the representation (4.19) of points in the corresponding Möbius gyrovector space $(\mathbb{R}_s^n, \oplus_M, \otimes)$. Naturally, this result suggests the following definition of gyrobarycentric coordinate representation of points in Möbius gyrovector spaces.

Definition 4.7. (Gyrobarycentric Coordinates in Möbius Gyrovector Spaces). *Let* $S = \{A_1, \ldots, A_N\}$ *be a pointwise independent set of* $N \geq 2$ *points in a Möbius gyrovector space* $\mathbb{R}_s^n = (\mathbb{R}_s^n, \oplus, \otimes)$. *The real* *numbers* $m_1, \ldots, m_N$ *are gyrobarycentric coordinates of a point* $P \in \mathbb{R}_s^n$ *with respect to* S *if*

$$\sum_{k=1}^{N} m_k(\gamma_{A_k}^2 - \tfrac{1}{2}) \neq 0 \tag{4.21}$$

and

$$P = \tfrac{1}{2} \otimes \frac{\sum_{k=1}^{N} m_k \gamma_{A_k}^2 A_k}{\sum_{k=1}^{N} m_k(\gamma_{A_k}^2 - \frac{1}{2})} \tag{4.22}$$

Gyrobarycentric coordinates are homogeneous in the sense that the gyrobarycentric coordinates $(m_1, \ldots, m_N)$ *of the point* P *in* (4.22) *are equivalent to the gyrobarycentric coordinates* $(\lambda m_1, \ldots, \lambda m_N)$ *for any* $\lambda \neq 0$. *Since in gyrobarycentric coordinates only ratios of coordinates are relevant, the gyrobarycentric coordinates* $(m_1, \ldots, m_N)$ *are also written as* $(m_1 :, \ldots, : m_N)$.

Finally, the gyrobarycentric coordinate representation (4.22) *of* P *is special if the gyrobarycentric coordinates are normalized by the condition*

$$\sum_{k=1}^{N} m_k = 1 \tag{4.23}$$

The pointwise independence of the set S in Def. 4.7 insures that a gyrobarycentric coordinate representation of a point with respect to the set S are unique.

4.4 Einstein Gyromidpoint

Definition 4.8. (Einstein Gyromidpoints). *Let* $A_1, A_2 \in \mathbb{R}_s^n$, *be two distinct points of an Einstein gyrovector space* $(\mathbb{R}_s^n, \oplus, \otimes)$. *The gyromidpoint* M_{12} *of points* A_1 *and* A_2 *is a point of gyrosegment* A_1A_2 *equigyrodistant from* A_1 *and* A_2, *that is,*

$$\|\ominus A_1 \oplus M_{12}\| = \|\ominus A_2 \oplus M_{12}\| \tag{4.24}$$

or, equivalently,

$$\gamma_{\ominus A_1 \oplus M_{12}} = \gamma_{\ominus A_2 \oplus M_{12}} \tag{4.25}$$

Let M_{12} be the gyromidpoint of gyrosegment A_1A_2 in an Einstein gyrovector space $(\mathbb{R}_s^n, \oplus, \otimes)$, shown in Fig. 2.1, p. 92, with gyrobarycentric coordinate representation

$$M_{12} = \frac{m_1\gamma_{A_1} A_1 + m_2\gamma_{A_2} A_2}{m_1\gamma_{A_1} + m_2\gamma_{A_2}} \tag{4.26}$$

with respect to the set $\{A_1, A_2\}$, where the gyrobarycentric coordinates m_1 and m_2 of M_{12} are to be determined in (4.31) below.

By the gyrocovariance property with respect to left gyrotranslations in the second identity in (4.11), p. 174, of Theorem 4.4, with X replaced by $\ominus X$, the point M_{12} in (4.26) obeys the identity

$$\gamma_{\ominus X\oplus M_{12}} = \frac{m_1\gamma_{\ominus X\oplus A_1} + m_2\gamma_{\ominus X\oplus A_2}}{m_0} \tag{4.27}$$

for any $X\in\mathbb{R}_s^n$. If we use the notation $\gamma_{12} = \gamma_{\ominus A_1\oplus A_2}$, etc., adopted in (2.133), p. 100, $m_0 \neq 0$ is determined from (4.10), p. 173, by the equation

$$m_0^2 = (m_1 + m_2)^2 + 2m_1m_2(\gamma_{12} - 1) \tag{4.28}$$

where $m_0 > 0$ if m_1 and m_2 are positive.

Following (4.27) with $X = A_1$ and with $X = A_2$, respectively, we have

$$\begin{aligned}\gamma_{\ominus A_1\oplus M_{12}} &= \frac{m_1 + m_2\gamma_{\ominus A_1\oplus A_2}}{m_0} = \frac{m_1 + m_2\gamma_{12}}{m_0}\\ \gamma_{\ominus A_2\oplus M_{12}} &= \frac{m_1\gamma_{\ominus A_2\oplus A_1} + m_2}{m_0} = \frac{m_1\gamma_{12} + m_2}{m_0}\end{aligned} \tag{4.29}$$

noting that $\gamma_{\ominus A_1\oplus A_1} = \gamma_0 = 1$.

Following (4.25) and (4.29), along with the normalization condition $m_1 + m_2 = 1$, we have

$$\begin{aligned}m_1 + m_2 &= 1\\ m_1 + m_2\gamma_{12} &= m_1\gamma_{12} + m_2\end{aligned} \tag{4.30}$$

thus obtaining a system of two equations for the two unknowns m_1 and m_2. The unique solution of (4.30) is $m_1 = m_2 = \frac{1}{2}$. Hence, the special gyrobarycentric coordinates $(m_1 : m_2)$ of M_{12} with respect to the set $\{A_1, A_2\}$ are

$$(m_1, m_2) = (\tfrac{1}{2}, \tfrac{1}{2}) \tag{4.31}$$

so that convenient gyrobarycentric coordinates $(m_1 : m_2)$ of the gyromidpoint M_{12} with respect to the set $\{A_1, A_2\}$ are

$$(m_1 : m_2) = (1 : 1) \tag{4.32}$$

Following (4.32) and (4.26), the gyromidpoint M_{12} of a gyrosegment A_1A_2 in an Einstein gyrovector space $(\mathbb{R}_s^n, \oplus, \otimes)$ is given by the equation

$$M_{12} = \frac{\gamma_{A_1}A_1 + \gamma_{A_2}A_2}{\gamma_{A_1} + \gamma_{A_2}} \tag{4.33}$$

Gyromidpoints of several gyrosegments in an Einstein gyrovector plane are shown in Fig. 4.1, p. 182.

We have thus established the following theorem:

Theorem 4.9. (Einstein Gyromidpoint). *Let $A_1, A_2 \in \mathbb{R}_s^n$, $n \geq 1$ be two points of an Einstein gyrovector space $(\mathbb{R}_s^n, \oplus, \otimes)$, and let M_{12} be their gyromidpoint. Then M_{12} has the gyrobarycentric coordinate representation*

$$M_{12} = \frac{\gamma_{A_1} A_1 + \gamma_{A_2} A_2}{\gamma_{A_1} + \gamma_{A_2}} \tag{4.34}$$

with respect to the set $\{A_1, A_2\}$, with gyrobarycentric coordinates

$$(m_1 : m_2) = (1 : 1) \tag{4.35}$$

4.5 Möbius Gyromidpoint

The transformation of an Einstein gyrobarycentric coordinate representation (4.18) of a point in an Einstein gyrovector space into a Möbius gyrobarycentric coordinate representation (4.19) of a corresponding point in a corresponding Möbius gyrovector space, is presented in Theorem 4.6, p. 177.

Following Theorem 4.6, the transformation of the gyromidpoint identity (4.33) from Einstein gyrovector spaces into Möbius gyrovector spaces results in the Möbius gyromidpoint identity

$$M_{12} = \tfrac{1}{2} \otimes \frac{\gamma_{A_1}^2 A_1 + \gamma_{A_2}^2 A_2}{\gamma_{A_1}^2 + \gamma_{A_2}^2 - 1} \tag{4.36}$$

The point M_{12} in (4.36) is the gyromidpoint of the gyrosegment $A_1 A_2$ in a Möbius gyrovector space $(\mathbb{R}_s^n, \oplus_{\mathrm{M}}, \otimes)$, shown in Fig. 4.3, p. 190.

Following (4.36) and Def. 4.7, convenient gyrobarycentric coordinates $(m_1 : m_2)$ of M_{12} with respect to the set $\{A_1, A_2\}$ are

$$(m_1 : m_2) = (1 : 1) \tag{4.37}$$

Gyromidpoints of several gyrosegments in a Möbius gyrovector plane are shown in Fig. 4.3, p. 190.

Formalizing the main result of this section, we have the following theorem:

Theorem 4.10. (Möbius Gyromidpoint). *Let $A_1, A_2 \in \mathbb{R}_s^n$, $n \geq 1$, be two points of a Möbius gyrovector space $(\mathbb{R}_s^n, \oplus, \otimes)$, and let M_{12} be their gyromidpoint. Then M_{12} has the gyrobarycentric coordinate representation*

$$M_{12} = \tfrac{1}{2} \otimes \frac{\gamma_{A_1}^2 A_1 + \gamma_{A_2}^2 A_2}{\gamma_{A_1}^2 + \gamma_{A_2}^2 - 1} \tag{4.38}$$

with respect to the set $\{A_1, A_2\}$, with gyrobarycentric coordinates

$$(m_1 : m_2) = (1 : 1) \tag{4.39}$$

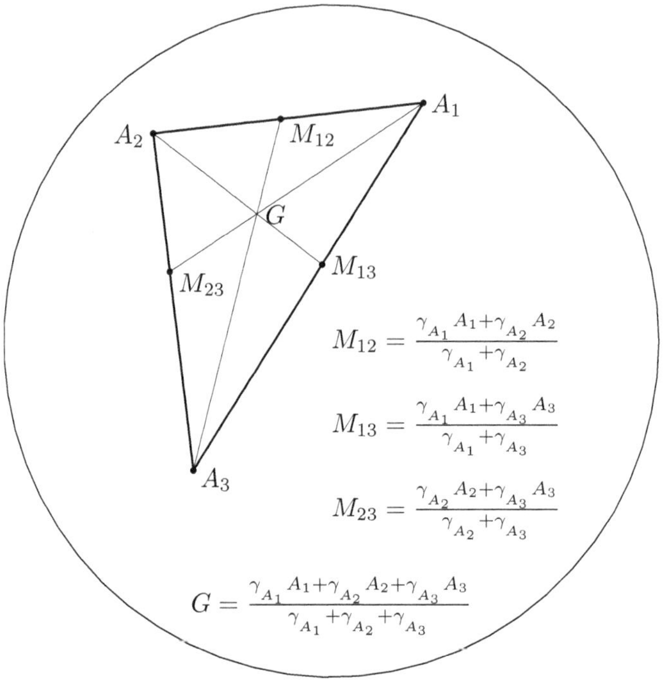

Fig. 4.1 The gyromidpoints M_{12}, M_{13}, and M_{23} of the sides, A_1A_2, A_1A_3 and A_2A_3, of gyrotriangle $A_1A_2A_3$ in an Einstein gyrovector space $(\mathbb{R}_s^n, \oplus_{\mathrm{E}}, \otimes_{\mathrm{E}})$ are shown here for $n = 2$, along with its gyromedians A_1M_{23}, A_2M_{13}, A_3M_{12}, and its gyrocentroid G.

4.6 Einstein Gyrotriangle Gyrocentroid

The hyperbolic triangle centroid is called, in gyrolanguage, the gyrotriangle gyrocentroid.

Definition 4.11. (Gyromedians, Gyrotriangle Gyrocentroids). *A gyromedian of a gyrotriangle is the gyrosegment joining a vertex of the gyrotriangle with the gyromidpoint of the opposing side, shown in Fig. 4.1. The gyrocentroid, G, of a gyrotriangle is the point of concurrency of the gyrotriangle gyromedians, shown in Fig. 4.1.*

Let $A_1A_2A_3$ be a gyrotriangle in an Einstein gyrovector space $(\mathbb{R}_s^n, \oplus, \otimes)$, $n \geq 2$, and let the gyromidpoints of its sides be M_{12}, M_{13} and M_{23}, as shown in Fig. 4.1. Hence, by (4.33), M_{12} and, in a similar way,

M_{13} and M_{23}, are given by the equations

$$\begin{aligned} M_{12} &= \frac{\gamma_{A_1} A_1 + \gamma_{A_2} A_2}{\gamma_{A_1} + \gamma_{A_2}} \\ M_{13} &= \frac{\gamma_{A_1} A_1 + \gamma_{A_3} A_3}{\gamma_{A_1} + \gamma_{A_3}} \\ M_{23} &= \frac{\gamma_{A_2} A_2 + \gamma_{A_3} A_3}{\gamma_{A_2} + \gamma_{A_3}} \end{aligned} \tag{4.40}$$

The three gyromedians of gyrotriangle $A_1A_2A_3$ in Fig. 4.1 are the gyrosegments A_1M_{23}, A_2M_{13}, and A_3M_{12}. Since gyrosegments in Einstein gyrovector spaces coincide with Euclidean segments, one can employ methods of linear algebra to determine the point of concurrency, that is the gyrocentroid, of the three gyromedians of gyrotriangle $A_1A_2A_3$ in Fig. 4.1.

The details of the use of methods of linear algebra for the determination of the gyrobarycentric coordinates of the gyrotriangle gyrocentroid in Einstein gyrovector spaces are presented below.

In order to determine the gyrobarycentric coordinates of the gyrotriangle gyrocentroid in Einstein gyrovector spaces we begin with some gyroalgebra that reduce the task we face to a problem in linear algebra.

Let the gyrocentroid G of gyrotriangle $A_1A_2A_3$ in an Einstein gyrovector space $(\mathbb{R}^n_s, \oplus, \otimes)$, shown in Fig. 4.1, be given by its gyrobarycentric coordinate representation with respect to the set $S = \{A_1, A_2, A_3\}$ of the gyrotriangle vertices,

$$G = \frac{m_1\gamma_{A_1} A_1 + m_2\gamma_{A_2} A_2 + m_3\gamma_{A_3} A_3}{m_1\gamma_{A_1} + m_2\gamma_{A_2} + m_3\gamma_{A_3}} \tag{4.41}$$

where the gyrobarycentric coordinates (m_1, m_2, m_3) of G in (4.41) are to be determined in (4.65), p. 188.

Left gyrotranslating gyrotriangle $A_1A_2A_3$ by $\ominus A_1$, the gyrotriangle becomes gyrotriangle $O(\ominus A_1 \oplus A_2)(\ominus A_1 \oplus A_3)$, where $O = \ominus A_1 \oplus A_1$ is the arbitrarily selected origin of the Einstein gyrovector space $\mathbb{R}^n_s$. The gyrotriangle side-gyromidpoints M_{12}, M_{13} and M_{23} become, respectively, $\ominus A_1 \oplus M_{12}$, $\ominus A_1 \oplus M_{13}$ and $\ominus A_1 \oplus M_{23}$. These are calculated in (4.42) below by means of the gyroalgebraic relations in (4.6) and in the first identity in (4.11) of Theorem 4.4, p. 173, and in terms of the standard gyrotriangle notation, shown in Fig. 2.3, p. 101 and in (2.133), p. 100.

$$\begin{aligned}
\ominus A_1 \oplus M_{12} &= \ominus A_1 \oplus \frac{\gamma_{A_1} A_1 + \gamma_{A_2} A_2}{\gamma_{A_1} + \gamma_{A_2}} = \frac{\gamma_{\ominus A_1 \oplus A_2} (\ominus A_1 \oplus A_2)}{1 + \gamma_{\ominus A_1 \oplus A_2}} = \frac{\gamma_{12} \mathbf{a}_{12}}{\gamma_{12} + 1} \\
\ominus A_1 \oplus M_{13} &= \ominus A_1 \oplus \frac{\gamma_{A_1} A_1 + \gamma_{A_3} A_3}{\gamma_{A_1} + \gamma_{A_3}} = \frac{\gamma_{\ominus A_1 \oplus A_3} (\ominus A_1 \oplus A_3)}{1 + \gamma_{\ominus A_1 \oplus A_3}} = \frac{\gamma_{13} \mathbf{a}_{13}}{\gamma_{13} + 1} \\
\ominus A_1 \oplus M_{23} &= \ominus A_1 \oplus \frac{\gamma_{A_2} A_2 + \gamma_{A_3} A_3}{\gamma_{A_2} + \gamma_{A_3}} \\
&= \frac{\gamma_{\ominus A_1 \oplus A_2} (\ominus A_1 \oplus A_2) + \gamma_{\ominus A_1 \oplus A_3} (\ominus A_1 \oplus A_3)}{\gamma_{\ominus A_1 \oplus A_2} + \gamma_{\ominus A_1 \oplus A_3}} = \frac{\gamma_{12} \mathbf{a}_{12} + \gamma_{13} \mathbf{a}_{13}}{\gamma_{12} + \gamma_{13}}
\end{aligned} \tag{4.42}$$

Note that, by Def. 4.1, p. 171, the set of points $S = \{A_1, A_2, A_3\}$ is pointwise independent in an Einstein gyrovector space $(\mathbb{R}^n_s, \oplus, \otimes)$. Hence, the two gyrovectors $\mathbf{a}_{12} = \ominus A_1 \oplus A_2$ and $\mathbf{a}_{13} = \ominus A_1 \oplus A_3$ in $\mathbb{R}^n_s \subset \mathbb{R}^n$ in (4.42), considered as vectors in $\mathbb{R}^n$, are linearly independent in $\mathbb{R}^n$.

Similarly to the gyroalgebra in (4.42), under a left gyrotranslation by $\ominus A_1$ the gyrocentroid G in (4.41) becomes

$$\begin{aligned}
\ominus A_1 \oplus G &= \frac{m_2 \gamma_{\ominus A_1 \oplus A_2} (\ominus A_1 \oplus A_2) + m_3 \gamma_{\ominus A_1 \oplus A_3} (\ominus A_1 \oplus A_3)}{m_1 + m_2 \gamma_{\ominus A_1 \oplus A_2} + m_3 \gamma_{\ominus A_1 \oplus A_3}} \\
&= \frac{m_2 \gamma_{12} \mathbf{a}_{12} + m_3 \gamma_{13} \mathbf{a}_{13}}{m_1 + m_2 \gamma_{12} + m_3 \gamma_{13}}
\end{aligned} \tag{4.43}$$

Left gyrotranslating the triangle $A_1 A_2 A_3$ in Fig. 4.1 by $\ominus A_1$ we obtain the left gyrotranslated gyrotriangle

$$O(\ominus A_1 \oplus A_2)(\ominus A_1 \oplus A_3) \tag{4.44}$$

with vertices $\ominus A_1 \oplus A_1 = O = \mathbf{0}$, $\mathbf{a}_{12} = \ominus A_1 \oplus A_2$ and $\mathbf{a}_{13} = \ominus A_1 \oplus A_3$.

The gyromedian of the left gyrotranslated gyrotriangle (4.44) that joins the vertex O with the gyromidpoint of its opposing side, as calculated in (4.42),

$$\ominus A_1 \oplus M_{23} = \frac{\gamma_{12} \mathbf{a}_{12} + \gamma_{13} \mathbf{a}_{13}}{\gamma_{12} + \gamma_{13}} \tag{4.45}$$

is contained in the Euclidean line

$$L_1 = O + (-O + \{\ominus A_1 \oplus M_{23}\}) t_1 = \frac{\gamma_{12} \mathbf{a}_{12} + \gamma_{13} \mathbf{a}_{13}}{\gamma_{12} + \gamma_{13}} t_1 \tag{4.46}$$

where $t_1 \in \mathbb{R}$ is the line parameter. This line passes through the point $O = \mathbf{0} \in \mathbb{R}_s^n \subset \mathbb{R}^n$ when $t_1 = 0$, and it passes through the point

$$\ominus A_1 \oplus M_{23} = \frac{\gamma_{12}\mathbf{a}_{12} + \gamma_{13}\mathbf{a}_{13}}{\gamma_{12} + \gamma_{13}} \in \mathbb{R}_s^n \subset \mathbb{R}^n \tag{4.47}$$

when $t_1 = 1$.

Similarly to (4.45) – (4.46), the gyromedian of the left gyrotranslated gyrotriangle $O(\ominus A_1 \oplus A_2)(\ominus A_1 \oplus A_3)$, (4.44), that joins the vertex

$$\ominus A_1 \oplus A_2 = \mathbf{a}_{12} \tag{4.48}$$

with the gyromidpoint of its opposing side, as calculated in (4.42),

$$\ominus A_1 \oplus M_{13} = \frac{\gamma_{13}\mathbf{a}_{13}}{\gamma_{13} + 1} \tag{4.49}$$

is contained in the Euclidean line

$$L_2 = \mathbf{a}_{12} + (-\mathbf{a}_{12} + \{\ominus A_1 \oplus M_{13}\})t_2 = \mathbf{a}_{12} + \left(-\mathbf{a}_{12} + \frac{\gamma_{13}\mathbf{a}_{13}}{\gamma_{13} + 1}\right) t_2 \tag{4.50}$$

where $t_2 \in \mathbb{R}$ is the line parameter. This line passes through the point $\mathbf{a}_{12} \in \mathbb{R}_s^n \subset \mathbb{R}^n$ when $t_2 = 0$, and it passes through the point $\ominus A_1 \oplus M_{13} = \gamma_{13}\mathbf{a}_{13}/(\gamma_{13} + 1) \in \mathbb{R}_s^n \subset \mathbb{R}^n$ when $t_2 = 1$.

Similarly to (4.45) – (4.46), and similarly to (4.49) – (4.50), the gyromedian of the left gyrotranslated gyrotriangle $O(\ominus A_1 \oplus A_2)(\ominus A_1 \oplus A_3)$ that joins the vertex

$$\ominus A_1 \oplus A_3 = \mathbf{a}_{13} \tag{4.51}$$

with the gyromidpoint of its opposing side, as calculated in (4.42),

$$\ominus A_1 \oplus M_{12} = \frac{\gamma_{12}\mathbf{a}_{12}}{\gamma_{12} + 1} \tag{4.52}$$

is contained in the Euclidean line

$$L_3 = \mathbf{a}_{13} + (-\mathbf{a}_{13} + \{\ominus A_1 \oplus M_{12}\})t_3 = \mathbf{a}_{13} + \left(-\mathbf{a}_{13} + \frac{\gamma_{12}\mathbf{a}_{12}}{\gamma_{12} + 1}\right) t_3 \tag{4.53}$$

where $t_3 \in \mathbb{R}$ is the line parameter. This line passes through the point $\mathbf{a}_{13} \in \mathbb{R}_s^n \subset \mathbb{R}^n$ when $t_3 = 0$, and it passes through the point $\ominus A_1 \oplus M_{12} = \gamma_{12}\mathbf{a}_{12}/(\gamma_{12} + 1) \in \mathbb{R}_s^n \subset \mathbb{R}^n$ when $t_3 = 1$.

Hence, if the gyrocentroid G exists, its left gyrotranslated gyrocentroid, $-\ominus A_1 \oplus G$, given by (4.43), is contained in each of the three Euclidean lines L_k, $k = 1, 2, 3$, in (4.46), (4.50) and (4.53). Formalizing, if G exists then the point P, (4.43),

$$P = \ominus A_1 \oplus G = \frac{m_2\gamma_{12}\mathbf{a}_{12} + m_3\gamma_{13}\mathbf{a}_{13}}{m_1 + m_2\gamma_{12} + m_3\gamma_{13}} \tag{4.54}$$

lies on each of the lines L_k, $k = 1, 2, 3$. Imposing the normalization condition $m_1 + m_2 + m_3 = 1$ of gyrobarycentric coordinates, (4.54) can be simplified by means of the resulting equation $m_1 = 1 - m_2 - m_3$, obtaining

$$P = \ominus A_1 \oplus G = \frac{m_2 \gamma_{12} \mathbf{a}_{12} + m_3 \gamma_{13} \mathbf{a}_{13}}{1 + m_2(\gamma_{12} - 1) + m_3(\gamma_{13} - 1)} \tag{4.55}$$

Since the point P lies on each of the three lines L_k, $k = 1, 2, 3$, there exist values $t_{k,0}$ of the line parameters t_k, $k = 1, 2, 3$, respectively, such that

$$\begin{aligned} P - \frac{\gamma_{12}\mathbf{a}_{12} + \gamma_{13}\mathbf{a}_{13}}{\gamma_{12} + \gamma_{13}} t_{1,0} \qquad\qquad &= 0 \\ P - \mathbf{a}_{12} - \left(-\mathbf{a}_{12} + \frac{\gamma_{13}\mathbf{a}_{13}}{\gamma_{13} + 1} \right) t_{2,0} &= 0 \\ P - \mathbf{a}_{13} - \left(-\mathbf{a}_{13} + \frac{\gamma_{12}\mathbf{a}_{12}}{\gamma_{12} + 1} \right) t_{3,0} &= 0 \end{aligned} \tag{4.56}$$

The kth equation in (4.56), $k = 1, 2, 3$, is equivalent to the condition that point P lies on line L_k.

The system of equations (4.56) was obtained by methods of gyroalgebra, and will be solved below by a common method of linear algebra.

Substituting P from (4.55) into (4.56), and rewriting each equation in (4.56) as a linear combination of $\mathbf{a}_{12}$ and $\mathbf{a}_{13}$ equals zero, one obtains the following linear homogeneous system of three gyrovector equations

$$\begin{aligned} c_{11}\mathbf{a}_{12} + c_{12}\mathbf{a}_{13} &= \mathbf{0} \\ c_{21}\mathbf{a}_{12} + c_{22}\mathbf{a}_{13} &= \mathbf{0} \\ c_{31}\mathbf{a}_{12} + c_{32}\mathbf{a}_{13} &= \mathbf{0} \end{aligned} \tag{4.57}$$

where each coefficient c_{ij}, $i = 1, 2, 3$, $j = 1, 2$, is a function of γ_{12}, γ_{13}, γ_{23}, and the five unknowns m_2, m_3, and $t_{k,0}$, $k = 1, 2, 3$.

Since the set $S = \{A_1, A_2, A_3\}$ is pointwise independent, the two gyrovectors $\mathbf{a}_{12} = \ominus A_1 \oplus A_2$ and $\mathbf{a}_{13} = \ominus A_1 \oplus A_3$ in $\mathbb{R}_s^n$, considered as vectors in $\mathbb{R}^n$, are linearly independent. Hence, each coefficient c_{ij} in (4.57) equals zero. Accordingly, the three gyrovector equations in (4.57) are equivalent to the following six scalar equations,

$$c_{11} = c_{12} = c_{21} = c_{22} = c_{31} = c_{32} = 0 \tag{4.58}$$

for the five unknowns m_2, m_3 and $t_{k,0}$, $k = 1, 2, 3$.

Explicitly, the six scalar equations in (4.58) are equivalent to the following six equations:

$$\begin{aligned}
&[1 + m_2(\gamma_{12} - 1) + m_3(\gamma_{13} - 1)]t_{1,0} - m_2(\gamma_{12} + \gamma_{13}) &&= 0\\
&[1 + m_2(\gamma_{12} - 1) + m_3(\gamma_{13} - 1)]t_{1,0} - m_3(\gamma_{12} + \gamma_{13}) &&= 0\\
&[1 + m_2(\gamma_{12} - 1) + m_3(\gamma_{13} - 1)]t_{2,0} - m_3(\gamma_{13} + 1) &&= 0\\
&[1 + m_2(\gamma_{12} - 1) + m_3(\gamma_{13} - 1)]t_{2,0} - m_3(\gamma_{13} - 1) + m_2 - 1 &&= 0\\
&[1 + m_2(\gamma_{12} - 1) + m_3(\gamma_{13} - 1)]t_{3,0} - m_2(\gamma_{12} + 1) &&= 0\\
&[1 + m_2(\gamma_{12} - 1) + m_3(\gamma_{13} - 1)]t_{3,0} - m_2(\gamma_{12} - 1) + m_3 - 1 &&= 0
\end{aligned} \tag{4.59}$$

The unique solution of (4.59) is given by

$$\begin{aligned}
t_{1,0} &= \frac{\gamma_{12} + \gamma_{13}}{\gamma_{12} + \gamma_{13} + 1}\\
t_{2,0} &= \frac{\gamma_{13} + 1}{\gamma_{12} + \gamma_{13} + 1}\\
t_{3,0} &= \frac{\gamma_{12} + 1}{\gamma_{12} + \gamma_{13} + 1}
\end{aligned} \tag{4.60}$$

and

$$m_2 = m_3 = \frac{1}{3} \tag{4.61}$$

so that by the normalization condition $m_1 + m_2 + m_3 = 1$, also $m_1 = 1/3$.

Hence, the special gyrobarycentric coordinates of the gyrocentroid of a gyrotriangle $A_1A_2A_3$ with respect to the pointwise independent set $\{A_1, A_2, A_3\}$ in an Einstein gyrovector space $(\mathbb{R}^n_s, \oplus, \otimes)$, Fig. 4.1, are given by $(m_1, m_2, m_3) = (\frac{1}{3}, \frac{1}{3}, \frac{1}{3})$, so that convenient gyrobarycentric coordinates of the gyrotriangle gyrocentroid are

$$(m_1 : m_2 : m_3) = (1, 1, 1) \tag{4.62}$$

Finally, following (4.62) and (4.41), the gyrocentroid of a gyrotriangle $A_1A_2A_3$ in an Einstein gyrovector space $(\mathbb{R}^n_s, \oplus, \otimes)$ or, equivalently, in the Cartesian-Beltrami-Klein ball model of hyperbolic geometry, is given by the equation in Fig. 4.1,

$$G = \frac{\gamma_{A_1} A_1 + \gamma_{A_2} A_2 + \gamma_{A_3} A_3}{\gamma_{A_1} + \gamma_{A_2} + \gamma_{A_3}} \tag{4.63}$$

Formalizing the main result of this section, we have the following theorem.

Theorem 4.12. (Einstein Gyrotriangle Gyrocentroid). *Let $S = \{A_1, A_2, A_3\}$ be a pointwise independent set of three points in an Einstein gyrovector space $(\mathbb{R}_s^n, \oplus, \otimes)$. $n \geq 2$, the gyrocentroid G, shown in Fig. 4.1, of gyrotriangle $A_1A_2A_3$ has the gyrobarycentric coordinate representation*

$$G = \frac{\gamma_{A_1} A_1 + \gamma_{A_2} A_2 + \gamma_{A_3} A_3}{\gamma_{A_1} + \gamma_{A_2} + \gamma_{A_3}} \tag{4.64}$$

with respect to the set $\{A_1, A_2, A_3\}$, with gyrobarycentric coordinates

$$(m_1 : m_2 : m_3) = (1 : 1 : 1) \tag{4.65}$$

4.7 Einstein Gyrotetrahedron Gyrocentroid

A hyperbolic tetrahedron centroid is called, in gyrolanguage, a gyrotetrahedron gyrocentroid.

Definition 4.13. (Gyrotetrahedra, Gyromedians and Gyrocentroids). *A gyrotetrahedron in an Einstein or a Möbius gyrovector space $\mathbb{R}_s^n$, $n \geq 3$, is a 3-gyrosimplex $A_1A_2A_3A_4$ with vertices A_k, $k = 1, 2, 3, 4$, in $\mathbb{R}_s^n$.*

A gyromedian of a gyrotetrahedron is the gyrosegment joining a vertex of the gyrotetrahedron with the gyrocentroid of the opposing side, shown in Fig. 4.2 and Fig. 4.4.

The gyrocentroid, G, of a gyrotetrahedron is the point of concurrency of the four gyrotetrahedron gyromedians, shown in Fig. 4.2 and Fig. 4.4.

The similarity between the gyrotriangle gyrocentroid G of gyrotriangle $A_1A_2A_3$ in (4.64) and the gyromidpoint M_{12} of gyrosegment A_1A_2 in (4.40) reveals a remarkable pattern. The extension to higher dimensions is now obvious. Indeed, the gyrocentroid G of a gyrotetrahedron $A_1A_2A_3A_4$ in an Einstein gyrovector space $\mathbb{R}_s^n$, $n \geq 3$, is given by the following theorem:

Theorem 4.14. (Einstein Gyrotetrahedron Gyrocentroid). *Let $S = \{A_1, A_2, A_3, A_4\}$ be a pointwise independent set of four points in an Einstein gyrovector space $(\mathbb{R}_s^n, \oplus, \otimes)$, $n \geq 3$. The gyrocentroid G of gyrotetrahedron $A_1A_2A_3A_4$ has the gyrobarycentric coordinate representation*

$$G = \frac{\gamma_{A_1} A_1 + \gamma_{A_2} A_2 + \gamma_{A_3} A_3 + \gamma_{A_4} A_4}{\gamma_{A_1} + \gamma_{A_2} + \gamma_{A_3} + \gamma_{A_4}} \tag{4.66}$$

with respect to the set $\{A_1, A_2, A_3\}$, shown in Fig. 4.2, with gyrobarycentric coordinates

$$(m_1 : m_2 : m_3 : m_4) = (1 : 1 : 1 : 1) \tag{4.67}$$

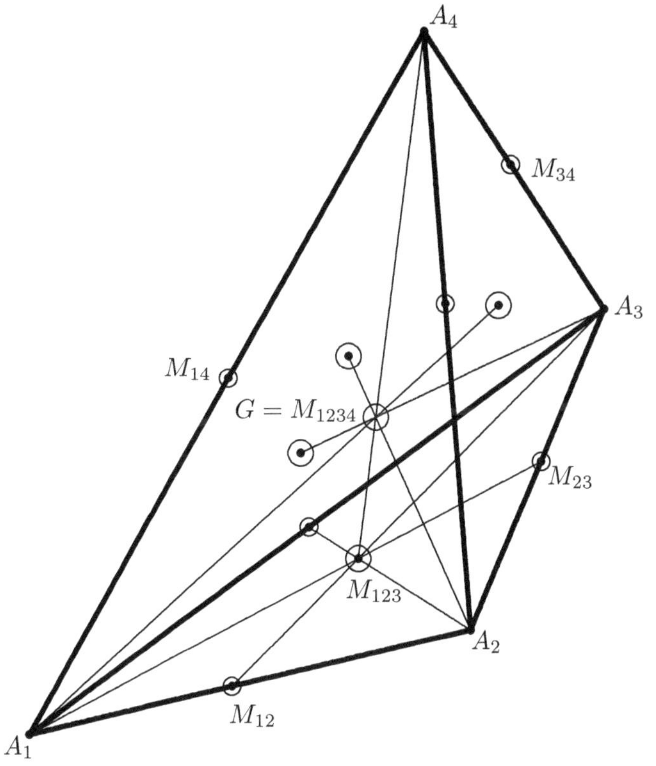

Fig. 4.2 A gyrotetrahedron $A_1A_2A_3A_4$ in an Einstein gyrovector space $(\mathbb{R}^n_s, \oplus, \otimes)$, $n = 3$, along with its four gyromedians and gyrocentroid $G = M_{1234}$. Also the edge gyromidpoints and the gyrocentroid $G = M_{123}$ of side $A_1A_2A_3$ of the gyrotetrahedron are shown.

Proof. The proof of Theorem 4.14 is similar to that of Theorem 4.12. □

The pattern in (4.66) that gyrobarycentric coordinate representations of gyrocentroids possess survives unimpaired for the gyrocentroids of n-dimensional gyrosimplices in Einstein gyrovector spaces $(\mathbb{R}^n_s, \oplus, \otimes)$ for any $n \geq 2$, as shown in [19, Eq. (10.31)].

4.8 Möbius Gyrotriangle Gyrocentroid

We determine the gyrotriangle gyrocentroid in Möbius gyrovector spaces by transforming the gyrobarycentric coordinate representation of the gyrotriangle gyrocentroid from Einstein to Möbius gyrovector spaces by means of Theorem 4.6, p. 177.

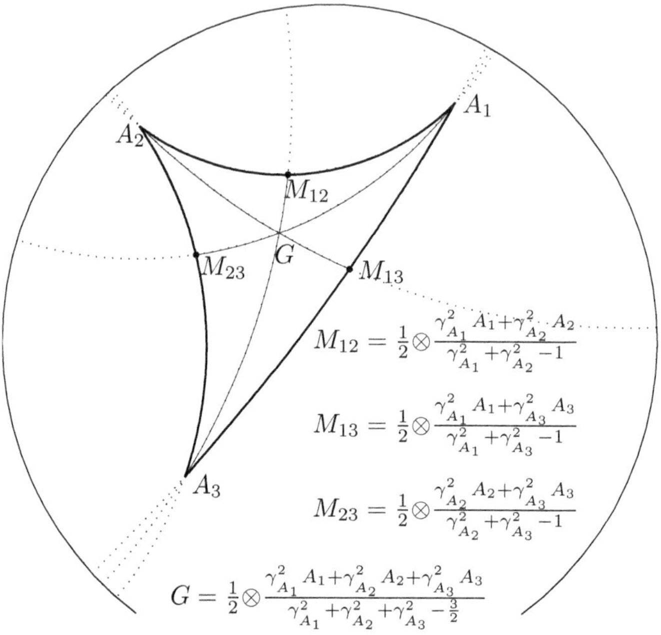

Fig. 4.3 The gyromidpoints M_{12}, M_{13}, and M_{23} of the three sides, A_1A_2, A_1A_3 and A_2A_3, of gyrotriangle $A_1A_2A_3$ in a Möbius gyrovector space $(\mathbb{R}_s^n, \oplus, \otimes)$ are shown here for $n = 2$, along with its gyromedians $A_1M_{23}, A_2M_{13}, A_3M_{12}$, and its gyrocentroid G.

Following Theorem 4.6, the gyromidpoints of the sides of gyrotriangle $A_1A_2A_3$ in a Möbius gyrovector space $(\mathbb{R}_s^n, \oplus, \otimes)$, $n \geq 2$, shown in Fig. 4.3, are determined by the transformation of (4.40), p. 183, from Einstein to Möbius gyrovector spaces, obtaining

$$\begin{aligned}
M_{12} &= \tfrac{1}{2}\otimes\frac{\gamma_{A_1}^2 A_1+\gamma_{A_2}^2 A_2}{\gamma_{A_1}^2+\gamma_{A_2}^2-1} \\
M_{13} &= \tfrac{1}{2}\otimes\frac{\gamma_{A_1}^2 A_1+\gamma_{A_3}^2 A_3}{\gamma_{A_1}^2+\gamma_{A_3}^2-1} \\
M_{23} &= \tfrac{1}{2}\otimes\frac{\gamma_{A_2}^2 A_2+\gamma_{A_3}^2 A_3}{\gamma_{A_2}^2+\gamma_{A_3}^2-1}
\end{aligned} \tag{4.68}$$

Similarly, the gyrocentroid G of gyrotriangle $A_1A_2A_3$ in a Möbius gyrovector space, shown in Fig. 4.3, is the transformation of (4.41), p. 183, from Einstein to Möbius gyrovector spaces, obtaining

$$G = \tfrac{1}{2}\otimes_{\mathrm{M}}\frac{\gamma_{A_1}^2 A_1+\gamma_{A_2}^2 A_2+\gamma_{A_3}^2 A_3}{\gamma_{A_1}^2+\gamma_{A_2}^2+\gamma_{A_3}^2-\frac{3}{2}} \tag{4.69}$$

Formalizing the main result of this section, we have the following theorem:

Theorem 4.15. (The Möbius Gyrotriangle Gyrocentroid). *Let $S = \{A_1, A_2, A_3\}$ be a pointwise independent set of three points in a Möbius gyrovector space $(\mathbb{R}^n_s, \oplus, \otimes)$, $n \geq 2$. The gyrocentroid G of gyrotriangle $A_1A_2A_3$, shown in Fig. 4.3, has the gyrobarycentric coordinate representation*

$$G = \tfrac{1}{2}\otimes \frac{\gamma^2_{A_1} A_1 + \gamma^2_{A_2} A_2 + \gamma^2_{A_3} A_3}{\gamma^2_{A_1} + \gamma^2_{A_2} + \gamma^2_{A_3} - \frac{3}{2}} \tag{4.70}$$

with respect to the set $\{A_1, A_2, A_3\}$, with gyrobarycentric coordinates

$$(m_1 : m_2 : m_3) = (1 : 1 : 1) \tag{4.71}$$

4.9 Möbius Gyrotetrahedron Gyrocentroid

We determine the gyrotetrahedron gyrocentroid in Möbius gyrovector spaces by transforming the gyrobarycentric coordinate representation of the gyrotetrahedron gyrocentroid from Einstein into Möbius gyrovector spaces by means of Theorem 4.6, p. 177. Accordingly, the transformation of Theorem 4.14, p. 188, from Einstein into Möbius gyrovector spaces results in the following theorem:

Theorem 4.16. (Möbius Gyrotetrahedron Gyrocentroid). *Let $S = \{A_1, A_2, A_3, A_4\}$ be a pointwise independent set of four points in a Möbius gyrovector space $(\mathbb{R}^n_s, \oplus, \otimes)$. $n \geq 3$, the gyrocentroid G of gyrotetrahedron $A_1A_2A_3A_4$ has the gyrobarycentric coordinate representation*

$$G = \tfrac{1}{2}\otimes \frac{\gamma^2_{A_1} A_1 + \gamma^2_{A_2} A_2 + \gamma^2_{A_3} A_3 + \gamma^2_{A_4} A_4}{\gamma^2_{A_1} + \gamma^2_{A_2} + \gamma^2_{A_3} + \gamma^2_{A_4} - 2} \tag{4.72}$$

with respect to the set $\{A_1, A_2, A_3\}$, shown in Fig. 4.4, with gyrobarycentric coordinates

$$(m_1 : m_2 : m_3 : m_4) = (1 : 1 : 1 : 1) \tag{4.73}$$

4.10 Foot of a Gyrotriangle Gyroaltitude

Let A_3P_3 be the gyroaltitude of gyrotriangle $A_1A_2A_3$ drawn from vertex A_3 to its foot P_3 on its opposite side A_1A_2 in an Einstein gyrovector space $(\mathbb{R}^n_s, \oplus, \otimes)$, $n \geq 2$, as shown in Fig. 4.5. Furthermore, let

$$P_3 = \frac{m_1\gamma_{A_1} A_1 + m_2\gamma_{A_2} A_2}{m_1\gamma_{A_1} + m_2\gamma_{A_2}} \tag{4.74}$$

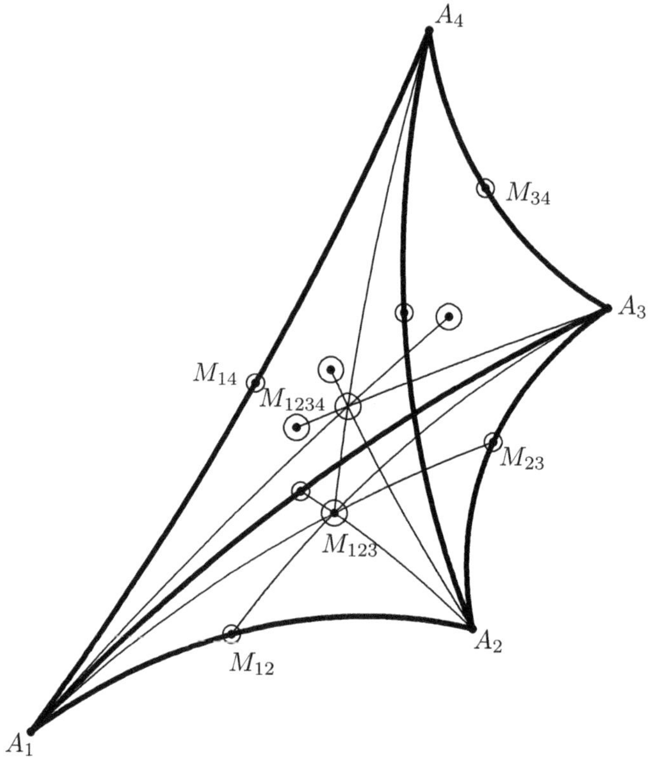

Fig. 4.4 A gyrotetrahedron $A_1A_2A_3A_4$ in a Möbius gyrovector space $(\mathbb{R}_s^n, \oplus, \otimes)$, $n = 3$, along with its four gyromedians and gyrocentroid $G_t = M_{1234}$. Also the edge gyromidpoints and the gyrocentroid $G = M_{123}$ of side $A_1A_2A_3$ of the gyrotetrahedron are shown.

be the gyrobarycentric coordinate representation of P_3 with respect to the set $\{A_1, A_2\}$, where the gyrobarycentric coordinates (m_1, m_2) are to be determined in (4.85) – (4.87).

Employing the gyroalgebraic relations (4.6) – (4.12) in Theorem 4.4, p. 173, we have from the second identity in (4.11) and from (4.10),

$$\gamma_{\ominus X \oplus P_3} = \frac{m_1 \gamma_{\ominus X \oplus A_1} + m_2 \gamma_{\ominus X \oplus A_2}}{m_0} \tag{4.75}$$

where

$$m_0^2 = m_1^2 + m_2^2 + 2m_1m_2\gamma_{12} \tag{4.76}$$

Hence, for $X = A_1$, $X = A_2$ and $X = A_3$ in (4.75) we have, respectively, in

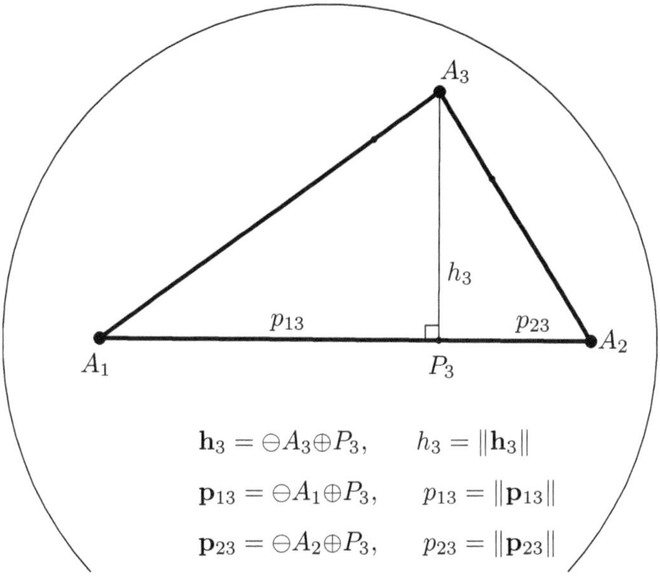

Fig. 4.5 The foot P_3 of gyroaltitude A_3P_3 of gyrotriangle $A_1A_2A_3$ in an Einstein gyrovector space $(\mathbb{R}^n_s, \oplus, \otimes)$.

the notation presented in Fig. 2.3, p. 101, in (2.133), p. 100, and in Fig. 4.5,

$$\begin{aligned}
\gamma_{p_{13}} = \gamma_{\ominus A_1 \oplus P_3} &= \frac{m_1 + m_2\gamma_{\ominus A_1 \oplus A_2}}{m_0} = \frac{m_1 + m_2\gamma_{12}}{m_0} \\
\gamma_{p_{23}} = \gamma_{\ominus A_2 \oplus P_3} &= \frac{m_1\gamma_{\ominus A_2 \oplus A_1} + m_2}{m_0} = \frac{m_1\gamma_{12} + m_2}{m_0} \\
\gamma_{h_3} = \gamma_{\ominus A_3 \oplus P_3} &= \frac{m_1\gamma_{\ominus A_3 \oplus A_1} + m_2\gamma_{\ominus A_3 \oplus A_2}}{m_0} = \frac{m_1\gamma_{13} + m_2\gamma_{23}}{m_0}
\end{aligned} \tag{4.77}$$

Applying the Einstein-Pythagoras identity (2.179), p. 113 to each of the two right gyroangled gyrotriangles $A_1P_3A_3$ and $A_2P_3A_3$ in Fig. 4.5 we have

$$\begin{aligned}
\gamma_{p_{13}}\gamma_{h_3} &= \gamma_{13} \\
\gamma_{p_{23}}\gamma_{h_3} &= \gamma_{23}
\end{aligned} \tag{4.78}$$

Substituting (4.76) – (4.77) into (4.78), we obtain a system of two equations for the two unknowns m_1 and m_2. This system does not possess a unique solution. Adding the normalization condition $m_1 + m_2 = 1$ results in the unique solution,

$$\begin{aligned} m_1 &= \frac{\gamma_{12}\gamma_{23} - \gamma_{13}}{(\gamma_{13} + \gamma_{23})(\gamma_{12} - 1)} \\ m_2 &= \frac{\gamma_{12}\gamma_{13} - \gamma_{23}}{(\gamma_{13} + \gamma_{23})(\gamma_{12} - 1)} \end{aligned} \tag{4.79}$$

as one can readily check. The unique special gyrobarycentric coordinates (m_1, m_2) of the point P_3 with respect to the set $S = \{A_1, A_2\}$ in Fig. 4.5 are thus determined by (4.79).

The special gyrobarycentric coordinates (m_1, m_2) in (4.79) suggest the following convenient gyrobarycentric coordinates $(m_1' : m_2')$ of the point P_3 with respect to the set $S = \{A_1, A_2\}$,

$$\begin{aligned} m_1' &= \gamma_{12}\gamma_{23} - \gamma_{13} \\ m_2' &= \gamma_{12}\gamma_{13} - \gamma_{23} \end{aligned} \tag{4.80}$$

so that the gyrobarycentric coordinate representation (4.74) of P_3 with respect to the set $S = \{A_1, A_2\}$ is given by

$$P_3 = \frac{(\gamma_{12}\gamma_{23} - \gamma_{13})\gamma_{A_1} A_1 + (\gamma_{12}\gamma_{13} - \gamma_{23})\gamma_{A_2} A_2}{(\gamma_{12}\gamma_{23} - \gamma_{13})\gamma_{A_1} + (\gamma_{12}\gamma_{13} - \gamma_{23})\gamma_{A_2}} \tag{4.81}$$

A different convenient gyrobarycentric coordinate set $(m_1'' : m_2'')$ of P_3 with respect to $\{A_1, A_2\}$ can be obtained from (4.80) by means of (2.143), p. 103,

$$\begin{aligned} (m_1'' : m_2'') &= (\gamma_{12}\gamma_{23} - \gamma_{13} : \gamma_{12}\gamma_{13} - \gamma_{23}) \frac{1}{\sqrt{\gamma_{12}^2 - 1}\sqrt{\gamma_{13}^2 - 1}\sqrt{\gamma_{23}^2 - 1}} \\ &= \left(\frac{\cos\alpha_2}{\sqrt{\gamma_{13}^2 - 1}} : \frac{\cos\alpha_1}{\sqrt{\gamma_{23}^2 - 1}} \right) \end{aligned} \tag{4.82}$$

The advantage of the gyrobarycentric coordinates $(m_1'' : m_2'')$ of P_3 with respect to $\{A_1, A_2\}$ in (4.82) rests on the observation that the sign of m_1'' (m_2'') equals the sign of $\cos\alpha_2$ $(\cos\alpha_1)$.

Another set of convenient gyrobarycentric coordinates $(m_1''' : m_2''')$ of P_3 with respect to $\{A_1, A_2\}$ results from (4.82) and (2.163), p. 109, obtaining the following gyrotrigonometric gyrobarycentric coordinates:

$$(m_1''' : m_2''') = (\sin\alpha_1 \cos\alpha_2 : \cos\alpha_1 \sin\alpha_2) = (\tan\alpha_1 : \tan\alpha_2) \tag{4.83}$$

where α_k, $k = 1, 2, 3$ are the gyroangles of gyrotriangle $A_1A_2A_3$ in Fig. 4.5, in the standard gyrotriangle notation.

Interestingly, the gyrotrigonometric gyrobarycentric coordinates $(m_1''' : m_2''')$ of a gyrotriangle gyroaltitude foot in (4.83) are identical in form with their Euclidean counterparts in (1.69), p. 21. We will find that this is always the case when gyrobarycentric coordinates are given trigonometrically (in Euclidean geometry) and gyrotrigonometry (in hyperbolic geometry).

Formalizing the main result of this section, we have the following theorem.

Theorem 4.17. (The Foot of an Einstein Gyrotriangle Gyroaltitude). *Let $S = \{A_1, A_2, A_3\}$ be a pointwise independent set of three points in an Einstein gyrovector space $(\mathbb{R}_s^n, \oplus, \otimes)$, and let P_3 be the foot of gyroaltitude A_3P_3, as shown in Fig. 4.5. Then the gyroaltitude foot has the gyrobarycentric coordinate representation*

$$P_3 = \frac{m_1\gamma_{A_1}A_1 + m_2\gamma_{A_2}A_2}{m_1\gamma_{A_1} + m_2\gamma_{A_2}} \tag{4.84}$$

with respect to the set $\{A_1, A_2\}$, with gyrobarycentric coordinates

$$(m_1 : m_2) = (\gamma_{12}\gamma_{23} - \gamma_{13} : \gamma_{12}\gamma_{13} - \gamma_{23}) \tag{4.85}$$

or, equivalently,

$$(m_1 : m_2) = \left(\frac{\cos\alpha_2}{\sqrt{\gamma_{13}^2 - 1}} : \frac{\cos\alpha_1}{\sqrt{\gamma_{23}^2 - 1}}\right) \tag{4.86}$$

or, equivalently, with gyrotrigonometric gyrobarycentric coordinates

$$(m_1 : m_2) = (\sin\alpha_1\cos\alpha_2 : \cos\alpha_1\sin\alpha_2) = (\cot\alpha_2 : \cot\alpha_1) \tag{4.87}$$

4.11 Einstein Point to Gyroline Gyrodistance

Guided by analogies with Euclidean geometry we say that the point P_3 in Fig. 4.5 is the gyroperpendicular projection of the point A_3 on the gyroline $L_{A_1A_2}$ that passes through the points A_1 and A_2. It is useful to reformulate Theorem 4.17 in terms of a point to gyroline gyroperpendicular projection, obtaining the following theorem:

Theorem 4.18. (Point to Gyroline Gyroperpendicular Projection, Einstein). *Let A_1 and A_2 be any two distinct points of an Einstein gyrovector space $(\mathbb{R}_s^n, \oplus, \otimes)$, and let $L_{A_1A_2}$ be the gyroline passing through these points. Furthermore, let A_3 be any point of the space that does not* *lie on $L_{A_1A_2}$, as shown in Fig. 4.5. Then, in the notation of Fig. 4.5 and Fig. 2.3, p. 101, the gyroperpendicular projection of the point A_3 on the gyroline $L_{A_1A_2}$ is the point P_3 on the gyroline given by its gyrobarycentric coordinate representation,* (4.84),

$$P_3 = \frac{m_1\gamma_{A_1} A_1 + m_2\gamma_{A_2} A_2}{m_1\gamma_{A_1} + m_2\gamma_{A_2}} \tag{4.88}$$

with respect to the set $\{A_1, A_2\}$, with gyrobarycentric coordinates

$$(m_1 : m_2) = (\gamma_{12}\gamma_{23} - \gamma_{13} : \gamma_{12}\gamma_{13} - \gamma_{23}) \tag{4.89}$$

or, equivalently,

$$(m_1 : m_2) = \left(\frac{\cos\alpha_2}{\sqrt{\gamma_{13}^2 - 1}} : \frac{\cos\alpha_1}{\sqrt{\gamma_{23}^2 - 1}} \right) \tag{4.90}$$

or, equivalently, with gyrotrigonometric gyrobarycentric coordinates

$$(m_1 : m_2) = (\sin\alpha_1\cos\alpha_2 : \cos\alpha_1\sin\alpha_2) = (\tan\alpha_1 : \tan\alpha_2) \tag{4.91}$$

Theorem 4.19. (Point to Gyroline Gyrodistance, Einstein). *Let A_1 and A_2 be any two distinct points of an Einstein gyrovector space $(\mathbb{R}^n_s, \oplus, \otimes)$, and let $L_{A_1A_2}$ be the gyroline passing through these points. Furthermore, let A_3 be any point of the space that does not lie on $L_{A_1A_2}$, as shown in Fig. 4.5. Then, in the notation of Fig. 4.5 and Fig. 2.3, p. 101, the gyrodistance $h_3 = \|\ominus A_3 \oplus P_3\|$ between the point P_3 and the gyroline on $L_{A_1A_2}$ is given by the equation*

$$h_3 = s\sqrt{\frac{1 + 2\gamma_{12}\gamma_{13}\gamma_{23} - \gamma_{12}^2 - \gamma_{13}^2 - \gamma_{23}^2}{2\gamma_{12}\gamma_{13}\gamma_{23} - \gamma_{13}^2 - \gamma_{23}^2}} \tag{4.92}$$

satisfying

$$\gamma_{h_3} = \sqrt{\frac{2\gamma_{12}\gamma_{13}\gamma_{23} - \gamma_{13}^2 - \gamma_{23}^2}{\gamma_{12}^2 - 1}} \tag{4.93}$$

Proof. By the third equation in (4.77) we have

$$\gamma_{h_3} = \frac{m_1\gamma_{13} + m_2\gamma_{23}}{m_0} \tag{4.94}$$

where, by (4.80),

$$\begin{aligned} m_1 &= \gamma_{12}\gamma_{23} - \gamma_{13} \\ m_2 &= \gamma_{12}\gamma_{13} - \gamma_{23} \end{aligned} \tag{4.95}$$

Substituting (4.95) into (4.76) yields

$$m_0^2 = (\gamma_{12}^2 - 1)(2\gamma_{12}\gamma_{13}\gamma_{23} - \gamma_{13}^2 - \gamma_{23}^2) \tag{4.96}$$

Substituting (4.95) – (4.96) into (4.94), we obtain the result (4.93) of the Theorem. The result (4.92) of the Theorem follows from (4.93). Indeed, by (2.11), p. 66, and (4.93) we have

$$h_3^2 = s^2 \frac{\gamma_{h_3}^2 - 1}{\gamma_{h_3}^2} = s^2 \frac{1 + 2\gamma_{12}\gamma_{13}\gamma_{23} - \gamma_{12}^2 - \gamma_{13}^2 - \gamma_{23}^2}{2\gamma_{12}\gamma_{13}\gamma_{23} - \gamma_{13}^2 - \gamma_{23}^2} \tag{4.97}$$

□

Theorem 4.19 is proved here by means of gyrobarycentric coordinates. In contrast, the same Theorem 2.39, p. 122, is proved by means of gyrotrigonometry.

As an immediate consequence of Theorem 2.39 we have the following Corollary.

Corollary 4.20. *Let $A_1, A_2, A_3 \in \mathbb{R}_s^n$ be any three points of an Einstein gyrovector space $\mathbb{R}_s^n$. Then, in the notation of* (2.133), *p. 100,*

$$1 + 2\gamma_{12}\gamma_{13}\gamma_{23} - \gamma_{12}^2 - \gamma_{13}^2 - \gamma_{23}^2 \geq 0 \tag{4.98}$$

where equality holds if and only if the three points are gyrocollinear.

4.12 Möbius Point to Gyroline Gyrodistance

In this section we transform the results of Sec. 4.11 from Einstein gyrovector spaces into Möbius gyrovector spaces.

Theorem 4.21. (Point to Gyroline Gyroperpendicular Projection, Möbius). *Let A_1 and A_2 be any two distinct points of a Möbius gyrovector space $(\mathbb{R}_s^n, \oplus, \otimes)$, and let $L_{A_1A_2}$ be the gyroline passing through these points. Furthermore, let A_3 be any point of the space that does not lie on $L_{A_1A_2}$, as shown in Fig. 4.6. Then, in the notation of Fig. 4.6 and Fig. 2.15, p. 140, the gyroperpendicular projection of the point A_3 on the gyroline $L_{A_1A_2}$ is the point P_3 on the gyroline given by its gyrobarycentric coordinate representation,* (4.19), *p. 177,*

$$P_3 = \tfrac{1}{2} \otimes \frac{m_1\gamma_{A_1}^2 A_1 + m_2\gamma_{A_2}^2 A_2}{m_1\gamma_{A_1}^2 + m_2\gamma_{A_2}^2 - \frac{1}{2}(m_1 + m_2)} \tag{4.99}$$

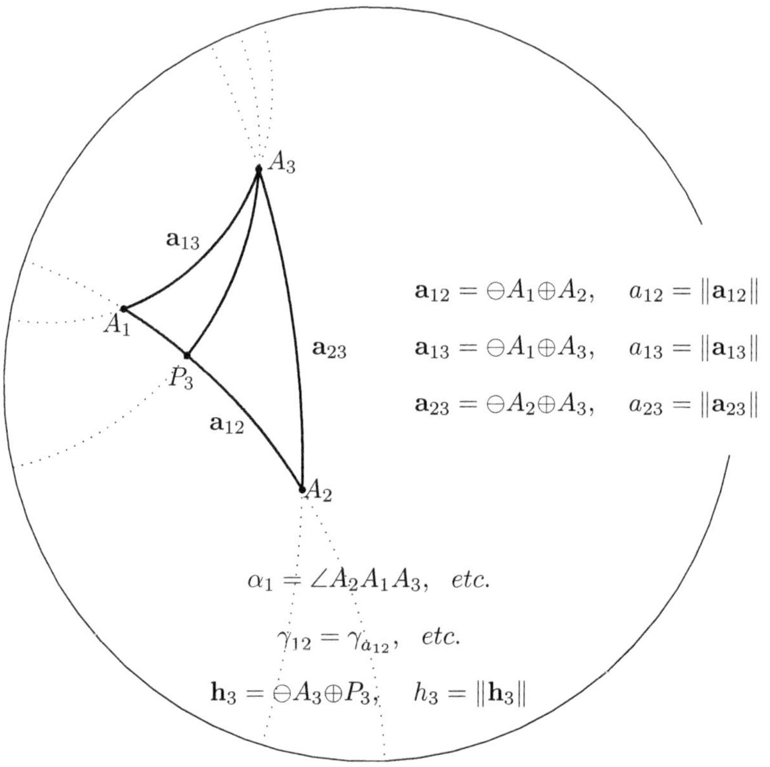

Fig. 4.6 The foot P_3 of gyroaltitude A_3P_3 of gyrotriangle $A_1A_2A_3$ in a Möbius gyrovector space $(\mathbb{R}^n_s, \oplus, \otimes)$.

with respect to the set $\{A_1, A_2\}$, *with gyrobarycentric coordinates*

$$(m_1 : m_2) = (2\gamma_{12}^2\gamma_{23}^2 - \gamma_{12}^2 - \gamma_{13}^2 - \gamma_{23}^2 + 1 : 2\gamma_{12}^2\gamma_{13}^2 - \gamma_{12}^2 - \gamma_{13}^2 - \gamma_{23}^2 + 1) \tag{4.100}$$

or, equivalently,

$$(m_1 : m_2) = \left(\frac{\cos\alpha_2}{\gamma_{13}\sqrt{\gamma_{13}^2 - 1}} : \frac{\cos\alpha_1}{\gamma_{23}\sqrt{\gamma_{23}^2 - 1}} \right) \tag{4.101}$$

or, equivalently, with gyrotrigonometric gyrobarycentric coordinates

$$(m_1 : m_2) = (\sin\alpha_1\cos\alpha_2 : \cos\alpha_1\sin\alpha_2) = (\tan\alpha_1 : \tan\alpha_2) \tag{4.102}$$

Proof. The gyrobarycentric coordinate representation (4.99) of P_3 in the Möbius gyrovector space $\mathbb{R}^n_s$ is the isomorphic image of its corresponding point P_3 in the corresponding Einstein gyrovector space $\mathbb{R}_s^n$ with the gyrobarycentric coordinate representation (4.88), according to Theorem 4.6, p. 177.

The gyrobarycentric coordinate set $(m_1 : m_2)$ in (4.100) is the isomorphic image of its corresponding gyrobarycentric coordinate set $(m_1 : m_2)$ in (4.89) according to the isomorphism between gamma factors in (2.280), p. 142.

The gyrobarycentric coordinate set $(m_1 : m_2)$ in (4.101) is the isomorphic image of its corresponding gyrobarycentric coordinate set $(m_1 : m_2)$ in (4.90) according to the isomorphism between gamma factors in (2.280), p. 142. Here we should note that gyroangles remain invariant under the isomorphism between Einstein and Möbius gyrovector spaces, as stated in Theorem 2.48, p. 143.

The gyrotrigonometric gyrobarycentric coordinate set $(m_1 : m_2)$ in (4.102) in the Möbius gyrovector space is identical with its isomorphic image (4.91) in the corresponding Einstein gyrovector space owing to the invariance of gyroangles under the isomorphism between Einstein and Möbius gyrovector spaces according to Theorem 2.48. □

Theorem 4.22. (Point to Gyroline Gyrodistance, Möbius). *Let A_1 and A_2 be any two distinct points of a Möbius gyrovector space $(\mathbb{R}_s^n, \oplus, \otimes)$, and let $L_{A_1A_2}$ be the gyroline passing through these points. Furthermore, let A_3 be any point of the space that does not lie on $L_{A_1A_2}$, as shown in Fig. 4.6. Then, in the notation of Fig. 4.6 and Fig. 2.15, p. 140, the gyrodistance $h_3 = \|\ominus A_3 \oplus P_3\|$ between the point A_3 and the gyroline $L_{A_1A_2}$ is given by the equation*

$$2\otimes h_3 = s\sqrt{\frac{4\gamma_{12}^2\gamma_{13}^2\gamma_{23}^2 - (\gamma_{12}^2+\gamma_{13}^2+\gamma_{23}^2-1)^2}{\gamma_{12}^2(\gamma_{12}^2+4\gamma_{13}^2\gamma_{23}^2-1)-(\gamma_{12}^2+\gamma_{13}^2+\gamma_{23}^2-1)^2}} \tag{4.103}$$

satisfying

$$\gamma_{2\otimes h_3} = \sqrt{\frac{\gamma_{12}^2(\gamma_{12}^2+4\gamma_{13}^2\gamma_{23}^2-1)-(\gamma_{12}^2+\gamma_{13}^2+\gamma_{23}^2-1)^2}{\gamma_{12}^2(\gamma_{12}^2-1)}} \tag{4.104}$$

Proof. Equations (4.103) – (4.104) result from the transformation of (4.92) – (4.93) from Einstein to Möbius gyrovector spaces by means of the isomorphism in (2.279), p. 142, as explained in Sec. 2.29. □

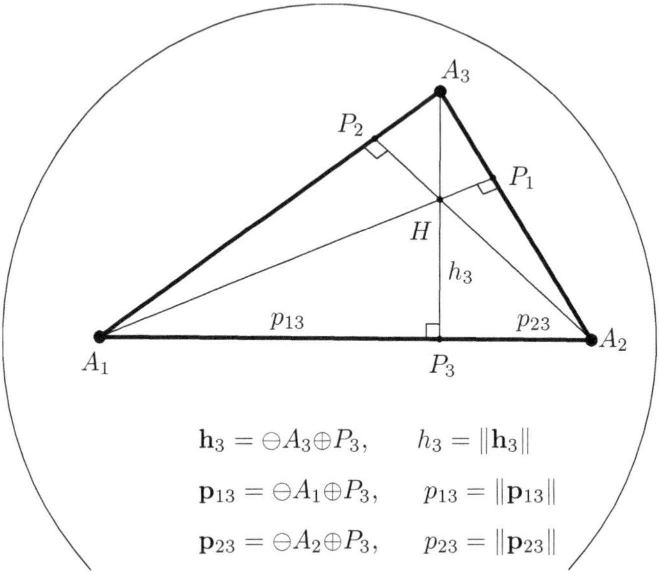

Fig. 4.7 The orthogyrocenter H of a gyrotriangle $A_1A_2A_3$ in an Einstein gyrovector space $(\mathbb{R}_s^n, \oplus, \otimes)$. Here the orthogyrocenter lies inside its gyrotriangle. There are gyrotriangles with their orthogyrocenters lying out of their gyrotriangles, and there are gyrotriangles that possess no orthogyrocenters, as shown in Figs. 4.8 – 4.11.

4.13 Einstein Gyrotriangle Orthogyrocenter

The hyperbolic triangle orthocenter, H, shown in Fig. 4.7, is called in gyrolanguage a gyrotriangle orthogyrocenter.

Definition 4.23. The orthogyrocenter, H, of a gyrotriangle is the point of concurrency of the gyrotriangle gyroaltitudes.

The three feet, P_1, P_2 and P_3 of the three gyroaltitudes of gyrotriangle $A_1A_2A_3$ in an Einstein gyrovector space $(\mathbb{R}_s^n, \oplus, \otimes)$, shown in Fig. 4.7 for $n = 2$, are given by

$$
\begin{aligned}
P_1 &= \frac{(\gamma_{13}\gamma_{23} - \gamma_{12})\gamma_{A_2} A_2 + (\gamma_{12}\gamma_{23} - \gamma_{13})\gamma_{A_3} A_3}{(\gamma_{13}\gamma_{23} - \gamma_{12})\gamma_{A_2} + (\gamma_{12}\gamma_{23} - \gamma_{13})\gamma_{A_3}} \\
P_2 &= \frac{(\gamma_{13}\gamma_{23} - \gamma_{12})\gamma_{A_1} A_1 + (\gamma_{12}\gamma_{13} - \gamma_{23})\gamma_{A_3} A_3}{(\gamma_{13}\gamma_{23} - \gamma_{12})\gamma_{A_1} + (\gamma_{12}\gamma_{13} - \gamma_{23})\gamma_{A_3}} \\
P_3 &= \frac{(\gamma_{12}\gamma_{23} - \gamma_{13})\gamma_{A_1} A_1 + (\gamma_{12}\gamma_{13} - \gamma_{23})\gamma_{A_2} A_2}{(\gamma_{12}\gamma_{23} - \gamma_{13})\gamma_{A_1} + (\gamma_{12}\gamma_{13} - \gamma_{23})\gamma_{A_2}}
\end{aligned}
\tag{4.105}
$$

The third equation in (4.105) is a copy of (4.81), which was established in Sec. 4.10. The first and second equations in (4.105) are obtained from the third one by cyclic permutations of the vertices of gyrotriangle $A_1A_2A_3$.

The gyroaltitudes of gyrotriangle $A_1A_2A_3$ in an Einstein gyrovector space $(\mathbb{R}_s^n, \oplus, \otimes)$, shown in Fig. 4.7 for $n = 2$, are the gyrosegments A_1P_1, A_2P_2, and A_1P_3. Since gyrosegments in Einstein gyrovector spaces coincide with Euclidean segments, one can employ methods of linear algebra to determine the point of concurrency, that is the orthogyrocenter, of the three gyroaltitudes of gyrotriangle $A_1A_2A_3$ in Fig. 4.7.

In order to determine the gyrobarycentric coordinates of the gyrotriangle orthogyrocenter in Einstein gyrovector spaces we begin with some gyroalgebraic manipulations that reduce the task we face to a problem in linear algebra.

Let the orthogyrocenter H of gyrotriangle $A_1A_2A_3$ in an Einstein gyrovector space $(\mathbb{R}_s^n, \oplus, \otimes)$, shown in Fig. 4.7, be given in terms of its gyrobarycentric coordinate representation with respect to the set $S = \{A_1, A_2, A_3\}$ of the gyrotriangle vertices by the equation

$$H = \frac{m_1\gamma_{A_1}A_1 + m_2\gamma_{A_2}A_2 + m_3\gamma_{A_3}A_3}{m_1\gamma_{A_1} + m_2\gamma_{A_2} + m_3\gamma_{A_3}} \tag{4.106}$$

where the gyrobarycentric coordinates (m_1, m_2, m_3) of H in (4.106) are to be determined in Theorem 4.24.

Left gyrotranslating gyrotriangle $A_1A_2A_3$ by $\ominus A_1$, the gyrotriangle becomes gyrotriangle $O(\ominus A_1 \oplus A_2)(\ominus A_1 \oplus A_3)$, where $O = \ominus A_1 \oplus A_1$ is the arbitrarily selected origin of the Einstein gyrovector space $\mathbb{R}_s^n$. The gyrotriangle gyroaltitude feet P_1, P_2 and P_3 become, respectively, $\ominus A_1 \oplus P_1$, $\ominus A_1 \oplus P_2$ and $\ominus A_1 \oplus P_3$. These are calculated in (4.107) below by means of the gyroalgebraic relations (4.6) and the first identity in (4.11) in Theorem 4.4, p. 173, and the standard gyrotriangle notation, shown in Fig. 2.3, p. 101 and in (2.133), p. 100:

$$\begin{aligned}
\ominus A_1 \oplus P_1 &= \ominus A_1 \oplus \frac{(\gamma_{13}\gamma_{23} - \gamma_{12})\gamma_{A_2}A_2 + (\gamma_{12}\gamma_{23} - \gamma_{13})\gamma_{A_3}A_3}{(\gamma_{13}\gamma_{23} - \gamma_{12})\gamma_{A_2} + (\gamma_{12}\gamma_{23} - \gamma_{13})\gamma_{A_3}} \\
&= \frac{(\gamma_{13}\gamma_{23} - \gamma_{12})\gamma_{\ominus A_1 \oplus A_2}(\ominus A_1 \oplus A_2) + (\gamma_{12}\gamma_{23} - \gamma_{13})\gamma_{\ominus A_1 \oplus A_3}(\ominus A_1 \oplus A_3)}{(\gamma_{13}\gamma_{23} - \gamma_{12})\gamma_{\ominus A_1 \oplus A_2} + (\gamma_{12}\gamma_{23} - \gamma_{13})\gamma_{\ominus A_1 \oplus A_3}} \\
&= \frac{(\gamma_{13}\gamma_{23} - \gamma_{12})\gamma_{12}\mathbf{a}_{12} + (\gamma_{12}\gamma_{23} - \gamma_{13})\gamma_{13}\mathbf{a}_{13}}{(\gamma_{13}\gamma_{23} - \gamma_{12})\gamma_{12} + (\gamma_{12}\gamma_{23} - \gamma_{13})\gamma_{13}}
\end{aligned} \tag{4.107a}$$

$$\ominus A_1 \oplus P_2 = \ominus A_1 \oplus \frac{(\gamma_{13}\gamma_{23} - \gamma_{12})\gamma_{_{A_1}} A_1 + (\gamma_{12}\gamma_{13} - \gamma_{23})\gamma_{_{A_3}} A_3}{(\gamma_{13}\gamma_{23} - \gamma_{12})\gamma_{_{A_1}} + (\gamma_{12}\gamma_{13} - \gamma_{23})\gamma_{_{A_3}}}$$

$$= \frac{(\gamma_{12}\gamma_{13} - \gamma_{23})\gamma_{_{\ominus A_1 \oplus A_3}}(\ominus A_1 \oplus A_3)}{(\gamma_{13}\gamma_{23} - \gamma_{12}) + (\gamma_{12}\gamma_{13} - \gamma_{23})\gamma_{_{\ominus A_1 \oplus A_3}}} \tag{4.107b}$$

$$= \frac{(\gamma_{12}\gamma_{13} - \gamma_{23})\gamma_{13}\mathbf{a}_{13}}{(\gamma_{13}\gamma_{23} - \gamma_{12}) + (\gamma_{12}\gamma_{13} - \gamma_{23})\gamma_{13}}$$

$$\ominus A_1 \oplus P_3 = \ominus A_1 \oplus \frac{(\gamma_{12}\gamma_{23} - \gamma_{13})\gamma_{_{A_1}} A_1 + (\gamma_{12}\gamma_{13} - \gamma_{23})\gamma_{_{A_2}} A_2}{(\gamma_{12}\gamma_{23} - \gamma_{13})\gamma_{_{A_1}} + (\gamma_{12}\gamma_{13} - \gamma_{23})\gamma_{_{A_2}}}$$

$$= \frac{(\gamma_{12}\gamma_{13} - \gamma_{23})\gamma_{_{\ominus A_1 \oplus A_2}}(\ominus A_1 \oplus A_2)}{(\gamma_{12}\gamma_{23} - \gamma_{13}) + (\gamma_{12}\gamma_{13} - \gamma_{23})\gamma_{_{\ominus A_1 \oplus A_2}}} \tag{4.107c}$$

$$= \frac{(\gamma_{12}\gamma_{13} - \gamma_{23})\gamma_{12}\mathbf{a}_{12}}{(\gamma_{12}\gamma_{23} - \gamma_{13}) + (\gamma_{12}\gamma_{13} - \gamma_{23})\gamma_{12}}$$

Note that, by Def. 4.1, p. 171, the set of points $S = \{A_1, A_2, A_3\}$ is pointwise independent in an Einstein gyrovector space $(\mathbb{R}_s^n, \oplus, \otimes)$. Hence, the two gyrovectors $\mathbf{a}_{12} = \ominus A_1 \oplus A_2$ and $\mathbf{a}_{13} = \ominus A_1 \oplus A_3$ in $\mathbb{R}_s^n \subset \mathbb{R}^n$ in (4.107) are linearly independent in $\mathbb{R}^n$.

Similarly to the gyroalgebra in (4.107), under a left gyrotranslation by $\ominus A_1$ the orthogyrocenter H in (4.106) becomes

$$\begin{aligned} \ominus A_1 \oplus H &= \frac{m_2\gamma_{_{\ominus A_1 \oplus A_2}}(\ominus A_1 \oplus A_2) + m_3\gamma_{_{\ominus A_1 \oplus A_3}}(\ominus A_1 \oplus A_3)}{m_1 + m_2\gamma_{_{\ominus A_1 \oplus A_2}} + m_3\gamma_{_{\ominus A_1 \oplus A_3}}} \\ &= \frac{m_2\gamma_{12}\mathbf{a}_{12} + m_3\gamma_{13}\mathbf{a}_{13}}{m_1 + m_2\gamma_{12} + m_3\gamma_{13}} \end{aligned} \tag{4.108}$$

Left gyrotranslating the gyrotriangle $A_1A_2A_3$ in Fig. 4.7 by $\ominus A_1$ we obtain the left gyrotranslated gyrotriangle

$$O(\ominus A_1 \oplus A_2)(\ominus A_1 \oplus A_3) \tag{4.109}$$

with vertices $\ominus A_1 \oplus A_1 = O = \mathbf{0}$, $\mathbf{a}_{12} = \ominus A_1 \oplus A_2$ and $\mathbf{a}_{13} = \ominus A_1 \oplus A_3$.

The gyroaltitude of the left gyrotranslated gyrotriangle (4.109) that joins the vertex O with the left gyrotranslated gyroaltitude foot $\ominus A_1 \oplus P_1$ on its opposing side, as calculated in (4.107a),

$$\ominus A_1 \oplus P_1 = \frac{(\gamma_{13}\gamma_{23} - \gamma_{12})\gamma_{12}\mathbf{a}_{12} + (\gamma_{12}\gamma_{23} - \gamma_{13})\gamma_{13}\mathbf{a}_{13}}{(\gamma_{13}\gamma_{23} - \gamma_{12})\gamma_{12} + (\gamma_{12}\gamma_{23} - \gamma_{13})\gamma_{13}} \tag{4.110}$$

is contained in the Euclidean line

$$L_1 = O + (-O + \{\ominus A_1 \oplus P_1\})t_1$$
$$= \frac{(\gamma_{13}\gamma_{23} - \gamma_{12})\gamma_{12}\mathbf{a}_{12} + (\gamma_{12}\gamma_{23} - \gamma_{13})\gamma_{13}\mathbf{a}_{13}}{(\gamma_{13}\gamma_{23} - \gamma_{12})\gamma_{12} + (\gamma_{12}\gamma_{23} - \gamma_{13})\gamma_{13}} t_1 \tag{4.111}$$

where $t_1 \in \mathbb{R}$ is the line parameter. This line passes through the point $O = \mathbf{0} \in \mathbb{R}^n_s \subset \mathbb{R}^n$ when $t_1 = 0$, and it passes through the point $\ominus A_1 \oplus P_1$ when $t_1 = 1$.

Similarly to (4.110) – (4.111), the gyroaltitude of the left gyrotranslated gyrotriangle $O(\ominus A_1 \oplus A_2)(\ominus A_1 \oplus A_3)$ that joins the vertex

$$\ominus A_1 \oplus A_2 = \mathbf{a}_{12} \tag{4.112}$$

with the left gyrotranslated gyroaltitude foot on its opposing side, $\ominus A_1 \oplus P_2$, as calculated in (4.107b),

$$\ominus A_1 \oplus P_2 = \frac{(\gamma_{12}\gamma_{13} - \gamma_{23})\gamma_{13}\mathbf{a}_{13}}{(\gamma_{13}\gamma_{23} - \gamma_{12}) + (\gamma_{12}\gamma_{13} - \gamma_{23})\gamma_{13}} \tag{4.113}$$

is contained in the Euclidean line

$$L_2 = \mathbf{a}_{12} + (-\mathbf{a}_{12} + \{\ominus A_1 \oplus P_2\})t_2$$
$$= \mathbf{a}_{12} + \left(-\mathbf{a}_{12} + \frac{(\gamma_{12}\gamma_{13} - \gamma_{23})\gamma_{13}\mathbf{a}_{13}}{(\gamma_{13}\gamma_{23} - \gamma_{12}) + (\gamma_{12}\gamma_{13} - \gamma_{23})\gamma_{13}}\right) t_2 \tag{4.114}$$

where $t_2 \in \mathbb{R}$ is the line parameter. This line passes through the point $\mathbf{a}_{12} \in \mathbb{R}^n_s \subset \mathbb{R}^n$ when $t_2 = 0$, and it passes through the point $\ominus A_1 \oplus P_2$ when $t_2 = 1$.

Similarly to (4.110) – (4.111), and similarly to (4.113) – (4.114), the gyroaltitude of the left gyrotranslated gyrotriangle $O(\ominus A_1 \oplus A_2)(\ominus A_1 \oplus A_3)$ that joins the vertex

$$\ominus A_1 \oplus A_3 = \mathbf{a}_{13} \tag{4.115}$$

with the left gyrotranslated gyroaltitude foot $\ominus A_1 \oplus P_3$ on its opposing side, as calculated in (4.107c),

$$\ominus A_1 \oplus P_3 = \frac{(\gamma_{12}\gamma_{13} - \gamma_{23})\gamma_{12}\mathbf{a}_{12}}{(\gamma_{12}\gamma_{23} - \gamma_{13}) + (\gamma_{12}\gamma_{13} - \gamma_{23})\gamma_{12}} \tag{4.116}$$

is contained in the Euclidean line

$$L_3 = \mathbf{a}_{13} + (-\mathbf{a}_{13} + \{\ominus A_1 \oplus P_3\})t_3$$
$$= \mathbf{a}_{13} + \left(-\mathbf{a}_{13} + \frac{(\gamma_{12}\gamma_{13} - \gamma_{23})\gamma_{12}\mathbf{a}_{12}}{(\gamma_{12}\gamma_{23} - \gamma_{13}) + (\gamma_{12}\gamma_{13} - \gamma_{23})\gamma_{12}}\right) t_3 \tag{4.117}$$

where $t_3 \in \mathbb{R}$ is the line parameter. This line passes through the point $\mathbf{a}_{13} \in \mathbb{R}^n_s \subset \mathbb{R}^n$ when $t_3 = 0$, and it passes through the point $\ominus A_1 \oplus P_3 \in \mathbb{R}^n_s \subset \mathbb{R}^n$ when $t_3 = 1$.

Hence, if the orthogyrocenter H exists, its left gyrotranslated orthogyrocenter, $-\ominus A_1 \oplus H$, given by (4.108), is contained in each of the three Euclidean lines L_k, $k = 1, 2, 3$, in (4.111), (4.114) and (4.117). Formalizing, if H exists then the point P, (4.108),

$$P = \ominus A_1 \oplus H = \frac{m_2 \gamma_{12} \mathbf{a}_{12} + m_3 \gamma_{13} \mathbf{a}_{13}}{m_1 + m_2 \gamma_{12} + m_3 \gamma_{13}} \tag{4.118}$$

lies on each of the lines L_k, $k = 1, 2, 3$. Imposing the normalization condition $m_1 + m_2 + m_3 = 1$ of gyrobarycentric coordinates, (4.118) can be simplified by means of the resulting equation $m_1 = 1 - m_2 - m_3$, obtaining

$$P = \ominus A_1 \oplus H = \frac{m_2 \gamma_{12} \mathbf{a}_{12} + m_3 \gamma_{13} \mathbf{a}_{13}}{1 + m_2(\gamma_{12} - 1) + m_3(\gamma_{13} - 1)} \tag{4.119}$$

Since the point P lies on each of the three lines L_k, $k = 1, 2, 3$, there exist values $t_{k,0}$ of the line parameters t_k, $k = 1, 2, 3$, respectively, such that

$$\begin{aligned}
&P - \frac{(\gamma_{13}\gamma_{23} - \gamma_{12})\gamma_{12}\mathbf{a}_{12} + (\gamma_{12}\gamma_{23} - \gamma_{13})\gamma_{13}\mathbf{a}_{13}}{(\gamma_{13}\gamma_{23} - \gamma_{12})\gamma_{12} + (\gamma_{12}\gamma_{23} - \gamma_{13})\gamma_{13}} t_{1,0} &= 0 \\
&P - \mathbf{a}_{12} - \left(-\mathbf{a}_{12} + \frac{(\gamma_{12}\gamma_{13} - \gamma_{23})\gamma_{13}\mathbf{a}_{13}}{(\gamma_{13}\gamma_{23} - \gamma_{12}) + (\gamma_{12}\gamma_{13} - \gamma_{23})\gamma_{13}} \right) t_{2,0} &= 0 \\
&P - \mathbf{a}_{13} - \left(-\mathbf{a}_{13} + \frac{(\gamma_{12}\gamma_{13} - \gamma_{23})\gamma_{12}\mathbf{a}_{12}}{(\gamma_{12}\gamma_{23} - \gamma_{13}) + (\gamma_{12}\gamma_{13} - \gamma_{23})\gamma_{12}} \right) t_{3,0} &= 0
\end{aligned} \tag{4.120}$$

The kth equation in (4.120), $k = 1, 2, 3$, is equivalent to the condition that point P lies on line L_k.

The system of equations (4.120) was obtained by methods of gyroalgebra, and will be solved below by a common method of linear algebra.

Substituting P from (4.119) into (4.120), and rewriting each equation in (4.120) as a linear combination of $\mathbf{a}_{12}$ and $\mathbf{a}_{13}$ equals zero, one obtains the following linear homogeneous system of three gyrovector equations:

$$\begin{aligned}
c_{11}\mathbf{a}_{12} + c_{12}\mathbf{a}_{13} &= \mathbf{0} \\
c_{21}\mathbf{a}_{12} + c_{22}\mathbf{a}_{13} &= \mathbf{0} \\
c_{31}\mathbf{a}_{12} + c_{32}\mathbf{a}_{13} &= \mathbf{0}
\end{aligned} \tag{4.121}$$

where each coefficient c_{ij}, $i = 1, 2, 3$, $j = 1, 2$, is a function of γ_{12}, γ_{13}, γ_{23}, and the five unknowns m_2, m_3, and $t_{k,0}$, $k = 1, 2, 3$.

Since the set $S = \{A_1, A_2, A_3\}$ is pointwise independent, the two gyrovectors $\mathbf{a}_{12} = \ominus A_1 \oplus A_2$ and $\mathbf{a}_{13} = \ominus A_1 \oplus A_3$ in $\mathbb{R}_s^n$, considered as vectors in $\mathbb{R}^n$, are linearly independent. Hence, each coefficient c_{ij} in (4.121) equals zero. Accordingly, the three gyrovector equations in (4.121) are equivalent to the following six scalar equations,

$$c_{11} = c_{12} = c_{21} = c_{22} = c_{31} = c_{32} = 0 \tag{4.122}$$

for the five unknowns m_2, m_3 and $t_{k,0}$, $k = 1, 2, 3$.

Explicitly, the six scalar equations in (4.122) are equivalent to the following six equations:

$$\begin{aligned}
&(2\gamma_{12}\gamma_{13}\gamma_{23} - \gamma_{12}^2 - \gamma_{13}^2)m_2 \\
&\qquad -(\gamma_{13}\gamma_{23} - \gamma_{12})(1 - m_2 - m_3 + \gamma_{12}m_2 + \gamma_{13}m_3)t_{1,0} = 0 \\
&(2\gamma_{12}\gamma_{13}\gamma_{23} - \gamma_{12}^2 - \gamma_{13}^2)m_3 \\
&\qquad -(\gamma_{12}\gamma_{23} - \gamma_{13})(1 - m_2 - m_3 + \gamma_{12}m_2 + \gamma_{13}m_3)t_{1,0} = 0 \\
&1 - m_2 - m_3 + \gamma_{13}m_3 - (1 - m_2 - m_3 + \gamma_{12}m_2 + \gamma_{13}m_3)t_{2,0} = 0 \\
&\gamma_{12}(\gamma_{13}^2 - 1)m_3 - (\gamma_{12}\gamma_{13} - \gamma_{23})(1 - m_2 - m_3 + \gamma_{12}m_2 + \gamma_{13}m_3)t_{2,0} = 0 \\
&\gamma_{13}(\gamma_{12}^2 - 1)m_2 - (\gamma_{12}\gamma_{13} - \gamma_{23})(1 - m_2 - m_3 + \gamma_{12}m_2 + \gamma_{13}m_3)t_{3,0} = 0 \\
&1 - m_2 - m_3 + \gamma_{12}m_2 - (1 - m_2 - m_3 + \gamma_{12}m_2 + \gamma_{13}m_3)t_{3,0} = 0
\end{aligned} \tag{4.123}$$

The unique solution of (4.123) is given by (4.124) and (4.126) below: The values of the line parameters are

$$\begin{aligned}
t_{1,0} &= \frac{(\gamma_{12}\gamma_{13} - \gamma_{23})(2\gamma_{12}\gamma_{13}\gamma_{23} - \gamma_{12}^2 - \gamma_{13}^2)}{\gamma_{12}\gamma_{13}(1 + 2\gamma_{12}\gamma_{13}\gamma_{23} - \gamma_{12}^2 - \gamma_{13}^2)} \\
t_{2,0} &= \frac{(\gamma_{13}^2 - 1)(\gamma_{12}\gamma_{23} - \gamma_{13})}{\gamma_{13}(1 + 2\gamma_{12}\gamma_{13}\gamma_{23} - \gamma_{12}^2 - \gamma_{13}^2)} \\
t_{3,0} &= \frac{(\gamma_{12}^2 - 1)(\gamma_{13}\gamma_{23} - \gamma_{12})}{\gamma_{12}(1 + 2\gamma_{12}\gamma_{13}\gamma_{23} - \gamma_{12}^2 - \gamma_{13}^2)}
\end{aligned} \tag{4.124}$$

where

$$1 + 2\gamma_{12}\gamma_{13}\gamma_{23} - \gamma_{12}^2 - \gamma_{13}^2 > 0 \tag{4.125}$$

by (2.152), p. 105.

The gyrobarycentric coordinates (m_1, m_2, m_3) are given by

$$
\begin{aligned}
m_1 &= \frac{1}{D}(\gamma_{12}\gamma_{23} - \gamma_{13})(\gamma_{13}\gamma_{23} - \gamma_{12}) \\
m_2 &= \frac{1}{D}(\gamma_{12}\gamma_{13} - \gamma_{23})(\gamma_{13}\gamma_{23} - \gamma_{12}) \\
m_3 &= \frac{1}{D}(\gamma_{12}\gamma_{23} - \gamma_{13})(\gamma_{12}\gamma_{23} - \gamma_{13})
\end{aligned}
\tag{4.126}
$$

satisfying the normalization condition

$$
m_1 + m_2 + m_3 = 1 \tag{4.127}
$$

where D is the determinant

$$
D = \begin{vmatrix} \gamma_{12}\gamma_{23} - \gamma_{13} & -(\gamma_{13}\gamma_{23} - \gamma_{12}) \\ \gamma_{12}\gamma_{13} - \gamma_{23} & (\gamma_{12}\gamma_{13} - \gamma_{23}) + (\gamma_{13}\gamma_{23} - \gamma_{12}) \end{vmatrix} \tag{4.128}
$$

or, equivalently,

$$
\begin{aligned}
D = {}& \gamma_{12}\gamma_{13}(1 - \gamma_{12} - \gamma_{13}) + \gamma_{12}\gamma_{23}(1 - \gamma_{12} - \gamma_{23}) + \gamma_{13}\gamma_{23}(1 - \gamma_{13} - \gamma_{23}) \\
& + \gamma_{12}\gamma_{13}\gamma_{23}(\gamma_{12} + \gamma_{13} + \gamma_{23})
\end{aligned}
\tag{4.129}
$$

Following (4.126), convenient gyrobarycentric coordinates of the gyrotriangle orthogyrocenter H are given by the equation

$$
(m_1 : m_2 : m_3) = (C_{12}C_{13} : C_{12}C_{23} : C_{13}C_{23}) \tag{4.130}
$$

or, equivalently, by the equation

$$
(m_1 : m_2 : m_3) = \left(\frac{C_{12}}{C_{23}} : \frac{C_{12}}{C_{13}} : 1\right) \tag{4.131}
$$

where

$$
\begin{aligned}
C_{12} &= \gamma_{13}\gamma_{23} - \gamma_{12} \\
C_{13} &= \gamma_{12}\gamma_{23} - \gamma_{13} \\
C_{23} &= \gamma_{12}\gamma_{13} - \gamma_{23}
\end{aligned}
\tag{4.132}
$$

Accordingly, the gyrobarycentric coordinate representation of the orthogyrocenter H of gyrotriangle $A_1A_2A_3$ with respect to the set of the gyrotriangle vertices is given by the equation

$$
H = \frac{C_{12}C_{13}\gamma_{A_1}A_1 + C_{12}C_{23}\gamma_{A_2}A_2 + C_{13}C_{23}\gamma_{A_3}A_3}{C_{12}C_{13}\gamma_{A_1} + C_{12}C_{23}\gamma_{A_2} + C_{13}C_{23}\gamma_{A_3}} \in \mathbb{R}^n \tag{4.133}
$$

Gyrotrigonometry enables the gyrobarycentric coordinate representation (4.133) of the gyrotriangle orthogyrocenter H with respect to the gyrotriangle set of vertices $S = \{A_1, A_2, A_3\}$ to be drastically simplified. Gamma factors of gyrotriangle side gyrolengths are related to its gyroangles by the equations, (2.157), p. 106,

$$\begin{aligned}\gamma_{23} &= \frac{\cos\alpha_1 + \cos\alpha_2\cos\alpha_3}{\sin\alpha_2\sin\alpha_3}\\ \gamma_{13} &= \frac{\cos\alpha_2 + \cos\alpha_1\cos\alpha_3}{\sin\alpha_1\sin\alpha_3}\\ \gamma_{12} &= \frac{\cos\alpha_3 + \cos\alpha_1\cos\alpha_2}{\sin\alpha_1\sin\alpha_2}\end{aligned} \tag{4.134}$$

It follows from (4.132) and (4.134) that

$$\begin{aligned}\frac{C_{12}}{C_{23}} &= \frac{\tan\alpha_1}{\tan\alpha_3}\\ \frac{C_{12}}{C_{13}} &= \frac{\tan\alpha_2}{\tan\alpha_3}\end{aligned} \tag{4.135}$$

so that the gyrobarycentric coordinates of H in (4.131) can be written as

$$(m_1 : m_2 : m_3) = \left(\frac{\tan\alpha_1}{\tan\alpha_3} : \frac{\tan\alpha_2}{\tan\alpha_3} : 1\right) \tag{4.136}$$

which are, in turn, equivalent to the gyrobarycentric coordinates

$$(m_1 : m_2 : m_3) = (\tan\alpha_1 : \tan\alpha_2 : \tan\alpha_3) \tag{4.137}$$

Interestingly, the gyrotrigonometric gyrobarycentric coordinates (4.137) of the gyrotriangle orthogyrocenter H are identical in form with trigonometric barycentric coordinates of the triangle orthocenter in Euclidean geometry, as we see from Table (1.1), p. 55.

Following (4.130) and the definition in (4.10), p. 173, of the constant m_0 of a point P with a gyrobarycentric representation, the constant m_0 of the gyrotriangle orthogyrocenter H in (4.133) with respect to the set of the gyrotriangle vertices is given by the equation

$$\begin{aligned}m_0^2 &= m_1^2 + m_2^2 + m_3^2 + 2m_1m_2\gamma_{12} + 2m_1m_3\gamma_{13} + 2m_2m_3\gamma_{23}\\ &= \tfrac{1}{2}F_1(F_1^2 + F_2)\end{aligned} \tag{4.138}$$

where

$$\begin{aligned}F_1 &= 1 + 2\gamma_{12}\gamma_{13}\gamma_{23} - \gamma_{12}^2 - \gamma_{13}^2 - \gamma_{23}^2\\ &= 2(\gamma_{12}\gamma_{13}\gamma_{23} - 1) - (\gamma_{12}^2 - 1) - (\gamma_{13}^2 - 1) - (\gamma_{23}^2 - 1)\\ F_2 &= 2(\gamma_{12}\gamma_{13}\gamma_{23} - 1)^2 - (\gamma_{12}^2 - 1)^2 - (\gamma_{13}^2 - 1)^2 - (\gamma_{23}^2 - 1)^2\end{aligned} \tag{4.139}$$

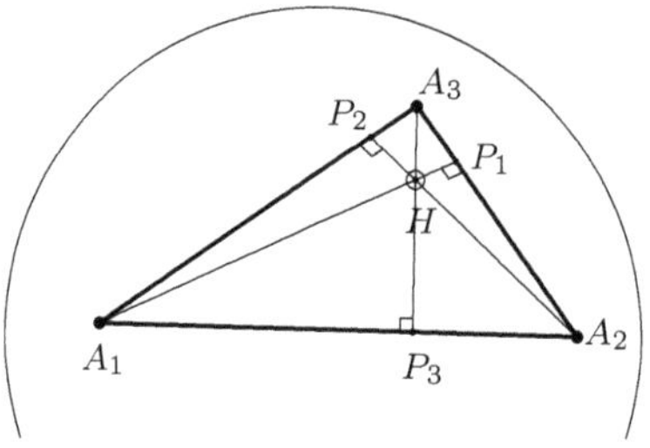

Fig. 4.8 The gyroaltitudes, and the orthogyrocenter H, of a gyrotriangle $A_1A_2A_3$ in an Einstein gyrovector space. Case I: The orthogyrocenter H lies inside the acute gyrotriangle. Gyrobarycentric coordinates $(m_1 : m_2 : m_3)$ of the orthogyrocenter H with respect to the set $\{A_1, A_2, A_3\}$ of the gyrotriangle vertices, given by (4.137), are all positive (or, equivalently, all negative) so that $m_0^2 > 0$ in (4.138), in agreement with Remark 4.5, p. 174.

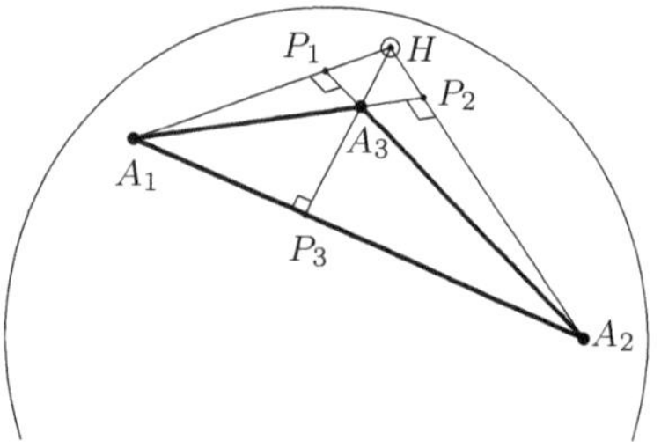

Fig. 4.9 The gyroaltitudes, and the orthogyrocenter H, of a gyrotriangle $A_1A_2A_3$ in an Einstein gyrovector space. Case II: The orthogyrocenter H lies outside the obtuse gyrotriangle. One of the gyrobarycentric coordinates $(m_1 : m_2 : m_3)$ of the orthogyrocenter H with respect to the set $\{A_1, A_2, A_3\}$ of the gyrotriangle vertices, given by (4.137), is negative and the other two are positive, in agreement with Remark 4.5.

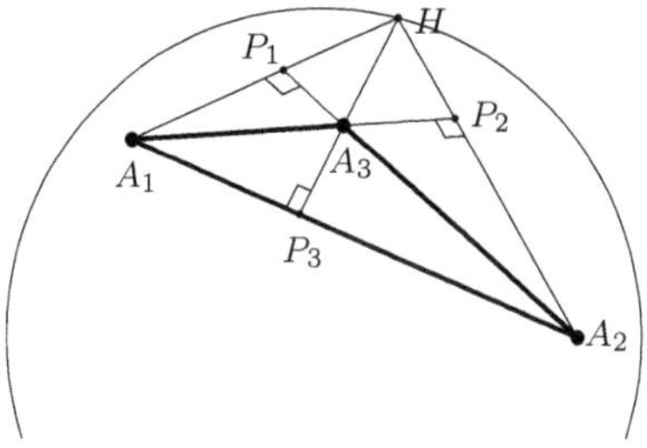

Fig. 4.10 A gyrotriangle $A_1A_2A_3$ that does not possess a orthogyrocenter H in an Einstein gyrovector plane $(\mathbb{R}_s^2, \oplus, \otimes)$. The point $H \in \mathbb{R}^n$ lies on the boundary of the ball $\mathbb{R}_s^n \subset \mathbb{R}^n$. accordingly $m_0^2 = 0$ in (4.138), in agreement with Remark 4.5.

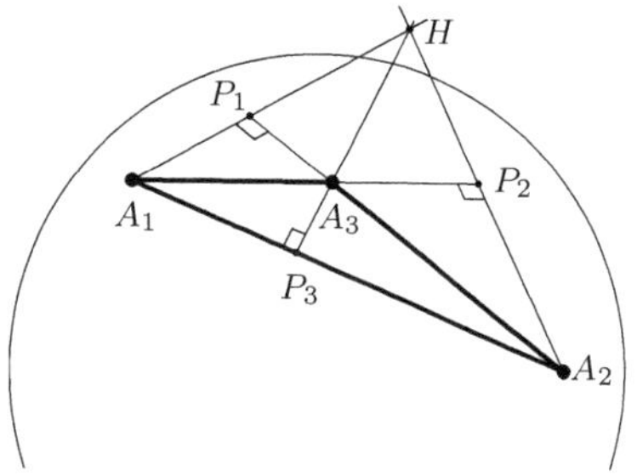

Fig. 4.11 A gyrotriangle $A_1A_2A_3$ that does not possess a orthogyrocenter H in an Einstein gyrovector plane $(\mathbb{R}_s^2, \oplus, \otimes)$. The point $H \in \mathbb{R}^n$ lies outside of the ball $\mathbb{R}_s^n \subset \mathbb{R}^n$. accordingly $m_0^2 < 0$ in (4.138), in agreement with Remark 4.5, p. 174.

Since $F_1 > 0$, by Corollary 4.20, p. 197, the constant m_0^2 in (4.138) is positive, zero, or negative if and only if $F_1^2 + F_2$ is positive, zero, or negative, respectively. According to Remark 4.5, p. 174, if $m_0^2 > 0$ then gyrotriangle $A_1A_2A_3$ possesses an orthogyrocenter H. The orthogyrocenter H lies in the interior of gyrotriangle $A_1A_2A_3$ if and only if gyrobarycentric coordinates of H are all positive or all negative. The gyrotriangle $A_1A_2A_3$ does not possess an orthogyrocenter H when $m_0^2 \le 0$ in (4.138). When $m_0^2 = 0$ the point H lies on the boundary of the ball $\mathbb{R}_s^n$, and when $m_0^2 < 0$ the point H lies outside of the ball, as shown in Figs. 4.8 – 4.11 for $n = 2$.

Formalizing the main result of this section, we have the following theorem.

Theorem 4.24. (Einstein Gyrotriangle Orthogyrocenter). *Let* $S = \{A_1, A_2, A_3\}$ *be a pointwise independent set of three points in an Einstein gyrovector space* $(\mathbb{R}_s^n, \oplus, \otimes)$, $n \ge 2$. *The orthogyrocenter* H, *shown in Figs. 4.8 – 4.11, of gyrotriangle* $A_1A_2A_3$ *has the gyrobarycentric coordinate representation*

$$H = \frac{m_1\gamma_{A_1}A_1 + m_2\gamma_{A_2}A_2 + m_3\gamma_{A_3}A_3}{m_1\gamma_{A_1} + m_2\gamma_{A_2} + m_3\gamma_{A_3}} \tag{4.140}$$

with respect to the set $\{A_1, A_2, A_3\}$, *with gyrotrigonometric gyrobarycentric coordinates*

$$(m_1 : m_2 : m_3) = (\tan\alpha_1 : \tan\alpha_2 : \tan\alpha_3) \tag{4.141}$$

The existence of the gyrotriangle orthogyrocenter H *is determined by the squared orthogyrocenter constant* m_0^2 *with respect to the set of the gyrotriangle vertices,*

$$\begin{aligned} m_0^2 &= \tan^2\alpha_1 + \tan^2\alpha_2 + \tan^2\alpha_3 \\ &\quad + 2(\tan\alpha_1\,\tan\alpha_2\,\gamma_{12} + \tan\alpha_1\,\tan\alpha_3\,\gamma_{13} + \tan\alpha_2\,\tan\alpha_3\,\gamma_{23}) \end{aligned} \tag{4.142}$$

The gyrotriangle orthogyrocenter H *exists if and only if* $m_0^2 > 0$. *Furthermore, the gyrotriangle orthogyrocenter* H *lies on the interior of its gyrotriangle if and only if* $\tan\alpha_1 > 0$, $\tan\alpha_2 > 0$ *and* $\tan\alpha_3 > 0$ *or, equivalently, if and only if the gyrotriangle is acute (Figs. 4.8 – 4.11).*

4.14 Möbius Gyrotriangle Orthogyrocenter

We determine the gyrotriangle orthogyrocenter in Möbius gyrovector spaces by transforming the gyrobarycentric coordinate representation of the gyrotriangle orthogyrocenter from Einstein to Möbius gyrovector spaces. In order to calculate this transformation we need the following Lemma:

Lemma 4.25. *Let* $A_{1,e}, A_{2,e} \in \mathbb{R}_s^n$ *be any two points of an Einstein gyrovector space* $(\mathbb{R}_s^n, \oplus_E, \otimes)$, *and let* $A_{1,m}, A_{2,m} \in \mathbb{R}_s^n$ *be their respective image in* *a corresponding Möbius gyrovector space* $(\mathbb{R}^n_s, \oplus_M, \otimes)$ *under isomorphism* (2.277), *p. 142. Furthermore, let*

$$\gamma_{12,e} = \gamma_{\ominus_E A_{1,e} \oplus_E A_{2,e}} \tag{4.143}$$

be the gamma factor of a gyrodifference in the Einstein gyrovector space.

Then, the image $\gamma_{12,m}$,

$$\gamma_{12,m} = \gamma_{\ominus_M A_{1,m} \oplus_M A_{2,m}} \tag{4.144}$$

of $\gamma_{12,e}$ *in the corresponding Möbius gyrovector space under isomorphism* (2.277) *is given by the equation*

$$\gamma_{12,e} = 2\gamma^2_{12,m} - 1 \tag{4.145}$$

Proof. The proof of (4.145) is given by the chain of equations (2.282), p. 143. □

In order to calculate the transformation of the gyrobarycentric coordinate representation of the gyrotriangle orthogyrocenter H from an Einstein gyrovector space to its corresponding Möbius gyrovector space, we rewrite the gyrobarycentric coordinate representation of H in an Einstein gyrovector space, introducing the subscript "e".

Thus, following (4.133), the gyrobarycentric coordinate representation of the gyrotriangle orthogyrocenter H_e in an Einstein gyrovector space $(\mathbb{R}^n_s, \oplus_E, \otimes)$ is given by

$$H_e = \frac{m_{1,e}\gamma_{A_{1,e}} A_{1,e} + m_{2,e}\gamma_{A_{2,e}} A_{2,e} + m_{3,e}\gamma_{A_{3,e}} A_{3,e}}{m_{1,e}\gamma_{A_{1,e}} + m_{2,e}\gamma_{A_{2,e}} + m_{3,e}\gamma_{A_{3,e}}} \tag{4.146}$$

in terms of its gyrobarycentric coordinates $(m_{1,e} : m_{2,e} : m_{3,e})$ with respect to the set of the gyrotriangle vertices $\{A_{1,e}, A_{2,e}, A_{3,e}\}$. These are

$$\begin{aligned} m_{1,e} &= C_{12,e}C_{13,e} \\ m_{2,e} &= C_{12,e}C_{23,e} \\ m_{3,e} &= C_{13,e}C_{23,e} \end{aligned} \tag{4.147}$$

where, by (4.132),

$$\begin{aligned} C_{12,e} &= \gamma_{13,e}\gamma_{23,e} - \gamma_{12,e} \\ C_{13,e} &= \gamma_{12,e}\gamma_{23,e} - \gamma_{13,e} \\ C_{23,e} &= \gamma_{12,e}\gamma_{13,e} - \gamma_{23,e} \end{aligned} \tag{4.148}$$

and where

$$\gamma_{ij,e} = \gamma_{\ominus_E A_{i,e} \oplus_E A_{j,e}} \tag{4.149}$$

$i, j = 1, 2, 3$, and $i < j$.

By isomorphism (2.277), p. 142, H_m is related to H_e by the equation

$$H_e = 2\otimes H_m \tag{4.150}$$

and

$$A_{k,e} = 2\otimes A_{k,m} \tag{4.151}$$

$k = 1, 2, 3$. It follows from (4.151) that

$$\gamma_{A_{k,e}} = 2\gamma^2_{A_{k,m}} - 1 \tag{4.152}$$

and

$$\gamma_{A_{k,e}} A_{k,e} = 2\gamma^2_{A_{k,m}} A_{k,m} \tag{4.153}$$

$k = 1, 2, 3$, as shown in (2.279) – (2.280), p. 142.

In (4.150) – (4.153) we have transformed several terms of (4.146) from an Einstein gyrovector space $(\mathbb{R}^n_s, \oplus_{\text{E}}, \otimes)$ into its corresponding Möbius gyrovector space $(\mathbb{R}^n_s, \oplus_{\text{M}}, \otimes)$. It remains to transform the gyrobarycentric coordinates $(m_{1,e} : m_{2,e} : m_{3,e})$ as well.

Following (4.148) and Lemma 4.25

$$\begin{aligned} C_{12,e} :&= \gamma_{13,e}\gamma_{23,e} - \gamma_{12,e} \\ &= (2\gamma^2_{13,m} - 1)(2\gamma^2_{23,m} - 1) - (2\gamma^2_{12,m} - 1) \\ &=: C_{12,m} \end{aligned} \tag{4.154}$$

etc., so that

$$\begin{aligned} C_{12,e} &= C_{12,m} = (2\gamma^2_{13,m} - 1)(2\gamma^2_{23,m} - 1) - (2\gamma^2_{12,m} - 1) \\ C_{13,e} &= C_{13,m} = (2\gamma^2_{12,m} - 1)(2\gamma^2_{23,m} - 1) - (2\gamma^2_{13,m} - 1) \\ C_{23,e} &= C_{23,m} = (2\gamma^2_{12,m} - 1)(2\gamma^2_{13,m} - 1) - (2\gamma^2_{23,m} - 1) \end{aligned} \tag{4.155}$$

where, according to (2.282), we use the notation

$$\gamma_{ij,m} = \gamma_{\ominus_{\text{M}} A_{i,m} \oplus_{\text{M}} A_{j,m}} \tag{4.156}$$

$i, j = 1, 2, 3$, and $i < j$.

Finally, by (4.147) and (4.155),

$$\begin{aligned} m_{1,e} &= C_{12,e}C_{13,e} = C_{12,m}C_{13,m} =: m_{1,m} \\ m_{2,e} &= C_{12,e}C_{23,e} = C_{12,m}C_{23,m} =: m_{2,m} \\ m_{3,e} &= C_{13,e}C_{23,e} = C_{13,m}C_{23,m} =: m_{3,m} \end{aligned} \tag{4.157}$$

We are now in the position to rewrite the Einstein orthogyrocenter equation (4.146) by means of corresponding terms that involve the subscript "m" rather than "e". We thus substitute into (4.146):

(1) H_e from (4.150);
(2) $\gamma_{A_{k,e}}$, $k = 1,2,3$, from (4.152); and
(3) $\gamma_{A_{k,e}} A_{k,e}$, $k = 1,2,3$, from (4.153);
(4) $m_{k,e}$, $k = 1,2,3$, from (4.157).

These substitutions result in the Möbius orthogyrocenter equation,

$$2\otimes_{\mathrm{M}} H_m = \frac{2m_{1,m}\gamma^2_{A_{1,m}} A_{1,m} + 2m_{2,m}\gamma^2_{A_{2,m}} A_{2,m} + 2m_{3,m}\gamma^2_{A_{3,m}} A_{3,m}}{2m_{1,m}(\gamma^2_{A_{1,m}} - 1) + 2m_{2,m}(\gamma^2_{A_{2,m}} - 1) + 2m_{3,m}(\gamma^2_{A_{3,m}} - 1)} \tag{4.158}$$

where H_m is the orthogyrocenter of a gyrotriangle $A_{1,m}A_{2,m}A_{3,m}$ in a Möbius gyrovector space $(\mathbb{R}^n_s, \oplus_{\mathrm{M}}, \otimes)$, and where the gyrobarycentric coordinates $(m_{1,m} : m_{2,m} : m_{3,m})$ are given by (4.157) and (4.155).

To formalize the result in (4.158) we slightly rearrange Identity (4.158) and omit the subscript "m", obtaining the following theorem.

Theorem 4.26. (Möbius Gyrotriangle Orthogyrocenter). *Let $A_1A_2A_3$ be a gyrotriangle in a Möbius gyrovector space $(\mathbb{R}^n_s, \oplus, \otimes)$, $n \geq 2$. The orthogyrocenter H (Fig. 4.12) of the gyrotriangle is given by the equation*

$$H = \tfrac{1}{2}\otimes \frac{m_1\gamma^2_{A_1} A_1 + m_2\gamma^2_{A_2} A_2 + m_3\gamma^2_{A_3} A_3}{m_1(\gamma^2_{A_1} - \frac{1}{2}) + m_2(\gamma^2_{A_2} - \frac{1}{2}) + m_3(\gamma^2_{A_3} - \frac{1}{2})} \tag{4.159}$$

where the gyrobarycentric coordinates $(m_1 : m_2 : m_3)$ are given by

$$\begin{aligned} m_1 &= C_{12}C_{13} \\ m_2 &= C_{12}C_{23} \\ m_3 &= C_{13}C_{23} \end{aligned} \tag{4.160}$$

where

$$\begin{aligned} C_{12} &= (2\gamma^2_{13} - 1)(2\gamma^2_{23} - 1) - (2\gamma^2_{12} - 1) \\ C_{13} &= (2\gamma^2_{12} - 1)(2\gamma^2_{23} - 1) - (2\gamma^2_{13} - 1) \\ C_{23} &= (2\gamma^2_{12} - 1)(2\gamma^2_{13} - 1) - (2\gamma^2_{23} - 1) \end{aligned} \tag{4.161}$$

and where

$$\gamma_{ij} = \gamma_{\ominus A_i \oplus A_j} \tag{4.162}$$

$i, j = 1,2,3$, and $i < j$.

Equivalently, gyrobarycentric coordinates $(m_1 : m_2 : m_3)$ of H are given by

$$(m_1 : m_2 : m_3) = (\tan\alpha_1 : \tan\alpha_2 : \tan\alpha_3) \tag{4.163}$$

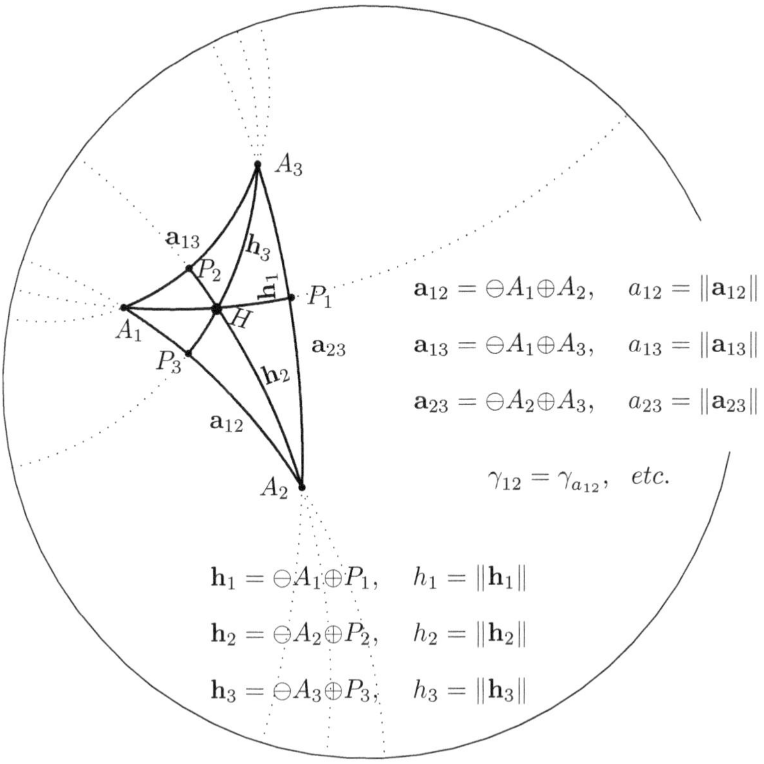

Fig. 4.12 A Möbius gyrotriangle $A_1A_2A_3$ in the Möbius gyrovector plane $(\mathbb{R}^2_s, \oplus, \otimes)$ and its gyroaltitudes are shown. The gyroaltitude $\mathbf{h}_3$ of gyrotriangle $A_1A_2A_3$ is the gyrosegment drawn perpendicularly from vertex A_3 to its opposite side A_1A_2. The gyrolines containing the gyrotriangle gyroaltitudes are concurrent at the point H, called the orthogyrocenter of the gyrotriangle. The position of the orthogyrocenter H relative to the gyrotriangle vertices as calculated in Identity (4.159) of Theorem 4.26 is shown here.

The existence of the gyrotriangle orthogyrocenter H is determined by the squared orthogyrocenter constant m_0^2 with respect to the set of the gyrotriangle vertices,

$$\begin{aligned} m_0^2 &= \tan^2\alpha_1 + \tan^2\alpha_2 + \tan^2\alpha_3 \\ &\quad + 2(\tan\alpha_1 \tan\alpha_2 \,(2\gamma_{12}^2 - 1) \\ &\quad + \tan\alpha_1 \tan\alpha_3 \,(2\gamma_{13}^2 - 1) \\ &\quad + \tan\alpha_2 \tan\alpha_3 \,(2\gamma_{23}^2 - 1)) \end{aligned} \tag{4.164}$$

The gyrotriangle orthogyrocenter H exists if and only if $m_0^2 > 0$. Furthermore, the gyrotriangle orthogyrocenter H lies on the interior of its gyrotriangle if and only if $\tan\alpha_1 > 0$, $\tan\alpha_2 > 0$ and $\tan\alpha_3 > 0$ or, equivalently, if and only if the gyrotriangle is acute (Fig. 4.12).

A Möbius gyrotriangle orthogyrocenter, with position calculated by Identity (4.159) of Theorem 4.26, is shown in Fig. 4.12.

4.15 Foot of a Gyrotriangle Gyroangle Bisector

A gyrotriangle $A_1A_2A_3$ and its gyroangle bisectors in an Einstein gyrovector space $(\mathbb{R}_s^n, \oplus, \otimes)$ are presented in Fig. 4.13, along with the standard notation in Fig. 2.4 and in (2.133), p. 100.

Let P_3 be a point on side A_1A_2 of gyrotriangle $A_1A_2A_3$ in an Einstein gyrovector space $(\mathbb{R}_s^n \oplus, \otimes)$, $n \geq 2$, such that A_3P_3 is a gyroangle bisector of gyroangle $\angle A_1A_3A_2$, as shown in Fig. 4.13 for $n = 2$. Then, the point P_3 is the foot of the gyroangle bisector A_3P_3 in gyrotriangle $A_1A_2A_3$.

Let P_3 be given in terms of its gyrobarycentric coordinates $(m_1 : m_2)$ with respect to the set $\{A_1, A_2\}$ by the equation

$$P_3 = \frac{m_1\gamma_{A_1}A_1 + m_2\gamma_{A_2}A_2}{m_1\gamma_{A_1} + m_2\gamma_{A_2}} \tag{4.165}$$

The gyrobarycentric coordinates m_1 and m_2 of P_3 in (4.165) are to be determined in (4.181) below in terms of gyroangles α_1 and α_2 of gyrotriangle $A_1A_2A_3$ and in (4.182) in terms of the side gyrolengths of the gyrotriangle embodied in the gamma factors γ_{13} and γ_{23}.

By Theorem 4.4, p. 173, the point P_3 in (4.165) gives rise to the equations in (4.166) – (4.168) below,

$$\ominus X \oplus P_3 = \frac{m_1\gamma_{\ominus X\oplus A_1}(\ominus X\oplus A_1) + m_2\gamma_{\ominus X\oplus A_2}(\ominus X\oplus A_2)}{m_1\gamma_{\ominus X\oplus A_1} + m_2\gamma_{\ominus X\oplus A_2}} \tag{4.166}$$

and

$$\begin{aligned} \gamma_{\ominus X\oplus P_3} &= \frac{m_1\gamma_{\ominus X\oplus A_1} + m_2\gamma_{\ominus X\oplus A_2}}{m_0} \\ \gamma_{\ominus X\oplus P_3}(\ominus X\oplus P_3) &= \frac{m_1\gamma_{\ominus X\oplus A_1}(\ominus X\oplus A_1) + m_2\gamma_{\ominus X\oplus A_2}(\ominus X\oplus A_2)}{m_0} \end{aligned} \tag{4.167}$$

for any $X \in \mathbb{R}_s^n$, where, in the notation of Fig. 4.13 for the gamma factor γ_{12}, the constant $m_0 > 0$ in (4.10), p. 173, specializes to

$$m_0^2 = (m_1 + m_2)^2 + 2m_1m_2(\gamma_{12} - 1) \tag{4.168}$$

in (4.167).

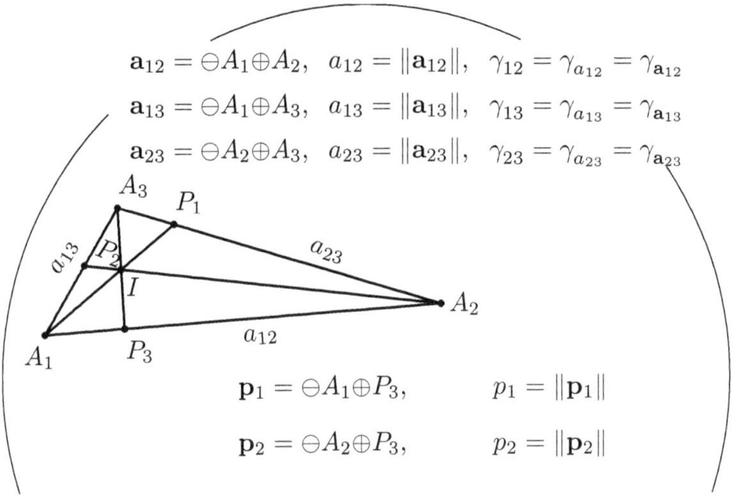

Fig. 4.13 The gyrotriangle gyroangle bisectors are concurrent. The point of concurrency, I, is called the ingyrocenter of the gyrotriangle. Let $A_1A_2A_3$ be a gyrotriangle in an Einstein gyrovector space $(\mathbb{R}_s^n, \oplus, \otimes)$. The gyroline A_kP_k is the gyroangle bisector from vertex A_k to the intersection point P_k with the opposite side, $k = 1, 2, 3$. The point P_k is the foot of the gyroangle bisector A_kP_k.

Using the notation in Fig. 4.13, it follows from (4.166) with $X = A_1$ that

$$\mathbf{p}_1 := \ominus A_1 \oplus P_3 = \frac{m_2 \gamma_{\ominus A_1 \oplus A_2} (\ominus A_1 \oplus A_2)}{m_1 + m_2 \gamma_{\ominus A_1 \oplus A_2}} = \frac{m_2 \gamma_{12} \mathbf{a}_{12}}{m_1 + m_2 \gamma_{12}} \tag{4.169}$$

and, similarly, with $X = A_2$,

$$\mathbf{p}_2 := \ominus A_2 \oplus P_3 = \frac{m_1 \gamma_{\ominus A_2 \oplus A_1} (\ominus A_2 \oplus A_1)}{m_1 \gamma_{\ominus A_2 \oplus A_1} + m_2} = \frac{m_1 \gamma_{21} \mathbf{a}_{21}}{m_1 \gamma_{21} + m_2} \tag{4.170}$$

Hence, by (4.169) – (4.170), in the notation of Fig. 4.13,

$$\begin{aligned} p_1 &:= \|\mathbf{p}_1\| = \frac{m_2 \gamma_{12} a_{12}}{m_1 + m_2 \gamma_{12}} \\ p_2 &:= \|\mathbf{p}_2\| = \frac{m_1 \gamma_{12} a_{12}}{m_1 + m_2 \gamma_{12}} \end{aligned} \tag{4.171}$$

noting that while, in general, $\mathbf{a}_{21} = \ominus A_2 \oplus A_1 \neq \ominus A_1 \oplus A_2 = \mathbf{a}_{12}$, we have $a_{21} = \|\ominus A_2 \oplus A_1\| = \|\ominus A_1 \oplus A_2\| = a_{12}$ and, hence, $\gamma_{21} = \gamma_{12}$.

Similarly, it follows from the first equation in (4.167) with $X = A_1$, and with $X = A_2$, respectively,

$$\begin{aligned} \gamma_{p_1} &= \gamma_{\ominus A_1 \oplus P_3} = \frac{m_1 + m_2\gamma_{\ominus A_1 \oplus A_2}}{m_0} = \frac{m_1 + m_2\gamma_{12}}{m_0} \\ \gamma_{p_2} &= \gamma_{\ominus A_2 \oplus P_3} = \frac{m_1\gamma_{\ominus A_2 \oplus A_1} + m_2}{m_0} = \frac{m_1\gamma_{12} + m_2}{m_0} \end{aligned} \tag{4.172}$$

It follows from (4.171) and (4.172) or, equivalently, from (4.165) and the second equation in (4.167) that

$$\begin{aligned} \gamma_{p_1} p_1 &= \frac{m_1 + m_2\gamma_{12}}{m_0} \frac{m_2\gamma_{12}a_{12}}{m_1 + m_2\gamma_{12}} = \frac{m_2}{m_0}\gamma_{12}a_{12} \\ \gamma_{p_2} p_2 &= \frac{m_1\gamma_{12} + m_2}{m_0} \frac{m_1\gamma_{12}a_{12}}{m_1\gamma_{12} + m_2} = \frac{m_1}{m_0}\gamma_{12}a_{12} \end{aligned} \tag{4.173}$$

implying

$$\frac{\gamma_{p_1} p_1}{\gamma_{p_2} p_2} = \frac{m_2}{m_1} \tag{4.174}$$

Applying the law of gyrosines (2.167), p. 110, to each of the two gyrotriangles $A_1A_3P_3$ and $A_2A_3P_3$ in Fig. 4.13, we have

$$\frac{\gamma_{p_1} p_1}{\sin \angle A_1A_3P_3} = \frac{\gamma_{13}a_{13}}{\sin \angle A_1P_3A_3} \tag{4.175}$$

and

$$\frac{\gamma_{p_2} p_2}{\sin \angle A_2A_3P_3} = \frac{\gamma_{23}a_{23}}{\sin \angle A_2P_3A_3} \tag{4.176}$$

By the gyroangle bisector definition, $\angle A_1A_3P_3 = \angle A_2A_3P_3$, so that

$$\sin \angle A_1A_3P_3 = \sin \angle A_2A_3P_3 \tag{4.177}$$

Gyroangles $\angle A_1P_3A_3$ and $\angle A_2P_3A_3$ are supplementary (their sum is π). Hence, they have equal gyrosines,

$$\sin \angle A_1P_3A_3 = \sin \angle A_2P_3A_3 \tag{4.178}$$

If follows from (4.175) – (4.178) immediately that

$$\frac{\gamma_{p_1} p_1}{\gamma_{p_2} p_2} = \frac{\gamma_{13}a_{13}}{\gamma_{23}a_{23}} \tag{4.179}$$

Hence, by (4.174) – (4.179), and by the law of gyrosines (2.167), p. 110,

$$\frac{m_2}{m_1} = \frac{\gamma_{13}a_{13}}{\gamma_{23}a_{23}} = \frac{\sin\alpha_2}{\sin\alpha_1} \tag{4.180}$$

so that gyrotrigonometric gyrobarycentric coordinates of point P_3 in Fig. 4.13 are given by the equation

$$(m_1 : m_2) = (\sin\alpha_2 : \sin\alpha_1) \tag{4.181}$$

Interestingly, the set of gyrotrigonometric gyrobarycentric coordinates of P_3 in (4.181) is identical, in form, with its Euclidean trigonometric counterpart in (1.100), p. 29.

It, finally, follows from (4.180) and (2.11), p. 66, that gyrobarycentric coordinates of point P_3 in Fig. 4.13 are given by the equation

$$(m_1 : m_2) = (\gamma_{23}a_{23} : \gamma_{13}a_{13}) = \left(\sqrt{\gamma_{23}^2 - 1} : \sqrt{\gamma_{13}^2 - 1}\right) \tag{4.182}$$

so that, by (4.182) and (4.165), we have

$$\begin{aligned} P_3 &= \frac{\gamma_{23}a_{23}\gamma_{A_1}A_1 + \gamma_{13}a_{13}\gamma_{A_2}A_2}{\gamma_{23}a_{23}\gamma_{A_1} + \gamma_{13}a_{13}\gamma_{A_2}} \\ &= \frac{\sqrt{\gamma_{23}^2 - 1}\gamma_{A_1}A_1 + \sqrt{\gamma_{13}^2 - 1}\gamma_{A_2}A_2}{\sqrt{\gamma_{23}^2 - 1}\gamma_{A_1} + \sqrt{\gamma_{13}^2 - 1}\gamma_{A_2}} \end{aligned} \tag{4.183}$$

Formalizing the main result of this section, we have the following theorem.

Theorem 4.27. (The Foot of an Einstein Gyrotriangle Gyroangle Bisector). *Let $S = \{A_1, A_2, A_3\}$ be a pointwise independent set of three points in an Einstein gyrovector space $(\mathbb{R}_s^n, \oplus, \otimes)$ $n \ge 2$, and let P_3 be the foot of gyroangle bisector A_3P_3 (Fig. 4.13). Then the foot has the gyrobarycentric coordinate representation*

$$P_3 = \frac{m_1\gamma_{A_1}A_1 + m_2\gamma_{A_2}A_2}{m_1\gamma_{A_1} + m_2\gamma_{A_2}} \tag{4.184}$$

with respect to the set $\{A_1, A_2\}$, with gyrobarycentric coordinates

$$(m_1 : m_2) = (\gamma_{23}a_{23} : \gamma_{13}a_{13}) \tag{4.185}$$

or, equivalently, with gyrotrigonometric gyrobarycentric coordinates

$$(m_1 : m_2) = (\sin\alpha_1 : \sin\alpha_2) \tag{4.186}$$

4.16 Einstein Gyrotriangle Ingyrocenter

The hyperbolic triangle incenter, I, shown in Fig. 4.13, is called in gyrolanguage a gyrotriangle ingyrocenter.

Definition 4.28. The ingyrocenter, I, of a gyrotriangle is the point of concurrency of the gyrotriangle gyroangle bisectors.

The three feet, P_1, P_2 and P_3 of the three gyroangle bisectors of gyrotriangle $A_1A_2A_3$ in an Einstein gyrovector space $(\mathbb{R}_s^n, \oplus, \otimes)$, shown in Fig. 4.13 for $n = 2$, are given by the equations

$$\begin{aligned}
P_1 &= \frac{\gamma_{13}a_{13}\gamma_{A_2}A_2 + \gamma_{12}a_{12}\gamma_{A_3}A_3}{\gamma_{13}a_{13}\gamma_{A_2} + \gamma_{12}a_{12}\gamma_{A_3}} \\
P_2 &= \frac{\gamma_{12}a_{12}\gamma_{A_3}A_3 + \gamma_{23}a_{23}\gamma_{A_1}A_1}{\gamma_{12}a_{12}\gamma_{A_3} + \gamma_{23}a_{23}\gamma_{A_1}} \\
P_3 &= \frac{\gamma_{23}a_{23}\gamma_{A_1}A_1 + \gamma_{13}a_{13}\gamma_{A_2}A_2}{\gamma_{23}a_{23}\gamma_{A_1} + \gamma_{13}a_{13}\gamma_{A_2}}
\end{aligned} \tag{4.187}$$

The third equation in (4.187) is a copy of (4.183). The first and second equations in (4.183) are obtained from the third one by cyclic permutations of the vertices of gyrotriangle $A_1A_2A_3$.

The gyroangle bisectors of gyrotriangle $A_1A_2A_3$ in an Einstein gyrovector space $(\mathbb{R}_s^n, \oplus, \otimes)$, shown in Fig. 4.13 for $n = 2$, are the gyrosegments A_1P_1, A_2P_2, and A_1P_3. Since gyrosegments in Einstein gyrovector spaces coincide with Euclidean segments, one can employ methods of linear algebra to determine the ingyrocenter, that is, the point of concurrency of the three gyroangle bisectors of gyrotriangle $A_1A_2A_3$ in Fig. 4.13.

In order to determine the gyrobarycentric coordinates of the gyrotriangle ingyrocenter in Einstein gyrovector spaces we begin with some gyroalgebraic manipulations that reduce the task we face to the task of solving a problem in linear algebra.

Let the ingyrocenter I of gyrotriangle $A_1A_2A_3$ in an Einstein gyrovector space $(\mathbb{R}_s^n, \oplus, \otimes)$, shown in Fig. 4.13, be given in terms of its gyrobarycentric coordinate representation with respect to the set $S = \{A_1, A_2, A_3\}$ of the gyrotriangle vertices by the equation

$$I = \frac{m_1\gamma_{A_1}A_1 + m_2\gamma_{A_2}A_2 + m_3\gamma_{A_3}A_3}{m_1\gamma_{A_1} + m_2\gamma_{A_2} + m_3\gamma_{A_3}} \tag{4.188}$$

The gyrobarycentric coordinates (m_1, m_2, m_3) of I in (4.188) are to be determined in (4.215) below.

Left gyrotranslating gyrotriangle $A_1A_2A_3$ by $\ominus A_1$, the gyrotriangle becomes gyrotriangle $O(\ominus A_1 \oplus A_2)(\ominus A_1 \oplus A_3)$, where $O = \ominus A_1 \oplus A_1$ is the arbitrarily selected origin of the Einstein gyrovector space $\mathbb{R}^n_s$. The gyrotriangle gyroangle bisector feet P_1, P_2 and P_3 become, respectively, $\ominus A_1 \oplus P_1$, $\ominus A_1 \oplus P_2$ and $\ominus A_1 \oplus P_3$.

The left gyrotranslated feet are calculated in (4.189) below by means of the gyroalgebraic relations (4.6) and the first identity in (4.11) in Theorem 4.4, p. 173, and the standard gyrotriangle notation, shown in Fig. 2.3, p. 101 and in (2.133), p. 100:

$$\begin{aligned}
\ominus A_1 \oplus P_1 &= \ominus A_1 \oplus \frac{\gamma_{13} a_{13} \gamma_{_{A_2}} A_2 + \gamma_{12} a_{12} \gamma_{_{A_3}} A_3}{\gamma_{13} a_{13} \gamma_{_{A_2}} + \gamma_{12} a_{12} \gamma_{_{A_3}}} \\
&= \frac{\gamma_{13} a_{13} \gamma_{_{\ominus A_1 \oplus A_2}} (\ominus A_1 \oplus A_2) + \gamma_{12} a_{12} \gamma_{_{\ominus A_1 \oplus A_3}} (\ominus A_1 \oplus A_3)}{\gamma_{13} a_{13} \gamma_{_{\ominus A_1 \oplus A_2}} + \gamma_{12} a_{12} \gamma_{_{\ominus A_1 \oplus A_3}}} \\
&= \frac{\gamma_{13} a_{13} \gamma_{12} \mathbf{a}_{12} + \gamma_{12} a_{12} \gamma_{13} \mathbf{a}_{13}}{\gamma_{13} a_{13} \gamma_{12} + \gamma_{12} a_{12} \gamma_{13}} = \frac{\gamma_{13} a_{13} \mathbf{a}_{12} + a_{12} \gamma_{13} \mathbf{a}_{13}}{\gamma_{13} a_{13} + a_{12} \gamma_{13}}
\end{aligned} \tag{4.189a}$$

$$\begin{aligned}
\ominus A_1 \oplus P_2 &= \ominus A_1 \oplus \frac{\gamma_{23} a_{23} \gamma_{_{A_1}} A_1 + \gamma_{12} a_{12} \gamma_{_{A_3}} A_3}{\gamma_{23} a_{23} \gamma_{_{A_1}} + \gamma_{12} a_{12} \gamma_{_{A_3}}} \\
&= \frac{\gamma_{12} a_{12} \gamma_{_{\ominus A_1 \oplus A_3}} (\ominus A_1 \oplus A_3)}{\gamma_{23} a_{23} + \gamma_{12} a_{12} \gamma_{_{\ominus A_1 \oplus A_3}}} \\
&= \frac{\gamma_{12} a_{12} \gamma_{13} \mathbf{a}_{13}}{\gamma_{23} a_{23} + \gamma_{12} a_{12} \gamma_{13}}
\end{aligned} \tag{4.189b}$$

$$\begin{aligned}
\ominus A_1 \oplus P_3 &= \ominus A_1 \oplus \frac{\gamma_{23} a_{23} \gamma_{_{A_1}} A_1 + \gamma_{13} a_{13} \gamma_{_{A_2}} A_2}{\gamma_{23} a_{23} \gamma_{_{A_1}} + \gamma_{13} a_{13} \gamma_{_{A_2}}} \\
&= \frac{\gamma_{13} a_{13} \gamma_{_{\ominus A_1 \oplus A_2}} (\ominus A_1 \oplus A_2)}{\gamma_{23} a_{23} + \gamma_{13} a_{13} \gamma_{_{\ominus A_1 \oplus A_2}}} \\
&= \frac{\gamma_{13} a_{13} \gamma_{12} \mathbf{a}_{12}}{\gamma_{23} a_{23} + \gamma_{13} a_{13} \gamma_{12}}
\end{aligned} \tag{4.189c}$$

Note that, by Def. 4.1, p. 171, the set of points $S = \{A_1, A_2, A_3\}$ is pointwise independent in an Einstein gyrovector space $(\mathbb{R}^n_s, \oplus, \otimes)$. Hence, the two gyrovectors $\mathbf{a}_{12} = \ominus A_1 \oplus A_2$ and $\mathbf{a}_{13} = \ominus A_1 \oplus A_3$ in $\mathbb{R}_s^n \subset \mathbb{R}^n$ in (4.189) are linearly independent in $\mathbb{R}^n$.

Similarly to the gyroalgebra in (4.189), under a left gyrotranslation by $\ominus A_1$ the ingyrocenter I in (4.188) becomes

$$\begin{aligned}\ominus A_1 \oplus I &= \ominus A_1 \oplus \frac{m_1 \gamma_{A_1} A_1 + m_2 \gamma_{A_2} A_2 + m_3 \gamma_{A_3} A_3}{m_1 \gamma_{A_1} + m_2 \gamma_{A_2} + m_3 \gamma_{A_3}} \\ &= \frac{m_2 \gamma_{\ominus A_1 \oplus A_2} (\ominus A_1 \oplus A_2) + m_3 \gamma_{\ominus A_1 \oplus A_3} (\ominus A_1 \oplus A_3)}{m_1 + m_2 \gamma_{\ominus A_1 \oplus A_2} + m_3 \gamma_{\ominus A_1 \oplus A_3}} \\ &= \frac{m_2 \gamma_{12} \mathbf{a}_{12} + m_3 \gamma_{13} \mathbf{a}_{13}}{m_1 + m_2 \gamma_{12} + m_3 \gamma_{13}}\end{aligned} \tag{4.190}$$

The gyroangle bisector of the left gyrotranslated gyrotriangle $O(\ominus A_1 \oplus A_2)(\ominus A_1 \oplus A_3)$ that joins the vertex

$$\ominus A_1 \oplus A_1 = O = \mathbf{0} \tag{4.191}$$

with the gyroangle bisector foot on its opposing side, $\oplus A_1 \oplus P_1$, as calculated in (4.189a),

$$\ominus A_1 \oplus P_1 = \frac{\gamma_{13} a_{13} \mathbf{a}_{12} + a_{12} \gamma_{13} \mathbf{a}_{13}}{\gamma_{13} a_{13} + a_{12} \gamma_{13}} \tag{4.192}$$

is contained in the Euclidean line

$$\begin{aligned}L_1 &= O + (-O + \{\ominus A_1 \oplus P_1\}) t_1 \\ &= \frac{\gamma_{13} a_{13} \mathbf{a}_{12} + a_{12} \gamma_{13} \mathbf{a}_{13}}{\gamma_{13} a_{13} + a_{12} \gamma_{13}} t_1\end{aligned} \tag{4.193}$$

where $t_1 \in \mathbb{R}$ is the line parameter. This line passes through the point $O = \mathbf{0} \in \mathbb{R}_s^n \subset \mathbb{R}^n$ when $t_1 = 0$, and it passes through the point $\ominus A_1 \oplus P_1$ when $t_1 = 1$.

Similarly to (4.191) – (4.193), the gyroangle bisector of the left gyrotranslated gyrotriangle $O(\ominus A_1 \oplus A_2)(\ominus A_1 \oplus A_3)$ that joins the vertex

$$\ominus A_1 \oplus A_2 = \mathbf{a}_{12} \tag{4.194}$$

with the gyroangle bisector foot on its opposing side, $\oplus A_1 \oplus P_2$, as calculated in (4.189b),

$$\ominus A_1 \oplus P_2 = \frac{\gamma_{12} a_{12} \gamma_{13} \mathbf{a}_{13}}{\gamma_{23} a_{23} + \gamma_{12} a_{12} \gamma_{13}} \tag{4.195}$$

is contained in the Euclidean line

$$\begin{aligned} L_2 &= \mathbf{a}_{12} + (-\mathbf{a}_{12} + \{\ominus A_1 \oplus P_2\})t_2 \\ &= \mathbf{a}_{12} + \left(-\mathbf{a}_{12} + \frac{\gamma_{12}a_{12}\gamma_{13}\mathbf{a}_{13}}{\gamma_{23}a_{23} + \gamma_{12}a_{12}\gamma_{13}}\right)t_2 \end{aligned} \tag{4.196}$$

where $t_2 \in \mathbb{R}$ is the line parameter. This line passes through the point $\mathbf{a}_{12} \in \mathbb{R}_s^n \subset \mathbb{R}^n$ when $t_2 = 0$, and it passes through the point $\ominus A_1 \oplus P_2$ when $t_2 = 1$.

Similarly to (4.191) – (4.193), and (4.194) – (4.196), the gyroangle bisector of the left gyrotranslated gyrotriangle $O(\ominus A_1 \oplus A_2)(\ominus A_1 \oplus A_3)$ that joins the vertex

$$\ominus A_1 \oplus A_3 = \mathbf{a}_{13} \tag{4.197}$$

with the gyroangle bisector foot on its opposing side, $\oplus A_1 \oplus P_3$, as calculated in (4.189c),

$$\ominus A_1 \oplus P_3 = \frac{\gamma_{13}a_{13}\gamma_{12}\mathbf{a}_{12}}{\gamma_{23}a_{23} + \gamma_{13}a_{13}\gamma_{12}} \tag{4.198}$$

is contained in the Euclidean line

$$\begin{aligned} L_3 &= \mathbf{a}_{13} + (-\mathbf{a}_{13} + \{\ominus A_1 \oplus P_3\})t_3 \\ &= \mathbf{a}_{13} + \left(-\mathbf{a}_{13} + \frac{\gamma_{13}a_{13}\gamma_{12}\mathbf{a}_{12}}{\gamma_{23}a_{23} + \gamma_{13}a_{13}\gamma_{12}}\right)t_3 \end{aligned} \tag{4.199}$$

where $t_3 \in \mathbb{R}$ is the line parameter. This line passes through the point $\mathbf{a}_{13} \in \mathbb{R}_s^n \subset \mathbb{R}^n$ when $t_3 = 0$, and it passes through the point $\ominus A_1 \oplus P_3 \in \mathbb{R}_s^n \subset \mathbb{R}^n$ when $t_3 = 1$.

Hence, if the ingyrocenter I exists, its left gyrotranslated ingyrocenter, $-\ominus A_1 \oplus I$, given by (4.190), is contained in each of the three Euclidean lines L_k, $k = 1, 2, 3$, in (4.193), (4.196) and (4.199). Formalizing, if I exists then the point P, (4.190),

$$P = \ominus A_1 \oplus I = \frac{m_2\gamma_{12}\mathbf{a}_{12} + m_3\gamma_{13}\mathbf{a}_{13}}{m_1 + m_2\gamma_{12} + m_3\gamma_{13}} \tag{4.200}$$

lies on each of the lines L_k, $k = 1, 2, 3$. Imposing the normalization condition $m_1 + m_2 + m_3 = 1$ of gyrobarycentric coordinates, (4.200) can be simplified by means of the resulting equation $m_1 = 1 - m_2 - m_3$, obtaining

$$P = \ominus A_1 \oplus I = \frac{m_2\gamma_{12}\mathbf{a}_{12} + m_3\gamma_{13}\mathbf{a}_{13}}{1 + m_2(\gamma_{12} - 1) + m_3(\gamma_{13} - 1)} \tag{4.201}$$

Since the point P lies on each of the three lines L_k, $k = 1, 2, 3$, there exist values $t_{k,0}$ of the line parameters t_k, $k = 1, 2, 3$, respectively, such that

$$\begin{aligned} P - \frac{\gamma_{13}a_{13}\gamma_{12}\mathbf{a}_{12} + \gamma_{12}a_{12}\gamma_{13}\mathbf{a}_{13}}{\gamma_{13}a_{13}\gamma_{12} + \gamma_{12}a_{12}\gamma_{13}} t_{1,0} \qquad\qquad &= 0 \\ P - \mathbf{a}_{12} - \left(-\mathbf{a}_{12} + \frac{\gamma_{12}a_{12}\gamma_{13}\mathbf{a}_{13}}{\gamma_{23}a_{23} + \gamma_{12}a_{12}\gamma_{13}} \right) t_{2,0} &= 0 \\ P - \mathbf{a}_{13} - \left(-\mathbf{a}_{13} + \frac{\gamma_{13}a_{13}\gamma_{12}\mathbf{a}_{12}}{\gamma_{23}a_{23} + \gamma_{13}a_{13}\gamma_{12}} \right) t_{3,0} &= 0 \end{aligned} \tag{4.202}$$

The kth equation in (4.202), $k = 1, 2, 3$, is equivalent to the condition that point P lies on line L_k.

The system of equations (4.202) was obtained by methods of gyroalgebra, and will be solved below by a common method of linear algebra.

Substituting P from (4.201) into (4.202), and rewriting each equation in (4.202) as a linear combination of $\mathbf{a}_{12}$ and $\mathbf{a}_{13}$ equals zero, one obtains the following linear homogeneous system of three gyrovector equations

$$\begin{aligned} c_{11}\mathbf{a}_{12} + c_{12}\mathbf{a}_{13} &= \mathbf{0} \\ c_{21}\mathbf{a}_{12} + c_{22}\mathbf{a}_{13} &= \mathbf{0} \\ c_{31}\mathbf{a}_{12} + c_{32}\mathbf{a}_{13} &= \mathbf{0} \end{aligned} \tag{4.203}$$

where each coefficient c_{ij}, $i = 1, 2, 3$, $j = 1, 2$, is a function of γ_{12}, γ_{13}, γ_{23}, and the five unknowns m_2, m_3, and $t_{k,0}$, $k = 1, 2, 3$.

Since the set $S = \{A_1, A_2, A_3\}$ is pointwise independent, the two gyrovectors $\mathbf{a}_{12} = \ominus A_1 \oplus A_2$ and $\mathbf{a}_{13} = \ominus A_1 \oplus A_3$ in $\mathbb{R}^n_s$, considered as vectors in $\mathbb{R}^n$, are linearly independent. Hence, each coefficient c_{ij} in (4.203) equals zero. Accordingly, the three gyrovector equations in (4.203) are equivalent to the following six scalar equations,

$$c_{11} = c_{12} = c_{21} = c_{22} = c_{31} = c_{32} = 0 \tag{4.204}$$

for the five unknowns m_2, m_3 and $t_{k,0}$, $k = 1, 2, 3$.

Explicitly, the six scalar equations in (4.204) are equivalent to the following six equations:

$$\begin{aligned}
\gamma_{12}(a_{12}+a_{13})m_2 - (1-m_2-m_3+\gamma_{12}m_2+\gamma_{13}m_3)a_{13}t_{1,0} &= 0\\
\gamma_{13}(a_{12}+a_{13})m_3 - (1-m_2-m_3+\gamma_{12}m_2+\gamma_{13}m_3)a_{12}t_{1,0} &= 0\\
1-m_2-m_3+\gamma_{13}m_3 - (1-m_2-m_3+\gamma_{12}m_2+\gamma_{13}m_3)t_{2,0} &= 0\\
(\gamma_{12}\gamma_{13}a_{12}+\gamma_{23}a_{23})m_3 - (1-m_2-m_3+\gamma_{12}m_2+\gamma_{13}m_3)\gamma_{12}a_{12}t_{2,0} &= 0\\
(\gamma_{12}\gamma_{13}a_{13}+\gamma_{23}a_{23})m_2 - (1-m_2-m_3+\gamma_{12}m_2+\gamma_{13}m_3)\gamma_{13}a_{13}t_{3,0} &= 0\\
1-m_2-m_3+\gamma_{12}m_2 - (1-m_2-m_3+\gamma_{12}m_2+\gamma_{13}m_3)t_{3,0} &= 0
\end{aligned} \tag{4.205}$$

The unique solution of (4.205) is given by (4.206) and (4.208) below: The values of the line parameters are

$$\begin{aligned}
t_{1,0} &= \frac{1}{D'}\gamma_{12}\gamma_{13}(a_{12}+a_{13})\\
t_{2,0} &= \frac{1}{D'}(\gamma_{12}\gamma_{13}a_{12}+\gamma_{23}a_{23})\\
t_{3,0} &= \frac{1}{D'}(\gamma_{12}\gamma_{13}a_{13}+\gamma_{23}a_{23})
\end{aligned} \tag{4.206}$$

where

$$D' = \gamma_{12}\gamma_{13}a_{12}+\gamma_{12}\gamma_{13}a_{13}+\gamma_{23}a_{23} > 0 \tag{4.207}$$

The gyrobarycentric coordinates (m_1, m_2, m_3) are given by

$$\begin{aligned}
m_1 &= \frac{1}{D}\gamma_{23}a_{23}\\
m_2 &= \frac{1}{D}\gamma_{13}a_{13}\\
m_3 &= \frac{1}{D}\gamma_{12}a_{12}
\end{aligned} \tag{4.208}$$

satisfying the normalization condition $m_1+m_2+m_3 = 1$, where D is given by

$$\gamma_{12}a_{12}+\gamma_{13}a_{13}+\gamma_{23}a_{23} > 0 \tag{4.209}$$

Following (4.208), convenient gyrobarycentric coordinates of the gyrotriangle ingyrocenter I are given by the equation

$$(m_1 : m_2 : m_3) = (\gamma_{23}a_{23} : \gamma_{13}a_{13} : \gamma_{12}a_{12}) \tag{4.210}$$

or, equivalently, by the equation

$$(m_1 : m_2 : m_3) = \left(\frac{\gamma_{23}a_{23}}{\gamma_{12}a_{12}} : \frac{\gamma_{13}a_{13}}{\gamma_{12}a_{12}} : 1\right) = \left(\frac{\sin\alpha_1}{\sin\alpha_3} : \frac{\sin\alpha_2}{\sin\alpha_3} : 1\right) \tag{4.211}$$

as we see from the law of gyrosines (2.167), p. 110. Hence a convenient set of gyrotrigonometric gyrobarycentric coordinates of the gyrotriangle ingyrocenter I is given by the equation

$$(m_1 : m_2 : m_3) = (\sin\alpha_1 : \sin\alpha_2 : \sin\alpha_3) \tag{4.212}$$

The gyrobarycentric coordinates in (4.212) are positive for any gyrotriangle gyroangles α_k, $k = 1, 2, 3$. Hence, by Remark 4.5, p. 174, the gyrotriangle ingyrocenter always lies in the interior of its gyrotriangle, as shown, for instance, in Fig. 4.13, p. 215.

We have thus found that the ingyrocenter of gyrotriangle $A_1A_2A_3$ lies in the interior of gyrotriangle $A_1A_2A_3$, and it has the gyrobarycentric coordinate representation with respect to the set $\{A_1, A_2, A_3\}$ given by each equation in the following chain of equations,

$$\begin{aligned} I &= \frac{\gamma_{23}a_{23}\gamma_{A_1}A_1 + \gamma_{13}a_{13}\gamma_{A_2}A_2 + \gamma_{12}a_{12}\gamma_{A_3}A_3}{\gamma_{23}a_{23}\gamma_{A_1} + \gamma_{13}a_{13}\gamma_{A_2} + \gamma_{12}a_{12}\gamma_{A_3}} \\ &= \frac{\sqrt{\gamma_{23}^2-1}\gamma_{A_1}A_1 + \sqrt{\gamma_{13}^2-1}\gamma_{A_2}A_2 + \sqrt{\gamma_{12}^2-1}\gamma_{A_3}A_3}{\sqrt{\gamma_{23}^2-1}\gamma_{A_1} + \sqrt{\gamma_{13}^2-1}\gamma_{A_2} + \sqrt{\gamma_{12}^2-1}\gamma_{A_3}} \\ &= \frac{\sin\alpha_1\gamma_{A_1}A_1 + \sin\alpha_2\gamma_{A_2}A_2 + \sin\alpha_3\gamma_{A_3}A_3}{\sin\alpha_1\gamma_{A_1} + \sin\alpha_2\gamma_{A_2} + \sin\alpha_3\gamma_{A_3}} \in \mathbb{R}_s^n \end{aligned} \tag{4.213}$$

The first equation in (4.213) follows from (4.208). The second equation in (4.213) follows from the first by (2.11), p. 66, and the third equation in (4.213) follows from the first by the law of gyrosines (2.167), p. 110, according to which, by (2.11),

$$\begin{aligned} \frac{\sqrt{\gamma_{23}^2-1}}{\sqrt{\gamma_{12}^2-1}} &= \frac{\gamma_{23}a_{23}}{\gamma_{12}a_{12}} = \frac{\sin\alpha_1}{\sin\alpha_3} \\ \frac{\sqrt{\gamma_{13}^2-1}}{\sqrt{\gamma_{12}^2-1}} &= \frac{\gamma_{13}a_{13}}{\gamma_{12}a_{12}} = \frac{\sin\alpha_2}{\sin\alpha_3} \end{aligned} \tag{4.214}$$

Formalizing the main result of this section, we thus have the following theorem.

Theorem 4.29. *Let $A_1A_2A_3$ be a gyrotriangle in an Einstein gyrovector space $(\mathbb{R}_s^n \oplus, \otimes)$, $n \geq 2$, as shown in Fig. 4.13, p. 215, along with its standard notation. Then, gyrobarycentric coordinates of the gyrotriangle ingyrocenter, I, are given by each of the following three equations:*

$$
\begin{aligned}
(m_1 : m_2 : m_3) &= \left(\sqrt{\gamma_{23}^2 - 1} : \sqrt{\gamma_{13}^2 - 1} : \sqrt{\gamma_{12}^2 - 1}\right) \\
(m_1 : m_2 : m_3) &= (\gamma_{23}a_{23} : \gamma_{13}a_{13} : \gamma_{12}a_{12}) \\
(m_1 : m_2 : m_3) &= (\sin\alpha_1 : \sin\alpha_2 : \sin\alpha_3)
\end{aligned}
\tag{4.215}
$$

The gyrobarycentric coordinates m_1, m_1, m_3 in (4.215) of a gyrotriangle ingyrocenter I are all positive and, accordingly, the point I lies on the interior of its gyrotriangle, as remarked in Remark 4.5, p. 174.

Interestingly, in the Euclidean limit of large s, $s \to \infty$, the three sets of gyrobarycentric coordinates $(m_1 : m_2 : m_3)$ in Theorem 4.29, (4.215), exhibit the following distinct features:

The first set of gyrobarycentric coordinates of the gyrotriangle ingyrocenter in (4.215) reduces to $(m_1 : m_2 : m_3) = (0 : 0 : 0)$, which makes no sense in Euclidean geometry;

The second set of gyrobarycentric coordinates of the gyrotriangle ingyrocenter in (4.215) reduces to its Euclidean counterpart,

$$(m_1 : m_2 : m_3) = (a_{23} : a_{13} : a_{12}) \qquad \text{(Euclidean Geometry)} \tag{4.216}$$

noting that in the limit of large s, $s \to \infty$, gamma factors tend to 1, and gyrolengths tend to lengths. Equation (4.216) gives a well-known barycentric coordinates for the Euclidean triangle incenter [8,9], as determined directly in (1.153), p. 43. We should note that a_{23}, a_{13}, a_{12} in (4.215) are gyrotriangle side gyrolengths in $\mathbb{R}_s^n$ while, in contrast, a_{23}, a_{13}, a_{12} in (4.216) are triangle side lengths in $\mathbb{R}^n$.

The third set of gyrobarycentric coordinates of the gyrotriangle ingyrocenter in (4.215) appears in a gyrotrigonometric form. As such, it is identical, in form, with its Euclidean trigonometric counterpart. Indeed, in the limit $s \to \infty$ the third equation in (4.215), which is in a gyrotrigonometric form, remains intact in form in the transition from hyperbolic geometry to Euclidean geometry. It leads to a well-known barycentric coordinates of the Euclidean triangle incenter in a trigonometric form, [8,9], as shown in (1.109) – (1.111), p. 31,

$$(m_1 : m_2 : m_3) = (\sin\alpha_1 : \sin\alpha_2 : \sin\alpha_3) \qquad \text{(Euclidean Geometry)} \tag{4.217}$$

We should note that while the third equation in (4.215) and the equation in (4.217) are equal in form, they are different in context. The former involves gyrosines of gyrotriangle gyroangles while, in contrast, the latter involves sines of triangle angles.

By Theorem 4.29 and the ingyrocenter gyrobarycentric coordinate representation (4.188), p. 218, we have the following theorem:

Theorem 4.30. (Einstein Gyrotriangle Ingyrocenter). *Let $S = \{A_1, A_2, A_3\}$ be a pointwise independent set of three points in an Einstein gyrovector space $(\mathbb{R}^n_s, \oplus, \otimes)$, $n \geq 2$. The ingyrocenter I (Fig. 4.13, p. 215) of gyrotriangle $A_1A_2A_3$ has the gyrobarycentric coordinate representation*

$$I = \frac{m_1\gamma_{A_1}A_1 + m_2\gamma_{A_2}A_2 + m_3\gamma_{A_3}A_3}{m_1\gamma_{A_1} + m_2\gamma_{A_2} + m_3\gamma_{A_3}} \tag{4.218}$$

with respect to the set $\{A_1, A_2, A_3\}$, with gyrobarycentric coordinates given by

$$(m_1 : m_2 : m_3) = \left(\sqrt{\gamma_{23}^2 - 1} : \sqrt{\gamma_{13}^2 - 1} : \sqrt{\gamma_{12}^2 - 1}\right) \tag{4.219}$$

or, equivalently, by

$$(m_1 : m_2 : m_3) = (\gamma_{23}a_{23} : \gamma_{13}a_{13} : \gamma_{12}a_{12}) \tag{4.220}$$

or, equivalently, by the gyrotrigonometric gyrobarycentric coordinates

$$(m_1 : m_2 : m_3) = (\sin\alpha_1 : \sin\alpha_2 : \sin\alpha_3) \tag{4.221}$$

4.17 Ingyrocenter to Gyrotriangle Side Gyrodistance

The gyrodistance between a point and a gyroline in an Einstein gyrovector space is determined in Sec. 2.21. This result is used here to show that the gyrotriangle ingyrocenter is the point of the interior of the gyrotriangle that is equigyrodistant from the three gyrotriangle sides.

Let $A_1A_2A_3$ be a gyrotriangle in an Einstein gyrovector space $(\mathbb{R}^n_s, \oplus, \otimes)$, and let h_3 be the gyrodistance from the gyrotriangle ingyrocenter I to its side A_1A_2, as shown in Fig. 4.14. Following the notation in Fig. 4.14, let

$$\begin{aligned} \mathbf{a}_1 &= \ominus A_1 \oplus I\,, & a_1 &= \|\mathbf{a}_1\| \\ \mathbf{a}_2 &= \ominus A_2 \oplus I\,, & a_2 &= \|\mathbf{a}_2\| \end{aligned} \tag{4.222}$$

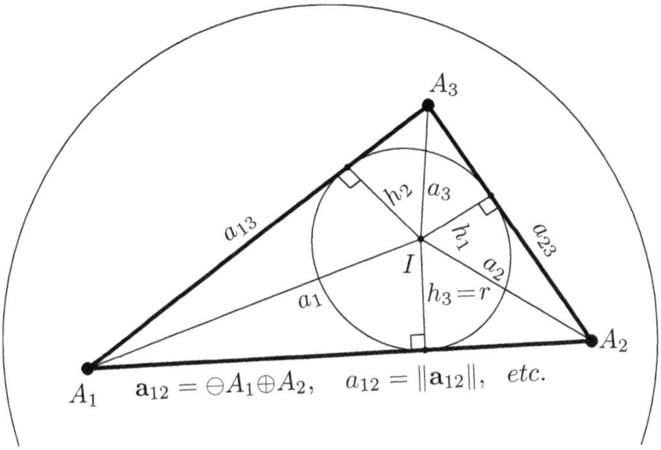

Fig. 4.14 The ingyrocircle and its ingyrocenter I of a gyrotriangle $A_1A_2A_3$ in an Einstein gyrovector plane $(\mathbb{R}_s^2, \oplus, \otimes)$. The gyrotriangle ingyrocenter is the point of the interior of the gyrotriangle that is equigyrodistant from the three gyrotriangle sides, that is, $h_1 = h_2 = h_3 =: r$, r being the gyrotriangle ingyroradius. The gyrotriangle gyroangle at vertex A_k is α_k, $k = 1, 2, 3$. A gyrocircle in an Einstein gyrovector plane $(\mathbb{R}_s^2, \oplus, \otimes)$ with gyrocenter at the center of the disk $\mathbb{R}_s^2$ has the shape of a Euclidean circle. It is, however, increasingly flattened as the gyrocircle gyrocenter moves away from the center of $\mathbb{R}_s^2$, as we see in this figure and in Fig. 4.17, p. 234.

Then, by Theorem 4.19, p. 196, in the notation of Fig. 4.14, along with the standard gyrotriangle notation, shown in Fig. 2.3, p. 101 and in (2.133), p. 100,

$$h_3^2 = s^2 \left(1 - \frac{\gamma_{12}^2 - 1}{2\gamma_{a_1}\gamma_{a_2}\gamma_{12} - \gamma_{a_1}^2 - \gamma_{a_2}^2}\right) \tag{4.223}$$

where, following the standard gyrotriangle notation, we use the notation

$$\gamma_{12} = \gamma_{a_{12}} = \gamma_{\mathbf{a}_{12}} = \gamma_{\ominus A_1 \oplus A_2} \tag{4.224}$$

as in Fig. 4.13, p. 215.

Applying Theorem 4.4, p. 173, about the gyrobarycentric coordinate representation of a point P, (4.6), to the gyrobarycentric coordinate representation of the gyrotriangle ingyrocenter I, (4.188), we obtain the following two equations.

$$\begin{aligned}\gamma_{a_1} = \gamma_{\mathbf{a}_1} = \gamma_{\ominus A_1 \oplus I} &= \frac{m_1\gamma_{\ominus A_1\oplus A_1} + m_2\gamma_{\ominus A_1\oplus A_2} + m_3\gamma_{\ominus A_1\oplus A_3}}{m_0} \\ &= \frac{m_1 + m_2\gamma_{12} + m_3\gamma_{13}}{m_0} \\ \gamma_{a_2} = \gamma_{\mathbf{a}_2} = \gamma_{\ominus A_2 \oplus I} &= \frac{m_1\gamma_{\ominus A_2\oplus A_1} + m_2\gamma_{\ominus A_2\oplus A_2} + m_3\gamma_{\ominus A_2\oplus A_3}}{m_0} \\ &= \frac{m_1\gamma_{12} + m_2 + m_3\gamma_{23}}{m_0}\end{aligned} \tag{4.225}$$

The two equations in (4.225) are obtained by an application of the second identity in (4.11), p. 174, with $X = \ominus A_1$ and with $X = \ominus A_2$, to the gyrobarycentric coordinate representation of I in (4.188), p. 218.

The constant m_0 in (4.225) is given by the equation

$$m_0^2 = m_1^2 + m_2^2 + m_3^2 + 2(m_1m_2\gamma_{12} + m_1m_3\gamma_{13} + m_2m_3\gamma_{23}) \tag{4.226}$$

according to (4.10), p. 173, where m_k, $k = 1, 2, 3$, are given by (4.210) or, equivalently, by (4.212).

Substituting (4.225) and (4.226) into (4.223), we obtain an expression for h_3^2/s^2 as an involved function f of $\gamma_{12}, \gamma_{13}, \gamma_{23}$ and m_1, m_2, m_3,

$$\frac{1}{s^2}h_3^2 = f(\gamma_{12}, \gamma_{13}, \gamma_{23}, m_1, m_2, m_3) \tag{4.227}$$

The function f is too involved and hence is not displayed here. We obtain it by computer assisted computation.

In order to simplify the function f in (4.227), let $\alpha_1, \alpha_2, \alpha_3$ be the gyroangles of gyrotriangle $A_1A_2A_3$ in Fig. 4.14.

The arguments $\gamma_{12}, \gamma_{13}, \gamma_{23}$ of the function f in (4.227) can be written as functions of $\alpha_1, \alpha_2, \alpha_3$ by the AAA to SSS conversion law in Theorem 2.26, p. 106.

Furthermore, the arguments m_1, m_2, m_3 of the function f in (4.227) can also be written as functions of $\alpha_1, \alpha_2, \alpha_3$, by (4.212), p. 224.

Having h_3^2/s^2 in (4.227) expressed as a function of the gyrotriangle gyroangles $\alpha_1, \alpha_2, \alpha_3$, the resulting seemingly involved function f can be remarkably simplified by employing a computer software, like Mathematica or

Maple, for symbolic manipulation, obtaining the following elegant results,

$$\begin{aligned}\frac{1}{s^2}h_3^2 &= \frac{\cos\frac{\alpha_1+\alpha_2+\alpha_3}{2}\cos\frac{-\alpha_1+\alpha_2+\alpha_3}{2}\cos\frac{\alpha_1-\alpha_2+\alpha_3}{2}\cos\frac{\alpha_1+\alpha_2-\alpha_3}{2}}{4\cos^2\frac{\alpha_1}{2}\cos^2\frac{\alpha_2}{2}\cos^2\frac{\alpha_3}{2}} \\ &= \frac{2\cos\frac{\alpha_1+\alpha_2+\alpha_3}{2}\cos\frac{-\alpha_1+\alpha_2+\alpha_3}{2}\cos\frac{\alpha_1-\alpha_2+\alpha_3}{2}\cos\frac{\alpha_1+\alpha_2-\alpha_3}{2}}{(1+\cos\alpha_1)(1+\cos\alpha_2)(1+\cos\alpha_3)} \\ &= \frac{2\cos\alpha_1\cos\alpha_2\cos\alpha_3+\cos^2\alpha_1+\cos^2\alpha_2+\cos^2\alpha_3-1}{2(1+\cos\alpha_1)(1+\cos\alpha_2)(1+\cos\alpha_3)}\end{aligned} \tag{4.228}$$

It would be interesting to compare trigonometric identities in (4.228) with (2.164), p. 109.

Equation 4.228 determines the gyrodistance h_3 of the gyrotriangle ingyrocenter I from the gyrotriangle side A_1A_2 of any gyrotriangle $A_1A_2A_3$ in an Einstein gyrovector space $(\mathbb{R}_s^n, \oplus.\otimes)$, shown in Fig. 4.14 for $n = 2$.

Each of the right-hand sides of (4.228) is a symmetric function of the gyrotriangle gyroangles, so that it is invariant under vertex cyclic permutations. Hence, also the left-hand side of (4.228) is invariant under vertex cyclic permutations, implying

$$h_1 = h_2 = h_3 =: r \tag{4.229}$$

where we use the notation in Fig. 4.14. According to (4.229), the gyrotriangle ingyrocenter I is the point of the interior of the gyrotriangle that is equigyrodistant from the three gyrotriangle sides, so that r is the ingyroradius of gyrotriangle $A_1A_2A_3$.

In the Euclidean limit of large s, $s \to \infty$, the left hand side of (4.228) tends to 0, so that Identity (4.228) reduces to the identity

$$\cos\frac{\alpha_1+\alpha_2+\alpha_3}{2} = 0 \tag{4.230}$$

which, for triangle angles, is equivalent to the triangle angle condition $\alpha_1 + \alpha_2 + \alpha_3 = \pi$ in Euclidean geometry.

4.18 Möbius Gyrotriangle Ingyrocenter

Let us transform the gyrobarycentric coordinates $(m_1 : m_2 : m_3)$ of the gyrotriangle ingyrocenter in Einstein gyrovector spaces, as given in the first equation in (4.215), into their counterparts in corresponding Möbius gyrovector spaces by means of the transformation formula in (2.281) – (2.283), p. 143. The transformation, from which we remove the common factor 2,

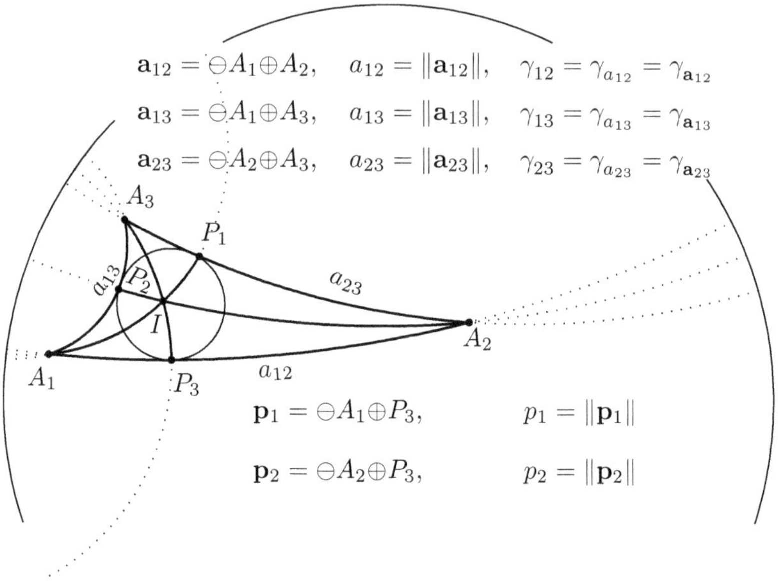

Fig. 4.15 A gyrotriangle $A_1A_2A_3$ and its ingyrocircle and ingyrocenter I, along with its standard index notation, is shown in a Möbius gyrovector space $(\mathbb{R}_s^n, \oplus, \otimes)$, $n = 2$. The gyroline A_kP_k is the gyroangle bisector from vertex A_k to the intersection point P_k with the opposite side, $k = 1, 2, 3$. The gyrotriangle gyroangle bisectors are concurrent. The point of concurrency, I, is the ingyrocenter of the gyrotriangle. A gyrocircle in a Möbius gyrovector plane $(\mathbb{R}_s^2, \oplus, \otimes)$ with gyrocenter at the center of the disk $\mathbb{R}_s^2$ has the shape of a Euclidean circle, and its gyrocircle gyrocenter coincides with the Euclidean circle center. In general, a gyrocircle in a Möbius gyrovector plane has the shape of a Euclidean circle, but its gyrocircle gyrocenter does not coincide with its Euclidean circle center, as we see in this figure and in Fig. 4.19, p. 241.

results in the following gyrobarycentric coordinates of the gyrotriangle ingyrocenter in Möbius gyrovector spaces, shown in Fig. 4.15,

$$(m_1 : m_2 : m_3) = \left(\gamma_{23}\sqrt{\gamma_{23}^2 - 1} : \gamma_{13}\sqrt{\gamma_{13}^2 - 1} : \gamma_{12}\sqrt{\gamma_{12}^2 - 1}\right) \qquad (4.231)$$

Accordingly, the transformation of the points P_1, P_2, P_3 in (4.187) from an Einstein gyrovector space into a corresponding Möbius gyrovector space, by means of the gyrobarycentric coordinates (4.231) and Def. 4.28, determines the points P_1, P_2, P_3, shown in Fig. 4.15, by the equations

$$
\begin{aligned}
P_1 &= \tfrac{1}{2}\otimes\frac{\gamma_{13}\sqrt{\gamma_{13}^2-1}\gamma_{A_2}^2 A_2+\gamma_{12}\sqrt{\gamma_{12}^2-1}\gamma_{A_3}^2 A_3}{\gamma_{13}\sqrt{\gamma_{13}^2-1}(\gamma_{A_2}^2-\frac{1}{2})+\gamma_{12}\sqrt{\gamma_{12}^2-1}(\gamma_{A_3}^2-\frac{1}{2})}\\
P_2 &= \tfrac{1}{2}\otimes\frac{\gamma_{23}\sqrt{\gamma_{23}^2-1}\gamma_{A_1}^2 A_1+\gamma_{12}\sqrt{\gamma_{12}^2-1}\gamma_{A_3}^2 A_3}{\gamma_{23}\sqrt{\gamma_{23}^2-1}(\gamma_{A_1}^2-\frac{1}{2})+\gamma_{12}\sqrt{\gamma_{12}^2-1}(\gamma_{A_3}^2-\frac{1}{2})}\\
P_3 &= \tfrac{1}{2}\otimes\frac{\gamma_{23}\sqrt{\gamma_{23}^2-1}\gamma_{A_1}^2 A_1+\gamma_{13}\sqrt{\gamma_{13}^2-1}\gamma_{A_2}^2 A_2}{\gamma_{23}\sqrt{\gamma_{23}^2-1}(\gamma_{A_1}^2-\frac{1}{2})+\gamma_{13}\sqrt{\gamma_{13}^2-1}(\gamma_{A_2}^2-\frac{1}{2})}
\end{aligned}
\tag{4.232}
$$

Equations (4.232) for P_1, P_2 and P_3 are obtained by transforming the gyrobarycentric coordinates $(m_1 : m_2 : m_3)$ in the first equation of (4.215) from an Einstein gyrovector space into the corresponding Möbius gyrovector space. However, it is simpler to transform the gyrobarycentric coordinates $(m_1 : m_2 : m_3)$ in the third equation of (4.215). The third equation of (4.215) presents gyrobarycentric coordinates in a gyrotrigonometric form the transformation of which is trivial.

Gyroangles are model independent, [3, Theorem 8.3], and, in particular, gyroangles remain invariant in the transition from Einstein to Möbius gyrovector spaces as stated in Theorem 2.48, p. 143. Hence, the transformation of the gyrobarycentric coordinates $(m_1 : m_2 : m_3)$ in the third equation of (4.215) from an Einstein gyrovector space into the corresponding Möbius gyrovector space keeps gyrotrigonometric gyrobarycentric coordinates invariant. Hence, in our comparative study of the standard Cartesian model of Euclidean geometry and the two Cartesian models of hyperbolic geometry that are regulated by Einstein and Möbius gyrovector spaces, gyrotrigonometric gyrobarycentric coordinates possess comparative advantage over other forms of gyrobarycentric coordinates.

We should note that while gyrobarycentric coordinates in a gyrotrigonometric form remain intact in the transformation between various models of hyperbolic geometry, the gyrobarycentric coordinates $(m_1 : m_2 : m_3)$ in the third equation of (4.215) should be (i) placed in formula (4.2) of Def. 4.2, p. 171, for Einstein gyrovector spaces; and (ii) placed in the different formula (4.22) of Def. 4.7, p. 178, for Möbius gyrovector spaces.

Accordingly, following this trivial transformation and Def. 4.7, p. 178, the transformation of the points P_1, P_2, P_3 in (4.187) from an Einstein gyrovector space into a corresponding Möbius gyrovector space gives the points P_1, P_2, P_3, shown in Fig. 4.15, by the equations,

$$P_1 = \tfrac{1}{2}\otimes\frac{\sin\alpha_2\gamma_{A_2}^2 A_1 + \sin\alpha_3\gamma_{A_3}^2 A_3}{\sin\alpha_2(\gamma_{A_2}^2 - \frac{1}{2}) + \sin\alpha_3(\gamma_{A_3}^2 - \frac{1}{2})}$$

$$P_2 = \tfrac{1}{2}\otimes\frac{\sin\alpha_1\gamma_{A_1}^2 A_1 + \sin\alpha_2\gamma_{A_2}^2 A_3}{\sin\alpha_1(\gamma_{A_1}^2 - \frac{1}{2}) + \sin\alpha_3(\gamma_{A_3}^2 - \frac{1}{2})} \qquad (4.233)$$

$$P_3 = \tfrac{1}{2}\otimes\frac{\sin\alpha_1\gamma_{A_1}^2 A_1 + \sin\alpha_2\gamma_{A_2}^2 A_2}{\sin\alpha_1(\gamma_{A_1}^2 - \frac{1}{2}) + \sin\alpha_2(\gamma_{A_2}^2 - \frac{1}{2})}$$

Accordingly, the gyrobarycentric coordinates $(m_1 : m_2 : m_3)$ of the gyrotriangle ingyrocenter in the Poincaré ball model in gyrotrigonometric form is given by the equation

$$(m_1 : m_2 : m_3) = (\sin\alpha_1 : \sin\alpha_2 : \sin\alpha_3) \qquad (4.234)$$

which is identical in form with well-known Euclidean barycentric coordinates of the triangle incenter, determined in (1.153), p. 43.

Hence, following (4.231) – (4.234) and Def. 4.7, the ingyrocenter I of gyrotriangle $A_1A_2A_3$ in a Möbius gyrovector space $(\mathbb{R}_s^n, \oplus, \otimes)$ is given by

$$I = \tfrac{1}{2}\otimes\frac{\sum_{k=1}^3 m_k\gamma_{A_k}^2 A_k}{\sum_{k=1}^3 m_k(\gamma_{A_k}^2 - \frac{1}{2})} \qquad (4.235)$$

where the gyrobarycentric coordinates in (4.235) are given by (4.231) or, equivalently, by (4.234). The points P_1, P_2, P_3, and I, as calculated in (4.233) and (4.235), are presented in Fig. 4.15, which demonstrates graphically that the point I coincides with the point of intersection of the gyrotriangle gyroangle bisectors.

Formalizing the main result of this section, we have the following theorem:

Theorem 4.31. (Möbius Ingyrocenter). *Let $S = \{A_1, A_2, A_3\}$ be a pointwise independent set of three points in a Möbius gyrovector space $(\mathbb{R}_s^n, \oplus, \otimes)$, $n \geq 2$. The ingyrocenter I of gyrotriangle $A_1A_2A_3$ (Fig. 4.15) has the gyrobarycentric coordinate representation*

$$I = \tfrac{1}{2}\otimes\frac{\sum_{k=1}^3 m_k\gamma_{A_k}^2 A_k}{\sum_{k=1}^3 m_k(\gamma_{A_k}^2 - \frac{1}{2})} \qquad (4.236)$$

with respect to the set $\{A_1, A_2, A_3\}$, with gyrotrigonometric gyrobarycentric coordinates

$$(m_1 : m_2 : m_3) = (\sin\alpha_1 : \sin\alpha_2 : \sin\alpha_3) \qquad (4.237)$$

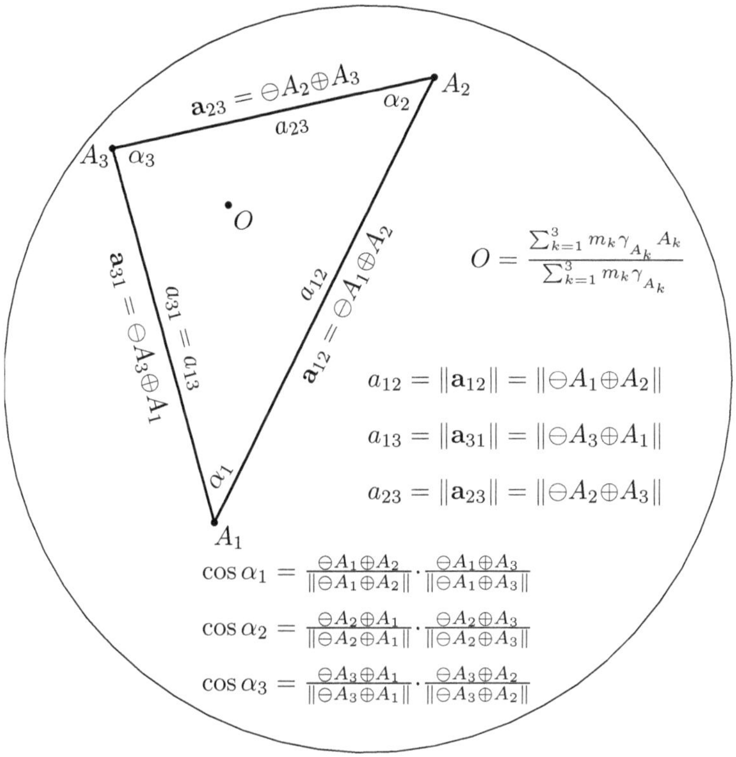

Fig. 4.16 The circumgyrocenter of gyrotriangle $A_1A_2A_3$ in an Einstein gyrovector space $(\mathbb{R}^n_s, \oplus, \otimes)$, $n = 2$, is shown along with its standard notation. Here $\|\ominus A_1 \oplus O\| = \|\ominus A_2 \oplus O\| = \|\ominus A_3 \oplus O\|$, where O is the gyrotriangle circumgyrocenter. The gyrotriangle circumgyrocircle is shown in Fig. 4.17, and gyrobarycentric coordinates of the gyrotriangle circumgyrocenter are determined in Theorem 4.33, p. 237.

4.19 Einstein Gyrotriangle Circumgyrocenter

Definition 4.32. The circumgyrocenter, O, of a gyrotriangle is the point in the gyrotriangle gyroplane equigyrodistant from the three gyrotriangle vertices.

Following Def. 4.32, the circumgyrocenter of a gyrotriangle, shown in Fig. 4.16, is the gyrocenter of the gyrotriangle circumgyrocircle, shown in Fig. 4.17.

Let O be the circumgyrocenter of gyrotriangle $A_1A_2A_3$ in an Einstein gyrovector space $(\mathbb{R}^n_s, \oplus, \otimes)$, shown in Fig. 4.16, and let $(m_1 : m_2 : m_3)$ be

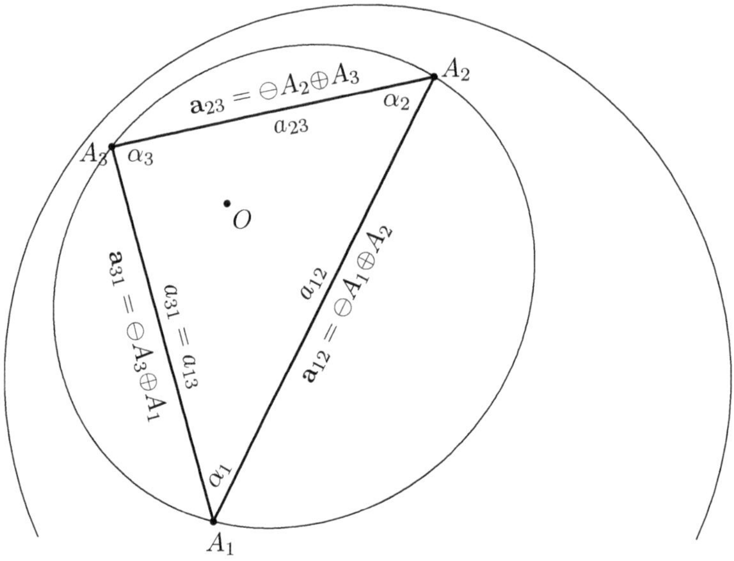

Fig. 4.17 The Einstein gyrotriangle circumgyrocircle with circumgyrocenter O. The circumgyrocircle of the gyrotriangle $A_1A_2A_3$ in Fig. 4.16 is shown. As remarked in Fig. 4.14, and as shown in this figure, a gyrocircle in an Einstein gyrovector space has the shape of a flattened Euclidean circle.

its gyrobarycentric coordinates with respect to the set $S = \{A_1, A_2, A_3\}$, so that

$$O = \frac{m_1\gamma_{A_1}A_1 + m_2\gamma_{A_2}A_2 + m_3\gamma_{A_3}A_3}{m_1\gamma_{A_1} + m_2\gamma_{A_2} + m_3\gamma_{A_3}} \tag{4.238}$$

The gyrobarycentric coordinates m_1, m_2 and m_3 are to be determined in (4.245) below, in terms of gamma factors of the gyrotriangle sides and, alternatively, in (4.251) in terms of the gyrotriangle gyroangles.

Then, by the second identity in (4.11), p. 174, for $X = \ominus A_1$, and by the standard gyrotriangle notation, shown in Fig. 4.16, in Fig. 2.3, p. 101 and in (2.133), p. 100, we have

$$\begin{aligned}\gamma_{\ominus A_1\oplus O} &= \frac{m_1\gamma_{\ominus A_1\oplus A_1} + m_2\gamma_{\ominus A_1\oplus A_2} + m_3\gamma_{\ominus A_1\oplus A_3}}{m_0} \\ &= \frac{m_1 + m_2\gamma_{12} + m_3\gamma_{13}}{m_0}\end{aligned} \tag{4.239}$$

where by (4.10), p. 173, the circumgyrocenter constant m_0 with respect to the set of the gyrotriangle vertices is given by the equation

$$m_0^2 = m_1^2 + m_2^2 + m_3^2 + 2(m_1m_2\gamma_{12} + m_1m_3\gamma_{13} + m_2m_3\gamma_{23}) \tag{4.240}$$

Hence, similarly, by the second identity in (4.11) for $X = \ominus A_1$, for $X = \ominus A_2$, and for $X = \ominus A_3$, we have

$$\begin{aligned} \gamma_{\ominus A_1 \oplus O} &= \frac{m_1 + m_2\gamma_{12} + m_3\gamma_{13}}{m_0} \\ \gamma_{\ominus A_2 \oplus O} &= \frac{m_1\gamma_{12} + m_2 + m_3\gamma_{23}}{m_0} \\ \gamma_{\ominus A_3 \oplus O} &= \frac{m_1\gamma_{13} + m_2\gamma_{23} + m_3}{m_0} \end{aligned} \tag{4.241}$$

The condition that the circumgyrocenter O is equigyrodistant from its gyrotriangle vertices A_1, A_2, and A_3 implies

$$\gamma_{\ominus A_1 \oplus O} = \gamma_{\ominus A_2 \oplus O} = \gamma_{\ominus A_3 \oplus O} \tag{4.242}$$

Equations (4.241) and (4.242), along with the normalization condition $m_1 + m_2 + m_3 = 1$, yield the following system of three equations for the three unknowns m_1, m_2, and m_3,

$$\begin{aligned} m_1 + m_2 + m_3 &= 1 \\ m_1 + m_2\gamma_{12} + m_3\gamma_{13} &= m_1\gamma_{13} + m_2\gamma_{23} + m_3 \\ m_1\gamma_{12} + m_2 + m_3\gamma_{23} &= m_1\gamma_{13} + m_2\gamma_{23} + m_3 \end{aligned} \tag{4.243}$$

which can be manipulated into the matrix equation,

$$\begin{pmatrix} 1 & 1 & 1 \\ 1 - \gamma_{13} & \gamma_{12} - \gamma_{23} & \gamma_{13} - 1 \\ \gamma_{12} - \gamma_{13} & 1 - \gamma_{23} & \gamma_{23} - 1 \end{pmatrix} \begin{pmatrix} m_1 \\ m_2 \\ m_3 \end{pmatrix} = \begin{pmatrix} 1 \\ 0 \\ 0 \end{pmatrix} \tag{4.244}$$

Solving (4.244) for the unknowns m_1, m_2, and m_3, we have

$$\begin{aligned} m_1 &= \frac{1}{D}(\ \ \gamma_{12} + \gamma_{13} - \gamma_{23} - 1)(\gamma_{23} - 1) \\ m_2 &= \frac{1}{D}(\ \ \gamma_{12} - \gamma_{13} + \gamma_{23} - 1)(\gamma_{13} - 1) \\ m_3 &= \frac{1}{D}(-\gamma_{12} + \gamma_{13} + \gamma_{23} - 1)(\gamma_{12} - 1) \end{aligned} \tag{4.245}$$

where D is the determinant of the 3×3 matrix in (4.244),

$$\begin{aligned} D &= 2(\gamma_{12}\gamma_{13} + \gamma_{12}\gamma_{23} + \gamma_{13}\gamma_{23}) \\ &- (\gamma_{12}^2 - 1) - (\gamma_{13}^2 - 1) - (\gamma_{23}^2 - 1) - 2(\gamma_{12} + \gamma_{13} + \gamma_{23}) \end{aligned} \tag{4.246}$$

Hence, the circumgyrocenter O of gyrotriangle $A_1A_2A_3$ is given by (4.238) where gyrobarycentric coordinates m_1, m_2, and m_3 are given by (4.245). Since in gyrobarycentric coordinates only ratios of coordinates are relevant, the gyrobarycentric coordinates, m_1, m_2, and m_3 in (4.245) can be simplified by removing their common factor $1/D$.

Gyrobarycentric coordinates, m_1, m_2, and m_3, of the circumgyrocenter O of gyrotriangle $A_1A_2A_3$ are thus given by the equations

$$\begin{aligned} m_1' &= (\ \ \gamma_{12} + \gamma_{13} - \gamma_{23} - 1)(\gamma_{23} - 1) \\ m_2' &= (\ \ \gamma_{12} - \gamma_{13} + \gamma_{23} - 1)(\gamma_{13} - 1) \\ m_3' &= (-\gamma_{12} + \gamma_{13} + \gamma_{23} - 1)(\gamma_{12} - 1) \end{aligned} \tag{4.247}$$

Hence, by (4.240) along with the gyrobarycentric coordinates in (4.247), we have

$$\begin{aligned} m_0^2 &= \{(\gamma_{12} + \gamma_{13} + \gamma_{23} - 1)^2 - 2(\gamma_{12}^2 + \gamma_{13}^2 + \gamma_{23}^2 - 1)\} \\ &\times\ (1 + 2\gamma_{12}\gamma_{13}\gamma_{23} - \gamma_{12}^2 - \gamma_{13}^2 - \gamma_{23}^2) \end{aligned} \tag{4.248}$$

Gyrotriangles for which the constant m_0^2 in (4.248) is nonpositive possess no circumgyrocenter.

Having the experience acquired from Theorem 4.29, we know that in order to capture trigonometric/gyrotrigonometric analogies with Euclidean geometry, gyrobarycentric coordinates of various gyrotriangle gyrocenters should be expressed in terms of gyrotriangle gyroangles, as in the third equation in (4.215) and as in (4.137). We therefore note that the gamma factors of gyrotriangle side gyrolengths are related to its gyroangles by (4.134). Substituting these from (4.134) into (4.247) we obtain

$$\begin{aligned} m_1' &= F \sin\left(\frac{-\alpha_1 + \alpha_2 + \alpha_3}{2}\right) \sin\alpha_1 \\ m_2' &= F \sin\left(\frac{\alpha_1 - \alpha_2 + \alpha_3}{2}\right) \sin\alpha_2 \\ m_3' &= F \sin\left(\frac{\alpha_1 + \alpha_2 - \alpha_3}{2}\right) \sin\alpha_3 \end{aligned} \tag{4.249}$$

where the common factor F in (4.249) is given by the equation

$$F = 2^3 \frac{\cos^2(\frac{\alpha_1+\alpha_2+\alpha_3}{2}) \cos(\frac{-\alpha_1+\alpha_2+\alpha_3}{2}) \cos(\frac{\alpha_1-\alpha_2+\alpha_3}{2}) \cos(\frac{\alpha_1+\alpha_2-\alpha_3}{2})}{\sin\alpha_1 \sin\alpha_2 \sin\alpha_3} \tag{4.250}$$

Since in gyrobarycentric coordinates only ratios of coordinates are relevant, the homogeneous gyrobarycentric coordinates, m_1', m_2', and m_3' in

(4.249) can be simplified by removing their common factor F. Hence, gyrobarycentric coordinates, m_1'', m_2'', and m_3'', of the circumgyrocenter O of gyrotriangle $A_1A_2A_3$, expressed in terms of the gyrotriangle gyroangles are given by the equations

$$\begin{aligned} m_1'' &= \sin\left(\frac{-\alpha_1+\alpha_2+\alpha_3}{2}\right)\sin\alpha_1 \\ m_2'' &= \sin\left(\frac{\alpha_1-\alpha_2+\alpha_3}{2}\right)\sin\alpha_2 \\ m_3'' &= \sin\left(\frac{\alpha_1+\alpha_2-\alpha_3}{2}\right)\sin\alpha_3 \end{aligned} \tag{4.251}$$

Interestingly, the gyrotrigonometric gyrobarycentric coordinates $(m_1'' : m_2'' : m_3'')$ in (4.251) of a gyrotriangle circumgyrocenter are identical in form with their Euclidean counterparts in (1.141), p. 38.

Formalizing the main result of this section, we have the following theorem.

Theorem 4.33. (Einstein Circumgyrocenter). *Let $S = \{A_1, A_2, A_3\}$ be a pointwise independent set of three points in an Einstein gyrovector space $(\mathbb{R}^n_s, \oplus, \otimes)$. The circumgyrocenter O (Fig. 4.16) of gyrotriangle $A_1A_2A_3$ has the gyrobarycentric coordinate representation*

$$O = \frac{m_1\gamma_{A_1}A_1 + m_2\gamma_{A_2}A_2 + m_3\gamma_{A_3}A_3}{m_1\gamma_{A_1} + m_2\gamma_{A_2} + m_3\gamma_{A_3}} \tag{4.252}$$

with respect to the set $\{A_1, A_2, A_3\}$, with gyrobarycentric coordinates $(m_1 : m_2 : m_3)$ given by

$$\begin{aligned} m_1 &= (\ \ \gamma_{12}+\gamma_{13}-\gamma_{23}-1)(\gamma_{23}-1) \\ m_2 &= (\ \ \gamma_{12}-\gamma_{13}+\gamma_{23}-1)(\gamma_{13}-1) \\ m_3 &= (-\gamma_{12}+\gamma_{13}+\gamma_{23}-1)(\gamma_{12}-1) \end{aligned} \tag{4.253}$$

or, equivalently, by the gyrotrigonometric gyrobarycentric coordinates

$$\begin{aligned} m_1 &= \sin\left(\frac{-\alpha_1+\alpha_2+\alpha_3}{2}\right)\sin\alpha_1 = \cos(\alpha_1+\tfrac{\delta}{2})\sin(\alpha_1) \\ m_2 &= \sin\left(\frac{\alpha_1-\alpha_2+\alpha_3}{2}\right)\sin\alpha_2 = \cos(\alpha_2+\tfrac{\delta}{2})\sin(\alpha_2) \\ m_3 &= \sin\left(\frac{\alpha_1+\alpha_2-\alpha_3}{2}\right)\sin\alpha_3 = \cos(\alpha_3+\tfrac{\delta}{2})\sin(\alpha_3) \end{aligned} \tag{4.254}$$

where

$$\delta = \pi - (\alpha_1+\alpha_2+\alpha_3) \tag{4.255}$$

is the defect of the reference gyrotriangle $A_1A_2A_3$. Clearly, the incorporation of δ in (4.254) yields simplicity and elegance.

In the special case when $\delta = 0$, (4.254) descends to its Euclidean counterpart, presented in (1.140), p. 38.

The circumgyrocenter constant m_0 with respect to the set $\{A_1, A_2, A_3\}$ is given by the equation

$$\begin{aligned} m_0^2 &= \{(\gamma_{12}+\gamma_{13}+\gamma_{23}-1)^2 - 2(\gamma_{12}^2+\gamma_{13}^2+\gamma_{23}^2-1)\} \\ &\times\ (1+2\gamma_{12}\gamma_{13}\gamma_{23}-\gamma_{12}^2-\gamma_{13}^2-\gamma_{23}^2) \end{aligned} \tag{4.256}$$

The gyrotriangle $A_1A_2A_3$ possesses a circumgyrocenter if and only if m_0^2 is positive, that is, equivalently (noting that the second factor in (4.256) is positive by inequality (2.146), p. 104), if and only if

$$(\gamma_{12}+\gamma_{13}+\gamma_{23}-1)^2 \ > \ 2(\gamma_{12}^2+\gamma_{13}^2+\gamma_{23}^2-1) \tag{4.257}$$

4.20 Einstein Gyrotriangle Circumgyroradius

Let $A_1A_2A_3$ be a gyrotriangle in an Einstein gyrovector space $(\mathbb{R}_s^n, \oplus, \otimes)$. The gyroradius R of the gyrotriangle circumgyrocircle is called the gyrotriangle circumgyroradius. As shown in Fig. 4.16, p. 233, the gyrotriangle circumgyroradius R of gyrotriangle $A_1A_2A_3$ is given by

$$R^2 = \|\ominus A_1 \oplus O\|^2 = \|\ominus A_2 \oplus O\|^2 = \|\ominus A_3 \oplus O\|^2 \tag{4.258}$$

so that, by (2.11), p. 66,

$$R^2 = \|\ominus A_1 \oplus O\|^2 = s^2 \frac{\gamma_{\ominus A_1 \oplus O}^2 - 1}{\gamma_{\ominus A_1 \oplus O}^2} \tag{4.259}$$

Substituting the first equation in (4.241) into (4.259), an elegant expression for R emerges,

$$R = \sqrt{2}s \sqrt{\frac{(\gamma_{12}-1)(\gamma_{13}-1)(\gamma_{23}-1)}{1+2\gamma_{12}\gamma_{13}\gamma_{23}-\gamma_{12}^2-\gamma_{13}^2-\gamma_{23}^2}} \tag{4.260}$$

A natural generalization of (4.260) into the circumgyroradius R of gyrosimplices of any dimension $n \geq 2$ appears in [19, Eq. 10.155].

4.21 Möbius Gyrotriangle Circumgyrocenter

Let us transform the gyrobarycentric coordinates $(m_1 : m_2 : m_3)$ of the gyrotriangle circumgyrocenter in Einstein gyrovector spaces $(\mathbb{R}_s^n, \oplus = \oplus_{\text{E}}, \otimes)$, as given in (4.253), into their counterparts in corresponding Möbius gyrovector spaces $(\mathbb{R}^n_s, \oplus = \oplus_{\mathrm{M}}, \otimes)$ by means of the transformation formula in (2.281), p. 143. The transformation, from which we remove a common factor 4, results in the gyrobarycentric coordinates $(m_1 : m_2 : m_3)$ of the gyrotriangle ingyrocenter in Möbius gyrovector spaces, shown in Fig. 4.15, p. 230,

$$\begin{aligned} m_1 &= (\ \ \gamma_{12}^2 + \gamma_{13}^2 - \gamma_{23}^2 - 1)(\gamma_{23}^2 - 1) \\ m_2 &= (\ \ \gamma_{12}^2 - \gamma_{13}^2 + \gamma_{23}^2 - 1)(\gamma_{13}^2 - 1) \\ m_3 &= (-\gamma_{12}^2 + \gamma_{13}^2 + \gamma_{23}^2 - 1)(\gamma_{12}^2 - 1) \end{aligned} \tag{4.261}$$

Similarly, the transformation of m_0^2 in (4.256) from an Einstein gyrovector space $(\mathbb{R}^n_s, \oplus = \oplus_{\mathrm{E}}, \otimes)$ into its counterpart, m_0^2, in a corresponding Möbius gyrovector space $(\mathbb{R}^n_s, \oplus = \oplus_{\mathrm{M}}, \otimes)$ by means of the transformation formula in (2.281), p. 143, gives

$$\begin{aligned} m_0^2 = 16\ &\{4\gamma_{12}^2\gamma_{13}^2\gamma_{23}^2 - (\gamma_{12}^2 + \gamma_{13}^2 + \gamma_{23}^2 - 1)^2\} \\ &\times\{4(\gamma_{12}^2\gamma_{13}^2 + \gamma_{12}^2\gamma_{23}^2 + \gamma_{13}^2\gamma_{23}^2 + 1) - (\gamma_{12}^2 + \gamma_{13}^2 + \gamma_{23}^2 + 1)^2\} \end{aligned} \tag{4.262}$$

The constant m_0^2 is presented in (4.262) as a product of three factors, the first of which is 16. The second factor is positive for any gyrotriangle. Hence, m_0^2 is positive if and only if the third factor on the right-hand side of (4.262) is positive.

As an alternative to (4.261), instead of transforming the circumgyrocenter gyrobarycentric coordinates in (4.253) from Einstein to Möbius gyrovector space, as we did in (4.261), we can transform the circumgyrocenter gyrotrigonometric gyrobarycentric coordinates in (4.254). The latter transformation is simpler than the former since the latter is just the trivial transformation, that is, a transformation of each gyrobarycentric coordinate into itself. Formalizing, we thus have the following theorem:

Theorem 4.34. (The Möbius Circumgyrocenter). *Let $S = \{A_1, A_2, A_3\}$ be a pointwise independent set of three points in a Möbius gyrovector space $(\mathbb{R}^n_s, \oplus, \otimes)$, $n \geq 2$. The circumgyrocenter O of gyrotriangle $A_1A_2A_3$, (Figs. 4.18–4.19) has the gyrobarycentric coordinate representation*

$$O = \tfrac{1}{2}\otimes\frac{\sum_{k=1}^{3} m_k\gamma_{A_k}^2 A_k}{\sum_{k=1}^{3} m_k(\gamma_{A_k}^2 - \frac{1}{2})} \tag{4.263}$$

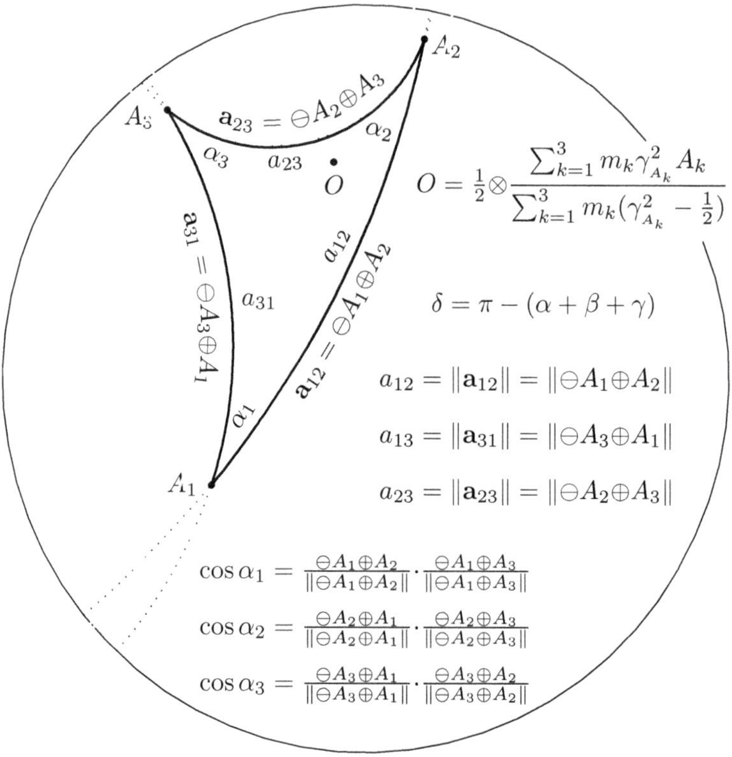

Fig. 4.18 The Möbius gyrotriangle circumgyrocenter O. The gyrotriangle $A_1A_2A_3$ in a Möbius gyrovector space $(\mathbb{R}_s^n, \oplus, \otimes)$ is presented for $n = 2$, along with its circumgyrocenter O. The circumgyrocenter O, determined by Theorem 4.34, is equigyrodistant from the three gyrotriangle vertices A_1, A_2 and A_3, that is, $\|\ominus A_1 \oplus O\| = \|\ominus A_2 \oplus O\| = \|\ominus A_3 \oplus O\|$. The gyrotriangle circumgyrocircle is shown in Fig. 4.19.

with respect to the set $\{A_1, A_2, A_3\}$, *with gyrobarycentric coordinates* $(m_1 : m_2 : m_3)$ *given by*

$$
\begin{aligned}
m_1 &= 4(\ \ \gamma_{12}^2 + \gamma_{13}^2 - \gamma_{23}^2 - 1)(\gamma_{23}^2 - 1) \\
m_2 &= 4(\ \ \gamma_{12}^2 - \gamma_{13}^2 + \gamma_{23}^2 - 1)(\gamma_{13}^2 - 1) \qquad (4.264) \\
m_3 &= 4(-\gamma_{12}^2 + \gamma_{13}^2 + \gamma_{23}^2 - 1)(\gamma_{12}^2 - 1)
\end{aligned}
$$

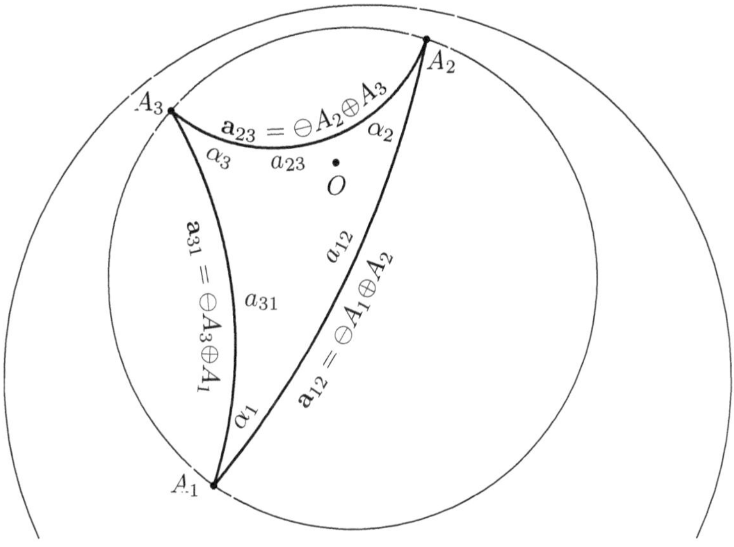

Fig. 4.19 The Möbius gyrotriangle circumgyrocircle with circumgyrocenter O. The circumgyrocircle of the gyrotriangle $A_1A_2A_3$ in Fig. 4.18 is shown. As remarked in Fig. 4.15, and as shown in this figure, a gyrocircle in a Möbius gyrovector space has the shape of a Euclidean circle, but in general, the gyrocircle gyrocenter is different from the corresponding circle center.

or, equivalently, with gyrotrigonometric gyrobarycentric coordinates $(m_1 : m_2 : m_3)$ *given by*

$$\begin{aligned}
m_1 &= \sin\left(\frac{-\alpha_1+\alpha_2+\alpha_3}{2}\right)\sin\alpha_1 = \cos(\alpha_1+\tfrac{\delta}{2})\sin(\alpha_1) \\
m_2 &= \sin\left(\frac{\alpha_1-\alpha_2+\alpha_3}{2}\right)\sin\alpha_2 = \cos(\alpha_2+\tfrac{\delta}{2})\sin(\alpha_2) \\
m_3 &= \sin\left(\frac{\alpha_1+\alpha_2-\alpha_3}{2}\right)\sin\alpha_3 = \cos(\alpha_3+\tfrac{\delta}{2})\sin(\alpha_3)
\end{aligned} \tag{4.265}$$

where $\delta = \pi - (\alpha_1+\alpha_2+\alpha_3)$ *is the defect of gyrotriangle* $A_1A_2A_3$.

The circumgyrocenter constant m_0 *with respect to the set* $\{A_1, A_2, A_3\}$ *is given by* (4.262), *and the gyrotriangle* $A_1A_2A_3$ *possesses a circumgyrocenter if and only if* m_0^2 *is positive.*

4.22 Comparative Study of Gyrotriangle Gyrocenters

We are now in the position to complete the comparative study of the classical triangle centers, initiated in Table 1.1, p. 55.

In this chapter we have introduced the notion of gyrobarycentric coordinates in the Cartesian-Beltrami-Klein model $\mathbb{R}_s^n$ and in the Cartesian-Poincaré model $\mathbb{R}_s^n$ of hyperbolic geometry and determined several gyrobarycentric coordinate representations, including those of the four hyperbolic counterparts of the classical triangle centers. Using the standard index notation for a gyrotriangle $A_1A_2A_3$, Fig. 2.3, p. 101, we expressed gyrobarycentric coordinates $\{m_1 : m_2 : m_3\}$ (i) in terms of gyrotriangle side-gyrolengths, a_{12}, a_{13}, a_{23}, and (ii) in terms of gyrotriangle gyroangles $\alpha_1, \alpha_2, \alpha_3$. The resulting gyrotrigonometric gyrobarycentric coordinates, in which gyrobarycentric coordinates are expressed in terms of the gyroangles of the reference gyrotriangle prove useful in the discovery of analogies with Euclidean geometry. Indeed, we have found in this chapter that gyrotrigonometric gyrobarycentric coordinates of gyrotriangle gyrocenters survive unimpaired the transition from hyperbolic to Euclidean geometry.

Tables of the gyrotrigonometric gyrobarycentric coordinate representations of the hyperbolic counterparts of the four classic triangle centers in the (i) Cartesian-Beltrami-Klein ball model and in the (ii) Cartesian-Poincaré ball model of hyperbolic geometry are presented in Tables 4.1 and 4.2.

The advantage of expressing gyrobarycentric coordinates gyrotrigonometrically is obvious from the tables. Gyrotrigonometric gyrobarycentric coordinates of a gyrotriangle gyrocenter in the Cartesian-Beltrami-Klein model $\mathbb{R}_s^n$ of hyperbolic geometry remain invariant in form in the transition to both the Cartesian-Poincaré model $\mathbb{R}_s^n$ of hyperbolic geometry and the standard Cartesian model $\mathbb{R}^n$ of Euclidean geometry.

It should be noted that while gyrotrigonometric gyrobarycentric coordinates m_k, $k = 1, 2, 3$, remain invariant in form in transitions between the two hyperbolic models and the Euclidean model, gyrotrigonometric gyrobarycentric coordinate representations are distinctive to each model, as illustrated by the gyrotriangle ingyrocenter in (4.266) below.

$$I = \frac{\sin\alpha_1 A_1 + \sin\alpha_2 A_2 + \sin\alpha_3 A_3}{\sin\alpha_1 + \sin\alpha_2 + \sin\alpha_3} \qquad \text{(Euclidean)}$$

$$I = \frac{\sin\alpha_1\gamma_{A_1} A_1 + \sin\alpha_2\gamma_{A_2} A_2 + \sin\alpha_3\gamma_{A_3} A_3}{\sin\alpha_1\gamma_{A_1} + \sin\alpha_2\gamma_{A_2} + \sin\alpha_3\gamma_{A_3}} \qquad \text{(Einsteinian)}$$

$$I = \tfrac{1}{2}\otimes\frac{\sin\alpha_1\gamma_{A_1}^2 A_1 + \sin\alpha_2\gamma_{A_2}^2 A_2 + \sin\alpha_3\gamma_{A_3}^2 A_3}{\sin\alpha_1(\gamma_{A_1}^2 - \frac{1}{2}) + \sin\alpha_2(\gamma_{A_2}^2 - \frac{1}{2}) + \sin\alpha_3(\gamma_{A_3}^2 - \frac{1}{2})} \qquad \text{(Möbius)}$$

$$(4.266)$$

Table 4.1 Gyrotrigonometric gyrobarycentric coordinates of the gyro-counterparts of the classical triangle centers and the triangle altitude foot in the Cartesian-Beltrami-Klein model of hyperbolic geometry, regulated algebraically by the Einstein gyrovector space $(\mathbb{R}^3, \oplus, \otimes)$, with the gyrobarycentric coordinate representation of points $P \in \mathbb{R}_s^n$ with respect to a pointwise independent set $S = \{A_1, A_2, A_3\} \subset \mathbb{R}_s^n$ given by (4.2), p. 171.
Note that
(i) on the one hand, gyroangles are different from angles and the gyrobarycentric coordinate representation definition, Def. 4.2, p. 171, is different from the barycentric coordinate representation definition, Def. 1.5, p. 9. But,
(ii) on the other hand, the gyrobarycentric coordinates m_k, $k = 1.2.3$, of each point in this table are identical in form to the barycentric coordinates m_k, $k = 1.2.3$, of their corresponding Euclidean points in Table 1.1, p. 55.

Gyrocenter	Symbol	Gyrotrigonometric Gyrobarycentric Coordinates in Einstein Gyrovector Spaces
Gyrocentroid	G, (4.63), p. 187 Fig. 4.1, p. 182	$\begin{pmatrix} m_1 \\ m_2 \\ m_3 \end{pmatrix} = \begin{pmatrix} 1 \\ 1 \\ 1 \end{pmatrix}$
Orthogyrocenter	H, (4.137), p. 207 Fig. 4.7, p. 200	$\begin{pmatrix} m_1 \\ m_2 \\ m_3 \end{pmatrix} = \begin{pmatrix} \tan\alpha_1 \\ \tan\alpha_2 \\ \tan\alpha_3 \end{pmatrix}$
Ingyrocenter	I, (4.213), p. 224 Fig. 4.13, p. 215	$\begin{pmatrix} m_1 \\ m_2 \\ m_3 \end{pmatrix} = \begin{pmatrix} \sin\alpha_1 \\ \sin\alpha_2 \\ \sin\alpha_3 \end{pmatrix}$
Circumgyrocenter	O, (4.254), p. 237 Fig. 4.16, p. 233	$\begin{pmatrix} m_1 \\ m_2 \\ m_3 \end{pmatrix} = \begin{pmatrix} \sin\frac{-\alpha_1+\alpha_2+\alpha_3}{2}\sin\alpha_1 \\ \sin\frac{\alpha_1-\alpha_2+\alpha_3}{2}\sin\alpha_2 \\ \sin\frac{\alpha_1+\alpha_2-\alpha_3}{2}\sin\alpha_3 \end{pmatrix}$
Gyroaltitude Foot	P_3, (4.91), p. 196 Fig. 4.7, p. 200	$\begin{pmatrix} m_1 \\ m_2 \\ m_3 \end{pmatrix} = \begin{pmatrix} \tan\alpha_1 \\ \tan\alpha_2 \\ 0 \end{pmatrix}$

The first equation in (4.266) presents the incenter trigonometric barycentric coordinate representation (1.111), p. 31, of a triangle $A_1A_2A_3$ in a Euclidean vector space $\mathbb{R}^n$ with respect to the triangle vertices. Here, the barycentric coordinate representation follows from Def. 1.5, p. 9, with trigonometric barycentric coordinates $m_k = \sin\alpha_k$, $k = 1.2.3$. The measure

Table 4.2 Gyrotrigonometric gyrobarycentric coordinates of the gyro-counterparts of the classical triangle centers and the triangle altitude foot in the Cartesian-Poincaré model of hyperbolic geometry, regulated algebraically by the Möbius gyrovector space $(\mathbb{R}^3, \oplus, \otimes)$, with the gyrobarycentric coordinate representation of points $P \in \mathbb{R}_s^n$ with respect to a pointwise independent set $S = \{A_1, A_2, A_3\} \subset \mathbb{R}_s^n$ given by (4.19), p. 177.
Note that
(i) on the one hand, gyroangles are different from angles and the gyrobarycentric coordinate representation definition, Def. 4.7, p. 178, is different from the barycentric coordinate representation definition, Def. 1.5, p. 9. But,
(ii) on the other hand, the gyrobarycentric coordinates m_k, $k = 1.2.3$, of each point in this table are identical in form to the barycentric coordinates m_k, $k = 1.2.3$, of their corresponding Euclidean points in Table 1.1, p. 55.

Gyrocenter	Symbol	Gyrotrigonometric Gyrobarycentric Coordinates in Möbius Gyrovector Spaces
Gyrocentroid	G, (4.73), p. 191 Fig. 4.3, p. 190	$\begin{pmatrix} m_1 \\ m_2 \\ m_3 \end{pmatrix} = \begin{pmatrix} 1 \\ 1 \\ 1 \end{pmatrix}$
Orthogyrocenter	H, (4.163), p. 212 Fig. 4.12, p. 213	$\begin{pmatrix} m_1 \\ m_2 \\ m_3 \end{pmatrix} = \begin{pmatrix} \tan\alpha_1 \\ \tan\alpha_2 \\ \tan\alpha_3 \end{pmatrix}$
Ingyrocenter	I, (4.237), p. 232 Fig. 4.15, p. 230	$\begin{pmatrix} m_1 \\ m_2 \\ m_3 \end{pmatrix} = \begin{pmatrix} \sin\alpha_1 \\ \sin\alpha_2 \\ \sin\alpha_3 \end{pmatrix}$
Circumgyrocenter	O, (4.265), p. 241 Fig. 4.18, p. 240	$\begin{pmatrix} m_1 \\ m_2 \\ m_3 \end{pmatrix} = \begin{pmatrix} \sin\frac{-\alpha_1+\alpha_2+\alpha_3}{2}\sin\alpha_1 \\ \sin\frac{\alpha_1-\alpha_2+\alpha_3}{2}\sin\alpha_2 \\ \sin\frac{\alpha_1+\alpha_2-\alpha_3}{2}\sin\alpha_3 \end{pmatrix}$
Gyroaltitude Foot	P_3, (4.102), p. 198 Fig. 4.6, p. 198	$\begin{pmatrix} m_1 \\ m_2 \\ m_3 \end{pmatrix} = \begin{pmatrix} \tan\alpha_1 \\ \tan\alpha_2 \\ 0 \end{pmatrix}$

of its angles α_k is given by the equation

$$\cos\alpha_1 = \frac{-A_1 + A_2}{\|-A_1 + A_2\|} \cdot \frac{-A_1 + A_3}{\|-A_1 + A_3\|}, \qquad (Euclidean\ Geometry) \tag{4.267}$$

etc.

The second equation in (4.266) presents the ingyrocenter gyrotrigonometric gyrobarycentric coordinate representation (4.213), p. 224, of a gyrotriangle $A_1A_2A_3$ in an Einstein gyrovector space $\mathbb{R}_s^n$ with respect to the gyrotriangle vertices. Here, the gyrobarycentric coordinate representation follows from Def. 4.2, p. 171, with gyrotrigonometric gyrobarycentric coordinates $m_k = \sin\alpha_k$, $k = 1.2.3$. The measure of its gyroangles α_k is given by the equation

$$\cos\alpha_1 = \frac{\ominus A_1\oplus A_2}{\|\ominus A_1\oplus A_2\|}\cdot\frac{\ominus A_1\oplus A_3}{\|\ominus A_1\oplus A_3\|}, \qquad (Einstein,\ \oplus = \oplus_{\rm E}) \qquad (4.268)$$

etc., where $\oplus = \oplus_{\rm E}$ represents Einstein addition in $\mathbb{R}_s^n$.

The third equation in (4.266) presents the ingyrocenter gyrotrigonometric gyrobarycentric coordinate representation (4.236), p. 232, of a gyrotriangle $A_1A_2A_3$ in a Möbius gyrovector space $\mathbb{R}_s^n$ with respect to the gyrotriangle vertices. Here, the gyrobarycentric coordinate representation follows from Def. 4.7, p. 178, with gyrotrigonometric gyrobarycentric coordinates $m_k = \sin\alpha_k$, $k = 1.2.3$. The measure of its gyroangles α_k is given by the equation

$$\cos\alpha_1 = \frac{\ominus A_1\oplus A_2}{\|\ominus A_1\oplus A_2\|}\cdot\frac{\ominus A_1\oplus A_3}{\|\ominus A_1\oplus A_3\|}, \qquad (\text{Möbius},\ \oplus = \oplus_{\rm M}) \qquad (4.269)$$

etc., where $\oplus = \oplus_{\rm M}$ represents Möbius addition in $\mathbb{R}_s^n$.

As suggested by Table 1.1, p. 55, and Tables 4.1 and 4.2, in general, and the representations in (4.266), in particular,

(1) once we determine a gyrotrigonometric gyrobarycentric coordinate representation of a gyrotriangle gyrocenter in the Cartesian-Beltrami-Klein model of hyperbolic geometry, regulated by Einstein gyrovector spaces,
(2) we can straightforwardly transform it into a corresponding gyrotriangle gyrocenter in the Cartesian-Poincaré model of hyperbolic geometry, regulated by möbius gyrovector spaces, and
(3) into a corresponding triangle center in the Cartesian model of Euclidean geometry, regulated by Euclidean vector spaces.

4.23 Exercises

(1) Show that the pointwise independence of the set S in Def. 4.2, p. 171, insures that the gyrobarycentric coordinate representation of a point with respect to the set S is unique.
(2) Prove that the relativistic momentum $\sum_{k=1}^{N} m_k\gamma_{A_k} A_k$ of the particle system S in Sec. 4.2, p. 175, with respect to the CM frame of S vanishes. Hint: Employ the first identity in (4.11), p. 174, to show that

$$\sum_{k=1}^{N} m_k\gamma_{\ominus P\oplus A_k}(\ominus P\oplus A_k) = 0 \qquad (4.270)$$

where $P \in \mathbb{R}_s^n$ is the relativistically admissible velocity of the CM frame of the particle system S, given by (4.6), p. 173.

(3) Prove Theorem 4.14, p. 188. (The proof of Theorem 4.14 is similar to that of Theorem 4.12, p. 188.)

(4) Prove that $D > 0$, where D is the determinant in (4.128), p. 206.

(5) Show in detail the steps leading to (4.79), p. 194.

(6) Prove that $(m_1 : m_2 : m_3)$ in (4.163), p. 212, form a set of gyrobarycentric coordinates with respect to the set $S = \{A_1, A_2, A_3\}$ of the Einstein orthogyrocenter H in Theorem 4.26, p. 212.

(7) Calculate explicitly the system of six scalar equations in (4.58), p. 186, for the unknowns m_2, m_3 and $t_{k,0}$, $k = 1, 2, 3$, and show that its unique solution is given by (4.60) and (4.61).

(8) Verify Theorem 4.14, p. 188, in a way similar to the proof of Theorem 4.12.

(9) Substitute the first equation in (4.241), p. 235, into (4.259), p. 238, to obtain the expression (4.260), p. 238, for the gyrotriangle circumgyroradius R.

Chapter 5

Hyperbolic Incircles and Excircles

An ingyrocircle of a gyrotriangle is a gyrocircle lying inside the gyrotriangle, tangent to each of its sides, shown in Fig. 4.14, p. 227 and in Figs. 5.1 – 5.2, pp. 248 – 249. The gyrocenter and gyroradius of the ingyrocircle of a gyrotriangle are called the gyrotriangle ingyrocenter and ingyroradius. Similarly, an exgyrocircle of a gyrotriangle is a gyrocircle lying outside the gyrotriangle, tangent to one of its sides and tangent to the extensions of the other two, as shown in Figs. 5.1 – 5.2. The gyrocenters and gyroradii of the exgyrocircles of a gyrotriangle are called the gyrotriangle exgyrocenters and exgyroradii. The ingyrocenter and exgyrocenters of a gyrotriangle are equigyrodistant from the triangles sides.

5.1 Einstein Gyrotriangle Ingyrocenter and Exgyrocenters

Let $A_1A_2A_3$ be a gyrotriangle in an Einstein gyrovector space $(\mathbb{R}^n_s, \oplus, \otimes)$, and let E be a point equigyrodistant from the gyrotriangle sides, so that E is the ingyrocenter or an exgyrocenter of the gyrotriangle. Furthermore, let

$$E = \frac{m_1\gamma_{A_1}A_1 + m_2\gamma_{A_2}A_2 + m_3\gamma_{A_3}A_3}{m_1\gamma_{A_1} + m_2\gamma_{A_2} + m_3\gamma_{A_3}} \tag{5.1}$$

be the gyrobarycentric coordinate representation of E with respect to the set $S = \{A_1, A_2, A_3\}$ of the gyrotriangle vertices, where the gyrobarycentric coordinates $(m_1 : m_2 : m_3)$ of E in (5.1) are to be determined in (5.10), p. 250.

Applying the second identity in the results (4.11), p. 174, of Theorem 4.4 to the gyrobarycentric coordinate representation of E in (5.1) we obtain the

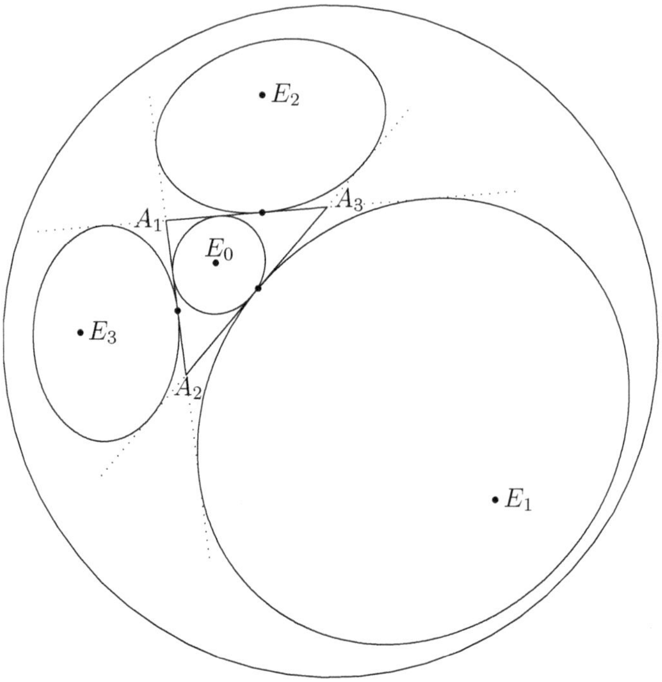

Fig. 5.1 The in-exgyrocenters E_k, $k = 0, 1, 2, 3$, with gyrobarycentric coordinate representations given by Theorem 5.1, p. 251, and the in-exgyrocircles of a gyrotriangle $A_1A_2A_3$ in an Einstein gyrovector plane $(\mathbb{R}_s^2, \oplus, \otimes)$.

following gamma factors,

$$\begin{aligned}\gamma_{\ominus A_1 \oplus E} &= \frac{m_1\gamma_{\ominus A_1\oplus A_1} + m_2\gamma_{\ominus A_1\oplus A_2} + m_3\gamma_{\ominus A_1\oplus A_3}}{m_0} = \frac{m_1 + m_2\gamma_{12} + m_3\gamma_{13}}{m_0} \\ \gamma_{\ominus A_2 \oplus E} &= \frac{m_1\gamma_{\ominus A_2\oplus A_1} + m_2\gamma_{\ominus A_2\oplus A_2} + m_3\gamma_{\ominus A_2\oplus A_3}}{m_0} = \frac{m_1\gamma_{12} + m_2 + m_3\gamma_{23}}{m_0} \\ \gamma_{\ominus A_3 \oplus E} &= \frac{m_1\gamma_{\ominus A_3\oplus A_1} + m_2\gamma_{\ominus A_3\oplus A_2} + m_3\gamma_{\ominus A_3\oplus A_3}}{m_0} = \frac{m_1\gamma_{13} + m_2\gamma_{23} + m_3}{m_0}\end{aligned} \tag{5.2}$$

where we use the standard gyrotriangle index notation, Fig. 2.3, p. 101, and where m_0 is the constant of the gyrobarycentric coordinate representation of E in (5.1) which, according to (4.10), p. 173, is given by the equation

$$m_0^2 = m_1^2 + m_2^2 + m_3^2 + 2m_1m_2\gamma_{12} + 2m_1m_3\gamma_{13} + 2m_2m_3\gamma_{23} \tag{5.3}$$

By Theorem 4.19, p. 196, the gyrodistance r_{12} from E to the gyroline $L_{A_1A_2}$ that contains side A_1A_2 of gyrotriangle $A_1A_2A_3$ satisfies the equation

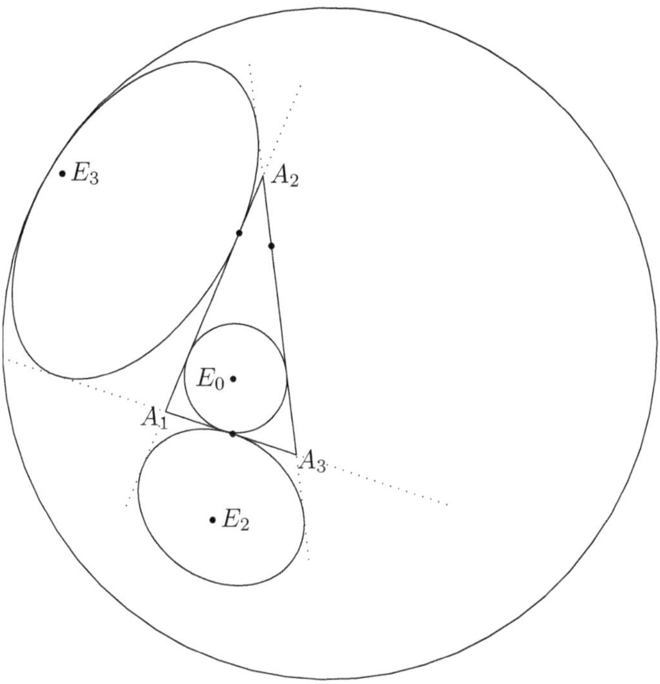

Fig. 5.2 As opposed to Fig. 5.1, here the gyrotriangle $A_1A_2A_3$ in an Einstein gyrovector plane $(\mathbb{R}_s^2, \oplus, \otimes)$ possesses only two exgyrocenters.

$$\gamma_{r_{12}}^2 = \frac{2\gamma_{12}\gamma_{\ominus A_1\oplus E}\gamma_{\ominus A_2\oplus E} - \gamma_{\ominus A_1\oplus E}^2 - \gamma_{\ominus A_2\oplus E}^2}{\gamma_{12}^2 - 1} \tag{5.4a}$$

Similarly, the gyrodistance r_{13} from E to the gyroline $L_{A_1A_3}$ that contains side A_1A_3 of gyrotriangle $A_1A_2A_3$ satisfies the equation

$$\gamma_{r_{13}}^2 = \frac{2\gamma_{13}\gamma_{\ominus A_1\oplus E}\gamma_{\ominus A_3\oplus E} - \gamma_{\ominus A_1\oplus E}^2 - \gamma_{\ominus A_3\oplus E}^2}{\gamma_{13}^2 - 1} \tag{5.4b}$$

and the gyrodistance r_{23} from E to the gyroline $L_{A_2A_3}$ that contains side A_2A_3 of gyrotriangle $A_1A_2A_3$ satisfies the equation

$$\gamma_{r_{23}}^2 = \frac{2\gamma_{23}\gamma_{\ominus A_2\oplus E}\gamma_{\ominus A_3\oplus E} - \gamma_{\ominus A_2\oplus E}^2 - \gamma_{\ominus A_3\oplus E}^2}{\gamma_{23}^2 - 1} \tag{5.4c}$$

By definition, the gyrodistances from E to each of the three gyrolines $L_{A_1A_2}$, $L_{A_1A_3}$ and $L_{A_2A_3}$ are equal. Hence,

$$\begin{aligned} \gamma_{r_{12}}^2 &= \gamma_{r_{13}}^2 \\ \gamma_{r_{12}}^2 &= \gamma_{r_{23}}^2 \end{aligned} \tag{5.5}$$

Substituting successively (5.4) and (5.2) in (5.5), along with the convenient normalization condition

$$m_1^2 + m_2^2 + m_3^2 = 1 \tag{5.6}$$

we obtain from (5.5) – (5.6) the following system of three equations for the three unknowns m_1^2, m_2^2 and m_3^2:

$$\begin{aligned} m_1^2(\gamma_{12}^2 - 1) - m_3^2(\gamma_{23}^2 - 1) &= 0 \\ m_2^2(\gamma_{12}^2 - 1) - m_3^2(\gamma_{13}^2 - 1) &= 0 \\ m_1^2 + m_2^2 + m_3^2 &= 1 \end{aligned} \tag{5.7}$$

The solution of the system (5.7) turns out to be

$$\begin{aligned} m_1^2 &= \frac{1}{D}(\gamma_{23}^2 - 1) \\ m_2^2 &= \frac{1}{D}(\gamma_{13}^2 - 1) \\ m_3^2 &= \frac{1}{D}(\gamma_{12}^2 - 1) \end{aligned} \tag{5.8}$$

where, according to (5.8) and (5.6),

$$D = \gamma_{12}^2 + \gamma_{13}^2 + \gamma_{23}^2 - 3 \ > \ 0 \tag{5.9}$$

and where the gyrobarycentric coordinates are normalized by (5.6).

It is now useful to drop the normalization condition (5.6), thus replacing special gyrobarycentric coordinates by convenient ones, in which a nonzero common factor of the coordinates is irrelevant, obtaining

$$\begin{aligned} m_1^2 &= s^2(\gamma_{23}^2 - 1) = \gamma_{23}^2 a_{23}^2 \\ m_2^2 &= s^2(\gamma_{13}^2 - 1) = \gamma_{13}^2 a_{13}^2 \\ m_3^2 &= s^2(\gamma_{12}^2 - 1) = \gamma_{12}^2 a_{12}^2 \end{aligned} \tag{5.10}$$

noting the frequently used identity (2.11), p. 66.

Owing to the homogeneity of gyrobarycentric coordinates, common factors are irrelevant. Hence, the gyrobarycentric coordinates $(m_1 : m_2 : m_3)$ in (5.10) give four distinct gyrobarycentric coordinate sets that correspond to the four different possible locations of the point E.

These are the ingyrocenter E_0 of gyrotriangle $A_1A_2A_3$, and the gyrotriangle A_k-exgyrocenter E_k opposite to vertex A_k, $k = 1, 2, 3$:

$$\begin{aligned}
E_0 :&\qquad (m_1 : m_2 : m_3) = (\ \ \gamma_{23}a_{23} : \ \ \gamma_{13}a_{13} : \ \ \gamma_{12}a_{12}) \\
E_1 :&\qquad (m_1 : m_2 : m_3) = (-\gamma_{23}a_{23} : \ \ \gamma_{13}a_{13} : \ \ \gamma_{12}a_{12}) \\
E_2 :&\qquad (m_1 : m_2 : m_3) = (\ \ \gamma_{23}a_{23} : -\gamma_{13}a_{13} : \ \ \gamma_{12}a_{12}) \\
E_3 :&\qquad (m_1 : m_2 : m_3) = (\ \ \gamma_{23}a_{23} : \ \ \gamma_{13}a_{13} : -\gamma_{12}a_{12})
\end{aligned} \tag{5.11}$$

By the law of gyrosines (2.167), p. 110, and owing to their homogeneity, the gyrobarycentric coordinates in the first equation in (5.11) can be written as

$$\begin{aligned}
(m_1 : m_2 : m_3) &= \left(\frac{\gamma_{23}a_{23}}{\gamma_{12}a_{12}} : \frac{\gamma_{13}a_{13}}{\gamma_{12}a_{12}} : 1 \right) \\
&= \left(\frac{\sin\alpha_1}{\sin\alpha_3} : \frac{\sin\alpha_2}{\sin\alpha_3} : 1 \right) \\
&= (\sin\alpha_1 : \sin\alpha_2 : \sin\alpha_3)
\end{aligned} \tag{5.12}$$

Similarly, all the barycentric coordinates in (5.11) can be expressed gyrotrigonometrically in terms of the gyrotriangle gyroangles.

Substituting (5.11) in (5.1) and, similarly, substituting (5.12) in (5.1) we obtain the following theorem:

Theorem 5.1. (In-Exgyrocenters Gyrobarycentric Representations in Einstein Gyrovector Spaces). *Let $A_1A_2A_3$ be a gyrotriangle with ingyrocenter E_0 and exgyrocenters E_k, $k = 1, 2, 3$, in an Einstein gyrovector space $(\mathbb{R}^n_s, \oplus, \otimes)$, Fig. 5.1. Then the gyrobarycentric coordinate representations of the gyrotriangle in-exgyrocenters E_k, $k = 0, 1, 2, 3$, are given by the equations*

$$\begin{aligned}
E_0 &= \frac{\gamma_{23}a_{23}\gamma_{A_1}A_1 + \gamma_{13}a_{13}\gamma_{A_2}A_2 + \gamma_{12}a_{12}\gamma_{A_3}A_3}{\gamma_{23}a_{23}\gamma_{A_1} + \gamma_{13}a_{13}\gamma_{A_2} + \gamma_{12}a_{12}\gamma_{A_3}} \\
E_1 &= \frac{-\gamma_{23}a_{23}\gamma_{A_1}A_1 + \gamma_{13}a_{13}\gamma_{A_2}A_2 + \gamma_{12}a_{12}\gamma_{A_3}A_3}{-\gamma_{23}a_{23}\gamma_{A_1} + \gamma_{13}a_{13}\gamma_{A_2} + \gamma_{12}a_{12}\gamma_{A_3}} \\
E_2 &= \frac{\gamma_{23}a_{23}\gamma_{A_1}A_1 - \gamma_{13}a_{13}\gamma_{A_2}A_2 + \gamma_{12}a_{12}\gamma_{A_3}A_3}{\gamma_{23}a_{23}\gamma_{A_1} - \gamma_{13}a_{13}\gamma_{A_2} + \gamma_{12}a_{12}\gamma_{A_3}} \\
E_3 &= \frac{\gamma_{23}a_{23}\gamma_{A_1}A_1 + \gamma_{13}a_{13}\gamma_{A_2}A_2 - \gamma_{12}a_{12}\gamma_{A_3}A_3}{\gamma_{23}a_{23}\gamma_{A_1} + \gamma_{13}a_{13}\gamma_{A_2} - \gamma_{12}a_{12}\gamma_{A_3}}
\end{aligned} \tag{5.13}$$

and their gyrotrigonometric gyrobarycentric coordinate representations are given by the equations

$$\begin{aligned}
E_0 &= \frac{\sin\alpha_1\gamma_{_{A_1}}A_1 + \sin\alpha_2\gamma_{_{A_2}}A_2 + \sin\alpha_3\gamma_{_{A_3}}A_3}{\sin\alpha_1\gamma_{_{A_1}} + \sin\alpha_2\gamma_{_{A_2}} + \sin\alpha_3\gamma_{_{A_3}}} \\
E_1 &= \frac{-\sin\alpha_1\gamma_{_{A_1}}A_1 + \sin\alpha_2\gamma_{_{A_2}}A_2 + \sin\alpha_3\gamma_{_{A_3}}A_3}{-\sin\alpha_1\gamma_{_{A_1}} + \sin\alpha_2\gamma_{_{A_2}} + \sin\alpha_3\gamma_{_{A_3}}} \\
E_2 &= \frac{\sin\alpha_1\gamma_{_{A_1}}A_1 - \sin\alpha_2\gamma_{_{A_2}}A_2 + \sin\alpha_3\gamma_{_{A_3}}A_3}{\sin\alpha_1\gamma_{_{A_1}} - \sin\alpha_2\gamma_{_{A_2}} + \sin\alpha_3\gamma_{_{A_3}}} \\
E_3 &= \frac{\sin\alpha_1\gamma_{_{A_1}}A_1 + \sin\alpha_2\gamma_{_{A_2}}A_2 - \sin\alpha_3\gamma_{_{A_3}}A_3}{\sin\alpha_1\gamma_{_{A_1}} + \sin\alpha_2\gamma_{_{A_2}} - \sin\alpha_3\gamma_{_{A_3}}}
\end{aligned} \tag{5.14}$$

Each gyrobarycentric coordinate representation of a point $E \in \mathbb{R}_s^n$ in Theorem 5.1 has the constant m_0, (5.3), which determines whether the point E exists. By Remark 4.5, p. 174, the point E in Theorem 5.1 exists if and only if the squared constant m_0^2 of its gyrobarycentric coordinate representation is positive. The squared constants, m_0^2, of the in-exgyrocenter E_k, $k = 0, 1, 2, 3$, of gyrotriangle $A_1A_2A_3$ turn out, respectively, to be

$$\begin{aligned}
m_0^2 &= (\gamma_{12}^2 - 1) + (\gamma_{13}^2 - 1) + (\gamma_{23}^2 - 1) \\
&\quad + 2\gamma_{12}\gamma_{13}\gamma_{23}(a_{12}a_{13} + a_{12}a_{13} + a_{12}a_{13}) \\
m_0^2 &= (\gamma_{12}^2 - 1) + (\gamma_{13}^2 - 1) + (\gamma_{23}^2 - 1) \\
&\quad + 2\gamma_{12}\gamma_{13}\gamma_{23}(a_{12}a_{13} - a_{12}a_{13} - a_{12}a_{13}) \\
m_0^2 &= (\gamma_{12}^2 - 1) + (\gamma_{13}^2 - 1) + (\gamma_{23}^2 - 1) \\
&\quad + 2\gamma_{12}\gamma_{13}\gamma_{23}(-a_{12}a_{13} + a_{12}a_{13} - a_{12}a_{13}) \\
m_0^2 &= (\gamma_{12}^2 - 1) + (\gamma_{13}^2 - 1) + (\gamma_{23}^2 - 1) \\
&\quad + 2\gamma_{12}\gamma_{13}\gamma_{23}(-a_{12}a_{13} - a_{12}a_{13} + a_{12}a_{13})
\end{aligned} \tag{5.15}$$

The squared constant, m_0^2, in the first equation in (5.15) gives the constant of the gyrobarycentric coordinate representation of the gyrotriangle ingyrocenter E_0. It is clearly always positive so that any gyrotriangle $A_1A_2A_3$ in an Einstein gyrovector space $\mathbb{R}_s^n$ possesses an ingyrocenter. This is, however, not the case for the gyrotriangle exgyrocenters E_k, $k = 1, 2, 3$. Hence, some gyrotriangles do not possess three exgyrocircles.

The similarity between the results, (5.13) – (5.14), of Theorem 5.1 and the results, (1.155) – (1.156), of its Euclidean counterpart, Theorem 1.16, p. 44, is remarkable.

5.2 Einstein Ingyrocircle and Exgyrocircle Tangency Points

Let E be an in-exgyrocenter of a gyrotriangle $A_1A_2A_3$ in an Einstein gyrovector space $(\mathbb{R}^n_s, \oplus, \otimes)$, representing one of the four in-exgyrocenters E_k, $k = 0, 1, 2, 3$, Fig. 5.1, p. 248, and let $L_{A_1A_2}$ be the gyroline that passes through the gyrotriangle vertices A_1 and A_2. Then, the tangency point T_3 where the in-exgyrocircle with in-exgyrocenter E meets the gyroline $L_{A_1A_2}$ is the gyroperpendicular projection of the point E on the gyroline $L_{A_1A_2}$.

By Theorem 4.18, p. 195, the gyroperpendicular projection T_3 of the point E on the gyroline $L_{A_1A_2}$ possesses the gyrobarycentric coordinate representation

$$T_3 = \frac{m'_1\gamma_{A_1}A_1 + m'_2\gamma_{A_2}A_2}{m'_1\gamma_{A_1} + m'_2\gamma_{A_2}} \tag{5.16a}$$

with respect to the set $\{A_1, A_2\}$, with gyrobarycentric coordinates given by

$$\begin{aligned} m'_1 &= \gamma_{12}\gamma_{\ominus A_2\oplus E} - \gamma_{\ominus A_1\oplus E} \\ m'_2 &= \gamma_{12}\gamma_{\ominus A_1\oplus E} - \gamma_{\ominus A_2\oplus E} \end{aligned} \tag{5.16b}$$

Hence, by (5.16b) and (5.2), p. 248,

$$\begin{aligned} m'_1 &= \frac{m_1(\gamma_{12}^2 - 1) + m_3(\gamma_{12}\gamma_{23} - \gamma_{13})}{m_0} \\ m'_2 &= \frac{m_2(\gamma_{12}^2 - 1) + m_3(\gamma_{12}\gamma_{13} - \gamma_{23})}{m_0} \end{aligned} \tag{5.16c}$$

where m_0 is given by (5.3), p. 248,

$$m_0^2 = m_1^2 + m_2^2 + m_3^2 + 2m_1m_2\gamma_{12} + 2m_1m_3\gamma_{13} + 2m_2m_3\gamma_{23} \tag{5.16d}$$

Owing to the homogeneity of gyrobarycentric coordinates, the common factor $1/m_0$ in (5.16c) can be removed, replacing the gyrobarycentric coordinates (5.16c) by the convenient gyrobarycentric coordinates

$$\begin{aligned} m'_1 &= m_1(\gamma_{12}^2 - 1) + m_3(\gamma_{12}\gamma_{23} - \gamma_{13}) \\ m'_2 &= m_2(\gamma_{12}^2 - 1) + m_3(\gamma_{12}\gamma_{13} - \gamma_{23}) \end{aligned} \tag{5.16e}$$

(0) If E specializes to $E = E_0$ then the gyrobarycentric coordinates $(m_1 : m_2 : m_3)$ of E in (5.16) is given by the first equation in (5.11), p. 251, that is,

$$(m_1 : m_2 : m_3) = (\gamma_{23}a_{23} : \gamma_{13}a_{13} : \gamma_{12}a_{12}) \tag{5.17a}$$

Hence, if E specializes to $E = E_0$ then T_3 in (5.16) specializes to $T_3 = T_{03}$ given by the equation

$$T_{03} = \frac{m'_1\gamma_{A_1}A_1 + m'_2\gamma_{A_2}A_2}{m'_1\gamma_{A_1} + m'_2\gamma_{A_2}} \tag{5.17b}$$

where, by (5.16e) and (5.17a),

$$\begin{aligned} m'_1 &= \gamma_{23}a_{23}(\gamma_{12}^2 - 1) + \gamma_{12}a_{12}(\gamma_{12}\gamma_{23} - \gamma_{13}) \\ m'_2 &= \gamma_{13}a_{13}(\gamma_{12}^2 - 1) + \gamma_{12}a_{12}(\gamma_{12}\gamma_{13} - \gamma_{23}) \end{aligned} \tag{5.17c}$$

(1) If E specializes to $E = E_1$ then the gyrobarycentric coordinates $(m_1 : m_2 : m_3)$ of E in (5.16) is given by the second equation in (5.11), p. 251, that is,

$$(m_1 : m_2 : m_3) = (-\gamma_{23}a_{23} : \gamma_{13}a_{13} : \gamma_{12}a_{12}) \tag{5.18a}$$

Hence, if E specializes to $E = E_1$ then T_3 in (5.16) specializes to $T_3 = T_{13}$ given by the equation

$$T_{13} = \frac{m'_1\gamma_{A_1}A_1 + m'_2\gamma_{A_2}A_2}{m'_1\gamma_{A_1} + m'_2\gamma_{A_2}} \tag{5.18b}$$

where, by (5.16e) and (5.18a),

$$\begin{aligned} m'_1 &= -\gamma_{23}a_{23}(\gamma_{12}^2 - 1) + \gamma_{12}a_{12}(\gamma_{12}\gamma_{23} - \gamma_{13}) \\ m'_2 &= \quad\gamma_{13}a_{13}(\gamma_{12}^2 - 1) + \gamma_{12}a_{12}(\gamma_{12}\gamma_{13} - \gamma_{23}) \end{aligned} \tag{5.18c}$$

(2) If E specializes to $E = E_2$ then the gyrobarycentric coordinates $(m_1 : m_2 : m_3)$ of E in (5.16) is given by the third equation in (5.11), p. 251, that is,

$$(m_1 : m_2 : m_3) = (\gamma_{23}a_{23} : -\gamma_{13}a_{13} : \gamma_{12}a_{12}) \tag{5.19a}$$

Hence, if E specializes to $E = E_2$ then T_3 in (5.16) specializes to $T_3 = T_{23}$ given by the equation

$$T_{23} = \frac{m'_1\gamma_{A_1}A_1 + m'_2\gamma_{A_2}A_2}{m'_1\gamma_{A_1} + m'_2\gamma_{A_2}} \tag{5.19b}$$

where, by (5.16e) and (5.19a),

$$\begin{aligned} m'_1 &= \quad\gamma_{23}a_{23}(\gamma_{12}^2 - 1) + \gamma_{12}a_{12}(\gamma_{12}\gamma_{23} - \gamma_{13}) \\ m'_2 &= -\gamma_{13}a_{13}(\gamma_{12}^2 - 1) + \gamma_{12}a_{12}(\gamma_{12}\gamma_{13} - \gamma_{23}) \end{aligned} \tag{5.19c}$$

(3) If E specializes to $E = E_3$ then the gyrobarycentric coordinates $(m_1 : m_2 : m_3)$ of E in (5.16) is given by the fourth equation in (5.11), p. 251, that is,

$$(m_1 : m_2 : m_3) = (\gamma_{23}a_{23} : \gamma_{13}a_{13} : -\gamma_{12}a_{12}) \tag{5.20a}$$

Hence, if E specializes to $E = E_3$ then T_3 in (5.16) specializes to $T_3 = T_{33}$ given by the equation

$$T_{33} = \frac{m'_1\gamma_{_{A_1}}A_1 + m'_2\gamma_{_{A_2}}A_2}{m'_1\gamma_{_{A_1}} + m'_2\gamma_{_{A_2}}} \tag{5.20b}$$

where, by (5.16e) and (5.20a),

$$\begin{aligned} m'_1 &= \gamma_{23}a_{23}(\gamma_{12}^2 - 1) - \gamma_{12}a_{12}(\gamma_{12}\gamma_{23} - \gamma_{13}) \\ m'_2 &= \gamma_{13}a_{13}(\gamma_{12}^2 - 1) - \gamma_{12}a_{12}(\gamma_{12}\gamma_{13} - \gamma_{23}) \end{aligned} \tag{5.20c}$$

The results in items (0)–(3) present gyrobarycentric coordinate representations for several in-exgyrocircle tangency points. Gyrobarycentric coordinate representations for the remaining tangency points can be obtained by cyclic permutations of the reference gyrotriangle, as shown in Sec. 5.4

5.3 Useful Gyrotriangle Gyrotrigonometric Relations

The AAA to SSS conversion law (2.157), p. 106, results in the gyrotriangle-gyrotrigonometric relations (2.163) – (2.164), p. 109, that prove useful. These relations are, therefore, presented below in (5.21) – (5.23) in the standard gyrotriangle index notation for gyrotriangle $A_1A_2A_3$.

$$\begin{aligned} \gamma_{12} &= \frac{\cos\alpha_3 + \cos\alpha_1\cos\alpha_2}{\sin\alpha_1\sin\alpha_2} \\ \gamma_{13} &= \frac{\cos\alpha_2 + \cos\alpha_1\cos\alpha_3}{\sin\alpha_1\sin\alpha_3} \\ \gamma_{23} &= \frac{\cos\alpha_1 + \cos\alpha_2\cos\alpha_3}{\sin\alpha_2\sin\alpha_3} \end{aligned} \tag{5.21}$$

and

$$\begin{aligned} \sqrt{\gamma_{12}^2 - 1} = \gamma_{12}\frac{a_{12}}{s} &= 2\frac{\sqrt{F(\alpha_1,\alpha_2,\alpha_3)}}{\sin\alpha_1\sin\alpha_2} \\ \sqrt{\gamma_{13}^2 - 1} = \gamma_{13}\frac{a_{13}}{s} &= 2\frac{\sqrt{F(\alpha_1,\alpha_2,\alpha_3)}}{\sin\alpha_1\sin\alpha_3} \\ \sqrt{\gamma_{23}^2 - 1} = \gamma_{23}\frac{a_{23}}{s} &= 2\frac{\sqrt{F(\alpha_1,\alpha_2,\alpha_3)}}{\sin\alpha_2\sin\alpha_3} \end{aligned} \tag{5.22}$$

where we note the identity (2.11), p. 66, and where

$$F(\alpha_1,\alpha_2,\alpha_3) = \frac{1}{4}(2\cos\alpha_1\cos\alpha_2\cos\alpha_3 + \cos^2\alpha_1 + \cos^2\alpha_2 + \cos^2\alpha_3 - 1) =$$

$$\cos\frac{\alpha_1+\alpha_2+\alpha_3}{2}\cos\frac{\alpha_1-\alpha_2-\alpha_3}{2}\cos\frac{-\alpha_1+\alpha_2-\alpha_3}{2}\cos\frac{-\alpha_1-\alpha_2+\alpha_3}{2}$$

$$\begin{aligned} &= \sin\tfrac{\delta}{2}\sin(\alpha_1+\tfrac{\delta}{2})\sin(\alpha_2+\tfrac{\delta}{2})\sin(\alpha_3+\tfrac{\delta}{2}) \\ &= \frac{1}{4}\frac{(1+2\gamma_{12}\gamma_{13}\gamma_{23}-\gamma_{12}^2-\gamma_{13}^2-\gamma_{23}^2)^2}{(\gamma_{12}^2-1)(\gamma_{13}^2-1)(\gamma_{23}^2-1)} \end{aligned} \tag{5.23}$$

where $\delta = \pi - (\alpha_1+\alpha_2+\alpha_3)$ is the defect of gyrotriangle $A_1A_2A_3$.

The SSS to AAA conversion law, Theorem 2.23, results in the law of gyrocosines (2.143), p. 103. The law of gyrocosines, in turn, along with (5.22), gives rise to the following useful gyrotriangle gyrotrigonometric relations:

$$\begin{aligned} \gamma_{12}\gamma_{13} - \gamma_{23} &= \frac{4F(\alpha_1,\alpha_2,\alpha_3)\cos\alpha_1}{\sin^2\alpha_1\sin\alpha_2\sin\alpha_3} \\ \gamma_{12}\gamma_{23} - \gamma_{13} &= \frac{4F(\alpha_1,\alpha_2,\alpha_3)\cos\alpha_2}{\sin\alpha_1\sin^2\alpha_2\sin\alpha_3} \\ \gamma_{13}\gamma_{23} - \gamma_{12} &= \frac{4F(\alpha_1,\alpha_2,\alpha_3)\cos\alpha_3}{\sin\alpha_1\sin\alpha_2\sin^2\alpha_3} \end{aligned} \tag{5.24}$$

If $\alpha_1 = \pi/2$ then gyrotriangle $A_1A_2A_3$ is right-gyroangled, satisfying the hyperbolic Pythagorean Theorem $\gamma_{12}\gamma_{13} = \gamma_{23}$, as we see from the first equation in (5.24) and as shown in Fig. 2.5, p. 114.

5.4 The Tangency Points Expressed Gyrotrigonometrically

(0) Following (5.17) the tangency point T_{03} has the gyrobarycentric coordinate representation

$$T_{03} = \frac{m_1\gamma_{A_1}A_1 + m_2\gamma_{A_2}A_2}{m_1\gamma_{A_1} + m_2\gamma_{A_2}} \tag{5.25a}$$

where

$$\begin{aligned} m_1 &= \gamma_{23}a_{23}(\gamma_{12}^2-1) + \gamma_{12}a_{12}(\gamma_{12}\gamma_{23}-\gamma_{13}) \\ m_2 &= \gamma_{13}a_{13}(\gamma_{12}^2-1) + \gamma_{12}a_{12}(\gamma_{12}\gamma_{13}-\gamma_{23}) \end{aligned} \tag{5.25b}$$

Substituting from (5.22) – (5.24) into (5.25b), and using the abbreviation $F = F(\alpha_1, \alpha_2, \alpha_3)$, we obtain the gyrobarycentric coordinate set $(m_1 : m_2)$ in gyrotrigonometric form,

$$m_1 = \frac{8F^{3/2}}{\sin^2\alpha_1 \sin^2\alpha_2 \sin\alpha_3} \frac{1+\cos\alpha_2}{\sin\alpha_2} = \frac{8F^{3/2}}{\sin^2\alpha_1 \sin^2\alpha_2 \sin\alpha_3} \cot\frac{\alpha_2}{2}$$

$$m_2 = \frac{8F^{3/2}}{\sin^2\alpha_1 \sin^2\alpha_2 \sin\alpha_3} \frac{1+\cos\alpha_1}{\sin\alpha_1} = \frac{8F^{3/2}}{\sin^2\alpha_1 \sin^2\alpha_2 \sin\alpha_3} \cot\frac{\alpha_1}{2} \tag{5.25c}$$

Since gyrobarycentric coordinates are homogeneous, a nonzero common factor of a system of gyrobarycentric coordinates is irrelevant. Hence, it follows from (5.25c) that a convenient gyrobarycentric coordinate set $(m_1 : m_2)$ for the tangency point T_{03} in (5.25a) is

$$(m_1 : m_2) = \left(\cot\frac{\alpha_2}{2} : \cot\frac{\alpha_1}{2}\right) = \left(\tan\frac{\alpha_1}{2} : \tan\frac{\alpha_2}{2}\right) \tag{5.25d}$$

thus obtaining the elegant gyrotrigonometric gyrobarycentric coordinate representation

$$T_{03} = \frac{\tan\frac{\alpha_1}{2}\gamma_{A_1} A_1 + \tan\frac{\alpha_2}{2}\gamma_{A_2} A_2}{\tan\frac{\alpha_1}{2}\gamma_{A_1} + \tan\frac{\alpha_2}{2}\gamma_{A_2}} \tag{5.25e}$$

(1) Following (5.18) the tangency point T_{13} has the gyrobarycentric coordinate representation

$$T_{13} = \frac{m_1\gamma_{A_1} A_1 + m_2\gamma_{A_2} A_2}{m_1\gamma_{A_1} + m_2\gamma_{A_2}} \tag{5.26a}$$

where

$$\begin{aligned} m_1 &= -\gamma_{23}a_{23}(\gamma_{12}^2 - 1) + \gamma_{12}a_{12}(\gamma_{12}\gamma_{23} - \gamma_{13}) \\ m_2 &= \gamma_{13}a_{13}(\gamma_{12}^2 - 1) + \gamma_{12}a_{12}(\gamma_{12}\gamma_{13} - \gamma_{23}) \end{aligned} \tag{5.26b}$$

Substituting from (5.22) – (5.24) into (5.26b), using the abbreviation $F = F(\alpha_1, \alpha_2, \alpha_3)$, we obtain the gyrobarycentric coordinate set $(m_1 : m_2)$ in gyrotrigonometric form,

$$m_1 = \frac{8F^{3/2}}{\sin^2\alpha_1 \sin^2\alpha_2 \sin\alpha_3} \frac{-1+\cos\alpha_2}{\sin\alpha_2} = \frac{-8F^{3/2}}{\sin^2\alpha_1 \sin^2\alpha_2 \sin\alpha_3} \tan\frac{\alpha_2}{2}$$

$$m_2 = \frac{8F^{3/2}}{\sin^2\alpha_1 \sin^2\alpha_2 \sin\alpha_3} \frac{1+\cos\alpha_1}{\sin\alpha_1} = \frac{8F^{3/2}}{\sin^2\alpha_1 \sin^2\alpha_2 \sin\alpha_3} \cot\frac{\alpha_1}{2} \tag{5.26c}$$

Since gyrobarycentric coordinates are homogeneous, a nonzero common factor of a system of gyrobarycentric coordinates is irrelevant. Hence, it follows from (5.26c) that a convenient gyrobarycentric coordinate set $(m_1 : m_2)$ for the tangency point T_{13} in (5.26a) is

$$(m_1 : m_2) = \left(-\tan\frac{\alpha_2}{2} : \cot\frac{\alpha_1}{2}\right) = \left(\tan\frac{\alpha_1}{2} : -\cot\frac{\alpha_2}{2}\right) \tag{5.26d}$$

thus obtaining the elegant gyrotrigonometric gyrobarycentric coordinate representation

$$T_{13} = \frac{\tan\frac{\alpha_1}{2}\gamma_{A_1}A_1 - \cot\frac{\alpha_2}{2}\gamma_{A_2}A_2}{\tan\frac{\alpha_1}{2}\gamma_{A_1} - \cot\frac{\alpha_2}{2}\gamma_{A_2}} \tag{5.26e}$$

(2) Following (5.19) the tangency point T_{23} has the gyrobarycentric coordinate representation

$$T_{23} = \frac{m_1\gamma_{A_1}A_1 + m_2\gamma_{A_2}A_2}{m_1\gamma_{A_1} + m_2\gamma_{A_2}} \tag{5.27a}$$

where

$$\begin{aligned} m_1 &= \gamma_{23}a_{23}(\gamma_{12}^2 - 1) + \gamma_{12}a_{12}(\gamma_{12}\gamma_{23} - \gamma_{13}) \\ m_2 &= -\gamma_{13}a_{13}(\gamma_{12}^2 - 1) + \gamma_{12}a_{12}(\gamma_{12}\gamma_{13} - \gamma_{23}) \end{aligned} \tag{5.27b}$$

Substituting from (5.22)–(5.24) into (5.27b), using the abbreviation $F = F(\alpha_1, \alpha_2, \alpha_3)$, we obtain the gyrobarycentric coordinate set $(m_1 : m_2)$ in gyrotrigonometric form,

$$\begin{aligned} m_1 &= \frac{8F^{3/2}}{\sin^2\alpha_1\sin^2\alpha_2\sin\alpha_3}\frac{1+\cos\alpha_2}{\sin\alpha_2} = \frac{8F^{3/2}}{\sin^2\alpha_1\sin^2\alpha_2\sin\alpha_3}\cot\frac{\alpha_2}{2} \\ m_2 &= \frac{8F^{3/2}}{\sin^2\alpha_1\sin^2\alpha_2\sin\alpha_3}\frac{-1+\cos\alpha_1}{\sin\alpha_1} = \frac{-8F^{3/2}}{\sin^2\alpha_1\sin^2\alpha_2\sin\alpha_3}\tan\frac{\alpha_1}{2} \end{aligned} \tag{5.27c}$$

Since gyrobarycentric coordinates are homogeneous, a nonzero common factor of a system of gyrobarycentric coordinates is irrelevant. Hence, it follows from (5.27c) that a convenient gyrobarycentric coordinate set $(m_1 : m_2)$ for the tangency point T_{23} in (5.27a) is

$$(m_1 : m_2) = \left(\cot\frac{\alpha_2}{2} : -\tan\frac{\alpha_1}{2}\right) = \left(\cot\frac{\alpha_1}{2} : -\tan\frac{\alpha_2}{2}\right) \tag{5.27d}$$

thus obtaining the elegant gyrotrigonometric gyrobarycentric coordinate representation

$$T_{23} = \frac{\cot\frac{\alpha_1}{2}\gamma_{A_1}A_1 - \tan\frac{\alpha_2}{2}\gamma_{A_2}A_2}{\cot\frac{\alpha_1}{2}\gamma_{A_1} - \tan\frac{\alpha_2}{2}\gamma_{A_2}} \tag{5.27e}$$

(3) Following (5.20) the tangency point T_{33} has the gyrobarycentric coordinate representation

$$T_{33} = \frac{m_1\gamma_{A_1}A_1 + m_2\gamma_{A_2}A_2}{m_1\gamma_{A_1} + m_2\gamma_{A_2}} \tag{5.28a}$$

where

$$\begin{aligned} m_1 &= \gamma_{23}a_{23}(\gamma_{12}^2 - 1) - \gamma_{12}a_{12}(\gamma_{12}\gamma_{23} - \gamma_{13}) \\ m_2 &= \gamma_{13}a_{13}(\gamma_{12}^2 - 1) - \gamma_{12}a_{12}(\gamma_{12}\gamma_{13} - \gamma_{23}) \end{aligned} \tag{5.28b}$$

Substituting from (5.22) – (5.24) into (5.28b), using the abbreviation $F = F(\alpha_1, \alpha_2, \alpha_3)$, we obtain the gyrobarycentric coordinate set $(m_1 : m_2)$ in gyrotrigonometric form,

$$\begin{aligned} m_1 &= \frac{8F^{3/2}}{\sin^2\alpha_1 \sin^2\alpha_2 \sin\alpha_3} \frac{1-\cos\alpha_2}{\sin\alpha_2} = \frac{8F^{3/2}}{\sin^2\alpha_1 \sin^2\alpha_2 \sin\alpha_3} \tan\frac{\alpha_2}{2} \\ m_2 &= \frac{8F^{3/2}}{\sin^2\alpha_1 \sin^2\alpha_2 \sin\alpha_3} \frac{1-\cos\alpha_1}{\sin\alpha_1} = \frac{8F^{3/2}}{\sin^2\alpha_1 \sin^2\alpha_2 \sin\alpha_3} \tan\frac{\alpha_1}{2} \end{aligned} \tag{5.28c}$$

Since gyrobarycentric coordinates are homogeneous, a nonzero common factor of a system of gyrobarycentric coordinates is irrelevant. Hence, it follows from (5.28c) that a convenient gyrobarycentric coordinate set $(m_1 : m_2)$ for the tangency point T_{33} in (5.28a) is

$$(m_1 : m_2) = \left(\tan\frac{\alpha_2}{2} : \tan\frac{\alpha_1}{2}\right) = \left(\cot\frac{\alpha_1}{2} : \cot\frac{\alpha_2}{2}\right) \tag{5.28d}$$

thus obtaining the elegant gyrotrigonometric gyrobarycentric coordinate representation

$$T_{33} = \frac{\cot\frac{\alpha_1}{2}\gamma_{A_1}A_1 + \cot\frac{\alpha_2}{2}\gamma_{A_2}A_2}{\cot\frac{\alpha_1}{2}\gamma_{A_1} + \cot\frac{\alpha_2}{2}\gamma_{A_2}} \tag{5.28e}$$

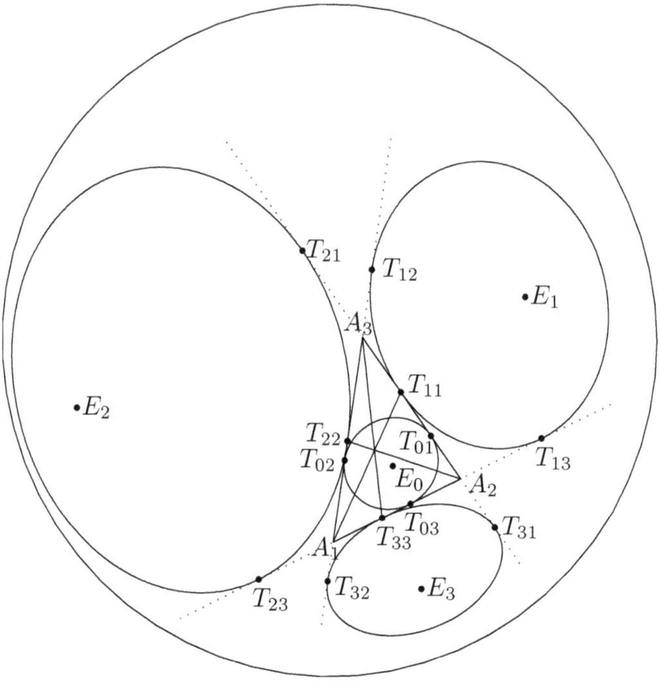

Fig. 5.3 The in-exgyrocircle tangency points, T_{ij}, of a gyrotriangle $A_1A_2A_3$ in an Einstein gyrovector plane $(\mathbb{R}_s^2, \oplus, \otimes)$. Here T_{ij} is the tangency point where the in-exgyrocircle with in-exgyrocenter E_i, $i = 0, 1, 2, 3$, meets the gyrotriangle side, or its extension, opposite to vertex A_j, $j = 1, 2, 3$. Gyrotrigonometric gyrobarycentric coordinate representations of the tangency points in any Einstein gyrovector space $(\mathbb{R}_s^n, \oplus, \otimes)$ are determined by Theorem 5.2, p. 260. Note that the gyrosegments A_kT_{kk}, $k = 1, 2, 3$, are concurrent. Guided by analogies with Euclidean triangles, the point of concurrency is called the Nagel gyropoint of the gyrotriangle $A_1A_2A_3$.

The results in (5.25), (5.26), (5.27) and (5.28) lead to the following theorem:

Theorem 5.2. (Einstein In-Exgyrocircle Tangency Points). *Let $A_1A_2A_3$ be a gyrotriangle in an Einstein gyrovector space $\mathbb{R}_s^n$ and let T_{ij}, $i = 0, 1, 2, 3$, $j = 1, 2, 3$, be the points of tangency where the gyrotriangle in-exgyrocircle with gyrocenter E_i meets the side opposite to A_j, or its extension, of the gyrotriangle, Fig. 5.3.*

Then, gyrotrigonometric gyrobarycentric coordinate representations of the tangency points T_{ij} with respect to the pointwise independent set $S = \{A_1, A_2, A_3\}$ are given by the equations listed below.

$$T_{01} = \frac{\tan\frac{\alpha_2}{2}\gamma_{A_2}A_2 + \tan\frac{\alpha_3}{2}\gamma_{A_3}A_3}{\tan\frac{\alpha_2}{2}\gamma_{A_2} + \tan\frac{\alpha_3}{2}\gamma_{A_3}}$$

$$T_{02} = \frac{\tan\frac{\alpha_1}{2}\gamma_{A_1}A_1 + \tan\frac{\alpha_3}{2}\gamma_{A_3}A_3}{\tan\frac{\alpha_1}{2}\gamma_{A_1} + \tan\frac{\alpha_3}{2}\gamma_{A_3}} \qquad (5.29a)$$

$$T_{03} = \frac{\tan\frac{\alpha_1}{2}\gamma_{A_1}A_1 + \tan\frac{\alpha_2}{2}\gamma_{A_2}A_2}{\tan\frac{\alpha_1}{2}\gamma_{A_1} + \tan\frac{\alpha_2}{2}\gamma_{A_2}}$$

$$T_{11} = \frac{\cot\frac{\alpha_2}{2}\gamma_{A_2}A_2 + \cot\frac{\alpha_3}{2}\gamma_{A_3}A_3}{\cot\frac{\alpha_2}{2}\gamma_{A_2} + \cot\frac{\alpha_3}{2}\gamma_{A_3}}$$

$$T_{12} = \frac{\tan\frac{\alpha_1}{2}\gamma_{A_1}A_1 - \cot\frac{\alpha_3}{2}\gamma_{A_3}A_3}{\tan\frac{\alpha_1}{2}\gamma_{A_1} - \cot\frac{\alpha_3}{2}\gamma_{A_3}} \qquad (5.29b)$$

$$T_{13} = \frac{\tan\frac{\alpha_1}{2}\gamma_{A_1}A_1 - \cot\frac{\alpha_2}{2}\gamma_{A_2}A_2}{\tan\frac{\alpha_1}{2}\gamma_{A_1} - \cot\frac{\alpha_2}{2}\gamma_{A_2}}$$

$$T_{21} = \frac{\tan\frac{\alpha_2}{2}\gamma_{A_2}A_2 - \cot\frac{\alpha_3}{2}\gamma_{A_3}A_3}{\tan\frac{\alpha_2}{2}\gamma_{A_2} - \cot\frac{\alpha_3}{2}\gamma_{A_3}}$$

$$T_{22} = \frac{\cot\frac{\alpha_1}{2}\gamma_{A_1}A_1 + \cot\frac{\alpha_3}{2}\gamma_{A_3}A_3}{\cot\frac{\alpha_1}{2}\gamma_{A_1} + \cot\frac{\alpha_3}{2}\gamma_{A_3}} \qquad (5.29c)$$

$$T_{23} = \frac{\cot\frac{\alpha_1}{2}\gamma_{A_1}A_1 - \tan\frac{\alpha_2}{2}\gamma_{A_2}A_2}{\cot\frac{\alpha_1}{2}\gamma_{A_1} - \tan\frac{\alpha_2}{2}\gamma_{A_2}}$$

$$T_{31} = \frac{\cot\frac{\alpha_2}{2}\gamma_{A_2}A_2 - \tan\frac{\alpha_3}{2}\gamma_{A_3}A_3}{\cot\frac{\alpha_2}{2}\gamma_{A_2} - \tan\frac{\alpha_3}{2}\gamma_{A_3}}$$

$$T_{32} = \frac{\cot\frac{\alpha_1}{2}\gamma_{A_1}A_1 - \tan\frac{\alpha_3}{2}\gamma_{A_3}A_3}{\cot\frac{\alpha_1}{2}\gamma_{A_1} - \tan\frac{\alpha_3}{2}\gamma_{A_3}} \qquad (5.29d)$$

$$T_{33} = \frac{\cot\frac{\alpha_1}{2}\gamma_{A_1}A_1 + \cot\frac{\alpha_2}{2}\gamma_{A_2}A_2}{\cot\frac{\alpha_1}{2}\gamma_{A_1} + \cot\frac{\alpha_2}{2}\gamma_{A_2}}$$

Proof. The proof of the third equation in each of (5.29a), (5.29b), (5.29c) and (5.29d) is established, respectively, in (5.25), (5.26), (5.27) and (5.28). The proof of the remaining equations in (5.29a), (5.29b), (5.29c) and (5.29d) is obtained from the former by invoking cyclicity, that is, by cyclic permutations of the vertices of the reference gyrotriangle. □

5.5 Möbius Gyrotriangle Ingyrocenter and Exgyrocenters

In this section we transform the results of Theorems 5.1, p. 251, and 5.2, p. 260, from Einstein into Möbius gyrovector spaces, obtaining the following theorems 5.3 and 5.4:

Theorem 5.3. **(In-Exgyrocenters Gyrobarycentric Representations in Möbius Gyrovector Spaces).** *Let $A_1A_2A_3$ be a gyrotriangle with ingyrocenter E_0 and exgyrocenters E_k, $k = 1, 2, 3$, in a Möbius gyrovector space $(\mathbb{R}^n_s, \oplus, \otimes)$, Fig. 5.4. Then gyrotrigonometric gyrobarycentric coordinate representations of the gyrotriangle in-exgyrocenters E_k, $k = 0, 1, 2, 3$, are given by the equations*

$$
\begin{aligned}
E_0 &= \tfrac{1}{2}\otimes\frac{\sin\alpha_1\gamma^2_{A_1}A_1 + \sin\alpha_2\gamma^2_{A_2}A_2 + \sin\alpha_3\gamma^2_{A_3}A_3}{\sin\alpha_1(\gamma^2_{A_1}-\frac{1}{2}) + \sin\alpha_2(\gamma^2_{A_2}-\frac{1}{2}) + \sin\alpha_3(\gamma^2_{A_3}-\frac{1}{2})} \\
E_1 &= \tfrac{1}{2}\otimes\frac{-\sin\alpha_1\gamma^2_{A_1}A_1 + \sin\alpha_2\gamma^2_{A_2}A_2 + \sin\alpha_3\gamma^2_{A_3}A_3}{-\sin\alpha_1(\gamma^2_{A_1}-\frac{1}{2}) + \sin\alpha_2(\gamma^2_{A_2}-\frac{1}{2}) + \sin\alpha_3(\gamma^2_{A_3}-\frac{1}{2})} \\
E_2 &= \tfrac{1}{2}\otimes\frac{\sin\alpha_1\gamma^2_{A_1}A_1 - \sin\alpha_2\gamma^2_{A_2}A_2 + \sin\alpha_3\gamma^2_{A_3}A_3}{\sin\alpha_1(\gamma^2_{A_1}-\frac{1}{2}) - \sin\alpha_2(\gamma^2_{A_2}-\frac{1}{2}) + \sin\alpha_3(\gamma^2_{A_3}-\frac{1}{2})} \\
E_3 &= \tfrac{1}{2}\otimes\frac{\sin\alpha_1\gamma^2_{A_1}A_1 + \sin\alpha_2\gamma^2_{A_2}A_2 - \sin\alpha_3\gamma^2_{A_3}A_3}{\sin\alpha_1(\gamma^2_{A_1}-\frac{1}{2}) + \sin\alpha_2(\gamma^2_{A_2}-\frac{1}{2}) - \sin\alpha_3(\gamma^2_{A_3}-\frac{1}{2})}
\end{aligned}
\tag{5.30}
$$

Proof. The in-exgyrocenters of a gyrotriangle $A_1A_2A_3$ in an Einstein gyrovector space $\mathbb{R}^n_s$ are given in (5.14), p. 252, in terms of the Einstein gyrobarycentric coordinate representation, (4.18), p. 177, of appropriate gyrobarycentric coordinates with respect to the vertices of the reference gyrotriangle $A_1A_2A_3$. Hence, by Theorem 4.6, p. 177, the in-exgyrocenters of the corresponding gyrotriangle $A_1A_2A_3$ in the isomorphic Möbius gyrovector space $\mathbb{R}^n_s$ are the ones given in (5.30) in terms of the Möbius

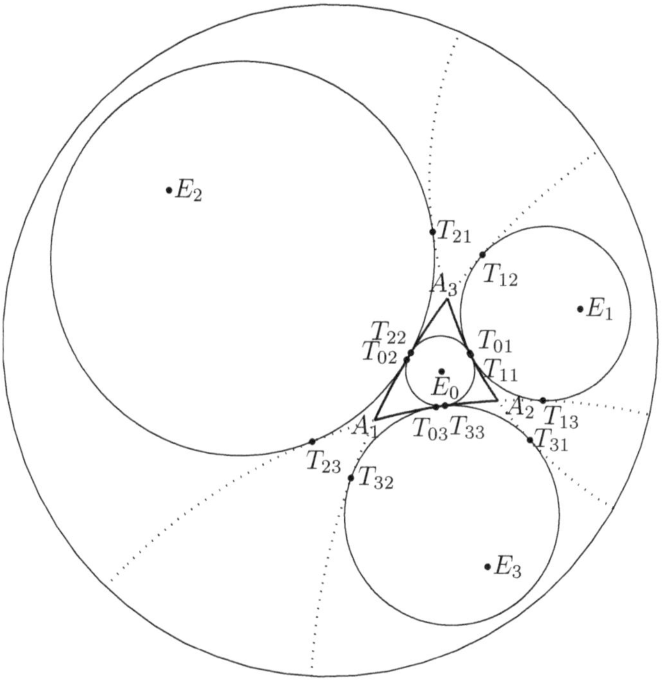

Fig. 5.4 The in-exgyrocircle tangency points, T_{ij}, $i = 0, 1, 2, 3$, $j = 1, 2, 3$, of a gyrotriangle $A_1A_2A_3$ in a Möbius gyrovector space $(\mathbb{R}^n_s, \oplus, \otimes)$. Gyrotrigonometric gyrobarycentric coordinate representations of the tangency points are determined by Theorem 5.4. Here T_{ij} is the tangency point where the gyrotriangle in-exgyrocircle with in-exgyrocenter E_i meets the gyrotriangle side opposite to vertex A_j, or its extension.

gyrobarycentric coordinate representation (4.19), p. 177, that involve isomorphic gyrobarycentric coordinates with respect to their respective sets of two points.

Finally, Einstein gyrobarycentric coordinates in (5.14) are gyrotrigonometric functions and hence, by Theorem 2.48, p. 143, their isomorphic images in the isomorphic Möbius gyrovector space survive unchanged in (5.30). □

Theorem 5.4. (Möbius In-Exgyrocircle Tangency Points). *Let $A_1A_2A_3$ be a gyrotriangle in a Möbius gyrovector space $\mathbb{R}^n_s$ and let T_{ij}, $i = 0, 1, 2, 3$, $j = 1, 2, 3$, be the points of tangency where the gyrotriangle in-exgyrocircle with gyrocenter E_i meets the side opposite to A_j, or its extension, of the gyrotriangle, Fig. 5.4.*

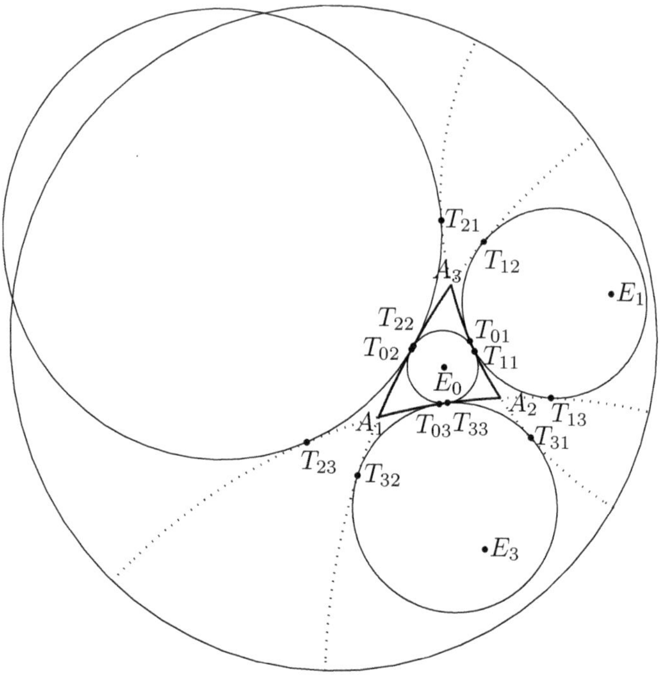

Fig. 5.5 The in-exgyrocircle tangency points, T_{ij}, $i = 0, 1, 2, 3$, $j = 1, 2, 3$, of a gyrotriangle $A_1A_2A_3$ in a Möbius gyrovector space $(\mathbb{R}^n_s, \oplus, \otimes)$. Here the exgyrocenter E_2 of the gyrotriangle does not exist. Yet, all the gyrotrigonometric gyrobarycentric coordinate representations of the tangency points of the gyrotriangle, determined by Theorem 5.4, p. 263, remain valid, as we see here graphically in a Möbius gyrovector plane $(\mathbb{R}^2_s, \oplus, \otimes)$. Indeed the Euclidean circle through the tangency points T_{21}, T_{22} and T_{23} does not lie entirely on the interior of the disk $\mathbb{R}^2_s$.

Then, gyrotrigonometric gyrobarycentric coordinate representations of the tangency points T_{ij} with respect to the pointwise independent set $S = \{A_1, A_2, A_3\}$ are given by the equations listed below.

$$T_{01} = \tfrac{1}{2}\otimes\frac{\tan\frac{\alpha_2}{2}\gamma^2_{A_2}A_2 + \tan\frac{\alpha_3}{2}\gamma^2_{A_3}A_3}{\tan\frac{\alpha_2}{2}(\gamma^2_{A_2}-\frac{1}{2}) + \tan\frac{\alpha_3}{2}(\gamma^2_{A_3}-\frac{1}{2})}$$

$$T_{02} = \tfrac{1}{2}\otimes\frac{\tan\frac{\alpha_1}{2}\gamma^2_{A_1}A_1 + \tan\frac{\alpha_3}{2}\gamma^2_{A_3}A_3}{\tan\frac{\alpha_1}{2}(\gamma^2_{A_1}-\frac{1}{2}) + \tan\frac{\alpha_3}{2}(\gamma^2_{A_3}-\frac{1}{2})} \tag{5.31a}$$

$$T_{03} = \tfrac{1}{2}\otimes\frac{\tan\frac{\alpha_1}{2}\gamma^2_{A_1}A_1 + \tan\frac{\alpha_2}{2}\gamma^2_{A_2}A_2}{\tan\frac{\alpha_1}{2}(\gamma^2_{A_1}-\frac{1}{2}) + \tan\frac{\alpha_2}{2}(\gamma^2_{A_2}-\frac{1}{2})}$$

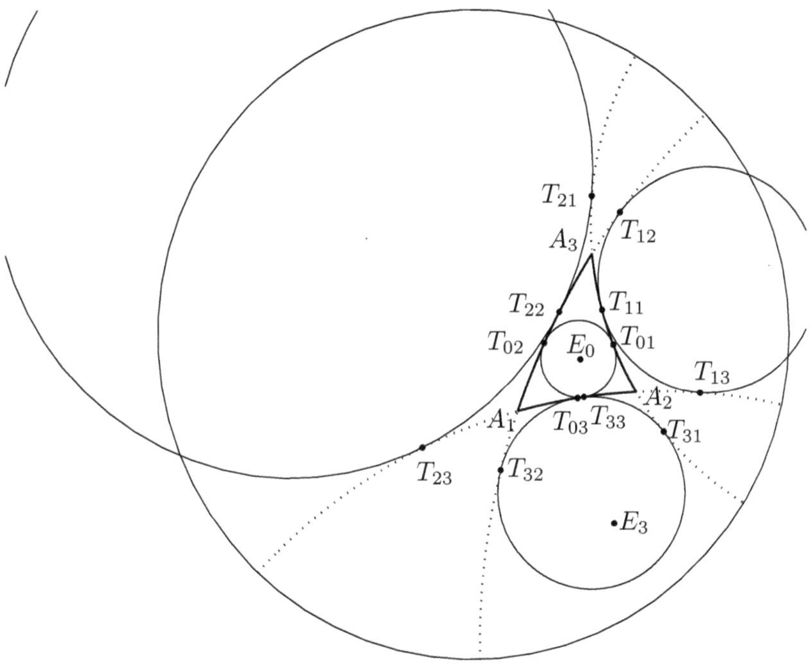

Fig. 5.6 The in-exgyrocircle tangency points, T_{ij}, $i = 0, 1, 2, 3$, $j = 1, 2, 3$, of a gyrotriangle $A_1A_2A_3$ in a Möbius gyrovector space $(\mathbb{R}^n_s, \oplus, \otimes)$. Here the exgyrocenters E_1 and E_2 of the gyrotriangle do not exist. Yet, all the gyrotrigonometric gyrobarycentric coordinate representations of the tangency points of the gyrotriangle, determined by Theorem 5.4, p. 263, remain valid, as we see here graphically.

$$T_{11} = \tfrac{1}{2}\otimes\frac{\cot\frac{\alpha_2}{2}\gamma^2_{A_2}A_2 + \cot\frac{\alpha_3}{2}\gamma^2_{A_3}A_3}{\cot\frac{\alpha_2}{2}(\gamma^2_{A_2}-\frac{1}{2}) + \cot\frac{\alpha_3}{2}(\gamma^2_{A_3}-\frac{1}{2})}$$

$$T_{12} = \tfrac{1}{2}\otimes\frac{\tan\frac{\alpha_1}{2}\gamma^2_{A_1}A_1 - \cot\frac{\alpha_3}{2}\gamma^2_{A_3}A_3}{\tan\frac{\alpha_1}{2}(\gamma^2_{A_1}-\frac{1}{2}) - \cot\frac{\alpha_3}{2}(\gamma^2_{A_3}-\frac{1}{2})} \tag{5.31b}$$

$$T_{13} = \tfrac{1}{2}\otimes\frac{\tan\frac{\alpha_1}{2}\gamma^2_{A_1}A_1 - \cot\frac{\alpha_2}{2}\gamma^2_{A_2}A_2}{\tan\frac{\alpha_1}{2}(\gamma^2_{A_1}-\frac{1}{2}) - \cot\frac{\alpha_2}{2}(\gamma^2_{A_2}-\frac{1}{2})}$$

$$T_{21} = \tfrac{1}{2} \otimes \frac{\tan \frac{\alpha_2}{2} \gamma_{A_2}^2 A_2 - \cot \frac{\alpha_3}{2} \gamma_{A_3}^2 A_3}{\tan \frac{\alpha_2}{2} (\gamma_{A_2}^2 - \frac{1}{2}) - \cot \frac{\alpha_3}{2} (\gamma_{A_3}^2 - \frac{1}{2})}$$

$$T_{22} = \tfrac{1}{2} \otimes \frac{\cot \frac{\alpha_1}{2} \gamma_{A_1}^2 A_1 + \cot \frac{\alpha_3}{2} \gamma_{A_3}^2 A_3}{\cot \frac{\alpha_1}{2} (\gamma_{A_1}^2 - \frac{1}{2}) + \cot \frac{\alpha_3}{2} (\gamma_{A_3}^2 - \frac{1}{2})} \qquad (5.31c)$$

$$T_{23} = \tfrac{1}{2} \otimes \frac{\cot \frac{\alpha_1}{2} \gamma_{A_1}^2 A_1 - \tan \frac{\alpha_2}{2} \gamma_{A_2}^2 A_2}{\cot \frac{\alpha_1}{2} (\gamma_{A_1}^2 - \frac{1}{2}) - \tan \frac{\alpha_2}{2} (\gamma_{A_2}^2 - \frac{1}{2})}$$

$$T_{31} = \tfrac{1}{2} \otimes \frac{\cot \frac{\alpha_2}{2} (\gamma_{A_2}^2 A_2 - \tan \frac{\alpha_3}{2} \gamma_{A_3}^2 A_3}{\cot \frac{\alpha_2}{2} (\gamma_{A_2}^2 - \frac{1}{2}) - \tan \frac{\alpha_3}{2} (\gamma_{A_3}^2 - \frac{1}{2})}$$

$$T_{32} = \tfrac{1}{2} \otimes \frac{\cot \frac{\alpha_1}{2} (\gamma_{A_1}^2 A_1 - \tan \frac{\alpha_3}{2} \gamma_{A_3}^2 A_3}{\cot \frac{\alpha_1}{2} (\gamma_{A_1}^2 - \frac{1}{2}) - \tan \frac{\alpha_3}{2} (\gamma_{A_3}^2 - \frac{1}{2})} \qquad (5.31d)$$

$$T_{33} = \tfrac{1}{2} \otimes \frac{\cot \frac{\alpha_1}{2} (\gamma_{A_1}^2 A_1 + \cot \frac{\alpha_2}{2} \gamma_{A_2}^2 A_2}{\cot \frac{\alpha_1}{2} (\gamma_{A_1}^2 - \frac{1}{2}) + \cot \frac{\alpha_2}{2} (\gamma_{A_2}^2 - \frac{1}{2})}$$

Proof. The following proof of this Theorem 5.4 is almost identical with that of Theorem 5.3.

The tangency points of the in-exgyrocircles of a gyrotriangle $A_1A_2A_3$ in an Einstein gyrovector space $\mathbb{R}_s^n$ are given in (5.29), p. 261, in terms of the Einstein gyrobarycentric coordinate representation, (4.18), p. 177, that involve isomorphic gyrobarycentric coordinates with respect to two vertices of the reference gyrotriangle $A_1A_2A_3$. Hence, by Theorem 4.6, p. 177, the in-exgyrocircle tangency points of the corresponding gyrotriangle $A_1A_2A_3$ in the isomorphic Möbius gyrovector space $\mathbb{R}_s^n$ are the ones given in (5.31) in terms of the Möbius gyrobarycentric coordinate representation (4.19), p. 177, that involve isomorphic gyrobarycentric coordinates with respect to their respective sets of two vertices of the gyrotriangle.

Finally, Einstein gyrobarycentric coordinates in (4.18) are gyrotrigonometric functions and hence, by Theorem 2.48, p. 143, their isomorphic images in the isomorphic Möbius gyrovector space survive unchanged in (5.31). □

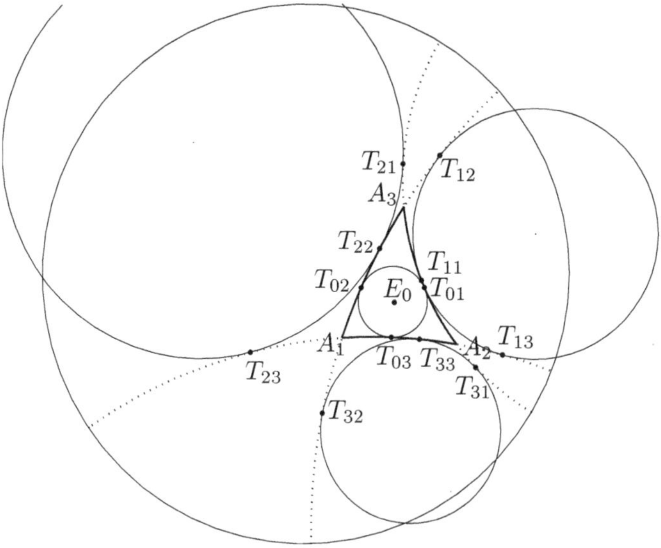

Fig. 5.7 The in-exgyrocircle tangency points, T_{ij}, $i = 0,1,2,3$, $j = 1,2,3$, of a gyrotriangle $A_1A_2A_3$ in a Möbius gyrovector space $(\mathbb{R}^n_s, \oplus, \otimes)$. Here all the exgyrocenters E_1, E_2 and E_3 of the gyrotriangle do not exist. Yet, all the gyrotrigonometric gyrobarycentric coordinate representations of the tangency points of the gyrotriangle, determined by Theorem 5.4, p. 263, remain valid, as we see here graphically.

5.6 From Gyrotriangle Tangency Points to Gyrotriangle Gyrocenters

The Möbius gyrotriangle tangency points T_{ij}, $i = 0,1,2,3$, $j = 1,2,3$, in Theorem 5.4, shown in Figs. 5.4–5.10, give rise to the following three gyrotriangle gyrocenters:

(1) *Gergonne Gyropoint* G_e. In the notation of Fig. 5.4, the gyrolines A_kT_{0k}, $k = 1,2,3$, are concurrent, Fig. 5.8. The resulting point of concurrency, called the gyrotriangle Gergonne gyropoint, G_e, possesses the gyrotrigonometric gyrobarycentric coordinate representation (see Exercise 2, p. 270)

$$G_e = \tfrac{1}{2}\otimes\frac{\tan\frac{\alpha_1}{2}\gamma^2_{A_1}A_1+\tan\frac{\alpha_2}{2}\gamma^2_{A_2}A_2+\tan\frac{\alpha_3}{2}\gamma^2_{A_3}A_3}{\tan\frac{\alpha_1}{2}(\gamma^2_{A_1}-\frac{1}{2})+\tan\frac{\alpha_2}{2}(\gamma^2_{A_2}-\frac{1}{2})+\tan\frac{\alpha_3}{2}(\gamma^2_{A_3}-\frac{1}{2})} \tag{5.32}$$

with respect to the vertices of the reference gyrotriangle $A_1A_2A_3$.

(2) *Nagel Gyropoint* N_a. In the notation of Fig. 5.4, the gyrolines A_kT_{kk}, $k = 1,2,3$, are concurrent, Fig. 5.9. The resulting point of

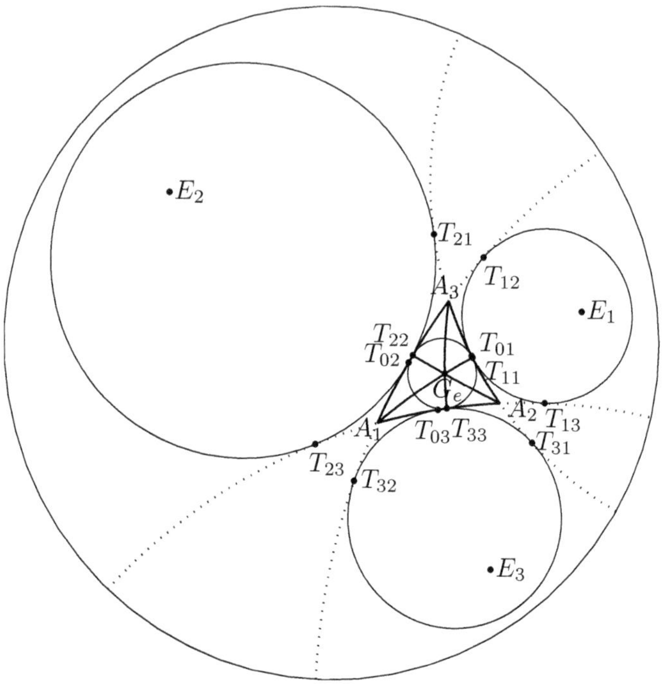

Fig. 5.8 Gergonne Gyropoint, G_e. In the notation of Fig. 5.4, p. 263, for the gyrotriangle in-exgyrocircle tangency points in a Möbius gyrovector plane, the gyrolines A_kT_{0k}, $k = 1, 2, 3$, are concurrent, and the resulting point of concurrency is the gyrotriangle Gergonne gyropoint, G_e. A gyrotrigonometric gyrobarycentric coordinate representation of the gyrotriangle Gergonne gyropoint, G_e, with respect to its vertices is given by (5.32).

concurrency, called the gyrotriangle Nagel gyropoint, N_a, possesses the gyrotrigonometric gyrobarycentric coordinate representation (see Exercise 3, p. 271)

$$N_a = \tfrac{1}{2}\otimes\frac{\cot\frac{\alpha_1}{2}\gamma^2_{A_1}A_1+\cot\frac{\alpha_2}{2}\gamma^2_{A_2}A_2+\cot\frac{\alpha_3}{2}\gamma^2_{A_3}A_3}{\cot\frac{\alpha_1}{2}(\gamma^2_{A_1}-\frac{1}{2})+\cot\frac{\alpha_2}{2}(\gamma^2_{A_2}-\frac{1}{2})+\cot\frac{\alpha_3}{2}(\gamma^2_{A_3}-\frac{1}{2})} \tag{5.33}$$

with respect to the vertices of the reference gyrotriangle $A_1A_2A_3$.

(3) A gyropoint P_u. In the notation of Fig. 5.4, the gyrolines E_kT_{kk}, $k = 1, 2, 3$, are concurrent, Fig. 5.10. The resulting point of concurrency, called the gyrotriangle gyropoint P_u, possesses the gyrotrigonometric gyrobarycentric coordinate representation (see Exercise 4, p. 271)

$$P_u = \tfrac{1}{2}\otimes\frac{m_1\gamma^2_{A_1}A_1+m_2\gamma^2_{A_2}A_2+m_3\gamma^2_{A_3}A_3}{m_1(\gamma^2_{A_1}-\frac{1}{2})+m_2(\gamma^2_{A_2}-\frac{1}{2})+m_3(\gamma^2_{A_3}-\frac{1}{2})} \tag{5.34a}$$

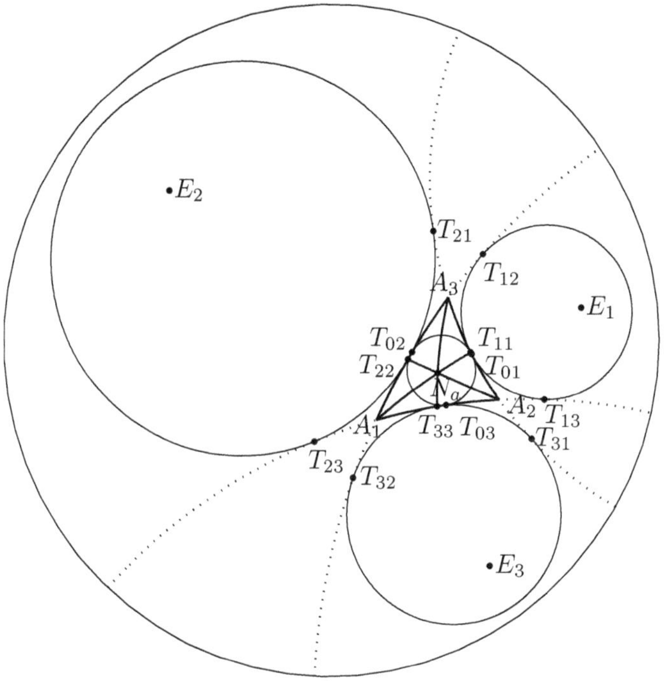

Fig. 5.9 Nagel Gyropoint, N_a. In the notation of Fig. 5.4, p. 263, for the gyrotriangle in-exgyrocircle tangency points in a Möbius gyrovector plane, the gyrolines A_kT_{kk}, $k = 1, 2, 3$, are concurrent, and the resulting point of concurrency is the gyrotriangle Nagel gyropoint, N_a. A gyrotrigonometric gyrobarycentric coordinate representation of the gyrotriangle Nagel gyropoint, N_a, with respect to its vertices is given by (5.33).

where gyrotrigonometric gyrobarycentric coordinates $(m_1 : m_2 : m_3)$ of P_u in (5.34a) are

$$\begin{aligned} m_1 &= \sin\alpha_1(1 + \cos\alpha_1 - \cos\alpha_2 - \cos\alpha_3) \\ m_2 &= \sin\alpha_2(1 - \cos\alpha_1 + \cos\alpha_2 - \cos\alpha_3) \\ m_3 &= \sin\alpha_3(1 - \cos\alpha_1 - \cos\alpha_2 + \cos\alpha_3) \end{aligned} \tag{5.34b}$$

with respect to the vertices of the reference gyrotriangle $A_1A_2A_3$.

5.7 Exercises

(1) Show that the law of gyrocosines (2.143), p. 103, along with (5.22), p. 255, implies the gyrotriangle gyrotrigonometric relations in (5.24), p. 256.

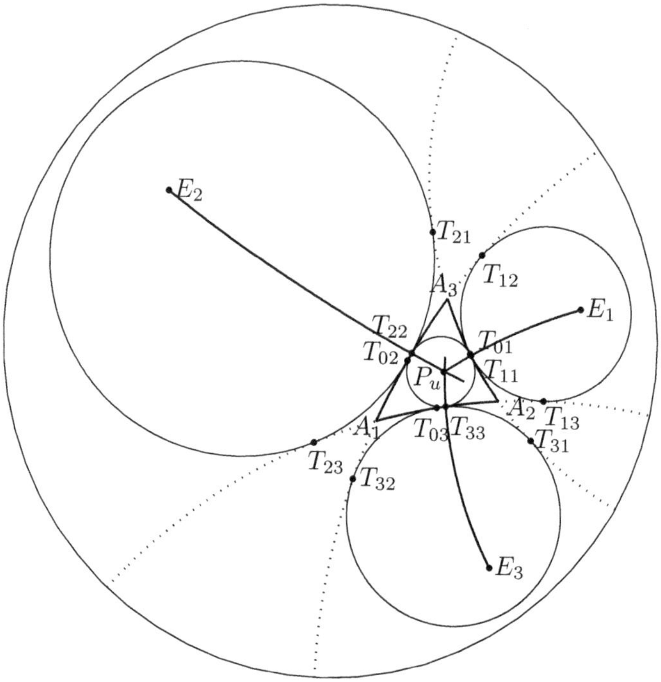

Fig. 5.10 P_u Gyropoint. In the notation of Fig. 5.4, p. 263, for the gyrotriangle in-exgyrocircle tangency points in a Möbius gyrovector plane, the gyrolines E_kT_{kk}, $k = 1, 2, 3$, are concurrent, and the resulting point of concurrency is the P_u gyropoint. A gyrotrigonometric gyrobarycentric coordinate representation of the gyrotriangle center P_u with respect to the vertices of its reference gyrotriangle is given by (5.34).

(2) Verify the gyrotrigonometric gyrobarycentric coordinate representation (5.32), p. 267, of the gyrotriangle Gergonne gyropoint G_e with respect to the vertices of its reference gyrotriangle in a Möbius gyrovector space $\mathbb{R}^n_s$.
Hint: It is too complicated to establish (5.32) in a Möbius gyrovector space $\mathbb{R}^n_s$ directly. Rather, one should exploit the comparative advantage that Einstein gyrovector spaces have over Möbius gyrovector spaces concerning gyroline intersections. In Einstein gyrovector spaces gyrolines are Euclidean straight lines so that gyroline intersection points can be determined by standard methods of linear algebra. Hence, one should first establish the counterpart of (5.32) in the context of Einstein gyrovector spaces, as in [18]. Once the Einstein counterpart of (5.32) has been determined (see [18, Chap. 7]) one may readily transform

the result into the context of both Möbius gyrovector spaces $\mathbb{R}^n_s$ and Euclidean vector spaces $\mathbb{R}^n$, thus solving simultaneously the present Exercise and Exercise 9, p. 61.

(3) Verify the gyrotrigonometric gyrobarycentric coordinate representation (5.33), p. 268, of the gyrotriangle Nagel gyropoint N_a with respect to the vertices of its reference gyrotriangle in a Möbius gyrovector space $\mathbb{R}^n_s$.
Hint: The Hint for solving Exercise 2 is applicable to this Exercise 3 as well.

(4) Verify the gyrotrigonometric gyrobarycentric coordinate representation (5.34), p. 268, of the gyrotriangle gyropoint P_u with respect to the vertices of its reference gyrotriangle in a Möbius gyrovector space $\mathbb{R}^n_s$.
Hint: The Hint for solving Exercise 2 is applicable to this Exercise 4 as well.

Chapter 6

Hyperbolic Tetrahedra

A gyrotetrahedron in an Einstein gyrovector space $(\mathbb{R}^n_s, \oplus, \otimes)$, $n \geq 3$, is a 3-dimensional gyrosimplex $A_1A_2A_3A_4$, Def. 4.3, p. 172, where $A_k \in \mathbb{R}^n_s$, $k = 1, 2, 3, 4$, are four pointwise independent points, called the vertices of the gyrotetrahedron. The four faces of a gyrotetrahedron $A_1A_2A_3A_4$ are the gyrotriangles $A_1A_2A_3$, $A_1A_2A_4$, $A_1A_3A_4$ and $A_2A_3A_4$, shown in Figs. 6.1 – 6.2. The gyrocentroid of a gyrotetrahedron, shown in Fig. 6.2, is presented in (4.64) – (4.65), p. 188,

6.1 Gyrotetrahedron Gyroaltitude

In this section we extend the study of gyrotriangle gyroaltitudes in an Einstein gyrovector space in Sec. 4.10, p. 191, from two dimensions to three dimensions.

A gyroline in an Einstein gyrovector space $(\mathbb{R}^n_s, \oplus, \otimes)$, $n \geq 2$, is the intersection of a Euclidean straight line in $\mathbb{R}^n$ and the ball $\mathbb{R}^n_s$. Similarly, a gyroplane in the Einstein gyrovector space $(\mathbb{R}^n_s, \oplus, \otimes)$, $n \geq 3$, is the intersection of a Euclidean plane in $\mathbb{R}^n$ and the ball $\mathbb{R}^n_s$. Any three nongyrocollinear points A, B, C of an Einstein gyrovector space determine uniquely a gyroplane π_{ABC}.

The orthogonal projection of a point A_4 on a gyroplane $\pi_{A_1A_2A_3}$ in an Einstein gyrovector space is a point P_4 at which the gyroline L passing through A_4 and orthogonal to the gyroplane $\pi_{A_1A_2A_3}$ intersects the gyroplane $\pi_{A_1A_2A_3}$, as shown in Fig. 6.3. The gyrosegment A_4P_4 is said to be the perpendicular dropped from the point A_4 on the gyroplane $\pi_{A_1A_2A_3}$.

Let $A_1A_2A_3A_4$ be a gyrotetrahedron with vertices A_1, A_2, A_3 and A_4, in an Einstein gyrovector space $(\mathbb{R}^n_s, \oplus, \otimes)$, $n \geq 3$, and let the point P_4 be the orthogonal projection of vertex A_4 onto its opposite face, $A_1A_2A_3$,

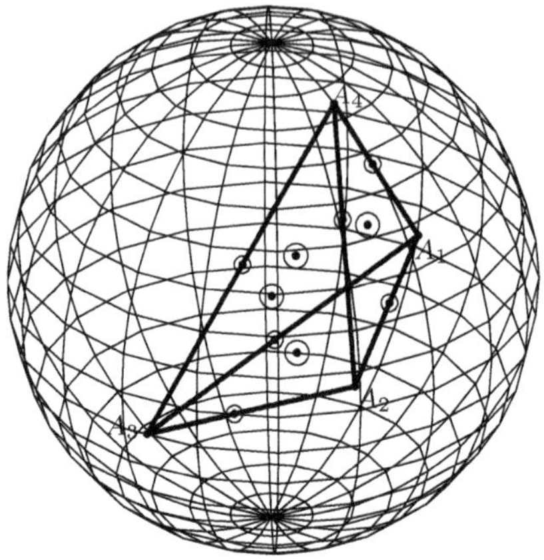

Fig. 6.1 A gyrotetrahedron $A_1A_2A_3A_4$ sitting inside an Einstein ball gyrovector space $\mathbb{R}_s^3$ is shown. The gyrotetrahedron is a gyrosimplex spanned by the set $S = \{A_1, A_2, A_3, A_4\}$ that forms a set of four pointwise independent points, called vertices. The faces of the gyrotetrahedron are gyrotriangles. For clarity, it is convenient to present the gyrotetrahedron without the ball in which it resides, as in Fig. 6.2.

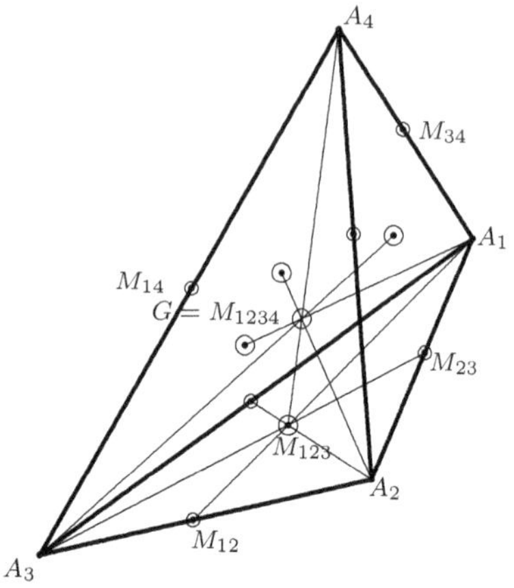

Fig. 6.2 Shown are the gyromidpoints of the 6 sides and the gyrocentroids of the 4 faces of a gyrotetrahedron. The gyroline joining a vertex of a gyrotetrahedron and the gyrocentroid of the opposite face is a gyrotetrahedron gyromedian. The four gyromedians of the gyrotetrahedron are concurrent. The point of concurrency being the gyrotetrahedron gyrocentroid G; see Sec. 4.7, p. 188.

(or its extension) of the gyrotetrahedron, as shown in Figs. 6.3–6.4 for $\mathbb{R}_s^n = \mathbb{R}_s^3$. Furthermore, let $(m_1 : m_2 : m_3)$ be a gyrobarycentric coordinate set of P_4 with respect to the set $\{A_1, A_2, A_3\}$ in the Einstein gyrovector space as presented in Def. 4.2, p. 171. Then,

$$P_4 = \frac{m_1\gamma_{A_1}A_1 + m_2\gamma_{A_2}A_2 + m_3\gamma_{A_3}A_3}{m_1\gamma_{A_1} + m_2\gamma_{A_2} + m_3\gamma_{A_3}} \tag{6.1}$$

where the gyrobarycentric coordinates m_1, m_2 and m_3 of P_4 are to be determined in (6.16), p. 277, in terms of the side gyrolengths of the gyrotetrahedron.

Hence, by the first identity in (4.11), p. 174, it follows from (6.1) that

$$\ominus X \oplus P_4 = \frac{m_1\gamma_{\ominus X\oplus A_1}(\ominus X\oplus A_1) + m_2\gamma_{\ominus X\oplus A_2}(\ominus X\oplus A_2) + m_3\gamma_{\ominus X\oplus A_3}(\ominus X\oplus A_3)}{m_1\gamma_{\ominus X\oplus A_1} + m_2\gamma_{\ominus X\oplus A_2} + m_3\gamma_{\ominus X\oplus A_3}} \tag{6.2}$$

so that, in particular, for $X = A_1$, $X = A_2$ and $X = A_3$ we have, respectively,

$$\begin{aligned}
\ominus A_1\oplus P_4 &= \frac{m_2\gamma_{\ominus A_1\oplus A_2}(\ominus A_1\oplus A_2) + m_3\gamma_{\ominus A_1\oplus A_3}(\ominus A_1\oplus A_3)}{m_1 + m_2\gamma_{\ominus A_1\oplus A_2} + m_3\gamma_{\ominus A_1\oplus A_3}} \\
\ominus A_2\oplus P_4 &= \frac{m_1\gamma_{\ominus A_1\oplus A_2}(\ominus A_2\oplus A_1) + m_3\gamma_{\ominus A_2\oplus A_3}(\ominus A_2\oplus A_3)}{m_1\gamma_{\ominus A_1\oplus A_2} + m_2 + m_3\gamma_{\ominus A_2\oplus A_3}} \\
\ominus A_3\oplus P_4 &= \frac{m_1\gamma_{\ominus A_1\oplus A_3}(\ominus A_3\oplus A_1) + m_2\gamma_{\ominus A_2\oplus A_3}(\ominus A_3\oplus A_2)}{m_1\gamma_{\ominus A_1\oplus A_3} + m_2\gamma_{\ominus A_2\oplus A_3} + m_3}
\end{aligned} \tag{6.3}$$

Along with the notation in Figs. 6.3 – 6.4 we use the notation

$$\begin{aligned}
\mathbf{a}_{12} &= \ominus A_1\oplus A_2, & a_{12} &= \|\mathbf{a}_{12}\|, & \gamma_{21} &= \gamma_{12} = \gamma_{a_{12}} \\
\mathbf{a}_{13} &= \ominus A_1\oplus A_3, & a_{13} &= \|\mathbf{a}_{13}\|, & \gamma_{31} &= \gamma_{13} = \gamma_{a_{13}} \\
\mathbf{a}_{14} &= \ominus A_1\oplus A_4, & a_{14} &= \|\mathbf{a}_{14}\|, & \gamma_{41} &= \gamma_{14} = \gamma_{a_{14}} \\
\mathbf{a}_{23} &= \ominus A_2\oplus A_3, & a_{23} &= \|\mathbf{a}_{23}\|, & \gamma_{32} &= \gamma_{23} = \gamma_{a_{23}} \\
\mathbf{a}_{24} &= \ominus A_2\oplus A_4, & a_{24} &= \|\mathbf{a}_{24}\|, & \gamma_{42} &= \gamma_{24} = \gamma_{a_{24}} \\
\mathbf{a}_{34} &= \ominus A_3\oplus A_4, & a_{34} &= \|\mathbf{a}_{34}\|, & \gamma_{43} &= \gamma_{34} = \gamma_{a_{34}}
\end{aligned} \tag{6.4}$$

and

$$\begin{aligned}
\mathbf{p}_1 &= \ominus A_1\oplus P_4, & p_1 &= \|\mathbf{p}_1\| \\
\mathbf{p}_2 &= \ominus A_2\oplus P_4, & p_2 &= \|\mathbf{p}_2\| \\
\mathbf{p}_3 &= \ominus A_3\oplus P_4, & p_3 &= \|\mathbf{p}_3\|
\end{aligned} \tag{6.5}$$

and

$$\mathbf{h}_k = \ominus A_k\oplus P_k, \qquad h_k = \|\mathbf{h}_k\| \tag{6.6}$$

$k = 1, 2, 3, 4$.

Following the notation in Figs. 6.3 – 6.4 and in (6.4) – (6.6), the application of Einstein-Pythagoras Identity, (2.179), p. 113, to the three right

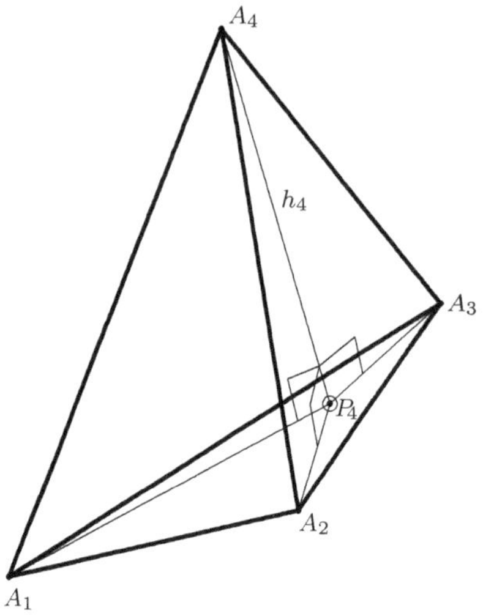

Fig. 6.3 A Gyrotetrahedron Gyroaltitude. The perpendicular projection of vertex A_4 of a gyrotetrahedron $A_1A_2A_3A_4$ in an Einstein gyrovector space $\mathbb{R}^n_s$, $n \geq 3$, on its opposite face $A_1A_2A_3$, Fig. 6.3, or its extension, Fig. 6.4. The gyrotetrahedron gyroaltitude drawn from A_4 is $\mathbf{h}_4 = \ominus P_4 \oplus A_4$.

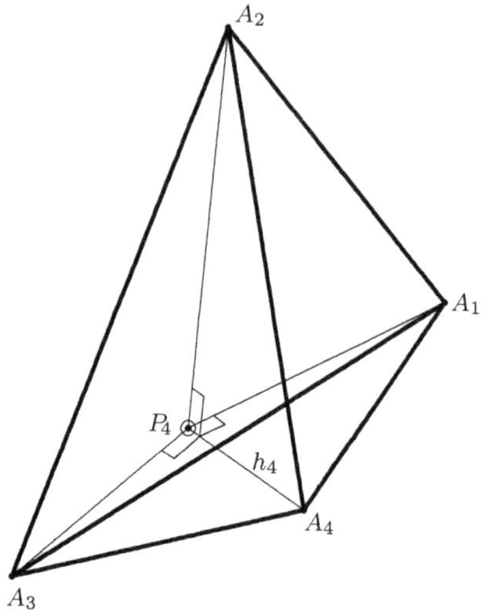

Fig. 6.4 and its gyrolength, also called gyroaltitude, is given by $h_4 = \|\ominus P_4 \oplus A_4\|$. The perpendicular projection P_4 of vertex A_4 on the gyroplane of the opposing face $A_1A_2A_3$ is the gyroaltitude foot. The gyrodistance between the point A_4 and the gyroplane that passes through the points A_1, A_2 and A_3 is h_4.

gyrotriangles $A_1P_4A_4$, $A_2P_4A_4$ and $A_3P_4A_4$ in Fig. 6.3 gives rise, respectively, to the equations

$$\begin{aligned} \gamma_{p_1}\gamma_{h_4} &= \gamma_{14} \\ \gamma_{p_2}\gamma_{h_4} &= \gamma_{24} \\ \gamma_{p_3}\gamma_{h_4} &= \gamma_{34} \end{aligned} \tag{6.7}$$

By the second identity in (4.11), p. 174, it follows from (6.1) that

$$\gamma_{\ominus X \oplus P_4} = \frac{m_1\gamma_{\ominus X \oplus A_1} + m_2\gamma_{\ominus X \oplus A_2} + m_3\gamma_{\ominus X \oplus A_3}}{m_0} \tag{6.8}$$

for any $X \in \mathbb{R}^n_s$, where

$$m_0^2 = (m_1+m_2+m_3)^2 + 2m_1m_2(\gamma_{12}-1) + 2m_1m_3(\gamma_{13}-1) + 2m_2m_3(\gamma_{23}-1) \tag{6.9}$$

For $X = A_1$, $X = A_2$ and $X = A_3$, respectively, (6.8) along with the notation in (6.5) gives the equations

$$\begin{aligned}\gamma_{p_1} &= \frac{m_1 + m_2\gamma_{12} + m_3\gamma_{13}}{m_0} \\ \gamma_{p_2} &= \frac{m_1\gamma_{12} + m_2 + m_3\gamma_{23}}{m_0} \\ \gamma_{p_3} &= \frac{m_1\gamma_{13} + m_2\gamma_{23} + m_3}{m_0}\end{aligned} \tag{6.10}$$

and Einstein-Pythagoras Identities (6.7) give

$$\frac{m_0\gamma_{p_1}}{\gamma_{14}} = \frac{m_0\gamma_{p_2}}{\gamma_{24}} = \frac{m_0\gamma_{p_3}}{\gamma_{34}} \tag{6.11}$$

Eliminating $m_0\gamma_{p_1}$, $m_0\gamma_{p_2}$ and $m_0\gamma_{p_3}$ between (6.10) and (6.11), we have

$$\frac{m_1 + m_2\gamma_{12} + m_3\gamma_{13}}{\gamma_{14}} = \frac{m_1\gamma_{12} + m_2 + m_3\gamma_{23}}{\gamma_{24}} = \frac{m_1\gamma_{13} + m_2\gamma_{23} + m_3}{\gamma_{34}} \tag{6.12}$$

Equation (6.12) and the gyrobarycentric coordinates normalization condition, $m_1 + m_2 + m_3 = 1$, form a system of three equations for the three unknowns m_1, m_2 and m_3, which can be written as the matrix equation

$$\begin{pmatrix} 1 & 1 & 1 \\ \frac{1}{\gamma_{14}} - \frac{\gamma_{12}}{\gamma_{24}} & \frac{\gamma_{12}}{\gamma_{14}} - \frac{1}{\gamma_{24}} & \frac{\gamma_{13}}{\gamma_{14}} - \frac{\gamma_{23}}{\gamma_{24}} \\ \frac{1}{\gamma_{14}} - \frac{\gamma_{13}}{\gamma_{34}} & \frac{\gamma_{12}}{\gamma_{14}} - \frac{\gamma_{23}}{\gamma_{34}} & \frac{\gamma_{13}}{\gamma_{14}} - \frac{1}{\gamma_{34}} \end{pmatrix} \begin{pmatrix} m_1 \\ m_2 \\ m_3 \end{pmatrix} = \begin{pmatrix} 1 \\ 0 \\ 0 \end{pmatrix} \tag{6.13}$$

The 3×3 matrix M in (6.13) is invertible, having the determinant

$$det(M) = \frac{K}{\gamma_{14}\gamma_{24}\gamma_{34}} > 0 \tag{6.14}$$

where K is given by the equation

$$\begin{aligned}K = \gamma_{14}(&\ \ \gamma_{12} + \gamma_{13} - \gamma_{23} - 1)(\gamma_{23} - 1) \\ + \gamma_{24}(&\ \ \gamma_{12} - \gamma_{13} + \gamma_{23} - 1)(\gamma_{13} - 1) \\ + \gamma_{34}(&-\gamma_{12} + \gamma_{13} + \gamma_{23} - 1)(\gamma_{12} - 1)\end{aligned} \tag{6.15}$$

Solving the matrix equation (6.13) we have

$$\begin{aligned}m_1 &= \frac{1}{K}\{\gamma_{23}(\gamma_{12}\gamma_{34} + \gamma_{13}\gamma_{24}) - (\gamma_{23}^2 - 1)\gamma_{14} - \gamma_{12}\gamma_{24} - \gamma_{13}\gamma_{34}\} \\ m_2 &= \frac{1}{K}\{\gamma_{13}(\gamma_{12}\gamma_{34} + \gamma_{23}\gamma_{14}) - (\gamma_{13}^2 - 1)\gamma_{24} - \gamma_{12}\gamma_{14} - \gamma_{23}\gamma_{34}\} \\ m_3 &= \frac{1}{K}\{\gamma_{12}(\gamma_{13}\gamma_{24} + \gamma_{23}\gamma_{14}) - (\gamma_{12}^2 - 1)\gamma_{34} - \gamma_{13}\gamma_{14} - \gamma_{23}\gamma_{24}\} \\ m_1 &+ m_2 + m_3 = 1\end{aligned} \tag{6.16}$$

thus determining the special gyrobarycentric coordinates (m_1, m_2, m_3) of the point P_4 with respect to the set $\{A_1, A_2, A_3\}$ in (6.1). Owing to their homogeneity, convenient gyrobarycentric coordinates $(m_1 : m_2 : m_3)$ for P_4 with respect to $\{A_1, A_2, A_3\}$ is obtained from (6.16) by omitting the nonzero common factor $1/K$, as we do in (6.17) – (6.17c) below.

Substituting the gyrobarycentric coordinates m_1, m_2 and m_3 from (6.16) into (6.1) we have

$$P_4 = \frac{N_{4123}}{D_{4123}} \tag{6.17a}$$

where

$$\begin{aligned} N_{4123} := &\ m_1\gamma_{A_1}A_1 + m_2\gamma_{A_2}A_2 + m_3\gamma_{A_3}A_3 \\ = &\ \left\{\gamma_{23}(\gamma_{12}\gamma_{34} + \gamma_{13}\gamma_{24}) - (\gamma_{23}^2 - 1)\gamma_{14} - \gamma_{12}\gamma_{24} - \gamma_{13}\gamma_{34}\right\}\gamma_{A_1}A_1 \\ &+ \left\{\gamma_{13}(\gamma_{12}\gamma_{34} + \gamma_{23}\gamma_{14}) - (\gamma_{13}^2 - 1)\gamma_{24} - \gamma_{12}\gamma_{14} - \gamma_{23}\gamma_{34}\right\}\gamma_{A_2}A_2 \\ &+ \left\{\gamma_{12}(\gamma_{13}\gamma_{24} + \gamma_{23}\gamma_{14}) - (\gamma_{12}^2 - 1)\gamma_{34} - \gamma_{13}\gamma_{14} - \gamma_{23}\gamma_{24}\right\}\gamma_{A_3}A_3 \end{aligned} \tag{6.17b}$$

and

$$\begin{aligned} D_{4123} := &\ m_1\gamma_{A_1} + m_2\gamma_{A_2} + m_3\gamma_{A_3} \\ = &\ \left\{\gamma_{23}(\gamma_{12}\gamma_{34} + \gamma_{13}\gamma_{24}) - (\gamma_{23}^2 - 1)\gamma_{14} - \gamma_{12}\gamma_{24} - \gamma_{13}\gamma_{34}\right\}\gamma_{A_1} \\ &+ \left\{\gamma_{13}(\gamma_{12}\gamma_{34} + \gamma_{23}\gamma_{14}) - (\gamma_{13}^2 - 1)\gamma_{24} - \gamma_{12}\gamma_{14} - \gamma_{23}\gamma_{34}\right\}\gamma_{A_2} \\ &+ \left\{\gamma_{12}(\gamma_{13}\gamma_{24} + \gamma_{23}\gamma_{14}) - (\gamma_{12}^2 - 1)\gamma_{34} - \gamma_{13}\gamma_{14} - \gamma_{23}\gamma_{24}\right\}\gamma_{A_3} \end{aligned} \tag{6.17c}$$

In (6.17) we obtain a homogeneous gyrobarycentric coordinate representation of the orthogonal projection P_4 of vertex A_4 onto its opposite face $A_1A_2A_3$ (or its extension) of gyrotetrahedron $A_1A_2A_3A_4$ of Fig. 6.3 with respect to the set $\{A_1, A_2, A_3\}$.

Following (6.8) – (6.9) with $X = A_4$, and (6.16) along with the notation in (6.4) we have, by computer algebra,

$$\gamma_{h_4} = \gamma_{\ominus A_4 \oplus P_4} = \frac{N'_{4123}}{D'_{4123}} \tag{6.18a}$$

where

$$\begin{aligned} N'_{4123} := &\ m_1\gamma_{\ominus A_4\oplus A_1} + m_2\gamma_{\ominus A_4\oplus A_2} + m_3\gamma_{\ominus A_4\oplus A_3} \\ = &\ \left\{\gamma_{23}(\gamma_{12}\gamma_{34} + \gamma_{13}\gamma_{24}) - (\gamma_{23}^2 - 1)\gamma_{14} - \gamma_{12}\gamma_{24} - \gamma_{13}\gamma_{34}\right\}\gamma_{14} \\ &+ \left\{\gamma_{13}(\gamma_{12}\gamma_{34} + \gamma_{23}\gamma_{14}) - (\gamma_{13}^2 - 1)\gamma_{24} - \gamma_{12}\gamma_{14} - \gamma_{23}\gamma_{34}\right\}\gamma_{24} \\ &+ \left\{\gamma_{12}(\gamma_{13}\gamma_{24} + \gamma_{23}\gamma_{14}) - (\gamma_{12}^2 - 1)\gamma_{34} - \gamma_{13}\gamma_{14} - \gamma_{23}\gamma_{24}\right\}\gamma_{34} \end{aligned} \tag{6.18b}$$

and where $D'_{4123} > 0$ and

$$
\begin{aligned}
(D'_{4123})^2 &= m_0^2 \\
&= (m_1 + m_2 + m_3)^2 \\
&+ 2m_1m_2(\gamma_{12} - 1) + 2m_1m_3(\gamma_{13} - 1) + 2m_2m_3(\gamma_{23} - 1)
\end{aligned}
\tag{6.18c}
$$

Hence, by (6.18) – (6.18c) and computer algebra,

$$
1 \le \gamma_{h_4}^2 = \gamma_{\ominus A_4 \oplus P_4}^2 = \frac{N''_{4123}}{D''_{4123}}
\tag{6.19a}
$$

where

$$
\begin{aligned}
N''_{4123} &= 2\gamma_{12}\gamma_{14}(\gamma_{23}\gamma_{34} - \gamma_{24}) \\
&+ 2\gamma_{13}\gamma_{34}(\gamma_{12}\gamma_{24} - \gamma_{14}) \\
&+ 2\gamma_{23}\gamma_{24}(\gamma_{13}\gamma_{14} - \gamma_{34}) \\
&- \gamma_{14}^2(\gamma_{23}^2 - 1) \\
&- \gamma_{24}^2(\gamma_{13}^2 - 1) \\
&- \gamma_{34}^2(\gamma_{12}^2 - 1)
\end{aligned}
\tag{6.19b}
$$

and where

$$
D''_{4123} = 1 + 2\gamma_{12}\gamma_{13}\gamma_{23} - \gamma_{12}^2 - \gamma_{13}^2 - \gamma_{23}^2 = \tfrac{1}{s^4} S_{A_1A_2A_3}^2
\tag{6.19c}
$$

Here, by (2.194), p. 117, $S_{A_1A_2A_3}$ is the gyrotriangle constant of the gyrotetrahedron face $A_1A_2A_3$ opposite to its vertex A_4.

Employing (6.19) and (2.11), p. 66, we have

$$
\begin{aligned}
\gamma_{h_4}^2 h_4^2 &= \gamma_{\ominus A_4 \oplus P_4}^2 \|\ominus A_4 \oplus P_4\|^2 \\
&= s^2(\gamma_{\ominus A_4 \oplus P_4}^2 - 1) \\
&= s^2 \frac{N''_{4123} - D''_{4123}}{D''_{4123}}
\end{aligned}
\tag{6.20}
$$

so that, by (6.19c) and (6.20),

$$
\begin{aligned}
S_{A_1A_2A_3A_4}^2 :&= S_{A_1A_2A_3}^2 \gamma_{\ominus A_4 \oplus P_4}^2 \|\ominus A_4 \oplus P_4\|^2 \\
&= s^4 D''_{4123} \gamma_{h_4}^2 h_4^2 \\
&= s^6 (N''_{4123} - D''_{4123})
\end{aligned}
\tag{6.21}
$$

Substituting (6.19b) – (6.19c) into (6.21) we have

$$
\begin{aligned}
S^2_{A_1A_2A_3A_4} &= s^6\{[2\gamma_{12}\gamma_{14}\gamma_{23}\gamma_{34} + 2\gamma_{12}\gamma_{13}\gamma_{24}\gamma_{34} + 2\gamma_{13}\gamma_{14}\gamma_{23}\gamma_{24}] \\
&\quad - [2\gamma_{12}\gamma_{13}\gamma_{23} + 2\gamma_{12}\gamma_{14}\gamma_{24} + 2\gamma_{13}\gamma_{14}\gamma_{34} + 2\gamma_{23}\gamma_{24}\gamma_{34}] \\
&\quad - [\gamma_{14}^2\gamma_{23}^2 + \gamma_{13}^2\gamma_{24}^2 + \gamma_{12}^2\gamma_{34}^2] \\
&\quad + [\gamma_{12}^2 + \gamma_{13}^2 + \gamma_{14}^2 + \gamma_{23}^2 + \gamma_{24}^2 + \gamma_{34}^2] \\
&\quad - 1\} \\
&= -s^6 \begin{vmatrix} 1 & \gamma_{12} & \gamma_{13} & \gamma_{14} \\ \gamma_{12} & 1 & \gamma_{23} & \gamma_{24} \\ \gamma_{13} & \gamma_{23} & 1 & \gamma_{34} \\ \gamma_{14} & \gamma_{24} & \gamma_{34} & 1 \end{vmatrix} > 0
\end{aligned}
\tag{6.22}
$$

Interestingly, the determinantal pattern (i) in (6.22) and (ii) in (2.195), p. 117, and (iii) in (7.11), p. 313, survives in higher dimensional gyrosimplices, as shown in [19, Sect. 10.5], where the matrix of the determinant is referred to as a *gamma matrix assigned to a gyrosimplex.*

Any gyrotetrahedron is invariant under any permutation of its vertices. As an example, a cyclic permutation of the vertices of gyrotetrahedron $A_1A_2A_3A_4$ in Fig. 6.3 into the gyrotetrahedron $A_3A_4A_1A_2$ is shown in Fig. 6.4.

Interestingly, $S_{A_1A_2A_3A_4}$ is a *gyrotetrahedron constant* of a gyrotetrahedron $A_1A_2A_3A_4$ in the sense that the right-hand side of (6.22) is independent of permutations of the indices 1,2,3,4. In fact, the right-hand side of (6.22) is presented as the sum of five partial sums each of which is invariant under any permutation of the indices 1,2,3,4. As such, the gyrotetrahedron constant $S_{A_1A_2A_3A_4}$ is analogous to the gyrotriangle constant $S_{A_1A_2A_3}$ in (2.194), p. 117.

Formalizing the result in (6.17) and previous related results, we have the following theorem.

Theorem 6.1. *Let $A_1A_2A_3A_4$ be a gyrotetrahedron in an Einstein gyrovector space, and let P_k, $k = 1, 2, 3, 4$, be the orthogonal projection of vertex A_k onto its opposite face, as shown in Figs. 6.3 – 6.4. Furthermore, let h_k be the gyrotetrahedron gyroaltitude drawn from vertex A_k, as shown in Figs. 6.3 – 6.4 and expressed in* (6.6). *Then, the gyrotetrahedron constant $S_{A_1A_2A_3A_4}$, given by* (6.22), *can be expressed in four different, but* *equivalent, ways:*

$$\begin{aligned} S_{A_1A_2A_3A_4} &= S_{A_2A_3A_4}\gamma_{h_1}h_1 \\ S_{A_1A_2A_3A_4} &= S_{A_1A_3A_4}\gamma_{h_2}h_2 \\ S_{A_1A_2A_3A_4} &= S_{A_1A_2A_4}\gamma_{h_3}h_3 \\ S_{A_1A_2A_3A_4} &= S_{A_1A_2A_3}\gamma_{h_4}h_4 \end{aligned} \tag{6.23}$$

Proof. The fourth equation in (6.23) has already been proved. It is identical with (6.21), where the gyrotetrahedron constant $S_{A_1A_2A_3A_4}$ is given by (6.22). It follows from (6.22) that the gyrotetrahedron constant $S_{A_1A_2A_3A_4}$ is invariant under any permutation of its indices $1, 2, 3, 4$. Hence, by the fourth equation in (6.23),

$$S_{A_1A_2A_3A_4} = S_{A_2A_3A_4A_1} = S_{A_2A_3A_4}\gamma_{h_1}h_1 \tag{6.24}$$

thus obtaining the first equation in (6.23) from the fourth equation in (6.23) by the cyclic permutation $(1, 2, 3, 4) \to (2, 3, 4, 1)$.

Similarly, the fourth equation in (6.23) with the cyclic permutation $(1, 2, 3, 4) \to (3, 4, 1, 2)$ gives

$$S_{A_1A_2A_3A_4} = S_{A_3A_4A_1A_2} = S_{A_3A_4A_1}\gamma_{h_2}h_2 = S_{A_1A_3A_4}\gamma_{h_2}h_2 \tag{6.25}$$

noting that the gyrotriangle constant $S_{A_1A_3A_4}$ is invariant under index permutations. In (6.25) we have thus shown that the second equation in (6.23) follows from the fourth equation in (6.23) by the cyclic permutation $(1, 2, 3, 4) \to (3, 4, 1, 2)$.

Similarly, the third equation in (6.23) follows from the fourth equation in (6.23) with the cyclic permutation $(1, 2, 3, 4) \to (4, 1, 2, 3)$. □

In the Euclidean limit of large s, $s \to \infty$, gamma factors reduce to 1 and, accordingly, in each of the four equation in (6.23) the gyrotetrahedron constant $S_{A_1A_2A_3A_4}$ reduces to $3V$, where V is the volume of the corresponding tetrahedron.

In the study of the gyrotriangle constant, defined in (2.193), p. 117, we were able to find a factor, symmetric in the gyrotriangle sides, that converts the gyrotriangle constant into the gyrotriangle gyroarea, in (2.198), p. 118, which obeys a gyrotriangle gyroarea addition rule, Sec. 2.20, p. 119. The question as to whether there exists, similarly, a symmetric factor that converts the gyrotetrahedron constant into an appropriate gyrotetrahedron gyrovolume that obeys some gyrotetrahedron gyrovolume addition rule remains open.

Figure 4.7, p. 200, indicates that the three gyroaltitudes of a gyrotriangle are concurrent. In contrast, the four gyroaltitudes of a gyrotetrahedron

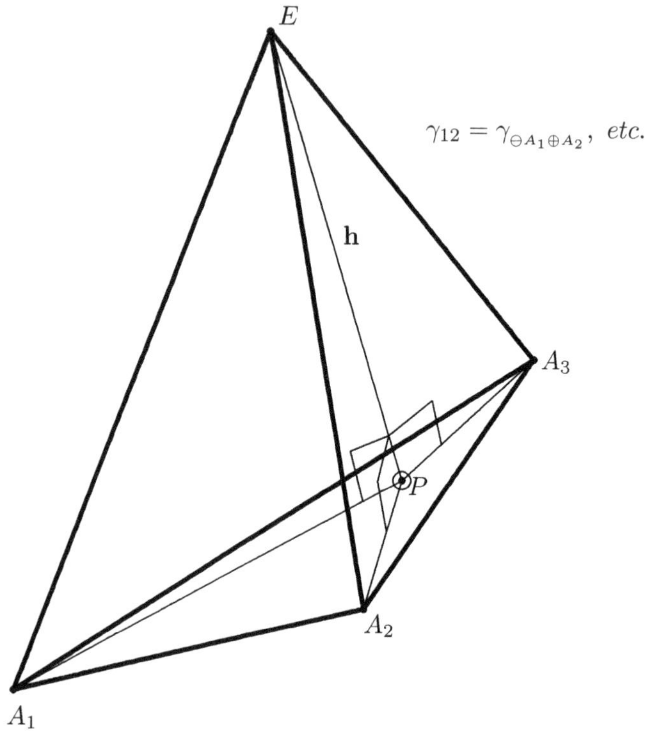

Fig. 6.5 The perpendicular projection P of a point E on a gyroplane $\pi_{A_1A_2A_3}$ in an Einstein gyrovector space $(\mathbb{R}^n_s, \oplus, \otimes)$, $n \geq 3$. The gyrodistance between the point E and the gyroplane $\pi_{A_1A_2A_3}$ is $h = \|\mathbf{h}\| = \|\ominus P \oplus E\|$.

are not concurrent in general. Indeed, it is well known in Euclidean geometry that the four altitudes of a tetrahedron are not concurrent in general; see, for instance [58].

6.2 Point Gyroplane Relations

In this section we determine in two theorems the perpendicular projection of a point on a gyroplane and the gyrodistance between a point and a gyroplane in an Einstein gyrovector space $(\mathbb{R}^n_s, \oplus, \otimes)$, $n \geq 3$, Fig. 6.5.

Theorem 6.2. (Perpendicular Projection of a Point on a Gyroplane). *Let A_1, A_2 and A_3 be any three pointwise independent points of an Einstein gyrovector space $(\mathbb{R}^n_s, \oplus, \otimes)$, $n \geq 3$, let $\pi_{A_1A_2A_3}$ be the gyroplane* *passing through these points, and let $E\in\mathbb{R}_s^n$ be a point not on the gyroplane, Fig. 6.5.*

Then, the perpendicular projection $P\in\mathbb{R}_s^n$ of E on the gyroplane is given by its gyrobarycentric coordinate representation

$$P = \frac{m_1\gamma_{A_1}A_1 + m_2\gamma_{A_2}A_2 + m_3\gamma_{A_3}A_3}{m_1\gamma_{A_1} + m_2\gamma_{A_2} + m_3\gamma_{A_3}} \tag{6.26a}$$

with respect to the set $S = \{A_1, A_2, A_3\}$, where the gyrobarycentric coordinates m_1, m_2 and m_3 of P are given by the equations

$$\begin{aligned}
m_1 &= \gamma_{23}(\gamma_{12}\gamma_{\ominus A_3\oplus E} + \gamma_{13}\gamma_{\ominus A_2\oplus E}) \\
&\quad - (\gamma_{23}^2 - 1)\gamma_{\ominus A_1\oplus E} - \gamma_{12}\gamma_{\ominus A_2\oplus E} - \gamma_{13}\gamma_{\ominus A_3\oplus E} \\
m_2 &= \gamma_{13}(\gamma_{12}\gamma_{\ominus A_3\oplus E} + \gamma_{23}\gamma_{\ominus A_1\oplus E}) \\
&\quad - (\gamma_{13}^2 - 1)\gamma_{\ominus A_2\oplus E} - \gamma_{12}\gamma_{\ominus A_1\oplus E} - \gamma_{23}\gamma_{\ominus A_3\oplus E} \\
m_3 &= \gamma_{12}(\gamma_{13}\gamma_{\ominus A_2\oplus E} + \gamma_{23}\gamma_{\ominus A_1\oplus E}) \\
&\quad - (\gamma_{12}^2 - 1)\gamma_{\ominus A_3\oplus E} - \gamma_{13}\gamma_{\ominus A_1\oplus E} - \gamma_{23}\gamma_{\ominus A_2\oplus E}
\end{aligned} \tag{6.26b}$$

Proof. The result (6.26) of the Theorem is established in (6.17), p. 278. Indeed, the derivation of (6.26) from the result (6.17) for P_4 in (6.1), p. 274 is a matter of notation: The point P_4 in (6.1) is renamed as P in (6.26) and the point A_4 in Figs. 6.1 – 6.4 and in (6.17) is renamed as E, Fig. 6.5. In accordance with renaming A_4 as E it becomes useful to replace the notation $\gamma_{k4} = \gamma_{\ominus A_k\oplus A_4}$, $k = 1, 2, 3$, in (6.17) by the notation $\gamma_{\ominus A_k\oplus E}$ in (6.26). □

Theorem 6.3. (Gyrodistance Between a Point and a Gyroplane). *Let A_1, A_2 and A_3 be any three pointwise independent points of an Einstein gyrovector space $(\mathbb{R}_s^n, \oplus, \otimes)$, $n \geq 3$, let $\pi_{A_1A_2A_3}$ be the gyroplane passing through these points, and let $E\in\mathbb{R}_s^n$ be a point not on the gyroplane, Fig. 6.5.*

Then, the gyrodistance, h, between the point E and the gyroplane is given by the equation

$$\gamma_h^2 = \frac{N_{E123}}{D_{E123}} \tag{6.27a}$$

where

$$
\begin{aligned}
N_{E123} &= 2\gamma_{12}\gamma_{\ominus A_1\oplus E}(\gamma_{23}\gamma_{\ominus A_3\oplus E} - \gamma_{\ominus A_2\oplus E}) \\
&+ 2\gamma_{13}\gamma_{\ominus A_3\oplus E}(\gamma_{12}\gamma_{\ominus A_2\oplus E} - \gamma_{\ominus A_1\oplus E}) \\
&+ 2\gamma_{23}\gamma_{\ominus A_2\oplus E}(\gamma_{13}\gamma_{\ominus A_1\oplus E} - \gamma_{\ominus A_3\oplus E}) \\
&- \gamma_{\ominus A_1\oplus E}^2(\gamma_{23}^2 - 1) \\
&- \gamma_{\ominus A_2\oplus E}^2(\gamma_{13}^2 - 1) \\
&- \gamma_{\ominus A_3\oplus E}^2(\gamma_{12}^2 - 1)
\end{aligned}
\tag{6.27b}
$$

and

$$
D_{E123} = 1 + 2\gamma_{12}\gamma_{13}\gamma_{23} - \gamma_{12}^2 - \gamma_{13}^2 - \gamma_{23}^2 \tag{6.27c}
$$

Proof. The result (6.27) of the Theorem is established in (6.19), p. 279. Indeed, the derivation of (6.27) from (6.19) is a matter of notation: The altitude gyrolength h_4 in (6.19) and in Figs. 6.3 – 6.4 becomes h in (6.27) and in Fig. 6.5, and the point A_4 in Figs. 6.1 – 6.4 and in (6.19) is renamed as E, in (6.27a) and in Fig. 6.5. In accordance with renaming A_4 as E it becomes useful to replace the notation $\gamma_{k4} = \gamma_{\ominus A_k\oplus A_4}$, $k = 1, 2, 3$, in (6.19) by the notation $\gamma_{\ominus A_k\oplus E}$ in (6.27). □

It should be noted that an explicit expression for the gyrodistance h in (6.27a) can readily be obtained from (6.27a) by employing Identity (2.11), p. 66.

6.3 Gyrotetrahedron Ingyrocenter and Exgyrocenters

Let $A_1A_2A_3A_4$ be a gyrotetrahedron in an Einstein gyrovector space $(\mathbb{R}_s^n, \oplus, \otimes)$, $n \geq 3$, and let E be a point equigyrodistant from the gyrotetrahedron faces, so that E is the ingyrocenter or an exgyrocenter of the gyrotetrahedron. Furthermore, let

$$
E = \frac{m_1\gamma_{A_1}A_1 + m_2\gamma_{A_2}A_2 + m_3\gamma_{A_3}A_3 + m_4\gamma_{A_4}A_4}{m_1\gamma_{A_1} + m_2\gamma_{A_2} + m_3\gamma_{A_3} + m_4\gamma_{A_4}} \tag{6.28}
$$

be the gyrobarycentric coordinate representation of E with respect to the set $S = \{A_1, A_2, A_3, A_4\}$ of the gyrotetrahedron vertices, where the gyrobarycentric coordinates $(m_1 : m_2 : m_3 : m_4)$ of E in (6.28) are to be determined in (6.38), p. 287.

Applying the second identity in the results (4.11), p. 174, of Theorem 4.4 to the gyrobarycentric coordinate representation of E in (6.28) we obtain the following gamma factors, where we use the gyrotetrahedron index notation in Figs. 6.3 – 6.4, p. 276, and in (6.4), p. 275,

$$\gamma_{\ominus A_1 \oplus E} = \frac{m_1\gamma_{\ominus A_1\oplus A_1} + m_2\gamma_{\ominus A_1\oplus A_2} + m_3\gamma_{\ominus A_1\oplus A_3} + m_4\gamma_{\ominus A_1\oplus A_4}}{m_0}$$

$$= \frac{m_1 + m_2\gamma_{12} + m_3\gamma_{13} + m_4\gamma_{14}}{m_0} \tag{6.29a}$$

$$\gamma_{\ominus A_2 \oplus E} = \frac{m_1\gamma_{\ominus A_2\oplus A_1} + m_2\gamma_{\ominus A_2\oplus A_2} + m_3\gamma_{\ominus A_2\oplus A_3} + m_4\gamma_{\ominus A_2\oplus A_4}}{m_0}$$

$$= \frac{m_1\gamma_{12} + m_2 + m_3\gamma_{23} + m_4\gamma_{24}}{m_0} \tag{6.29b}$$

$$\gamma_{\ominus A_3 \oplus E} = \frac{m_1\gamma_{\ominus A_3\oplus A_1} + m_2\gamma_{\ominus A_3\oplus A_2} + m_3\gamma_{\ominus A_3\oplus A_3} + m_3\gamma_{\ominus A_3\oplus A_4}}{m_0}$$

$$= \frac{m_1\gamma_{13} + m_2\gamma_{23} + m_3 + m_4\gamma_{34}}{m_0} \tag{6.29c}$$

$$\gamma_{\ominus A_4 \oplus E} = \frac{m_1\gamma_{\ominus A_4\oplus A_1} + m_2\gamma_{\ominus A_4\oplus A_2} + m_3\gamma_{\ominus A_4\oplus A_3} + m_4\gamma_{\ominus A_4\oplus A_4}}{m_0}$$

$$= \frac{m_1\gamma_{14} + m_2\gamma_{24} + m_3\gamma_{34} + m_4}{m_0} \tag{6.29d}$$

where $m_0 > 0$ is given by the equation

$$\begin{aligned} m_0^2 = m_1^2 + m_2^2 + m_3^2 + m_4^2 + 2m_1m_2\gamma_{12} + 2m_1m_3\gamma_{13} + 2m_1m_4\gamma_{14} \\ + 2m_2m_3\gamma_{23} + 2m_2m_4\gamma_{24} + 2m_3m_4\gamma_{34} \end{aligned} \tag{6.29e}$$

Note that if $m_0^2 \le 0$, that is, if $m_0 = 0$ or m_0 is purely imaginary, then the point $E \in \mathbb{R}_s^n$ does not exist.

Now, let $h = h_4$ be the gyrodistance between the point E and face $A_1A_2A_3$ of the gyrotetrahedron $A_1A_2A_3A_4$, as shown in Fig. 6.5. Then, by Theorem 6.3, p. 283, h_4 satisfies the equation

$$\gamma_{h_4}^2 = \frac{N_{E123}}{D_{E123}} \tag{6.30}$$

where

$$
\begin{aligned}
N_{E123} &= 2\gamma_{12}\gamma_{\ominus A_1\oplus E}(\gamma_{23}\gamma_{\ominus A_3\oplus E} - \gamma_{\ominus A_2\oplus E}) \\
&\quad + 2\gamma_{13}\gamma_{\ominus A_3\oplus E}(\gamma_{12}\gamma_{\ominus A_2\oplus E} - \gamma_{\ominus A_1\oplus E}) \\
&\quad + 2\gamma_{23}\gamma_{\ominus A_2\oplus E}(\gamma_{13}\gamma_{\ominus A_1\oplus E} - \gamma_{\ominus A_3\oplus E}) \\
&\quad - \gamma_{\ominus A_1\oplus E}^2(\gamma_{23}^2 - 1) \\
&\quad - \gamma_{\ominus A_2\oplus E}^2(\gamma_{13}^2 - 1) \\
&\quad - \gamma_{\ominus A_3\oplus E}^2(\gamma_{12}^2 - 1)
\end{aligned}
\tag{6.31}
$$

and

$$
D_{E123} = 1 + 2\gamma_{12}\gamma_{13}\gamma_{23} - \gamma_{12}^2 - \gamma_{13}^2 - \gamma_{23}^2 \tag{6.32}
$$

Substituting successively (6.29) into (6.31) and, then, substituting (6.31) – (6.32) in (6.30) one may express $\gamma_{h_4}^2$ in (6.30) in terms of the gamma factors γ_{ij}, $1 \le i < j \le 4$, and the unknown gyrobarycentric coordinates m_k, $k = 1, 2, 3, 4$, of E in (6.28), obtaining the first of the following four equations:

$$
\begin{aligned}
\gamma_{h_4}^2 &= f_4(m_k, \gamma_{ij}) \\
\gamma_{h_3}^2 &= f_3(m_k, \gamma_{ij}) \\
\gamma_{h_2}^2 &= f_2(m_k, \gamma_{ij}) \\
\gamma_{h_1}^2 &= f_1(m_k, \gamma_{ij})
\end{aligned}
\tag{6.33}
$$

In (6.33) h_k is the gyrodistance between the point E and the gyrotetrahedron face opposite to vertex A_k, $k = 1, 2, 3, 4$. The first equation in (6.33) is obtained as explained above, and the remaining three equations in (6.33) are obtained from the first by index cyclic permutations. The functions f_k, $k = 1, 2, 3, 4$, are too involved and hence are not listed here; see Exercises 1 –2, p. 307.

By definition, the gyrodistances from E to each of the four faces of the gyrotetrahedron $A_1A_2A_3A_4$ are equal. Hence,

$$
\begin{aligned}
\gamma_{h_4}^2 &= \gamma_{h_1}^2 \\
\gamma_{h_4}^2 &= \gamma_{h_2}^2 \\
\gamma_{h_4}^2 &= \gamma_{h_3}^2
\end{aligned}
\tag{6.34}
$$

Substituting (6.33) into (6.34), along with the convenient normalization condition

$$m_1^2 + m_2^2 + m_3^2 + m_4^2 = 1 \tag{6.35}$$

we obtain from (6.34) – (6.35) the following system of four equations for the four unknowns m_1^2, m_2^2, m_3^2 and m_4^2:

$$\begin{aligned}
&m_1^2(1 + 2\gamma_{12}\gamma_{13}\gamma_{23} - \gamma_{12}^2 - \gamma_{13}^2 - \gamma_{23}^2) \\
&-m_4^2(1 + 2\gamma_{23}\gamma_{24}\gamma_{34} - \gamma_{23}^2 - \gamma_{24}^2 - \gamma_{34}^2) = 0 \\
&m_2^2(1 + 2\gamma_{12}\gamma_{13}\gamma_{23} - \gamma_{12}^2 - \gamma_{13}^2 - \gamma_{23}^2) \\
&-m_4^2(1 + 2\gamma_{13}\gamma_{14}\gamma_{34} - \gamma_{13}^2 - \gamma_{14}^2 - \gamma_{34}^2) = 0 \\
&m_3^2(1 + 2\gamma_{12}\gamma_{13}\gamma_{23} - \gamma_{12}^2 - \gamma_{13}^2 - \gamma_{23}^2) \\
&-m_4^2(1 + 2\gamma_{12}\gamma_{14}\gamma_{24} - \gamma_{12}^2 - \gamma_{14}^2 - \gamma_{24}^2) = 0 \\
&m_1^2 + m_2^2 + m_3^2 + m_4^2 = 1
\end{aligned} \tag{6.36}$$

The unique solution of the system (6.36) turns out to be

$$\begin{aligned}
m_1^2 &= \frac{1}{D}(1 + 2\gamma_{23}\gamma_{24}\gamma_{34} - \gamma_{23}^2 - \gamma_{24}^2 - \gamma_{34}^2) \\
m_2^2 &= \frac{1}{D}(1 + 2\gamma_{13}\gamma_{14}\gamma_{34} - \gamma_{13}^2 - \gamma_{14}^2 - \gamma_{34}^2) \\
m_3^2 &= \frac{1}{D}(1 + 2\gamma_{12}\gamma_{14}\gamma_{24} - \gamma_{12}^2 - \gamma_{14}^2 - \gamma_{24}^2) \\
m_4^2 &= \frac{1}{D}(1 + 2\gamma_{12}\gamma_{13}\gamma_{23} - \gamma_{12}^2 - \gamma_{13}^2 - \gamma_{23}^2)
\end{aligned} \tag{6.37}$$

where $D > 0$ is determined from (6.37) by (6.35). Since nonzero common factors of a gyrobarycentric coordinate set of a point are irrelevant, the common factor $1/D$ in (6.37) can be omitted. Hence, a convenient gyrobarycentric coordinate set $(m_1 : m_2 : m_3 : m_4)$ for the point E in (6.28) is given by the equations

$$\begin{aligned}
m_1^2 &= 1 + 2\gamma_{23}\gamma_{24}\gamma_{34} - \gamma_{23}^2 - \gamma_{24}^2 - \gamma_{34}^2 \\
m_2^2 &= 1 + 2\gamma_{13}\gamma_{14}\gamma_{34} - \gamma_{13}^2 - \gamma_{14}^2 - \gamma_{34}^2 \\
m_3^2 &= 1 + 2\gamma_{12}\gamma_{14}\gamma_{24} - \gamma_{12}^2 - \gamma_{14}^2 - \gamma_{24}^2 \\
m_4^2 &= 1 + 2\gamma_{12}\gamma_{13}\gamma_{23} - \gamma_{12}^2 - \gamma_{13}^2 - \gamma_{23}^2
\end{aligned} \tag{6.38}$$

Selecting the positive sign for each of the gyrobarycentric coordinates m_k, $k = 1, 2, 3, 4$, in (6.38) results in the gyrobarycentric coordinate set of the gyrotetrahedron ingyrocenter $E = E_0$, which is the gyrocenter of the gyrotetrahedron ingyrosphere, with respect to the gyrotetrahedron vertices $\{A_1, A_2, A_3, A_4\}$. Other choices of signs for m_k give gyrobarycentric coordinate representations for the gyrotetrahedron exgyrocenters; see Exercise 3, p. 307.

Each of the four equations in (6.38) can be simplified as shown for the fourth equation in (6.38) in the chain of equations below, which are numbered for subsequent derivation.

$$
\begin{aligned}
m_4 &\overbrace{=\!=}^{(1)} \pm\sqrt{1 + 2\gamma_{12}\gamma_{13}\gamma_{23} - \gamma_{12}^2 - \gamma_{13}^2 - \gamma_{23}^2)} \\
&\overbrace{=\!=}^{(2)} \pm\sqrt{\gamma_{12}^2 - 1}\sqrt{\gamma_{13}^2 - 1}\sin\alpha_1 \\
&\overbrace{=\!=}^{(3)} \pm\frac{1}{s^2}\gamma_{12}a_{12}\gamma_{13}a_{13}\sin\alpha_1 \\
&\overbrace{=\!=}^{(4)} \pm\frac{1}{s^2}\gamma_{12}a_{12}\gamma_{h_3}h_3 \\
&\overbrace{=\!=}^{(5)} \pm\frac{1}{s^2}S_{A_1A_2A_3}
\end{aligned}
\tag{6.39}
$$

Derivation of the numbered equalities in (2.51) follows:

(1) This equation is the fourth equation in (6.38).
(2) Follows from (1) by the first equation in (2.145), p. 104, applied to gyrotriangle $ABC = A_1A_2A_3$, written in the gyrotriangle index notation presented in Fig. 2.3, p. 101.
(3) Follows from (2) by (2.11), p. 66.
(4) Follows from (3) by an elementary gyrotrigonometric relation, shown in Fig. 2.8, p. 122, where h_3 is the gyroaltitude gyrolength of gyrotriangle $A_1A_2A_3$ drawn from vertex A_3 of the gyrotriangle.
(5) Follows from (4) by Def. 2.35, p. 117, where $S_{A_1A_2A_3}$ is the gyrotriangle constant of the gyrotriangle $A_1A_2A_3$.

It is clear from (6.39) that the gyrotriangle constant $S_{A_1A_2A_3}$ of the gyrotriangle $A_1A_2A_3$ is a symmetric function of its indices $1, 2, 3$.

Following (6.39) and its consequent relations obtained by vertex cyclic permutations, and noting that a nonzero common factor of a gyrobarycentric coordinate set of a point can be omitted, the equations in (6.38) can be written as

$$\begin{aligned} m_1^2 &= S^2_{A_2A_3A_4} \\ m_2^2 &= S^2_{A_3A_4A_1} = S^2_{A_1A_3A_4} \\ m_3^2 &= S^2_{A_4A_1A_2} = S^2_{A_1A_2A_4} \\ m_4^2 &= S^2_{A_1A_2A_3} \end{aligned} \tag{6.40}$$

The similarity between (6.40) and its two-dimensional counterpart (5.10), p. 250, is remarkable.

Owing to the homogeneity of gyrobarycentric coordinates, common factors are irrelevant. Accordingly, by selecting various signs for m_k, $k = 1, 2, 3, 4$, in (6.40) the gyrobarycentric coordinates $(m_1 : m_2 : m_3 : m_4)$ in (6.40) give at most eight distinct gyrobarycentric coordinate sets that correspond to eight possible locations (counting multiplicities) of the point E.

The points in these eight locations, denoted E_k, $k = 0, \ldots, 7$, are the ingyrocenter E_0 of gyrotetrahedron $A_1A_2A_3A_4$, Figs. 6.6 – 6.7, p. 297; the gyrotetrahedron *near* A_k-exgyrocenters E_k opposite to vertices A_k, respectively, $k = 1, 2, 3, 4$, Figs. 6.7 – 6.9, p. 298; and *far* A_j-exgyrocenters E_j, $j = 5, 6, 7$, in front of some edges and opposite to some other edges of the gyrotetrahedron, Fig. 6.10, p. 301.

The eight in-exgyrocenters E_k, $k = 0, \ldots, 7$, are in general distinct. Each of them is associated with a gyrobarycentric coordinate representation constant m_0, given by (6.29e), p. 285. Accordingly, any of these in-exgyrocenters exists if and only if its associated constant m_0 satisfies $m_0^2 > 0$ (that is, m_0 is real and nonzero).

The gyrobarycentric coordinates of the in-exgyrocenters E_k are listed in (6.41) below.

$$
\begin{aligned}
E_0 &: (m_1 : m_2 : m_3 : m_4) = (\ \ S_{A_2A_3A_4} : \ \ S_{A_3A_4A_1} : \ \ S_{A_4A_1A_2} : \ \ S_{A_1A_2A_3}) \\
E_1 &: (m_1 : m_2 : m_3 : m_4) = (-S_{A_2A_3A_4} : \ \ S_{A_3A_4A_1} : \ \ S_{A_4A_1A_2} : \ \ S_{A_1A_2A_3}) \\
E_2 &: (m_1 : m_2 : m_3 : m_4) = (\ \ S_{A_2A_3A_4} : -S_{A_3A_4A_1} : \ \ S_{A_4A_1A_2} : \ \ S_{A_1A_2A_3}) \\
E_3 &: (m_1 : m_2 : m_3 : m_4) = (\ \ S_{A_2A_3A_4} : \ \ S_{A_3A_4A_1} : -S_{A_4A_1A_2} : \ \ S_{A_1A_2A_3}) \\
E_4 &: (m_1 : m_2 : m_3 : m_4) = (\ \ S_{A_2A_3A_4} : \ \ S_{A_3A_4A_1} : \ \ S_{A_4A_1A_2} : -S_{A_1A_2A_3}) \\
E_5 &: (m_1 : m_2 : m_3 : m_4) = (\ \ S_{A_2A_3A_4} : \ \ S_{A_3A_4A_1} : -S_{A_4A_1A_2} : -S_{A_1A_2A_3}) \\
E_6 &: (m_1 : m_2 : m_3 : m_4) = (\ \ S_{A_2A_3A_4} : -S_{A_3A_4A_1} : \ \ S_{A_4A_1A_2} : -S_{A_1A_2A_3}) \\
E_7 &: (m_1 : m_2 : m_3 : m_4) = (\ \ S_{A_2A_3A_4} : -S_{A_3A_4A_1} : -S_{A_4A_1A_2} : \ \ S_{A_1A_2A_3})
\end{aligned}
\tag{6.41}
$$

Among the points E_k, $k = 0, \ldots, 7$, in (6.41) there is a single one, E_0, the gyrobarycentric coordinates of which are all positive. As such, the point E_0 lies in the interior of its reference gyrotetrahedron. Hence, this point is identified as the gyrocenter of the gyrotetrahedron ingyrosphere, called the gyrotetrahedron ingyrocenter.

Accordingly, for instance, the gyrobarycentric coordinate representation of the gyrotetrahedron ingyrocenter with respect to the gyrotetrahedron vertices is given by

$$
E_0 = \frac{S_{A_2A_3A_4}\gamma_{A_1} A_1 + S_{A_1A_3A_4}\gamma_{A_2} A_2 + S_{A_1A_2A_4}\gamma_{A_3} A_3 + S_{A_1A_2A_3}\gamma_{A_4} A_4}{S_{A_2A_3A_4}\gamma_{A_1} + S_{A_1A_3A_4}\gamma_{A_2} + S_{A_1A_2A_4}\gamma_{A_3} + S_{A_1A_2A_3}\gamma_{A_4}}
\tag{6.42}
$$

Formalizing the results of this section, we obtain the following theorem:

Theorem 6.4. (Gyrotetrahedron In-Exgyrocenters, Einstein). *The in-exgyrocenters E_k, $k = 0, \ldots, 7$ of a gyrotetrahedron $A_1A_2A_3A_4$ in an Einstein gyrovector space $(\mathbb{R}^n_s, \oplus, \otimes)$, $n \geq 3$, Figs. 6.6 – 6.10, pp. 297 – 301, possess, when exist, the gyrobarycentric coordinate representations with respect to the set $S = \{A_1, A_2, A_3, A_4\}$ listed below:*

$$
E_0 = \frac{m_1\gamma_{A_1} A_1 + m_2\gamma_{A_2} A_2 + m_3\gamma_{A_3} A_3 + m_4\gamma_{A_4} A_4}{m_1\gamma_{A_1} + m_2\gamma_{A_2} + m_3\gamma_{A_3} + m_4\gamma_{A_4}}
\tag{6.43a}
$$

$$
E_1 = \frac{-m_1\gamma_{A_1} A_1 + m_2\gamma_{A_2} A_2 + m_3\gamma_{A_3} A_3 + m_4\gamma_{A_4} A_4}{-m_1\gamma_{A_1} + m_2\gamma_{A_2} + m_3\gamma_{A_3} + m_4\gamma_{A_4}}
\tag{6.43b}
$$

$$E_2 = \frac{m_1\gamma_{_{A_1}}A_1 - m_2\gamma_{_{A_2}}A_2 + m_3\gamma_{_{A_3}}A_3 + m_4\gamma_{_{A_4}}A_4}{m_1\gamma_{_{A_1}} - m_2\gamma_{_{A_2}} + m_3\gamma_{_{A_3}} + m_4\gamma_{_{A_4}}} \tag{6.43c}$$

$$E_3 = \frac{m_1\gamma_{_{A_1}}A_1 + m_2\gamma_{_{A_2}}A_2 - m_3\gamma_{_{A_3}}A_3 + m_4\gamma_{_{A_4}}A_4}{m_1\gamma_{_{A_1}} + m_2\gamma_{_{A_2}} - m_3\gamma_{_{A_3}} + m_4\gamma_{_{A_4}}} \tag{6.43d}$$

$$E_4 = \frac{m_1\gamma_{_{A_1}}A_1 + m_2\gamma_{_{A_2}}A_2 + m_3\gamma_{_{A_3}}A_3 - m_4\gamma_{_{A_4}}A_4}{m_1\gamma_{_{A_1}} + m_2\gamma_{_{A_2}} + m_3\gamma_{_{A_3}} - m_4\gamma_{_{A_4}}} \tag{6.43e}$$

$$E_5 = \frac{m_1\gamma_{_{A_1}}A_1 + m_2\gamma_{_{A_2}}A_2 - m_3\gamma_{_{A_3}}A_3 - m_4\gamma_{_{A_4}}A_4}{m_1\gamma_{_{A_1}} + m_2\gamma_{_{A_2}} - m_3\gamma_{_{A_3}} - m_4\gamma_{_{A_4}}} \tag{6.43f}$$

$$E_6 = \frac{m_1\gamma_{_{A_1}}A_1 - m_2\gamma_{_{A_2}}A_2 + m_3\gamma_{_{A_3}}A_3 - m_4\gamma_{_{A_4}}A_4}{m_1\gamma_{_{A_1}} - m_2\gamma_{_{A_2}} + m_3\gamma_{_{A_3}} - m_4\gamma_{_{A_4}}} \tag{6.43g}$$

$$E_7 = \frac{m_1\gamma_{_{A_1}}A_1 - m_2\gamma_{_{A_2}}A_2 - m_3\gamma_{_{A_3}}A_3 + m_4\gamma_{_{A_4}}A_4}{m_1\gamma_{_{A_1}} - m_2\gamma_{_{A_2}} - m_3\gamma_{_{A_3}} + m_4\gamma_{_{A_4}}} \tag{6.43h}$$

with gyrobarycentric coordinates m_k, $k = 1, 2, 3, 4$ *given by*

$$\begin{aligned} m_1 &= S_{A_2A_3A_4} \\ m_2 &= S_{A_3A_4A_1} = S_{A_1A_3A_4} \\ m_3 &= S_{A_4A_1A_2} = S_{A_1A_2A_4} \\ m_4 &= S_{A_1A_2A_3} \end{aligned} \tag{6.43i}$$

The right-hand sides of (6.43i) *are the gyrotriangle constants of the faces of the gyrotetrahedron listed below:*

$$S_{A_1A_2A_3} = \gamma_{12}a_{12}\gamma_{h_{43}}h_{43} = \gamma_{13}a_{13}\gamma_{h_{42}}h_{42} = \gamma_{23}a_{23}\gamma_{h_{41}}h_{41} \tag{6.44a}$$

where h_{4k}, $k = 1, 2, 3$, *is the gyroaltitude drawn from vertex* A_k *of the gyrotriangle* $A_1A_2A_3$ *that forms the gyrotetrahedron face opposite to vertex* A_4.

$$S_{A_2A_3A_4} = \gamma_{23}a_{23}\gamma_{h_{14}}h_{14} = \gamma_{24}a_{24}\gamma_{h_{13}}h_{13} = \gamma_{34}a_{34}\gamma_{h_{12}}h_{12} \tag{6.44b}$$

where h_{1k}, $k = 2, 3, 4$, *is the gyroaltitude drawn from vertex* A_k *of the gyrotriangle* $A_2A_3A_4$ *that forms the gyrotetrahedron face opposite to vertex* A_1.

$$S_{A_3A_4A_1} = \gamma_{34}a_{34}\gamma_{h_{21}}h_{21} = \gamma_{13}a_{13}\gamma_{h_{24}}h_{24} = \gamma_{14}a_{14}\gamma_{h_{23}}h_{23} \tag{6.44c}$$

where h_{2k}, $k = 1, 3, 4$, is the gyroaltitude drawn from vertex A_k of the gyrotriangle $A_3A_4A_1$ that forms the gyrotetrahedron face opposite to vertex A_2.

$$S_{A_4A_1A_2} = \gamma_{14}a_{14}\gamma_{h_{32}}h_{32} = \gamma_{24}a_{24}\gamma_{h_{31}}h_{31} = \gamma_{12}a_{12}\gamma_{h_{34}}h_{34} \tag{6.44d}$$

where h_{3k}, $k = 1, 2, 4$, is the gyroaltitude drawn from vertex A_k of the gyrotriangle $A_4A_1A_2$ that forms the gyrotetrahedron face opposite to vertex A_3.

Proof. The result (6.43) of the Theorem is established in (6.40) – (6.42), where the gyrotriangle constants $S_{A_1A_2A_3}$, *etc.*, that are involved, listed in (6.44), are given by (2.193), p. 117. □

The in-exgyrocenters of a gyrotetrahedron in an Einstein gyrovector space are shown in Figs. 6.6 – 6.10, pp. 297 – 301. As these figures indicate, the gyrotetrahedron in-exgyrocenters E_k, $k = 0, \ldots, 7$, in (6.41) split into three sets:

(1) The gyrotetrahedron ingyrocenter E_0, Fig. 6.6, p. 297. It meets each of the gyrotetrahedron faces.
(2) The gyrotetrahedron near exgyrocenters E_k, $k = 1, 2, 3, 4$, Figs. 6.7 – 6.9, pp. 298 – 300. Each of these meets one face of the gyrotetrahedron and the gyroplanar extension of each of the remaining gyrotetrahedron three faces.
(3) The gyrotetrahedron far exgyrocenters E_k, $k = 5, 6, 7$, Fig. 6.10, p. 301. Each of these meets none of the gyrotetrahedron faces. Rather, each of these meets the gyroplanar extension of each of the gyrotetrahedron four faces.

6.4 In-Exgyrosphere Tangency Points

Let E_0 be the ingyrocenter of a gyrotetrahedron $A_1A_2A_3A_4$ in an Einstein gyrovector space $(\mathbb{R}^n_s, \oplus, \otimes)$, $n \geq 3$. Then, by (6.29), p. 285, with $E = E_0$

we have the following gamma factors:

$$\begin{aligned}\gamma_{\ominus A_1 \oplus E_0} &= \frac{m_1 + m_2\gamma_{12} + m_3\gamma_{13} + m_4\gamma_{14}}{m_0} \\ \gamma_{\ominus A_2 \oplus E_0} &= \frac{m_1\gamma_{12} + m_2 + m_3\gamma_{23} + m_4\gamma_{24}}{m_0} \\ \gamma_{\ominus A_3 \oplus E_0} &= \frac{m_1\gamma_{13} + m_2\gamma_{23} + m_3 + m_4\gamma_{34}}{m_0} \\ \gamma_{\ominus A_4 \oplus E_0} &= \frac{m_1\gamma_{14} + m_2\gamma_{24} + m_3\gamma_{34} + m_4}{m_0}\end{aligned} \tag{6.45}$$

where $m_0 > 0$ is given by (6.29e), and where m_k, $k = 1, 2, 3, 4$, are the gyrobarycentric coordinates of E_0 with respect to the gyrotetrahedron vertices $\{A_1, A_2, A_3, A_4\}$, given by (6.43i).

Substituting m_k from (6.43i) into (6.45) we obtain the equations

$$\begin{aligned}\gamma_{\ominus A_1 \oplus E_0} &= \frac{S_{A_2A_3A_4} + S_{A_3A_4A_1}\gamma_{12} + S_{A_4A_1A_2}\gamma_{13} + S_{A_1A_2A_3}\gamma_{14}}{m_0} \\ \gamma_{\ominus A_2 \oplus E_0} &= \frac{S_{A_2A_3A_4}\gamma_{12} + S_{A_3A_4A_1} + S_{A_4A_1A_2}\gamma_{23} + S_{A_1A_2A_3}\gamma_{24}}{m_0} \\ \gamma_{\ominus A_3 \oplus E_0} &= \frac{S_{A_2A_3A_4}\gamma_{13} + S_{A_3A_4A_1}\gamma_{23} + S_{A_4A_1A_2} + S_{A_1A_2A_3}\gamma_{34}}{m_0} \\ \gamma_{\ominus A_4 \oplus E_0} &= \frac{S_{A_2A_3A_4}\gamma_{14} + S_{A_3A_4A_1}\gamma_{24} + S_{A_4A_1A_2}\gamma_{34} + S_{A_1A_2A_3}}{m_0}\end{aligned} \tag{6.46}$$

An explicit presentation of $m_0 > 0$ is not needed since the common factor $1/m_0$ in (6.46) will be omitted in the transition from (6.47b) to (6.47c) below.

Let T_{04} be the point of tangency where the gyrotetrahedron ingyrosphere meets the gyroplane $\pi_{A_1A_2A_3}$ of the face $A_1A_2A_3$ opposite to vertex A_4 of the gyrotetrahedron, Fig. 6.6, p. 297. Then, T_{04} turns out to be the perpendicular projection of E_0 on the gyroplane $\pi_{A_1A_2A_3}$. As such, it follows from Theorem 6.2, p. 282, that T_{04} possesses the gyrobarycentric coordinate representation

$$T_{04} = \frac{m_1\gamma_{A_1}A_1 + m_2\gamma_{A_2}A_2 + m_3\gamma_{A_3}A_3}{m_1\gamma_{A_1} + m_2\gamma_{A_2} + m_3\gamma_{A_3}} \tag{6.47a}$$

with respect to the set $S = \{A_1, A_2, A_3\}$, where the gyrobarycentric coordinates m_k, $k = 1, 2, 3$ of T_{04} are given by the equations

$$
\begin{aligned}
m_1 &= \gamma_{23}(\gamma_{12}\gamma_{\ominus A_3 \oplus E_0} + \gamma_{13}\gamma_{\ominus A_2 \oplus E_0}) \\
&\quad - (\gamma_{23}^2 - 1)\gamma_{\ominus A_1 \oplus E_0} - \gamma_{12}\gamma_{\ominus A_2 \oplus E_0} - \gamma_{13}\gamma_{\ominus A_3 \oplus E_0} \\
m_2 &= \gamma_{13}(\gamma_{12}\gamma_{\ominus A_3 \oplus E_0} + \gamma_{23}\gamma_{\ominus A_1 \oplus E_0}) \\
&\quad - (\gamma_{13}^2 - 1)\gamma_{\ominus A_2 \oplus E_0} - \gamma_{12}\gamma_{\ominus A_1 \oplus E_0} - \gamma_{23}\gamma_{\ominus A_3 \oplus E_0} \\
m_3 &= \gamma_{12}(\gamma_{13}\gamma_{\ominus A_2 \oplus E_0} + \gamma_{23}\gamma_{\ominus A_1 \oplus E_0}) \\
&\quad - (\gamma_{12}^2 - 1)\gamma_{\ominus A_3 \oplus E_0} - \gamma_{13}\gamma_{\ominus A_1 \oplus E_0} - \gamma_{23}\gamma_{\ominus A_2 \oplus E_0}
\end{aligned}
\tag{6.47b}
$$

Substituting gamma factors from (6.46) into (6.47b) and omitting the common factor $1/m_0$ we obtain the following gyrobarycentric coordinates:

$$
\begin{aligned}
m_1 &= \{\gamma_{24}(\gamma_{13}\gamma_{23} - \gamma_{12}) + \gamma_{34}(\gamma_{12}\gamma_{23} - \gamma_{13}) - \gamma_{14}(\gamma_{23}^2 - 1)\}S_{A_1A_2A_3} \\
&\quad + \{1 + 2\gamma_{12}\gamma_{13}\gamma_{23} - \gamma_{12}^2 - \gamma_{13}^2 - \gamma_{23}^2\}S_{A_2A_3A_4} \\
m_2 &= \{\gamma_{14}(\gamma_{13}\gamma_{23} - \gamma_{12}) + \gamma_{34}(\gamma_{12}\gamma_{13} - \gamma_{23}) - \gamma_{24}(\gamma_{13}^2 - 1)\}S_{A_1A_2A_3} \\
&\quad + \{1 + 2\gamma_{12}\gamma_{13}\gamma_{23} - \gamma_{12}^2 - \gamma_{13}^2 - \gamma_{23}^2\}S_{A_3A_4A_1} \\
m_3 &= \{\gamma_{14}(\gamma_{12}\gamma_{23} - \gamma_{12}) + \gamma_{24}(\gamma_{12}\gamma_{13} - \gamma_{23}) - \gamma_{24}(\gamma_{12}^2 - 1)\}S_{A_1A_2A_3} \\
&\quad + \{1 + 2\gamma_{12}\gamma_{13}\gamma_{23} - \gamma_{12}^2 - \gamma_{13}^2 - \gamma_{23}^2\}S_{A_4A_1A_2}
\end{aligned}
\tag{6.47c}
$$

noting that, by (6.39), p. 288, the gyrotriangle constant of a gyrotriangle $A_1A_2A_3$ in an Einstein gyrovector space is given by the equation

$$
S_{A_1A_2A_3} = s^2\sqrt{1 + 2\gamma_{12}\gamma_{13}\gamma_{23} - \gamma_{12}^2 - \gamma_{13}^2 - \gamma_{23}^2)}
\tag{6.47d}
$$

so that it is invariant under vertex permutations.

Formalizing the result of this section we obtain the following theorem:

Theorem 6.5. (Gyrotetrahedron Ingyrosphere Tangency Points, Einstein). *Let $A_1A_2A_3A_4$ be a gyrotetrahedron in an Einstein gyrovector space $\mathbb{R}_s^n$, $n \geq 3$, and let T_{0k}, $k = 1, 2, 3, 4$, be the tangency point where the gyrotetrahedron ingyrosphere meets the gyroplane of its face opposite to vertex A_k. Then, T_{04} possesses the following gyrobarycentric coordinate representation*

$$
T_{04} = \frac{m_1\gamma_{A_1}A_1 + m_2\gamma_{A_2}A_2 + m_3\gamma_{A_3}A_3}{m_1\gamma_{A_1} + m_2\gamma_{A_2} + m_3\gamma_{A_3}}
\tag{6.48}
$$

with respect to the set $S = \{A_1, A_2, A_3\}$, *where the gyrobarycentric coordinates* m_k, $k = 1, 2, 3$, *are given by* (6.47c). *The other three tangency points* T_{0k}, $k = 1, 2, 3$, *are obtained from* T_{04} *by cyclic permutations for the gyrotetrahedron vertices.*

The in-exgyrosphere tangency points T_{ik} are the points where the in-exgyrosphere with in-exgyrocenter E_i, $i = 0, \ldots, 7$, meets the gyrotetrahedron face opposite to vertex A_k, $k = 1, 2, 3, 4$, or its gyroplanar extension. The tangency points T_{ik} are presented graphically in Figs. 6.6 – 6.10, pp. 297 – 301. They are determined in a way similar to that of determining T_{04} in Theorem 6.5; see Exercise 5, p. 307.

6.5 Gyrotrigonometric Gyrobarycentric Coordinates for the Gyrotetrahedron In-Exgyrocenters

In order to emphasize that gyrotriangle $A_1A_2A_3$ is the face opposite to vertex A_4 of a gyrotetrahedron $A_1A_2A_3A_4$ in an Einstein gyrovector space $\mathbb{R}^n_s$, $n \ge 3$, the gyroangle of gyrotriangle $A_1A_2A_3$ with vertex A_k, $k = 1, 2, 3$, of the gyrotriangle is denoted α_{4k}. Accordingly, α_{ik} is the gyroangle with vertex A_k in the gyrotriangle that forms the gyrotetrahedron face opposite to the gyrotetrahedron vertex A_i, $i = 1, 2, 3, 4$.

Following (2.164), p. 109, and the α_{ik} notation for the gyroangles of a gyrotetrahedron faces, let

$$
\begin{aligned}
F_4 &= F(\alpha_{41}, \alpha_{42}, \alpha_{43}) \\
&:= \cos\frac{\alpha_{41} + \alpha_{42} + \alpha_{43}}{2} \cos\frac{\alpha_{41} - \alpha_{42} - \alpha_{43}}{2} \\
&\quad\times \cos\frac{-\alpha_{41} + \alpha_{42} - \alpha_{43}}{2} \cos\frac{-\alpha_{41} - \alpha_{42} + \alpha_{43}}{2} \\
&= \frac{1}{4}\frac{(1 + 2\gamma_{12}\gamma_{13}\gamma_{23} - \gamma_{12}^2 - \gamma_{13}^2 - \gamma_{23}^2)^2}{(\gamma_{12}^2 - 1)(\gamma_{13}^2 - 1)(\gamma_{23}^2 - 1)}
\end{aligned}
\tag{6.49}
$$

so that, by (2.163), p. 109,

$$
\begin{aligned}
\gamma_{12}^2 - 1 &= \frac{4F_4}{\sin^2\alpha_{41}\sin^2\alpha_{42}} \\
\gamma_{13}^2 - 1 &= \frac{4F_4}{\sin^2\alpha_{41}\sin^2\alpha_{43}} \\
\gamma_{23}^2 - 1 &= \frac{4F_4}{\sin^2\alpha_{42}\sin^2\alpha_{43}}
\end{aligned}
\tag{6.50}
$$

Hence, by (6.49) – (6.50),

$$(1+2\gamma_{12}\gamma_{13}\gamma_{23}-\gamma_{12}^2-\gamma_{13}^2-\gamma_{23}^2)^2 = 4F_4(\gamma_{12}^2-1)(\gamma_{13}^2-1)(\gamma_{23}^2-1)$$
$$= \frac{(4F_4)^4}{\sin^4\alpha_{41}\sin^4\alpha_{42}\sin^4\alpha_{43}} \qquad (6.51)$$

so that, by (6.39), p. 288,

$$\frac{1}{s^2}S_{A_1A_2A_3} = \sqrt{1+2\gamma_{12}\gamma_{13}\gamma_{23}-\gamma_{12}^2-\gamma_{13}^2-\gamma_{23}^2}$$
$$= \frac{4F_4}{\sin\alpha_{41}\sin\alpha_{42}\sin\alpha_{43}} \qquad (6.52)$$

We now extend the definition of F_4 in (6.49) by cyclic permutations of the indices $(1,2,3,4)$, obtaining

$$\begin{aligned} F_4 &= F_4(\alpha_{41},\alpha_{42},\alpha_{43}) \\ F_1 &= F_1(\alpha_{12},\alpha_{13},\alpha_{14}) \\ F_2 &= F_2(\alpha_{23},\alpha_{24},\alpha_{21}) \\ F_3 &= F_3(\alpha_{34},\alpha_{31},\alpha_{32}) \end{aligned} \qquad (6.53)$$

where, for instance, α_{32} is the gyroangle with vertex A_2 of gyrotriangle $A_4A_1A_2$ that forms the face opposite to vertex A_3 of gyrotetrahedron $A_1A_2A_3A_4$.

We are now in the position to rewrite Theorem 6.4, p. 290, in a gyrotrigonometric form of gyrobarycentric coordinates.

Theorem 6.6. (Gyrotetrahedron In-Exgyrocenters, Gyrotrigonometric Form, Einstein). *The in-exgyrocenters E_k, $k=0,\ldots,7$ of a gyrotetrahedron $A_1A_2A_3A_4$ in an Einstein gyrovector space $(\mathbb{R}_s^n,\oplus,\otimes)$, $n\geq 3$, Figs. 6.6 – 6.10, pp. 297 – 301, possess, when exist, the gyrobarycentric coordinate representations with respect to the set $S=\{A_1,A_2,A_3,A_4\}$ listed below:*

$$E_0 = \frac{m_1\gamma_{A_1}A_1+m_2\gamma_{A_2}A_2+m_3\gamma_{A_3}A_3+m_4\gamma_{A_4}A_4}{m_1\gamma_{A_1}+m_2\gamma_{A_2}+m_3\gamma_{A_3}+m_4\gamma_{A_4}} \qquad (6.54a)$$

$$E_1 = \frac{-m_1\gamma_{A_1}A_1+m_2\gamma_{A_2}A_2+m_3\gamma_{A_3}A_3+m_4\gamma_{A_4}A_4}{-m_1\gamma_{A_1}+m_2\gamma_{A_2}+m_3\gamma_{A_3}+m_4\gamma_{A_4}} \qquad (6.54b)$$

$$E_2 = \frac{m_1\gamma_{A_1}A_1-m_2\gamma_{A_2}A_2+m_3\gamma_{A_3}A_3+m_4\gamma_{A_4}A_4}{m_1\gamma_{A_1}-m_2\gamma_{A_2}+m_3\gamma_{A_3}+m_4\gamma_{A_4}} \qquad (6.54c)$$

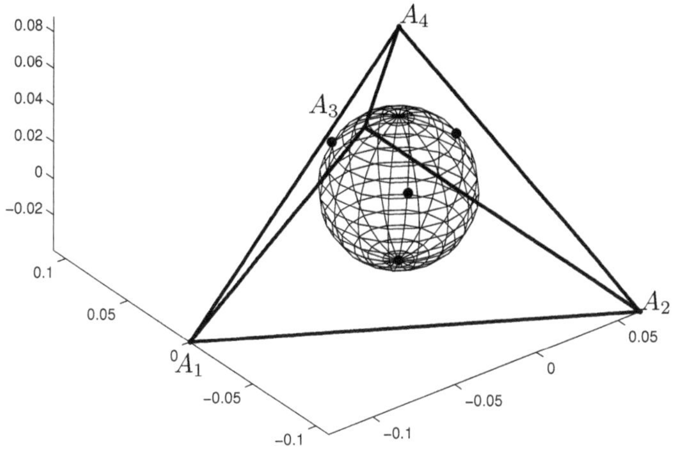

Fig. 6.6 A gyrotetrahedron in an Einstein gyrovector space $(\mathbb{R}^3_{s=1}, \oplus, \otimes)$ and its ingyrosphere. By left gyrotranslating the gyrotetrahedron vertices A_k, $k = 1, 2, 3, 4$, appropriately, the gyrotetrahedron is placed in a special position so that its ingyrocenter coincides with the origin $\mathbf{0}$ of the gyrovector space. Since the origin of any Einstein gyrovector space is conformal, an Einstein sphere with gyrocenter at the origin turns out to be a Euclidean sphere with center at the origin. The tangency points T_{0k} where the gyrotetrahedron ingyrosphere meets the gyrotetrahedron face opposite to vertex A_k, $k = 1, 2, 3, 4$, determined by Theorem 6.5, p. 294, are shown.

$$E_3 = \frac{m_1\gamma_{A_1}A_1 + m_2\gamma_{A_2}A_2 - m_3\gamma_{A_3}A_3 + m_4\gamma_{A_4}A_4}{m_1\gamma_{A_1} + m_2\gamma_{A_2} - m_3\gamma_{A_3} + m_4\gamma_{A_4}} \tag{6.54d}$$

$$E_4 = \frac{m_1\gamma_{A_1}A_1 + m_2\gamma_{A_2}A_2 + m_3\gamma_{A_3}A_3 - m_4\gamma_{A_4}A_4}{m_1\gamma_{A_1} + m_2\gamma_{A_2} + m_3\gamma_{A_3} - m_4\gamma_{A_4}} \tag{6.54e}$$

$$E_5 = \frac{m_1\gamma_{A_1}A_1 + m_2\gamma_{A_2}A_2 - m_3\gamma_{A_3}A_3 - m_4\gamma_{A_4}A_4}{m_1\gamma_{A_1} + m_2\gamma_{A_2} - m_3\gamma_{A_3} - m_4\gamma_{A_4}} \tag{6.54f}$$

$$E_6 = \frac{m_1\gamma_{A_1}A_1 - m_2\gamma_{A_2}A_2 + m_3\gamma_{A_3}A_3 - m_4\gamma_{A_4}A_4}{m_1\gamma_{A_1} - m_2\gamma_{A_2} + m_3\gamma_{A_3} - m_4\gamma_{A_4}} \tag{6.54g}$$

$$E_7 = \frac{m_1\gamma_{A_1}A_1 - m_2\gamma_{A_2}A_2 - m_3\gamma_{A_3}A_3 + m_4\gamma_{A_4}A_4}{m_1\gamma_{A_1} - m_2\gamma_{A_2} - m_3\gamma_{A_3} + m_4\gamma_{A_4}} \tag{6.54h}$$

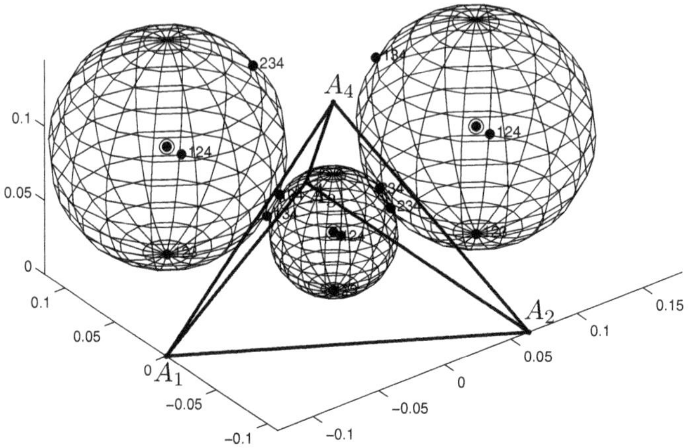

Fig. 6.7 A gyrotetrahedron and its ingyrosphere in an Einstein gyrovector space $(\mathbb{R}^3_{s=1}, \oplus, \otimes)$ is shown here along with two of its four near exgyrospheres that correspond to the gyrotetrahedron near exgyrocenters E_1 and E_2 in (6.41), p. 290. The tangency points T_{jk} where the gyrotetrahedron in-exgyrospheres gyrocentered at E_j, $j = 0, 1, 2$, meet the gyroplane of the gyrotetrahedron face opposite to vertex A_k, $k = 1, 2, 3, 4$, determined by exercise 5, p. 307, are shown. The gyrospheres in this figure are nearly coincident with (Euclidean) spheres since they are located near the origin of the gyrovector space ball $\mathbb{R}^3_{s=1}$. For clarity, the gyrotetrahedron face $A_1A_2A_3$ lies on the plane of the two horizontal axes.

These are gyrobarycentric coordinate representations of E_k, $k = 0, 1, \ldots, 7$, in an Einstein gyrovector space, with respect to the set $S = \{A_1, A_2, A_3, A_4\}$, with gyrotrigonometric gyrobarycentric coordinates given by

$$
\begin{aligned}
m_1 &= \frac{F_1}{\sin\alpha_{12}\sin\alpha_{13}\sin\alpha_{14}} \\
m_2 &= \frac{F_2}{\sin\alpha_{23}\sin\alpha_{24}\sin\alpha_{21}} \\
m_3 &= \frac{F_3}{\sin\alpha_{34}\sin\alpha_{31}\sin\alpha_{32}} \\
m_4 &= \frac{F_4}{\sin\alpha_{41}\sin\alpha_{42}\sin\alpha_{43}}
\end{aligned}
\qquad (6.54\text{i})
$$

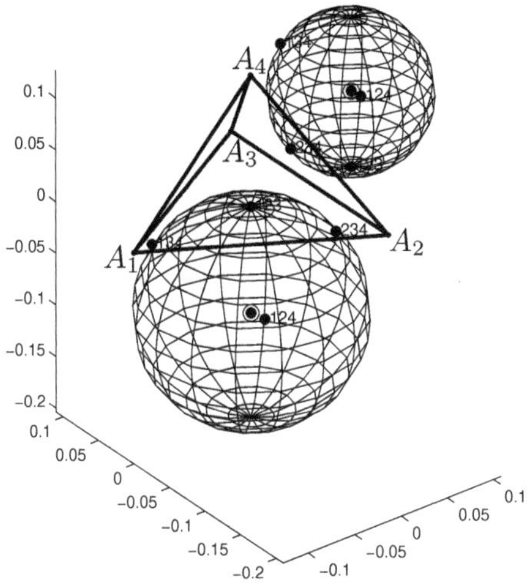

Fig. 6.8 Same gyrotetrahedron as in Fig. 6.7 along with two of its four near exgyrospheres that correspond to the near exgyrocenters E_1 and E_4 in (6.41), p. 290. The tangency points T_{jk} where the gyrotetrahedron near exgyrospheres gyrocentered at E_j, $j = 1, 4$, meet the gyroplane of the gyrotetrahedron face opposite to vertex A_k, $k = 1, 2, 3, 4$, determined by Exercise 5, p. 307, are shown.

Proof. By the fourth equation in (6.43i), p. 291, and by (6.52) we have the equation

$$m_4 = S_{A_1A_2A_3} = 4s^2 \frac{F_4}{\sin\alpha_{41}\sin\alpha_{42}\sin\alpha_{43}} \tag{6.55}$$

Rewriting (6.55) with cyclic permutations of the indices $(1, 2, 3, 4)$ and omitting the common factor $4s^2$ we find that the gyrobarycentric coordinates $(m_1 : m_2 : m_3 : m_4)$ in (6.54i) are equivalent to the gyrobarycentric coordinates $(m_1 : m_2 : m_3 : m_4)$ in (6.43i) of the gyrotetrahedron in-exgyrocenters E_k, $k = 0 \ldots 7$, in (6.43), p. 290, and in (6.54). □

The advantage of having in Theorem 6.6 gyrobarycentric coordinates expressed gyrotrigonometrically is that their transformation from Einstein to corresponding Möbius gyrovector spaces is trivial, giving rise to the following theorem:

Theorem 6.7. (Gyrotetrahedron In-Exgyrocenters, Gyrotrigonometric Form, Möbius). *The in-exgyrocenters E_k, $k = 0, \ldots, 7$ of a gyrotetrahedron $A_1A_2A_3A_4$ in a Möbius gyrovector space $(\mathbb{R}^n_s, \oplus, \otimes)$, $n \geq 3$,*

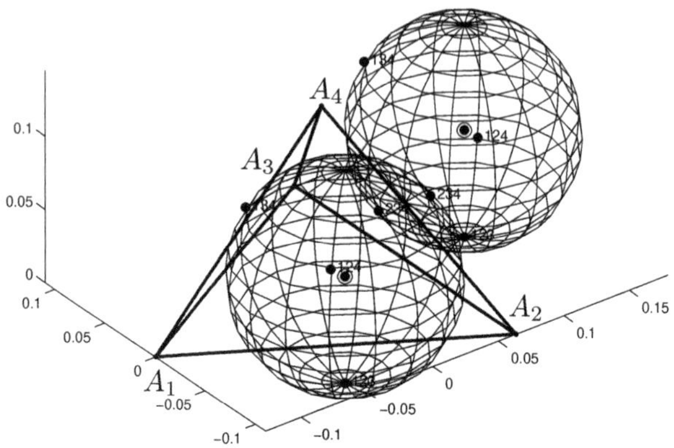

Fig. 6.9 The gyrotetrahedron in Figs. 6.7 – 6.8 is shown here along with two of its four near exgyrospheres that correspond to the gyrotetrahedron exgyrocenters E_1 and E_3 in (6.41), p. 290. The tangency points T_{jk} where the gyrotetrahedron near exgyrospheres gyrocentered at E_j, $j = 1, 3$, meet the gyroplane of the gyrotetrahedron face opposite to vertex A_k, $k = 1, 2, 3, 4$, determined by Exercise 5, p. 307, are shown.

Fig. 6.11, p. 302, possess, when exist, the gyrobarycentric coordinate representations with respect to the set $S = \{A_1, A_2, A_3, A_4\}$ listed below:

$$E_0 = \frac{m_1\gamma_{A_1}^2 A_1 + m_2\gamma_{A_2}^2 A_2 + m_3\gamma_{A_3}^2 A_3 + m_4\gamma_{A_4}^2 A_4}{m_1(\gamma_{A_1}^2 - \frac{1}{2}) + m_2(\gamma_{A_2}^2 - \frac{1}{2}) + m_3(\gamma_{A_3}^2 - \frac{1}{2}) + m_4(\gamma_{A_4}^2 - \frac{1}{2})} \tag{6.56a}$$

$$E_1 = \frac{-m_1\gamma_{A_1}^2 A_1 + m_2\gamma_{A_2}^2 A_2 + m_3\gamma_{A_3}^2 A_3 + m_4\gamma_{A_4}^2 A_4}{-m_1(\gamma_{A_1}^2 - \frac{1}{2}) + m_2(\gamma_{A_2}^2 - \frac{1}{2}) + m_3(\gamma_{A_3}^2 - \frac{1}{2}) + m_4(\gamma_{A_4}^2 - \frac{1}{2})} \tag{6.56b}$$

$$E_2 = \frac{m_1\gamma_{A_1}^2 A_1 - m_2\gamma_{A_2}^2 A_2 + m_3\gamma_{A_3}^2 A_3 + m_4\gamma_{A_4}^2 A_4}{m_1(\gamma_{A_1}^2 - \frac{1}{2}) - m_2(\gamma_{A_2}^2 - \frac{1}{2}) + m_3(\gamma_{A_3}^2 - \frac{1}{2}) + m_4(\gamma_{A_4}^2 - \frac{1}{2})} \tag{6.56c}$$

$$E_3 = \frac{m_1\gamma_{A_1}^2 A_1 + m_2\gamma_{A_2}^2 A_2 - m_3\gamma_{A_3}^2 A_3 + m_4\gamma_{A_4}^2 A_4}{m_1(\gamma_{A_1}^2 - \frac{1}{2}) + m_2(\gamma_{A_2}^2 - \frac{1}{2}) - m_3(\gamma_{A_3}^2 - \frac{1}{2}) + m_4(\gamma_{A_4}^2 - \frac{1}{2})} \tag{6.56d}$$

$$E_4 = \frac{m_1\gamma_{A_1}^2 A_1 + m_2\gamma_{A_2}^2 A_2 + m_3\gamma_{A_3}^2 A_3 - m_4\gamma_{A_4}^2 A_4}{m_1(\gamma_{A_1}^2 - \frac{1}{2}) + m_2(\gamma_{A_2}^2 - \frac{1}{2}) + m_3(\gamma_{A_3}^2 - \frac{1}{2}) - m_4(\gamma_{A_4}^2 - \frac{1}{2})} \tag{6.56e}$$

$$E_5 = \frac{m_1\gamma_{A_1}^2 A_1 + m_2\gamma_{A_2}^2 A_2 - m_3\gamma_{A_3}^2 A_3 - m_4\gamma_{A_4}^2 A_4}{m_1(\gamma_{A_1}^2 - \frac{1}{2}) + m_2(\gamma_{A_2}^2 - \frac{1}{2}) - m_3(\gamma_{A_3}^2 - \frac{1}{2}) - m_4(\gamma_{A_4}^2 - \frac{1}{2})} \tag{6.56f}$$

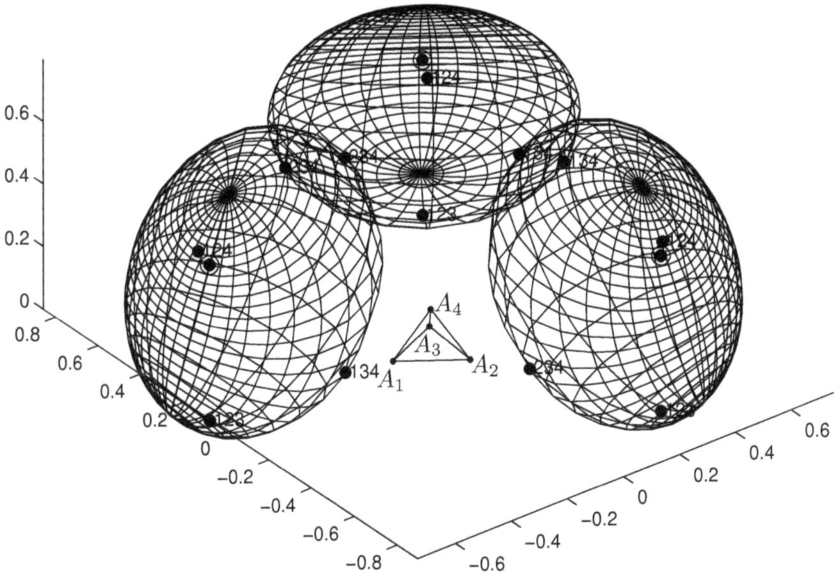

Fig. 6.10 The gyrotetrahedron $A_1A_2A_3A_4$ in Figs. 6.7 – 6.9 in an Einstein gyrovector space $(\mathbb{R}^3_{s=1}, \oplus, \otimes)$ is shown here along with its three far exgyrospheres, corresponding to the gyrotetrahedron three far exgyrocenters E_5, E_6 and E_7 in (6.41), p. 290. These are far exgyrocenters in the sense that they do not meet any face of the gyrotetrahedron. Rather, each of the far exgyrocenters meets all the gyroplanar extensions of the gyrotetrahedron faces. Contrasting the far exgyrocenters, each of the gyrotetrahedron exgyrocenters in Figs. 6.7 – 6.9, which is classified as "near", meets one face of the gyrotetrahedron and the gyroplanar extension of the remaining three faces. Each of the three far exgyrospheres is associated with two mutually opposite edges of the gyrotetrahedron. Accordingly, the six edges of a gyrotetrahedron admit three far exgyrospheres, or less in case some of these do not exist. It is clear in this figure that a gyrosphere in an Einstein gyrovector space is a flattened (Euclidean) sphere. The tangency points T_{jk} where the exgyrosphere gyrocentered at E_j, $j = 5, 6, 7$, meets the gyroplane of the gyrotetrahedron face opposite to vertex A_k, $k = 1, 2, 3, 4$, determined by Exercise 5, p. 307, are shown.

$$E_6 = \frac{m_1\gamma^2_{A_1}A_1 - m_2\gamma^2_{A_2}A_2 + m_3\gamma^2_{A_3}A_3 - m_4\gamma^2_{A_4}A_4}{m_1(\gamma^2_{A_1} - \frac{1}{2}) - m_2(\gamma^2_{A_2} - \frac{1}{2}) + m_3(\gamma^2_{A_3} - \frac{1}{2}) - m_4(\gamma^2_{A_4} - \frac{1}{2})} \tag{6.56g}$$

$$E_7 = \frac{m_1\gamma^2_{A_1}A_1 - m_2\gamma^2_{A_2}A_2 - m_3\gamma^2_{A_3}A_3 + m_4\gamma^2_{A_4}A_4}{m_1(\gamma^2_{A_1} - \frac{1}{2}) - m_2(\gamma^2_{A_2} - \frac{1}{2}) - m_3(\gamma^2_{A_3} - \frac{1}{2}) + m_4(\gamma^2_{A_4} - \frac{1}{2})} \tag{6.56h}$$

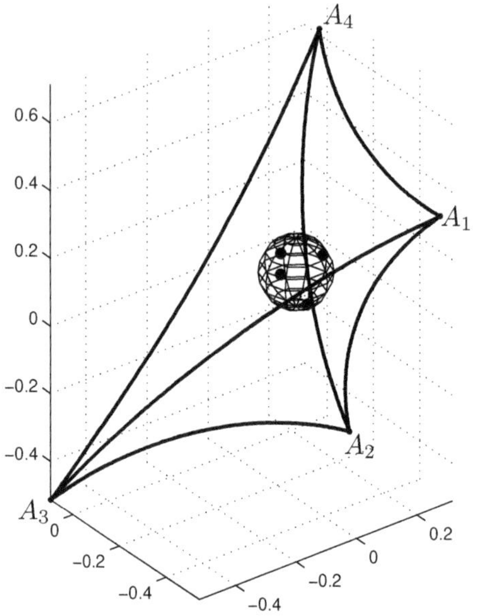

Fig. 6.11 A gyrotetrahedron in a Möbius gyrovector space $(\mathbb{R}_s^3, \oplus, \otimes)$ and its ingyrosphere. The tangency points T_{0k} where the gyrotetrahedron ingyrosphere meets the gyrotetrahedron face opposite to vertex A_k, $k = 1, 2, 3, 4$ are shown.

These are gyrobarycentric coordinate representations of E_k, $k = 0, 1, \dots, 7$, *in a Möbius gyrovector space, with respect to the set* $S = \{A_1, A_2, A_3, A_4\}$, *with gyrotrigonometric gyrobarycentric coordinates given by*

$$\begin{aligned} m_1 &= \frac{F_1}{\sin\alpha_{12}\sin\alpha_{13}\sin\alpha_{14}} \\ m_2 &= \frac{F_2}{\sin\alpha_{23}\sin\alpha_{24}\sin\alpha_{21}} \\ m_3 &= \frac{F_3}{\sin\alpha_{34}\sin\alpha_{31}\sin\alpha_{32}} \\ m_4 &= \frac{F_4}{\sin\alpha_{41}\sin\alpha_{42}\sin\alpha_{43}} \end{aligned} \tag{6.56i}$$

Proof. The proof of this theorem is similar to that of Theorem 5.3, p. 262, and Theorem 5.4, p. 263.

The gyrotetrahedron in-exgyrocenters E_k, $k = 0, \dots, 7$, in an Einstein gyrovector space $\mathbb{R}_s^n$ are given in Theorem 6.6 in terms of the gyrotrigonometric gyrobarycentric coordinate representations (6.54) with respect to the gyrotetrahedron vertices. Hence, by Theorem 4.6, p. 177, the gyrotetrahedron in-exgyrocenters E_k of the corresponding gyrotetrahedron $A_1A_2A_3A_4$ in the isomorphic Möbius gyrovector space $\mathbb{R}_s^n$ are the ones given in (6.56) in terms of the Möbius gyrotrigonometric gyrobarycentric coordinate representation (4.19), p. 177, that involve isomorphic gyrobarycentric coordinates with respect to the gyrotetrahedron vertices.

Finally, Einstein gyrobarycentric coordinates m_j, $j = 1, 2, 3, 4$, in (6.54) are gyrotrigonometric functions and hence, by Theorem 2.48, p. 143, their isomorphic images, also denoted m_j, in the isomorphic Möbius gyrovector space survive unchanged in (6.56). □

It should be remarked that

(1) the gyrobarycentric coordinates $(m_1 : m_2 : m_3 : m_4)$ of an Einstein gyrotetrahedron in-exgyrocenters in (6.54i) and
(2) the gyrobarycentric coordinates $(m_1 : m_2 : m_3 : m_4)$ of a Möbius gyrotetrahedron in-exgyrocenters in (6.56i)

are identical in form. However, they are different in context. Indeed,

(1) the gyrotrigonometric functions in (6.54i) are evaluated in terms of Einstein addition, as in Fig. 2.3, p. 101, while
(2) the gyrotrigonometric functions in (6.56i) are evaluated in terms of Möbius addition, as in Fig. 2.15, p. 140.

Owing to this identity in form but difference in context, gyrotrigonometric gyrobarycentric coordinate representations offer an attractive way to study Einstein gyrovector spaces and Möbius gyrovector spaces comparatively. An Einstein gyrotetrahedron ingyrosphere is shown in Fig. 6.6, p. 297, and a corresponding Möbius gyrotetrahedron ingyrosphere is shown in Fig. 6.11, p. 302.

6.6 Gyrotetrahedron Circumgyrocenter

Definition 6.8. The circumgyrocenter, O, of a gyrotetrahedron is the point in the interior of the gyrotetrahedron equigyrodistant from the four gyrotetrahedron vertices.

Let $A_1A_2A_3A_4$ be a gyrotetrahedron in an Einstein gyrovector space $(\mathbb{R}_s^n, \oplus, \otimes)$, $n \geq 3$, and let O be its circumgyrocenter. Furthermore, let

$$O = \frac{m_1\gamma_{A_1} A_1 + m_2\gamma_{A_2} A_2 + m_3\gamma_{A_3} A_3 + m_4\gamma_{A_4} A_4}{m_1\gamma_{A_1} + m_2\gamma_{A_2} + m_3\gamma_{A_3} + m_4\gamma_{A_4}} \tag{6.57}$$

be the gyrobarycentric coordinate representation of O with respect to the set $S = \{A_1, A_2, A_3, A_4\}$ of the gyrotetrahedron vertices, where the gyrobarycentric coordinates $(m_1 : m_2 : m_3 : m_4)$ of O in (6.57) are to be determined in (6.62), p. 305.

Following (6.28) and (6.29) with E replaced by O we have the gamma factors

$$\begin{aligned}
\gamma_{\ominus A_1 \oplus O} &= \frac{m_1 + m_2\gamma_{12} + m_3\gamma_{13} + m_4\gamma_{14}}{m_0} \\
\gamma_{\ominus A_2 \oplus O} &= \frac{m_1\gamma_{12} + m_2 + m_3\gamma_{23} + m_4\gamma_{24}}{m_0} \\
\gamma_{\ominus A_3 \oplus O} &= \frac{m_1\gamma_{13} + m_2\gamma_{23} + m_3 + m_4\gamma_{34}}{m_0} \\
\gamma_{\ominus A_4 \oplus O} &= \frac{m_1\gamma_{14} + m_2\gamma_{24} + m_3\gamma_{34} + m_4}{m_0}
\end{aligned} \tag{6.58}$$

where $m_0 > 0$ is given by, (6.29e),

$$\begin{aligned}
m_0^2 = m_1^2 + m_2^2 + m_3^2 + m_4^2 &+ 2m_1m_2\gamma_{12} + 2m_1m_3\gamma_{13} + 2m_1m_4\gamma_{14} \\
&+ 2m_2m_3\gamma_{23} + 2m_2m_4\gamma_{24} + 2m_3m_4\gamma_{34}
\end{aligned} \tag{6.59}$$

The condition that the circumgyrocenter O is equigyrodistant from its gyrotetrahedron vertices A_1, A_2, A_3, and A_4 implies

$$\gamma_{\ominus A_1 \oplus O} = \gamma_{\ominus A_2 \oplus O} = \gamma_{\ominus A_3 \oplus O} = \gamma_{\ominus A_4 \oplus O} \tag{6.60}$$

Equations (6.58) and (6.60), along with the normalization condition $m_1 + m_2 + m_3 + m_4 = 1$, yield the following system of four equations for the four unknowns m_1, m_2, m_3, and m_4,

$$\begin{aligned}
m_1 + m_2 + m_3 + m_4 &= 1 \\
m_1 + m_2\gamma_{12} + m_3\gamma_{13} + m_4\gamma_{14} &= m_1\gamma_{12} + m_2 + m_3\gamma_{23} + m_4\gamma_{24} \\
m_1 + m_2\gamma_{12} + m_3\gamma_{13} + m_4\gamma_{14} &= m_1\gamma_{13} + m_2\gamma_{23} + m_3 + m_4\gamma_{34} \\
m_1 + m_2\gamma_{12} + m_3\gamma_{13} + m_4\gamma_{14} &= m_1\gamma_{14} + m_2\gamma_{24} + m_3\gamma_{34} + m_4
\end{aligned} \tag{6.61}$$

Solving the linear system (6.61) for the unknowns m_k, $k = 1, 2, 3, 4$, we obtain the unique solution $(m_1 : m_2 : m_3 : m_4)$ listed below.

$$\begin{aligned} m_1 = \frac{1}{D}\{1 &+ 2\gamma_{23}\gamma_{24}\gamma_{34} - \gamma_{23}^2 - \gamma_{24}^2 - \gamma_{34}^2 \\ &- \gamma_{12}(\ \ \gamma_{23} + \gamma_{24} - \gamma_{34} - 1)(\gamma_{34} - 1) \\ &- \gamma_{13}(\ \ \gamma_{23} - \gamma_{24} + \gamma_{34} - 1)(\gamma_{24} - 1) \\ &- \gamma_{14}(-\gamma_{23} + \gamma_{24} + \gamma_{34} - 1)(\gamma_{23} - 1)\} \end{aligned} \tag{6.62a}$$

$$\begin{aligned} m_2 = \frac{1}{D}\{1 &+ 2\gamma_{13}\gamma_{14}\gamma_{34} - \gamma_{13}^2 - \gamma_{14}^2 - \gamma_{34}^2 \\ &- \gamma_{12}(\ \ \gamma_{13} + \gamma_{14} - \gamma_{34} - 1)(\gamma_{34} - 1) \\ &- \gamma_{23}(\ \ \gamma_{13} - \gamma_{14} + \gamma_{34} - 1)(\gamma_{14} - 1) \\ &- \gamma_{24}(-\gamma_{13} + \gamma_{14} + \gamma_{34} - 1)(\gamma_{13} - 1)\} \end{aligned} \tag{6.62b}$$

$$\begin{aligned} m_3 = \frac{1}{D}\{1 &+ 2\gamma_{12}\gamma_{14}\gamma_{24} - \gamma_{12}^2 - \gamma_{14}^2 - \gamma_{24}^2 \\ &- \gamma_{13}(\ \ \gamma_{12} + \gamma_{14} - \gamma_{24} - 1)(\gamma_{24} - 1) \\ &- \gamma_{23}(\ \ \gamma_{12} - \gamma_{14} + \gamma_{24} - 1)(\gamma_{14} - 1) \\ &- \gamma_{34}(-\gamma_{12} + \gamma_{14} + \gamma_{24} - 1)(\gamma_{12} - 1)\} \end{aligned} \tag{6.62c}$$

$$\begin{aligned} m_4 = \frac{1}{D}\{1 &+ 2\gamma_{12}\gamma_{13}\gamma_{23} - \gamma_{12}^2 - \gamma_{13}^2 - \gamma_{23}^2 \\ &- \gamma_{14}(\ \ \gamma_{12} + \gamma_{13} - \gamma_{23} - 1)(\gamma_{23} - 1) \\ &- \gamma_{24}(\ \ \gamma_{12} - \gamma_{13} + \gamma_{23} - 1)(\gamma_{13} - 1) \\ &- \gamma_{34}(-\gamma_{12} + \gamma_{13} + \gamma_{23} - 1)(\gamma_{12} - 1)\} \end{aligned} \tag{6.62d}$$

where $D \neq 0$ is determined by the first equation of the system (6.61) and by (6.62).

The gyrobarycentric coordinates $(m_1 : m_2 : m_3 : m_4)$ in (6.62) of the gyrotetrahedron circumgyrocenter O lead to the following theorem:

Theorem 6.9. (Gyrotetrahedron Circumgyrocenter, Einstein). *Let $A_1A_2A_3A_4$ be a gyrotetrahedron in an Einstein gyrovector space $\mathbb{R}_s^n$, $n \geq 3$, and let O be its circumgyrocenter, Fig. 6.12. Then, O possesses the gyrobarycentric coordinate representation*

$$O = \frac{m_1\gamma_{A_1}A_1 + m_2\gamma_{A_2}A_2 + m_3\gamma_{A_3}A_3 + m_4\gamma_{A_4}A_4}{m_1\gamma_{A_1} + m_2\gamma_{A_2} + m_3\gamma_{A_3} + m_4\gamma_{A_4}} \tag{6.63a}$$

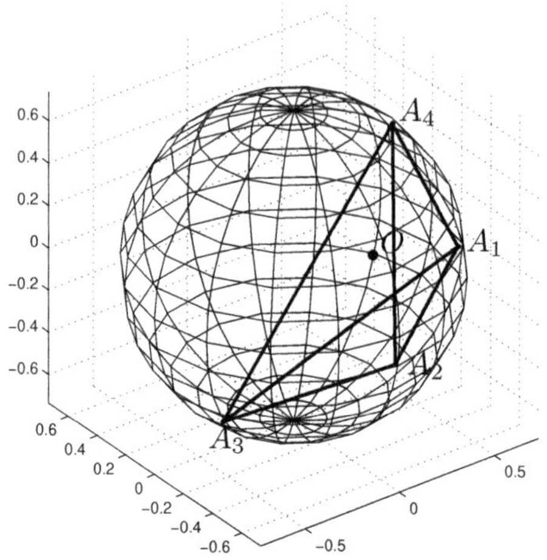

Fig. 6.12 A gyrotetrahedron circumgyrosphere in an Einstein gyrovector space $(\mathbb{R}^3_s, \oplus, \otimes)$. The gyrotetrahedron is left gyrotranslated to a position where its circumgyrocenter, that is, the center of its circumgyrosphere, is placed at the origin $\mathbf{0} = (0,0,0)$. Since the origin of an Einstein gyrovector space is conformal, a gyrosphere gyrocentered at the origin looks like a Euclidean sphere centered at the origin with radius equal to the gyrosphere gyroradius.

with respect to the set $S = \{A_1, A_2, A_3, A_4\}$, *with the gyrobarycentric coordinates* m_k, $k = 1, 2, 3, 4$, *listed below.*

$$\begin{aligned} m_1 = 1 &+ 2\gamma_{23}\gamma_{24}\gamma_{34} - \gamma_{23}^2 - \gamma_{24}^2 - \gamma_{34}^2 \\ &- \gamma_{12}(\ \ \gamma_{23} + \gamma_{24} - \gamma_{34} - 1)(\gamma_{34} - 1) \\ &- \gamma_{13}(\ \ \gamma_{23} - \gamma_{24} + \gamma_{34} - 1)(\gamma_{24} - 1) \\ &- \gamma_{14}(-\gamma_{23} + \gamma_{24} + \gamma_{34} - 1)(\gamma_{23} - 1) \end{aligned} \tag{6.63b}$$

$$\begin{aligned} m_2 = 1 &+ 2\gamma_{13}\gamma_{14}\gamma_{34} - \gamma_{13}^2 - \gamma_{14}^2 - \gamma_{34}^2 \\ &- \gamma_{12}(\ \ \gamma_{13} + \gamma_{14} - \gamma_{34} - 1)(\gamma_{34} - 1) \\ &- \gamma_{23}(\ \ \gamma_{13} - \gamma_{14} + \gamma_{34} - 1)(\gamma_{14} - 1) \\ &- \gamma_{24}(-\gamma_{13} + \gamma_{14} + \gamma_{34} - 1)(\gamma_{13} - 1) \end{aligned} \tag{6.63c}$$

$$\begin{aligned} m_3 = 1 &+ 2\gamma_{12}\gamma_{14}\gamma_{24} - \gamma_{12}^2 - \gamma_{14}^2 - \gamma_{24}^2 \\ &- \gamma_{13}(\ \ \gamma_{12} + \gamma_{14} - \gamma_{24} - 1)(\gamma_{24} - 1) \\ &- \gamma_{23}(\ \ \gamma_{12} - \gamma_{14} + \gamma_{24} - 1)(\gamma_{14} - 1) \\ &- \gamma_{34}(-\gamma_{12} + \gamma_{14} + \gamma_{24} - 1)(\gamma_{12} - 1) \end{aligned} \tag{6.63d}$$

$$\begin{aligned} m_4 = 1 &+ 2\gamma_{12}\gamma_{13}\gamma_{23} - \gamma_{12}^2 - \gamma_{13}^2 - \gamma_{23}^2 \\ &- \gamma_{14}(\ \ \gamma_{12} + \gamma_{13} - \gamma_{23} - 1)(\gamma_{23} - 1) \\ &- \gamma_{24}(\ \ \gamma_{12} - \gamma_{13} + \gamma_{23} - 1)(\gamma_{13} - 1) \\ &- \gamma_{34}(-\gamma_{12} + \gamma_{13} + \gamma_{23} - 1)(\gamma_{12} - 1) \end{aligned} \tag{6.63e}$$

Proof. The gyrobarycentric coordinates in (6.63) are obtained from the already established gyrobarycentric coordinates in (6.62) by omitting the irrelevant nonzero common factor $1/D$. □

A gyrotetrahedron with four equilateral gyrotriangular faces is *regular*. The gyrobarycentric coordinates m_k, $k = 1, 2, 3, 4$, in (6.63) of the circumgyrocenter of a regular gyrotetrahedron $A_1A_2A_3A_4$ in an Einstein gyrovector space are all equal. Hence, the gyrobarycentric coordinate representation (6.63a) of a regular gyrotetrahedron circumgyrocenter O descends to that of the gyrotetrahedron gyrocentroid G in (4.66), p. 188.

6.7 Exercises

(1) Derive the first equation in (6.33), p. 286, explicitly.
Hint: Substitute (6.29) into (6.31) and, then, substitute (6.31) – (6.32) into (6.30), p. 285, in order to express explicitly $\gamma_{h_4}^2$ in (6.30), p. 285, in terms of the gamma factors γ_{ij}, $1 \le i < j \le 4$, and the unknown gyrobarycentric coordinates m_k, $k = 1, 2, 3, 4$, of E in (6.28). The use of computer algebra software, like Mathematica, is recommended.

(2) Derive the last three equations in (6.33), p. 286, explicitly, from the first by index cyclic permutations.

(3) The gyrobarycentric coordinates of the gyrotetrahedron ingyrocenter E_0 are determined in (6.42), p. 290, by selecting the positive sign for each of the gyrobarycentric coordinates m_k, $k = 1, 2, 3, 4$, in (6.38), p. 287. Derive, in a similar way, gyrobarycentric coordinates for the gyrotetrahedron exgyrocenters.

(4) Transform Theorem 6.5, p. 294, from Einstein gyrovector spaces to corresponding Möbius gyrovector spaces by means of the isomorphisms between them studied in Sec. 2.29, p. 141.

(5) Let T_{ik} be the tangency points where the in-exgyrosphere with in-exgyrocenter E_i, $i = 0, \ldots, 7$, of a gyrotetrahedron $A_1A_2A_3A_4$ in an Einstein gyrovector space $(\mathbb{R}_s^n, \oplus, \otimes)$, $n \ge 3$, meets the gyrotetrahedron face opposite to vertex A_k, $k = 1, 2, 3, 4$, or its gyroplanar extension. The tangency point T_{04} is determined in Theorem 6.5, p. 294, by its

gyrobarycentric coordinate representation with respect to the vertices of its reference gyrotetrahedron. In a similar way of determining T_{04}, determine the remaining tangency points T_{ik}.

(6) Solve the system (6.61), p. 304.

(7) Substitute the gyrobarycentric coordinates (6.63), p. 305, into (6.59) to obtain explicitly the constant m_0 of the gyrobarycentric coordinate representation of the gyrotetrahedron circumgyrocenter O in (6.63a), p. 305.

Chapter 7

Comparative Patterns

A comparative study of Euclidean and hyperbolic geometry in various dimensions reveals in this book interesting, beautiful comparative patterns as, for instance, the patterns observed in Table 1.1, p. 55, Table 4.1, p. 243 and Table 4.2, p. 244. Comparative patterns are discovered in this book by a comparative study of triangle and tetrahedron centers in Euclidean and hyperbolic geometry by means of barycentric and gyrobarycentric calculus, sometimes employing distance and gyrodistance functions, as well as trigonometry and gyrotrigonometry. The algebraic tools for the comparative study come from vector spaces and gyrovector spaces. Owing to the obvious advantage of comparative studies, it is hoped that analytic Euclidean geometry will be studied comparatively, along with analytic hyperbolic geometry. The comparative study of analytic Euclidean and hyperbolic geometry, in turn, enables classical and relativistic mechanics to be studied comparatively as well, as emphasized in [18].

This chapter encourages explorers to extend the unfinished symphony of barycentric and gyrobarycentric calculus in the comparative study of analytic Euclidean and hyperbolic geometry.

7.1 Gyromidpoints and Gyrocentroids

Let A_1, A_2, A_3 and A_4 be four points in either Euclidean space $\mathbb{R}^n$, or Einstein gyrovector space $\mathbb{R}^n_s$, or Möbius gyrovector space $\mathbb{R}^n_s$, $n \geq 3$, depending on the context. Studying gyromidpoints and gyrocentroids comparatively, we present in items (1)–(3) below results that exhibit a remarkable comparative pattern. The remarkable pattern is that midpoints and centroids in the standard Cartesian model of Euclidean geometry and in two Cartesian models of hyperbolic geometry share barycentric coordinates.

(1) The Euclidean midpoint $M^{eu}_{A_1A_2}$ of A_1 and A_2 is given by its barycentric coordinate representation, (1.45), p. 17,

$$M^{eu}_{A_1A_2} = \frac{A_1 + A_2}{2} \tag{7.1a}$$

with respect to the set $S = \{A_1, A_2\}$, with barycentric coordinates

$$(m_1 : m_2) = (1 : 1) \tag{7.1b}$$

Similarly, the Euclidean centroid $M^{eu}_{A_1A_2A_3}$ of triangle $A_1A_2A_3$ is given by its barycentric coordinate representation, (1.48), p. 18,

$$M^{eu}_{A_1A_2A_3} = \frac{A_1 + A_2 + A_3}{3} \tag{7.1c}$$

with respect to the set $S = \{A_1, A_2, A_3\}$, with barycentric coordinates

$$(m_1 : m_2 : m_3) = (1 : 1 : 1) \tag{7.1d}$$

and the Euclidean centroid $M^{eu}_{A_1A_2A_3A_4}$ of tetrahedron $A_1A_2A_3A_4$ is given by its barycentric coordinate representation

$$M^{eu}_{A_1A_2A_3A_4} = \frac{A_1 + A_2 + A_3 + A_4}{4} \tag{7.1e}$$

with respect to the set $S = \{A_1, A_2, A_3, A_4\}$, with barycentric coordinates

$$(m_1 : m_2 : m_3 : m_4) = (1 : 1 : 1 : 1) \tag{7.1f}$$

in accordance with the generic barycentric coordinate representation (1.22), p. 9, in Euclidean geometry.

(2) The Einstein gyromidpoint $M^{en}_{A_1A_2}$ of points A_1 and A_2 is given by its gyrobarycentric coordinate representation, (4.26), p. 179,

$$M^{en}_{A_1A_2} = \frac{\gamma_{A_1} A_1 + \gamma_{A_2} A_2}{\gamma_{A_1} + \gamma_{A_2}} \tag{7.2a}$$

with respect to the set $S = \{A_1, A_2\}$, with gyrobarycentric coordinates

$$(m_1 : m_2) = (1 : 1) \tag{7.2b}$$

Similarly, the Einstein gyrocentroid $M^{en}_{A_1A_2A_3}$ of gyrotriangle $A_1A_2A_3$ is given by its gyrobarycentric coordinate representation (4.64), p. 188,

$$M^{en}_{A_1A_2A_3} = \frac{\gamma_{A_1} A_1 + \gamma_{A_2} A_2 + \gamma_{A_3} A_3}{\gamma_{A_1} + \gamma_{A_2} + \gamma_{A_3}} \tag{7.2c}$$

with respect to the set $S = \{A_1, A_2, A_3\}$, with gyrobarycentric coordinates

$$(m_1 : m_2 : m_3) = (1 : 1 : 1) \tag{7.2d}$$

Furthermore, the Einstein gyrocentroid $M^{en}_{A_1A_2A_3A_4}$ of gyrotetrahedron $A_1A_2A_3A_4$ is given by its gyrobarycentric coordinate representation, (4.66), p. 188,

$$M^{en}_{A_1A_2A_3A_4} = \frac{\gamma_{A_1}A_1 + \gamma_{A_2}A_2 + \gamma_{A_3}A_3 + \gamma_{A_4}A_4}{\gamma_{A_1} + \gamma_{A_2} + \gamma_{A_3} + \gamma_{A_4}} \tag{7.2e}$$

with respect to the set $S = \{A_1, A_2, A_3, A_4\}$, with gyrobarycentric coordinates

$$(m_1 : m_2 : m_3 : m_4) = (1 : 1 : 1 : 1) \tag{7.2f}$$

The gyrobarycentric coordinate representations (7.2a), (7.2c) and (7.2e) are in accordance with the generic gyrobarycentric coordinate representation (4.2), p. 171, in Einstein gyrovector spaces.

(3) The Möbius gyromidpoint $M^{mb}_{A_1A_2}$ of A_1 and A_2 is given by its gyrobarycentric coordinate representation, (4.38), p. 181,

$$M^{mb}_{A_1A_2} = \tfrac{1}{2}\otimes\frac{\gamma^2_{A_1}A_1 + \gamma^2_{A_2}A_2}{(\gamma^2_{A_1} - \frac{1}{2}) + (\gamma^2_{A_2} - \frac{1}{2})} \tag{7.3a}$$

with respect to the set $S = \{A_1, A_2\}$, with gyrobarycentric coordinates

$$(m_1 : m_2) = (1 : 1) \tag{7.3b}$$

Similarly, the Möbius gyrocentroid $M^{mb}_{A_1A_2A_3}$ of gyrotriangle $A_1A_2A_3$ is given by its gyrobarycentric coordinate representation, (4.70), p. 191,

$$M^{mb}_{A_1A_2A_3} = \tfrac{1}{2}\otimes\frac{\gamma^2_{A_1}A_1 + \gamma^2_{A_2}A_2 + \gamma^2_{A_3}A_3}{(\gamma^2_{A_1} - \frac{1}{2}) + (\gamma^2_{A_2} - \frac{1}{2}) + (\gamma^2_{A_3} - \frac{1}{2})} \tag{7.3c}$$

with respect to the set $S = \{A_1, A_2, A_3\}$, with gyrobarycentric coordinates

$$(m_1 : m_2 : m_3) = (1 : 1 : 1) \tag{7.3d}$$

Furthermore, the Möbius gyrocentroid $M^{mb}_{A_1A_2A_3A_4}$ of gyrotetrahedron $A_1A_2A_3A_4$ is given by its gyrobarycentric coordinate representation, (4.72), p. 191,

$$M^{mb}_{A_1A_2A_3A_4} = \tfrac{1}{2}\otimes\frac{\gamma^2_{A_1}A_1 + \gamma^2_{A_2}A_2 + \gamma^2_{A_3}A_3 + \gamma^2_{A_4}A_4}{(\gamma^2_{A_1} - \frac{1}{2}) + (\gamma^2_{A_2} - \frac{1}{2}) + (\gamma^2_{A_3} - \frac{1}{2}) + (\gamma^2_{A_4} - \frac{1}{2})} \tag{7.3e}$$

with respect to the set $S = \{A_1, A_2, A_3, A_4\}$, with gyrobarycentric coordinates

$$(m_1 : m_2 : m_3 : m_4) = (1 : 1 : 1 : 1) \tag{7.3f}$$

The gyrobarycentric coordinate representations (7.3a), (7.3c) and (7.3e) are in accordance with the generic gyrobarycentric coordinate representation (4.22), p. 179, in Möbius gyrovector spaces.

7.2 Two and Three Dimensional Ingyrocenters

It is interesting to compare the ingyrocenters of gyrotriangles, studied in Sec. 5.1, p. 247, with the ingyrocenters of gyrotetrahedra, studied in Sec. 6.3, p. 284.

Let A_1 and A_2 be two points in an Einstein gyrovector space $(\mathbb{R}^n_s, \oplus, \otimes)$, $n \geq 2$, with which we use the index notation

$$\begin{aligned} \mathbf{a}_{12} &= \ominus A_1 \oplus A_2 \\ a_{12} &= \|\mathbf{a}_{12}\| \\ \gamma_{12} &= \gamma_{\mathbf{a}_{12}} = \gamma_{a_{12}} \end{aligned} \tag{7.4}$$

so that the gyrolength of the gyrosegment A_1A_2 is a_{12} or, equivalently, the magnitude of the gyrovector $\mathbf{a}_{12}$ is a_{12}.

Definition 7.1. (The Gyrosegment Constant). *Let $\mathbf{a}_{12} = \ominus A_1 \oplus A_2$ be the gyrovector determined by two points, A_1 and A_2 in an Einstein gyrovector space $(\mathbb{R}^n_s, \oplus, \otimes)$, $n \geq 2$. Then, the gyrosegment constant, $S_{A_1A_2}$, of the gyrosegment A_1A_2 is given by*

$$S_{A_1A_2} = \gamma_{12}a_{12} \tag{7.5}$$

Definition 7.1 allows the gyrobarycentric coordinates of the gyrotriangle ingyrocenter with respect to the gyrotriangle vertices to be written in an elegant form that suggests a comparison with those of the gyrotetrahedron ingyrocenter.

Let E_0 be the ingyrocenter of a gyrotriangle $A_1A_2A_3$ in an Einstein gyrovector space $(\mathbb{R}^n_s, \oplus, \otimes)$, $n \geq 2$, given by its gyrobarycentric coordinate representation, (5.1),

$$E_0 = \frac{m_1\gamma_{A_1}A_1 + m_2\gamma_{A_2}A_2 + m_3\gamma_{A_3}A_3}{m_1\gamma_{A_1} + m_2\gamma_{A_2} + m_3\gamma_{A_3}} \tag{7.6}$$

with respect to the gyrotriangle vertices.

The gyrobarycentric coordinates m_k, $k = 1, 2, 3$, of the ingyrocenter E_0 of a gyrotriangle $A_1A_2A_3$ in (7.6) and in (5.11), p. 251, are given by (5.10). Following Def. 7.1, these in (5.10), p. 250, can be written as

$$\begin{aligned} m_3^2 &= s^2(\gamma_{12}^2 - 1) = \gamma_{12}^2 a_{12}^2 = S_{A_1A_2}^2 \\ m_1^2 &= s^2(\gamma_{23}^2 - 1) = \gamma_{23}^2 a_{23}^2 = S_{A_2A_3}^2 \\ m_2^2 &= s^2(\gamma_{13}^2 - 1) = \gamma_{13}^2 a_{13}^2 = S_{A_3A_1}^2 \end{aligned} \tag{7.7}$$

noting that $\gamma_{31} = \gamma_{13}$ and $S_{A_3A_1} = S_{A_1A_3}$, etc.

Let A_1, A_2 and A_3 be three points in an Einstein gyrovector space $(\mathbb{R}_s^n, \oplus, \otimes)$, $n \geq 3$, with which we use the index notation as illustrated in (7.4), and let $S_{A_1A_2A_3}$ be the gyrotriangle constant of gyrotriangle $A_1A_2A_3$. Then, by the gyrotriangle constant definition in Def. 2.35, p. 117, and by (6.39), p. 288,

$$S_{A_1A_2A_3} = \gamma_{12}a_{12}\gamma_{h_3}h_3 = s^2\sqrt{1 + 2\gamma_{12}\gamma_{13}\gamma_{23} - \gamma_{12}^2 - \gamma_{13}^2 - \gamma_{23}^2)} \tag{7.8}$$

Now, for comparison with (7.6)–(7.7), let E_0 be the ingyrocenter of a gyrotetrahedron $A_1A_2A_3A_4$ in an Einstein gyrovector space $(\mathbb{R}_s^n, \oplus, \otimes)$, $n \geq 3$, given by its gyrobarycentric coordinate representation, (6.28),

$$E_0 = \frac{m_1\gamma_{A_1}A_1 + m_2\gamma_{A_2}A_2 + m_3\gamma_{A_3}A_3 + m_4\gamma_{A_4}A_4}{m_1\gamma_{A_1} + m_2\gamma_{A_2} + m_3\gamma_{A_3} + m_4\gamma_{A_4}} \tag{7.9}$$

with respect to the gyrotetrahedron vertices. Then, by (7.8), the gyrobarycentric coordinates m_k, $k = 1, 2, 3, 4$, of the gyrotetrahedron ingyrocenter E_0 in (6.38), p. 287, and in (7.9) can be written as (noting that, being homogeneous, a nonzero common factor of a gyrobarycentric coordinate set is irrelevant)

$$\begin{aligned}
m_4^2 &= s^4(1 + 2\gamma_{12}\gamma_{13}\gamma_{23} - \gamma_{12}^2 - \gamma_{13}^2 - \gamma_{23}^2) = S_{A_1A_2A_3}^2 \\
m_1^2 &= s^4(1 + 2\gamma_{23}\gamma_{24}\gamma_{34} - \gamma_{23}^2 - \gamma_{24}^2 - \gamma_{34}^2) = S_{A_2A_3A_4}^2 \\
m_2^2 &= s^4(1 + 2\gamma_{13}\gamma_{14}\gamma_{34} - \gamma_{13}^2 - \gamma_{14}^2 - \gamma_{34}^2) = S_{A_3A_4A_1}^2 \\
m_3^2 &= s^4(1 + 2\gamma_{12}\gamma_{14}\gamma_{24} - \gamma_{12}^2 - \gamma_{14}^2 - \gamma_{24}^2) = S_{A_4A_1A_2}^2
\end{aligned} \tag{7.10}$$

A comparison between (7.7) and (7.10) reveals that the gyrotriangle constant $S_{A_1A_2A_3}$ is the natural extension of the gyrosegment constant $S_{A_1A_2}$ from a two-dimensional to a three-dimensional hyperbolic geometry.

Note that following the first equation in (7.7), $S_{A_1A_2}^2$ possesses the 2×2 *gamma matrix* representation

$$S_{A_1A_2}^2 = s^2(\gamma_{12}^2 - 1) = -s^2 \begin{vmatrix} 1 & \gamma_{12} \\ \gamma_{12} & 1 \end{vmatrix} \tag{7.11}$$

so that it shares a matrix pattern with (2.195), p. 117 and (6.22), p. 280.

7.3 Two and Three Dimensional Circumgyrocenters

It is interesting to compare the circumgyrocenters of gyrotriangles, studied in Sec. 4.19, p. 233, with the circumgyrocenters of gyrotetrahedra, studied in Sec. 6.6, p. 303.

Let O be the circumgyrocenter of a gyrotriangle $A_1A_2A_3$ in an Einstein gyrovector space $(\mathbb{R}^n_s, \oplus, \otimes)$, $n \geq 2$, given by its gyrobarycentric coordinate representation

$$O = \frac{m_1\gamma_{A_1}A_1 + m_2\gamma_{A_2}A_2 + m_3\gamma_{A_3}A_3}{m_1\gamma_{A_1} + m_2\gamma_{A_2} + m_3\gamma_{A_3}} \tag{7.12a}$$

with respect to the gyrotriangle vertices. Then, by Theorem 4.33, p. 237, barycentric coordinates m_k, $k = 1, 2, 3$, in (7.12a) are given by

$$\begin{aligned} m_1 &= (\ \ \gamma_{12} + \gamma_{13} - \gamma_{23} - 1)(\gamma_{23} - 1) \\ m_2 &= (\ \ \gamma_{12} - \gamma_{13} + \gamma_{23} - 1)(\gamma_{13} - 1) \\ m_3 &= (-\gamma_{12} + \gamma_{13} + \gamma_{23} - 1)(\gamma_{12} - 1) \end{aligned} \tag{7.12b}$$

Now, for comparison, let O be the circumgyrocenter of a gyrotetrahedron $A_1A_2A_3A_4$ in an Einstein gyrovector space $(\mathbb{R}^n_s, \oplus, \otimes)$, $n \geq 3$, given by its gyrobarycentric coordinate representation

$$O = \frac{m_1\gamma_{A_1}A_1 + m_2\gamma_{A_2}A_2 + m_3\gamma_{A_3}A_3 + m_4\gamma_{A_4}A_4}{m_1\gamma_{A_1} + m_2\gamma_{A_2} + m_3\gamma_{A_3} + m_4\gamma_{A_4}} \tag{7.13a}$$

with respect to the gyrotetrahedron vertices. Then, by Theorem 6.9, p. 305, barycentric coordinates m_k, $k = 1, 2, 3, 4$, in (7.12a) are given by

$$\begin{aligned} m_1 = 1 &+ 2\gamma_{23}\gamma_{24}\gamma_{34} - \gamma_{23}^2 - \gamma_{24}^2 - \gamma_{34}^2 \\ &- \gamma_{12}(\ \ \gamma_{23} + \gamma_{24} - \gamma_{34} - 1)(\gamma_{34} - 1) \\ &- \gamma_{13}(\ \ \gamma_{23} - \gamma_{24} + \gamma_{34} - 1)(\gamma_{24} - 1) \\ &- \gamma_{14}(-\gamma_{23} + \gamma_{24} + \gamma_{34} - 1)(\gamma_{23} - 1) \end{aligned} \tag{7.13b}$$

$$\begin{aligned} m_2 = 1 &+ 2\gamma_{13}\gamma_{14}\gamma_{34} - \gamma_{13}^2 - \gamma_{14}^2 - \gamma_{34}^2 \\ &- \gamma_{12}(\ \ \gamma_{13} + \gamma_{14} - \gamma_{34} - 1)(\gamma_{34} - 1) \\ &- \gamma_{23}(\ \ \gamma_{13} - \gamma_{14} + \gamma_{34} - 1)(\gamma_{14} - 1) \\ &- \gamma_{24}(-\gamma_{13} + \gamma_{14} + \gamma_{34} - 1)(\gamma_{13} - 1) \end{aligned} \tag{7.13c}$$

$$\begin{aligned} m_3 = 1 &+ 2\gamma_{12}\gamma_{14}\gamma_{24} - \gamma_{12}^2 - \gamma_{14}^2 - \gamma_{24}^2 \\ &- \gamma_{13}(\ \ \gamma_{12} + \gamma_{14} - \gamma_{24} - 1)(\gamma_{24} - 1) \\ &- \gamma_{23}(\ \ \gamma_{12} - \gamma_{14} + \gamma_{24} - 1)(\gamma_{14} - 1) \\ &- \gamma_{34}(-\gamma_{12} + \gamma_{14} + \gamma_{24} - 1)(\gamma_{12} - 1) \end{aligned} \tag{7.13d}$$

$$\begin{aligned} m_4 = 1 &+ 2\gamma_{12}\gamma_{13}\gamma_{23} - \gamma_{12}^2 - \gamma_{13}^2 - \gamma_{23}^2 \\ &- \gamma_{14}(\ \ \gamma_{12} + \gamma_{13} - \gamma_{23} - 1)(\gamma_{23} - 1) \\ &- \gamma_{24}(\ \ \gamma_{12} - \gamma_{13} + \gamma_{23} - 1)(\gamma_{13} - 1) \\ &- \gamma_{34}(-\gamma_{12} + \gamma_{13} + \gamma_{23} - 1)(\gamma_{12} - 1) \end{aligned} \tag{7.13e}$$

A comparison between (7.12) and (7.13) is interesting for exploration.

7.4 Tetrahedron Incenter and Excenters

Theorem 6.4, p. 290, admits a straightforward reduction from gyrotetrahedra into corresponding tetrahedra.

Let $A_1A_2A_3$ be a gyrotriangle with a gyrotriangle constant $S_{A_1A_2A_3}$ in an Einstein gyrovector space $(\mathbb{R}^n_s, \oplus, \otimes)$, given by (6.44a), p. 291. In the Euclidean limit of large s, $s \to \infty$, the gyrotriangle constant reduces to twice the area of a corresponding triangle in a corresponding Euclidean n-space $\mathbb{R}^n_s$, as explained below Def. 2.36, p. 118.

Let $S^e_{A_1A_2A_3}$ denote the area of a triangle $A_1A_2A_3$ in a Euclidean n-space $\mathbb{R}^n_s$. Then, in the Euclidean limit, Theorem 6.4, p. 290, gives rise to the following Theorem:

Theorem 7.2. **(Tetrahedron In-Excenters).** *The in-excenters E_k, $k = 0, \ldots, 7$ of a tetrahedron $A_1A_2A_3A_4$ in a Euclidean space $\mathbb{R}^n_s$, $n \geq 3$, possess, when exist, the barycentric coordinate representations with respect to the set $S = \{A_1, A_2, A_3, A_4\}$ listed below:*

$$E_0 = \frac{m_1A_1 + m_2A_2 + m_3A_3 + m_4A_4}{m_1 + m_2 + m_3 + m_4} \tag{7.14a}$$

$$E_1 = \frac{-m_1A_1 + m_2A_2 + m_3A_3 + m_4A_4}{-m_1 + m_2 + m_3 + m_4} \tag{7.14b}$$

$$E_2 = \frac{m_1A_1 - m_2A_2 + m_3A_3 + m_4A_4}{m_1 - m_2 + m_3 + m_4} \tag{7.14c}$$

$$E_3 = \frac{m_1A_1 + m_2A_2 - m_3A_3 + m_4A_4}{m_1 + m_2 - m_3 + m_4} \tag{7.14d}$$

$$E_4 = \frac{m_1A_1 + m_2A_2 + m_3A_3 - m_4A_4}{m_1 + m_2 + m_3 - m_4} \tag{7.14e}$$

$$E_5 = \frac{m_1A_1 + m_2A_2 - m_3A_3 - m_4A_4}{m_1 + m_2 - m_3 - m_4} \tag{7.14f}$$

$$E_6 = \frac{m_1A_1 - m_2A_2 + m_3A_3 - m_4A_4}{m_1 - m_2 + m_3 - m_4} \tag{7.14g}$$

$$E_7 = \frac{m_1A_1 - m_2A_2 - m_3A_3 + m_4A_4}{m_1 - m_2 - m_3 + m_4} \tag{7.14h}$$

with barycentric coordinates m_k, $k = 1, 2, 3, 4$ given by

$$\begin{aligned} m_1 &= S^e_{A_2A_3A_4} \\ m_2 &= S^e_{A_1A_3A_4} \\ m_3 &= S^e_{A_1A_2A_4} \\ m_4 &= S^e_{A_1A_2A_3} \end{aligned} \tag{7.14i}$$

A comparative study of Theorems 6.4, p. 290, and 7.2 suggests the following definition and a comparative consequence:

Definition 7.3. (Regular Tetrahedra and gyrotetrahedra). *A tetrahedron $A_1A_2A_3A_4$ in a Euclidean space $\mathbb{R}^n$, $n \geq 3$, is regular if the triangle constants $S^e_{A_1A_2A_3}$, $S^e_{A_1A_2A_4}$, $S^e_{A_1A_3A_4}$ and $S^e_{A_2A_3A_4}$ of its faces are equal (that is, equivalently, if its face areas are equal).*

A gyrotetrahedron $A_1A_2A_3A_4$ in an Einstein gyrovector space $\mathbb{R}^n_s$, $n \geq 3$, is regular if the gyrotriangle constants $S_{A_1A_2A_3}$, $S_{A_1A_2A_4}$, $S_{A_1A_3A_4}$ and $S_{A_2A_3A_4}$ of its faces are equal.

It is clear from (7.14) that the far excenters E_5, E_6 and E_7 of a regular tetrahedron $A_1A_2A_3A_4$ in $\mathbb{R}^n$, $n \geq 3$, do not exist. In full analogy, it is clear from (6.43), p. 290, that the far exgyrocenters E_5, E_6 and E_7 of a regular gyrotetrahedron $A_1A_2A_3A_4$ in an Einstein gyrovector space $\mathbb{R}^n$, $n \geq 3$, do not exist.

7.5 Comparative Study of the Pythagorean Theorem

Möbius gyrovector spaces have visual comparative advantage over Einstein gyrovector spaces in the comparative study of the law of cosines and its resulting Möbius-Pythagoras Theorem. This visual comparative advantage is convincingly seen in (7.15) – (7.16) and in Fig. 7.1 below.

Theorem 7.4. (Möbius Law of Gyrocosines) *Let ABC be a gyrotriangle in a Möbius gyrovector space $(\mathbb{R}^n_{s=1}, \oplus, \otimes)$ with vertices, $A, B, C \in \mathbb{R}^n_{s=1}$, with sides, $\mathbf{a}, \mathbf{b}, \mathbf{c}$, and side gyrolengths, a, b, c, given by*

$$\begin{aligned} \mathbf{a} &= \ominus B \oplus C, & a &= \|\mathbf{a}\| \\ \mathbf{b} &= \ominus C \oplus A, & b &= \|\mathbf{b}\| \\ \mathbf{c} &= \ominus A \oplus B, & c &= \|\mathbf{c}\| \end{aligned} \tag{7.15}$$

and with gyroangles α, β and γ at the vertices A, B and C, Fig. 2.15, p. 140. Then

$$c^2 = a^2 \oplus b^2 \ominus \frac{2ab\cos\gamma}{(1+a^2)(1+b^2) - 2ab\cos\gamma} \tag{7.16}$$

Proof. The proof of the hyperbolic law of cosines follows straightforwardly from the definition of gyroangle measure in Fig. 2.14, p. 139 (see Exercise 1, p. 319). □

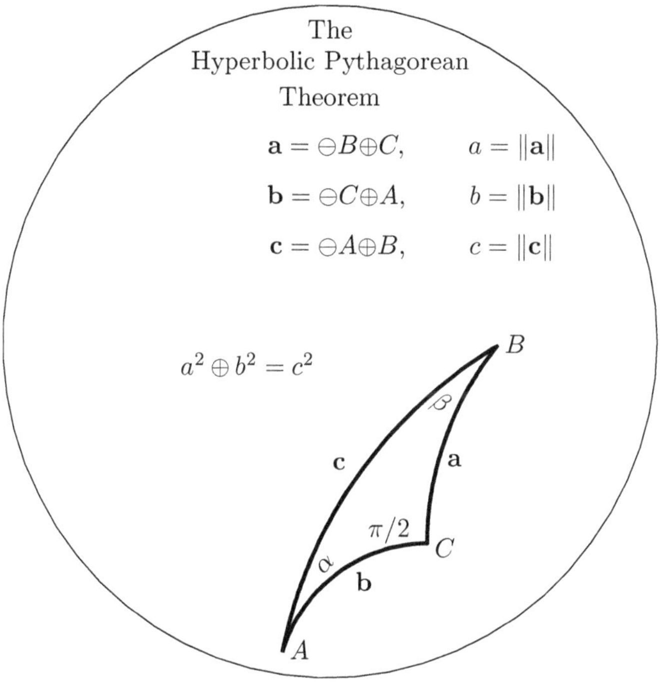

Fig. 7.1 The Hyperbolic Pythagorean Theorem in the Cartesian-Poincaré ball model of hyperbolic geometry and in its underlying Möbius gyrovector space $(\mathbb{R}^2_{s=1}, \oplus, \otimes)$.

We may note that Möbius addition $\oplus$ in (7.15) is a gyrogroup operation in a Möbius gyrovector space $(\mathbb{R}^n_{s=1}, \oplus, \otimes)$. In contrast, Möbius addition $\oplus$ in (7.16) is a commutative group operation in the one-dimensional Möbius gyrogroup $(\mathbb{R}_{s=1}, \oplus)$, $\mathbb{R}_{s=1}$ being the open unit interval $(-1, 1)$.

In the special case when $\gamma = \pi/2$, corresponding to a right gyroangled gyrotriangle, Fig. 7.1, the law of gyrocosines is of particular interest, giving rise to the hyperbolic Pythagorean theorem in the Poincaré ball model of hyperbolic geometry [48].

Theorem 7.5. (Möbius Hyperbolic Pythagorean Theorem) *Let ABC be a gyrotriangle in a Möbius gyrovector space $(\mathbb{R}^n_{s=1}, \oplus, \otimes)$ with the notation in Theorem 7.4. If $\gamma = \pi/2$, Fig. 7.1, then,*

$$a^2 \oplus b^2 = c^2 \tag{7.17}$$

Proof. The hyperbolic Pythagorean identity (7.17) follows from the law of gyrocosines (7.16) with $\gamma = \pi/2$. □

The Hyperbolic Pythagorean Theorem in the Cartesian-Poincaré ball model of hyperbolic geometry and in its underlying Möbius gyrovector space shares visual analogies with its Euclidean counterpart as seen from both Fig. 7.1 and the hyperbolic Pythagorean formula (7.17).

7.6 Hyperbolic Heron's Formula

Let us transform the gyrotriangular defect δ, (2.174), p. 112, from Einstein to Möbius gyrovector spaces by transforming gamma factors from Einstein to Möbius gyrovector spaces by means of the gamma transformation formula (4.145), p. 210. The resulting image of the transformation gives the gyrotriangular defect δ of a gyrotriangle ABC in Möbius gyrovector spaces by the equation

$$\tan\frac{\delta}{2} = \frac{\sqrt{4\gamma_a^2\gamma_b^2\gamma_c^2 - (\gamma_a^2+\gamma_b^2+\gamma_c^2-1)^2}}{\gamma_a^2+\gamma_b^2+\gamma_c^2-1} \tag{7.18}$$

Employing Identity (2.11), p. 66, the gyrotriangular defect formula (7.18) can be written as, [59],

$$\begin{aligned}\tan\frac{\delta}{2} = &\sqrt{a_s+b_s+c_s+a_sb_sc_s}\sqrt{-a_s+b_s+c_s-a_sb_sc_s}\ \times\\ &\sqrt{a_s-b_s+c_s-a_sb_sc_s}\sqrt{a_s+b_s-c_s-a_sb_sc_s}\ \times\\ &\frac{1}{2+a_s^2b_s^2c_s^2-a_s^2-b_s^2-c_s^2}\end{aligned} \tag{7.19}$$

The gyroarea $|ABC|$ of a gyrotriangle ABC in a Möbius gyrovector space is given in (2.197), p. 118, in terms of the gyrotriangle defect δ,

$$|ABC| = \tfrac{1}{2}s^2\tan\tfrac{\delta}{2} \tag{7.20}$$

Interestingly, (7.20) and (7.19) give rise to the hyperbolic formula of Heron in the following obvious theorem:

Theorem 7.6. (Möbius-Heron's Formula). *The gyroarea of gyrotriangle ABC in a Möbius gyrovector space is given by Möbius-Heron's formula*

$$\begin{aligned}|ABC| &= \tfrac{1}{2}s^2\tan\tfrac{\delta}{2}\\ &= \sqrt{a_s+b_s+c_s+a_sb_sc_s}\sqrt{-a_s+b_s+c_s-a_sb_sc_s}\\ &\times\sqrt{a_s-b_s+c_s-a_sb_sc_s}\sqrt{a_s+b_s-c_s-a_sb_sc_s}\\ &\times\tfrac{1}{2}s^2\frac{1}{2+a_s^2b_s^2c_s^2-a_s^2-b_s^2-c_s^2}\end{aligned} \tag{7.21}$$

where $a_s = a/s$, etc.

Remarkably, in the Euclidean limit, $s \to \infty$, of large s, when hyperbolic geometry in the ball $\mathbb{R}_s^n$ descends to Euclidean geometry in the space $\mathbb{R}^n$, Möbius-Heron's formula (7.21) of the gyrotriangle gyroarea in Möbius gyrovector spaces $\mathbb{R}_s^n$ descends to Heron's formula (1.76), p. 23, of the triangle area in Euclidean spaces $\mathbb{R}^n$.

7.7 Exercises

(1) Derive the Möbius law of gyrocosines (7.16), p. 316, from the definition of gyroangle measure in Fig. 2.14, p. 139 (see [2, Theorem 8.25, p. 258]).

(2) Transform the gyrotriangular defect δ, (2.174), p. 112, from Einstein to Möbius gyrovector spaces by transforming gamma factors from Einstein to Möbius gyrovector spaces by means of the gamma transformation formula (4.145), p. 210 and, hence, obtain the Möbius gyrotriangular defect formula (7.18), p. 318.

(3) Employ Identity (2.11), p. 66, to show that the Möbius gyrotriangular defect formula (7.18), p. 318, can be written as (7.19).

(4) Show that Möbius-Heron's formula (7.21), p. 318, tends to Heron's formula (1.76), p. 23, in the limit when $s \to \infty$. Note that in that limit $a/s = a_s \to 0$ and $a := a_{hyperbolic} \to a_{euclidean}$, *etc.*

Chapter 8

Gyrodistances and Distances

Results in hyperbolic geometry, associated with gyrobarycentric coordinates, along with their descent to well-known Euclidean counterparts, are presented in this chapter. In particular, attention is paid to the gyrodistance between two gyropoints which are given by their gyrobarycentric coordinates. The gyrodistance is calculated by applying the procedure described in the following Example, established in [19, Example 5.29, p. 167].

8.1 Gyrodistance Between Gyropoints

Example 8.1. (Gyropoint to gyropoint gyrodistance in a gyroplane). *Let $S = \{A_1, A_2, A_3\}$ be a gyrobarycentrically independent set of three gyropoints in an Einstein gyrovector space $(\mathbb{R}^n_s, \oplus, \otimes)$, $n \geq 2$, and let P and P' be two gyropoints in the gyroplane in $\mathbb{R}^n_s$ that passes through the gyropoints of S, given by their gyrobarycentric representations*

$$\begin{aligned} P &= \frac{m_1\gamma_{A_1}A_1 + m_2\gamma_{A_2}A_2 + m_3\gamma_{A_3}A_3}{m_1\gamma_{A_1} + m_2\gamma_{A_2} + m_3\gamma_{A_3}} \\ P' &= \frac{m'_1\gamma_{A_1}A_1 + m'_2\gamma_{A_2}A_2 + m'_3\gamma_{A_3}A_3}{m'_1\gamma_{A_1} + m'_2\gamma_{A_2} + m'_3\gamma_{A_3}} \end{aligned} \tag{8.1}$$

with respect to the set S. Then,

$$\begin{aligned} \gamma_{\ominus P' \oplus P} &= \frac{1}{m_P m_{P'}} \\ &\times \{(m_1 m'_2 + m'_1 m_2)\gamma_{12} + (m_1 m'_3 + m'_1 m_3)\gamma_{13} + (m_2 m'_3 + m'_2 m_3)\gamma_{23} \\ &\quad + m_1 m'_1 + m_2 m'_2 + m_3 m'_3\} \end{aligned} \tag{8.2}$$

where

$$m_P^2 = m_1^2 + m_2^2 + m_3^2 + 2(m_1m_2\gamma_{12} + m_1m_3\gamma_{13} + m_2m_3\gamma_{23})$$
$$m_{P'}^2 = (m_1')^2 + (m_2')^2 + (m_3')^2 + 2(m_1'm_2'\gamma_{12} + m_1'm_3'\gamma_{13} + m_2'm_3'\gamma_{23}) \tag{8.3}$$

By means of (8.2) *and* (2.11), *p. 66, the gyrodistance,* $\|\ominus P'\oplus P\|$, *between* P *and* P' *is given by the equation*

$$\|\ominus P'\oplus P\| = s\frac{\sqrt{\gamma_{\ominus P'\oplus P}^2 - 1}}{\gamma_{\ominus P'\oplus P}} \tag{8.4}$$

8.2 Ingyroradius – Circumgyroradius Ratio

Let $A_1A_2A_3$ be a gyrotriangle in an Einstein gyrovector space $(\mathbb{R}_s^n, \oplus, \otimes)$, $n \geq 2$, with gyroangles α_1, α_2 and α_3. Following (4.229) and the first equation in (4.228), p. 229, the ingyroradius r of gyrotriangle $A_1A_2A_3$ is given by

$$\begin{aligned} r^2 &= \frac{s^2}{4}\frac{\cos\frac{\alpha_1+\alpha_2+\alpha_3}{2}\cos\frac{-\alpha_1+\alpha_2+\alpha_3}{2}\cos\frac{\alpha_1-\alpha_2+\alpha_3}{2}\cos\frac{\alpha_1+\alpha_2-\alpha_3}{2}}{\cos^2\frac{\alpha_1}{2}\cos^2\frac{\alpha_2}{2}\cos^2\frac{\alpha_3}{2}} \\ &= \frac{s^2}{4}\frac{\sin\frac{\delta}{2}\sin(\alpha_1+\frac{\delta}{2})\sin(\alpha_2+\frac{\delta}{2})\sin(\alpha_3+\frac{\delta}{2})}{\cos^2\frac{\alpha_1}{2}\cos^2\frac{\alpha_2}{2}\cos^2\frac{\alpha_3}{2}} \end{aligned} \tag{8.5}$$

where

$$\delta = \pi - (\alpha_1 + \alpha_2 + \alpha_3) \tag{8.6}$$

is the defect of gyrotriangle $A_1A_2A_3$ in the notation of Fig. 4.17, p. 234.

By [19, Eq. (8.55), p. 296], the circumgyroradius R of gyrotriangle $AA_1A_2A_3$ is given by

$$\begin{aligned} R^2 &= s^2\frac{\cos\dfrac{\alpha_1+\alpha_2+\alpha_3}{2}}{\cos\dfrac{\alpha_1-\alpha_2-\alpha_3}{2}\cos\dfrac{-\alpha_1+\alpha_2-\alpha_3}{2}\cos\dfrac{-\alpha_1-\alpha_2+\alpha_3}{2}} \\ &= s^2\frac{\sin\frac{\delta}{2}}{\sin(\alpha_1+\frac{\delta}{2})\sin(\alpha_2+\frac{\delta}{2})\sin(\alpha_3+\frac{\delta}{2})} \end{aligned} \tag{8.7}$$

Eliminating s between (8.5) and (8.7) yields the most elegant equation

$$\frac{r}{R} = \frac{1}{2}\frac{\sin(\alpha_1+\frac{\delta}{2})\sin(\alpha_2+\frac{\delta}{2})\sin(\alpha_3+\frac{\delta}{2})}{\cos\frac{\alpha_1}{2}\cos\frac{\alpha_2}{2}\cos\frac{\alpha_3}{2}} \tag{8.8}$$

which descends to its Euclidean counterpart in the special case when $\delta = 0$.

Formalizing the result in (8.8), we obtain the following theorem.

Theorem 8.2. (Ingyroradius – Circumgyroradius Ratio). *Let $A_1A_2A_3$ be a gyrotriangle with gyroangles α_1, α_2 and α_3 in an Einstein gyrovector space $(\mathbb{R}^n_s, \oplus, \otimes)$, $n \geq 2$. The ratio r/R of the gyrotriangle ingyroradius r to the circumgyroradius R is given by the equation*

$$\frac{r}{R} = \frac{1}{2} \frac{\sin(\alpha_1 + \frac{\delta}{2}) \sin(\alpha_2 + \frac{\delta}{2}) \sin(\alpha_3 + \frac{\delta}{2})}{\cos \frac{\alpha_1}{2} \cos \frac{\alpha_2}{2} \cos \frac{\alpha_3}{2}} \tag{8.9}$$

In the Euclidean limit, $s \to \infty$, the gyrotriangle defect δ tends to zero and, hence, Theorem 8.2 descends to its Euclidean counterpart in the following corollary.

Corollary 8.3. (Inradius – Circumradius Ratio). *Let $A_1A_2A_3$ be a triangle with angles α_1, α_2 and α_3 in a Euclidean space $\mathbb{R}^n$, $n \geq 2$. The ratio r/R of the triangle inradius r to the circumradius R is given by the equation*

$$\frac{r}{R} = \frac{1}{2} \frac{\sin \alpha_1 \sin \alpha_2 \sin \alpha_3}{\cos \frac{\alpha_1}{2} \cos \frac{\alpha_2}{2} \cos \frac{\alpha_3}{2}} = 4 sin \tfrac{\alpha_1}{2} sin \tfrac{\alpha_2}{2} sin \tfrac{\alpha_3}{2} \tag{8.10}$$

Proof. In the Euclidean limit, $s \to \infty$, gyrotriangle defects δ tend to 0, so that (8.9) tends to (8.10). □

The result (8.10) of the Corollary is well-known in Euclidean geometry, found in [60, Eq. (5), p. 192]. its extension (8.9) to hyperbolic geometry is elegant and novel, demonstrating once again the advantage of the study of Euclidean and hyperbolic geometry comparatively.

8.3 Gyrotriangle Circumgyrocenter Gyrodistances from Gyrosides

As an application of Theorem 2.39, p. 122, it is determined in [18, Theorem 7.18, p. 191], shown in [18, Fig. 7.13, p. 189], that the gyrodistances h_1, h_2 and h_3 between a gyrotriangle circumgyrocenter, O, and its gyrosides A_2A_3, A_1A_3 and A_1A_2, respectively, shown in Fig. 4.16, p. 233, are given in the following Theorem.

Theorem 8.4. *Let $A_1A_2A_3$ be a gyrotriangle in an Einstein gyrovector space $(\mathbb{R}^n_s, \oplus, \otimes)$, $n \geq 2$, and let O be its circumgyrocenter (Fig. 4.17,* *p. 234). The gyrodistances h_k, $k = 1, 2, 3$ from O to the gyrotriangle gyroside opposite to gyrovertex A_k are related to the gyrotriangle circumgyroradius R and gyroangles $\alpha_1, \alpha_2, \alpha_3$ by the equations*

$$
\begin{aligned}
h_1^2 &= \frac{R^2 \cos\frac{-\alpha_1+\alpha_2-\alpha_3}{2} \cos\frac{-\alpha_1-\alpha_2+\alpha_3}{2} \sin^2\frac{-\alpha_1+\alpha_2+\alpha_3}{2}}{\sin\alpha_2 \sin\alpha_3} \\
h_2^2 &= \frac{R^2 \cos\frac{\alpha_1-\alpha_2-\alpha_3}{2} \cos\frac{-\alpha_1-\alpha_2+\alpha_3}{2} \sin^2\frac{\alpha_1-\alpha_2+\alpha_3}{2}}{\sin\alpha_1 \sin\alpha_3} \\
h_3^2 &= \frac{R^2 \cos\frac{\alpha_1-\alpha_2-\alpha_3}{2} \cos\frac{-\alpha_1+\alpha_2-\alpha_3}{2} \sin^2\frac{\alpha_1+\alpha_2-\alpha_3}{2}}{\sin\alpha_1 \sin\alpha_2}
\end{aligned}
\tag{8.11}
$$

As the study in Sect. 8.2 suggests, it is useful to involve the gyrotriangle defect δ in (8.11) in order to simplify Theorem 8.4 and its descent from hyperbolic geometry into Euclidean geometry. Accordingly, we rewrite Theorem 8.4 into which we insert δ from (8.6), obtaining the following elegant theorem.

Theorem 8.5. *Let $A_1A_2A_3$ be a gyrotriangle in an Einstein gyrovector space $(\mathbb{R}_s^n, \oplus, \otimes)$, $n \geq 2$, and let O be its circumgyrocenter. The gyrodistances h_k, $k = 1, 2, 3$, from O to the gyrotriangle gyroside opposite to gyrovertex A_k are related to the gyrotriangle circumgyroradius R and gyroangles $\alpha_1, \alpha_2, \alpha_3$ by the equations*

$$
\begin{aligned}
h_1^2 &= \frac{R^2 \cos\frac{-\alpha_1+\alpha_2-\alpha_3}{2} \cos\frac{-\alpha_1-\alpha_2+\alpha_3}{2} \sin^2\frac{-\alpha_1+\alpha_2+\alpha_3}{2}}{\sin\alpha_2 \sin\alpha_3} \\
&= \frac{R^2 \cos^2(\alpha_1+\frac{\delta}{2}) \sin(\alpha_2+\frac{\delta}{2}) \sin(\alpha_3+\frac{\delta}{2})}{\sin\alpha_2 \sin\alpha_3} \\
h_2^2 &= \frac{R^2 \cos\frac{\alpha_1-\alpha_2-\alpha_3}{2} \cos\frac{-\alpha_1-\alpha_2+\alpha_3}{2} \sin^2\frac{\alpha_1-\alpha_2+\alpha_3}{2}}{\sin\alpha_1 \sin\alpha_3} \\
&= \frac{R^2 \sin(\alpha_1+\frac{\delta}{2}) \cos^2(\alpha_2+\frac{\delta}{2}) \sin(\alpha_3+\frac{\delta}{2})}{\sin\alpha_1 \sin\alpha_3} \\
h_3^2 &= \frac{R^2 \cos\frac{\alpha_1-\alpha_2-\alpha_3}{2} \cos\frac{-\alpha_1+\alpha_2-\alpha_3}{2} \sin^2\frac{\alpha_1+\alpha_2-\alpha_3}{2}}{\sin\alpha_1 \sin\alpha_2} \\
&= \frac{R^2 \sin(\alpha_1+\frac{\delta}{2}) \sin(\alpha_2+\frac{\delta}{2}) \cos^2(\alpha_3+\frac{\delta}{2})}{\sin\alpha_1 \sin\alpha_2}
\end{aligned}
\tag{8.12}
$$

where δ is the defect of gyrotriangle $A_1A_2A_3$, given by (8.6).

In the Euclidean limit, $s \to \infty$, triangle defects tend to zero, $\delta \to 0$, so that in this limit Theorem 8.5 descends to the following theorem, which is well-known in Euclidean geometry [60, p. 191].

Theorem 8.6. *Let $A_1A_2A_3$ be a triangle in $\mathbb{R}^n$, $n \geq 2$, and let O be its circumcenter, shown in Fig. 1.10, p. 39. The distance h_k, $k = 1, 2, 3$, from O to the triangle side opposing vertex A_k is related to the triangle circumradius R and angle α_k by the equation*

$$h_k = R\cos\alpha_k \qquad (\textit{Euclidean Geometry}). \tag{8.13}$$

8.4 The Gyrodistance Between O and I

In this section we manipulate the gyrodistance $\|{\ominus}O{\oplus}I\|$ between the circumgyrocenter O and the ingyrocenter I of a gyrotriangle $A_1A_2A_3$ in Einstein gyrovector spaces $(\mathbb{R}^n_s, \oplus, \otimes)$, $n \geq 2$. In the Euclidean limit, $s \to \infty$, a gyrodistance in $\mathbb{R}^n_s$ tends to a corresponding distance in $\mathbb{R}^n$, that is,

$$\lim_{s\to\infty} \|{\ominus}O{\oplus}I\| = \|-O+I\| \tag{8.14}$$

where, contrasting the gyrodistance $\|{\ominus}O{\oplus}I\|$, the distance $\|-O+I\|$ is the Euclidean distance between the circumcenter O and the incenter I of triangle $A_1A_2A_3$ in corresponding ambient vector spaces $\mathbb{R}^n$.

The gyrodistance $\|{\ominus}O{\oplus}I\|$, expressed in terms of the circumgyroradius and gyroangles of the gyrotriangle, is presented in [18, Theorem 7.26, p. 215] in the following theorem.

Theorem 8.7. *Let α_k, $k = 1, 2, 3$, O and I be the gyroangles, circumgyrocenter and ingyrocenter of a gyrotriangle $A_1A_2A_3$ in an Einstein gyrovector space $(\mathbb{R}^n_s\oplus, \otimes)$, $n \geq 2$. Then, the gyrodistance $\|{\ominus}O{\oplus}I\|$ between O and I is given by the equation*

$$\begin{aligned} \|{\ominus}O{\oplus}I\|^2 &= \frac{4R^2\cos\frac{\alpha_1}{2}\cos\frac{\alpha_2}{2}\cos\frac{\alpha_3}{2}}{(\sin\alpha_1+\sin\alpha_2+\sin\alpha_3)^2} \\ &\times\Big\{\cos\frac{\alpha_1-\alpha_2-\alpha_3}{2}+\cos\frac{-\alpha_1+\alpha_2-\alpha_3}{2}+\cos\frac{-\alpha_1-\alpha_2+\alpha_3}{2} \\ &-\cos\frac{3\alpha_1-\alpha_2-\alpha_3}{2}-\cos\frac{-\alpha_1+3\alpha_2-\alpha_3}{2}-\cos\frac{-\alpha_1-\alpha_2+3\alpha_3}{2}\Big\} \end{aligned} \tag{8.15}$$

As the study in Sect. 8.2 and Sect. 8.3 suggests, it is useful to involve the gyrotriangle defect δ in (8.15), as shown in (8.16), in order to simplify Theorem 8.7 and its descent from hyperbolic geometry into Euclidean geometry. Accordingly, we rewrite Theorem 8.7 into which we insert δ from (8.6), obtaining the following elegant theorem.

Theorem 8.8. *Let α_k, $k=1,2,3$, R, O, and I be the gyroangles, circumgyroradius, circumgyrocenter and ingyrocenter of a gyrotriangle $A_1A_2A_3$ in an Einstein gyrovector space $(\mathbb{R}^n_s\oplus,\otimes)$, $n\geq 2$. Then, the squared gyrodistance between O and I is given by the equation*

$$\begin{aligned}\|\ominus O\oplus I\|^2 &= \frac{4R^2\cos\frac{\alpha_1}{2}\cos\frac{\alpha_2}{2}\cos\frac{\alpha_3}{2}}{(\sin\alpha_1+\sin\alpha_2+\sin\alpha_3)^2} \\ &\times\left\{\sin(\alpha_1+\tfrac{\delta}{2})+\sin(\alpha_2+\tfrac{\delta}{2})+\sin(\alpha_3+\tfrac{\delta}{2})\right. \\ &\left.-\sin(2\alpha_1+\tfrac{\delta}{2})-\sin(2\alpha_2+\tfrac{\delta}{2})-\sin(2\alpha_3+\tfrac{\delta}{2})\right\}\end{aligned}\tag{8.16}$$

If gyrotriangle $A_1A_2A_3$ in Theorem 8.8 is equilateral then $\alpha_1=\alpha_2=\alpha_3=:\alpha$ and $\delta=\pi-3\alpha$, so that

$$\sin(\alpha+\tfrac{\delta}{2})=\sin(2\alpha+\tfrac{\delta}{2})=\cos\tfrac{\alpha}{2}\tag{8.17}$$

and hence, by (8.16), $\|\ominus O\oplus I\|=0$ implying $I=O$. Hence, the ingyrocenter I and the circumgyrocenter O of an equilateral gyrotriangle coincide.

In the Euclidean limit, $s\to\infty$, the gyrotriangle defect δ tends to zero. Hence, we obtain the following corollary of Theorem 8.8:

Corollary 8.9. *Let α_k, $k=1,2,3$, R, O and I be the angles, circumradius, circumcenter and incenter of a triangle $A_1A_2A_3$ in a Euclidean space $\mathbb{R}^n$, $n\geq 2$. Then, the squared distance between O and I is given by the equation*

$$\begin{aligned}\|-O+I\|^2 &= \frac{4R^2\cos\frac{\alpha_1}{2}\cos\frac{\alpha_2}{2}\cos\frac{\alpha_3}{2}}{(\sin\alpha_1+\sin\alpha_2+\sin\alpha_3)^2} \\ &\times\{\sin\alpha_1+\sin\alpha_2+\sin\alpha_3-\sin 2\alpha_1-\sin 2\alpha_2-\sin 2\alpha_3\} \\ &= R^2\left(1-8\sin\frac{\alpha_1}{2}\sin\frac{\alpha_2}{2}\sin\frac{\alpha_3}{2}\right)\end{aligned}\tag{8.18}$$

The result in the extreme sides of (8.18) of the Corollary is well-known [60, p. 199]. One should note that the second equation in (8.18) is valid under the Euclidean triangle condition $\alpha_1+\alpha_2+\alpha_3=\pi$, since it obeys the condition $\delta=0$.

By means of the triangle inradius r in (8.10), p. 323, the extreme sides of (8.18) can be written as

$$\|-O+I\|^2 = R^2 - 2Rr \tag{8.19}$$

thus recovering Euler's equation [60, Eq. (15), p. 199], [61, p. 29].

8.5 Gyrotriangles with $O = I$

Gyrotrigonometric gyrobarycentric coordinates (i) of O are given in (4.254), p. 237; and (ii) of I are given in (4.221), p. 226. Clearly, $O = I$ if and only if the gyrobarycentric coordinates of O and I are equivalent, and they are equivalent if and only if $\alpha_1 = \alpha_2 = \alpha_3$. Formalizing this result, we have the following theorem.

Theorem 8.10. *Let α_k, $k = 1, 2, 3$, O, and I be the gyroangles, circumgyrocenter and ingyrocenter of a gyrotriangle $A_1A_2A_3$ in an Einstein gyrovector space $(\mathbb{R}^n_s \oplus, \otimes)$, $n \geq 2$. Then, $O = I$ if and only if $\alpha_1 = \alpha_2 = \alpha_3$.*

In the Euclidean limit $s \to \infty$, Theorem 8.10 descends to its Euclidean counterpart as follows.

Theorem 8.11. *Let α_k, $k = 1, 2, 3$, O, and I be the angles, circumcenter and incenter of a triangle $A_1A_2A_3$ in a vector space $\mathbb{R}^n$, $n \geq 2$, Then, $O = I$ if and only if $\alpha_1 = \alpha_2 = \alpha_3$.*

8.6 The Gyrodistance Between O and H

(1) Gyrotrigonometric gyrobarycentric coordinates of the circumgyrocenter O of a gyrotriangle $A_1A_2A_3$ with respect to the gyrotriangle in an Einstein gyrovector space $(\mathbb{R}^3_s, \oplus, \otimes)$ is given by Theorem 4.33, p. 237.
(2) Gyrotrigonometric gyrobarycentric coordinates of the orthogyrocenter H of a gyrotriangle $A_1A_2A_3$ with respect to the gyrotriangle in an Einstein gyrovector space $(\mathbb{R}^3_s, \oplus, \otimes)$ is given by Theorem 4.24, p. 209.

The theory of the gyrodistance between two gyropoints that are given by their gyrobarycentric coordinates is presented (i) in [18, Sec. 4.9]; and (ii) in [19, Sec. 5.12]. Following this theory and, in particular Example 8.1, p. 321, we obtain by computer assisted computation the following theorem (exercise).

Theorem 8.12. *Let α_k, $k = 1, 2, 3$, R, O and H be the gyroangles, circumgyroradius, circumgyrocenter and orthogyrocenter of a gyrotriangle*

$A_1A_2A_3$ that possesses a circumgyrocenter and an orthogyrocenter in an Einstein gyrovector space $(\mathbb{R}^n_s \oplus, \otimes)$, $n \geq 2$. Then, the squared gyrodistance between O and H is given by the equation

$$\|\ominus O \oplus H\|^2 = R^2(1 - 8K \cos\alpha_1 \cos\alpha_2 \cos\alpha_3)\,, \tag{8.20}$$

where

$$K = \frac{\{\sin(2\alpha_1 + \frac{\delta}{2}) + \sin(2\alpha_2 + \frac{\delta}{2}) + \sin(2\alpha_3 + \frac{\delta}{2})\}^2 - 9\sin^2\frac{\delta}{2}}{\{\sin(2\alpha_1 + \delta) + \sin(2\alpha_2 + \delta) + \sin(2\alpha_3 + \delta) + 3\sin\delta\}^2} > 0 \tag{8.21}$$

The inequality $K > 0$ in (8.21) stems from the *circumgyrocircle existence condition* discussed in [19, Eq. (8.34), p. 291].

If gyrotriangle $A_1A_2A_3$ in Theorem 8.12 is equilateral then $\alpha_1 = \alpha_2 = \alpha_3 =: \alpha$ and $\delta = \pi - 3\alpha$. These, in turn, imply that the factor

$$1 - 8K \cos\alpha_1 \cos\alpha_2 \cos\alpha_3$$

in (8.20) vanishes, as one can readily check, so that $\|\ominus O \oplus H\| = 0$. Hence, the circumgyrocenter O and the orthogyrocenter H of an equilateral gyrotriangle coincide.

In the Euclidean limit, $s \to \infty$, the gyrotriangle defect δ tends to zero, so that the coefficient K in (8.21) tends to 1. Hence, we obtain the following corollary of Theorem 8.12:

Corollary 8.13. *Let α_k, $k = 1, 2, 3$, R, O and H be the angles, circumradius, circumcenter and orthocenter of a triangle $A_1A_2A_3$ in a Euclidean space $\mathbb{R}^n$. Then, the squared distance between O and H is given by the equation*

$$\|-O + H\|^2 = R^2(1 - 8\cos\alpha_1 \cos\alpha_2 \cos\alpha_3) \tag{8.22}$$

The result (8.22) of the Corollary is well-known [60, Eq. (17), p. 199]. Note that the angles in (8.22) obey the triangle condition $\alpha_1 + \alpha_2 + \alpha_3 = \pi$, since they obey the Euclidean condition $\delta = 0$.

8.7 The Gyrodistance Between I and H

(1) Gyrotrigonometric gyrobarycentric coordinates of the ingyrocenter I of a gyrotriangle $A_1A_2A_3$ with respect to the gyrotriangle in an Einstein gyrovector space $(\mathbb{R}_s^3, \oplus, \otimes)$ is given by Theorem 4.30, p. 226.
(2) Gyrotrigonometric gyrobarycentric coordinates of the orthogyrocenter H of a gyrotriangle $A_1A_2A_3$ with respect to the gyrotriangle in an Einstein gyrovector space $(\mathbb{R}_s^3, \oplus, \otimes)$ is given by Theorem 4.24, p. 209.

The theory of the gyrodistance between two gyropoints that are given by their gyrobarycentric coordinates is presented (i) in [18, Sec. 4.9]; and (ii) in [19, Sec. 5.12]. Following this theory and, in particular Example 8.1, p. 321, we obtain by computer assisted computation the following theorem (exercise).

Theorem 8.14. *Let α_k, $k = 1, 2, 3$, R, I and H be the gyroangles, circumgyroradius, ingyrocenter and orthogyrocenter of a gyrotriangle $A_1A_2A_3$ in an Einstein gyrovector space $(\mathbb{R}_s^n \oplus, \otimes)$, $n \geq 2$. Then, the squared gyrodistance between I and H is given by the equation*

$$\|\ominus I \oplus H\|^2 = 2R^2 \left(\frac{\sin(\alpha_1 + \frac{\delta}{2}) \sin(\alpha_2 + \frac{\delta}{2}) \sin(\alpha_3 + \frac{\delta}{2})}{H_1} \right)^2 H_2 \qquad (8.23)$$

where δ is the defect (8.6) of gyrotriangle $A_1A_2A_3$, and where H_1 and H_2 are given gyrotrigonometrically by the following equations:

$$\begin{aligned} H_1 &= 2\cos\alpha_1 \cos\alpha_2 \cos\alpha_3 \\ &\quad + 3(\cos\alpha_1 \cos\alpha_2 + \cos\alpha_1 \cos\alpha_3 + \cos\alpha_2 \cos\alpha_3) \\ &\quad + \cos(2\alpha_1)(-1 + (1 + \cos\alpha_2)(1 + \cos\alpha_3)) \\ &\quad + \cos(2\alpha_2)(-1 + (1 + \cos\alpha_1)(1 + \cos\alpha_3)) \\ &\quad + \cos(2\alpha_3)(-1 + (1 + \cos\alpha_1)(1 + \cos\alpha_2)) \end{aligned} \qquad (8.24a)$$

and

$$\begin{aligned} H_2 &= 4\cos\alpha_1 \cos\alpha_2 \cos\alpha_3 \\ &\quad \times \{7 + 2\cos(2\alpha_2) + 2\cos(2\alpha_3) - 4\cos\alpha_3 - 4\cos\alpha_2(1 + \cos\alpha_3)\} \\ &\quad - 4\{-3 + \cos\alpha_3 + 2\cos\alpha_2 \cos\alpha_3 - \cos(3\alpha_1)\cos\alpha_2 \cos\alpha_3 \\ &\quad + 2\cos\alpha_2 \cos^2\alpha_3 + \cos(2\alpha_2)(-2 + \cos\alpha_3 - \cos(2\alpha_3)) \\ &\quad - 2\cos(2\alpha_3) + \cos(2\alpha_1)(-2 + \cos(2\alpha_2)(1 - \cos\alpha_3) + \cos\alpha_3 - \cos(2\alpha_3) \\ &\quad + \cos\alpha_2(1 + 2\cos\alpha_3 + \cos(2\alpha_3))))\} \end{aligned} \qquad (8.24b)$$

It is unknown whether (8.23) can be simplified by means of an appropriate introduction of δ into (8.23).

Under the Euclidean triangle condition $\delta = 0$, H_1 reduces to H_1^e given by

$$H_1^e = 2(1 - \sin^2 \tfrac{\alpha_1}{2} - \sin^2 \tfrac{\alpha_2}{2} - \sin^2 \tfrac{\alpha_3}{2})(\sin\alpha_1 + \sin\alpha_2 + \sin\alpha_3)^2 \quad (8.25)$$

and under the same Euclidean triangle condition $\delta = 0$, H_2 reduces to H_2^e given by

$$\begin{aligned} H_2^e = 4\,\{&3 + \cos(2\alpha_1) + \cos(2\alpha_2) + \cos(2\alpha_3) \\ &+ 3\cos(\alpha_1 + \alpha_2) + 3\cos(\alpha_1 + \alpha_3) + 3\cos(\alpha_2 + \alpha_3) \\ &+ \cos(\alpha_1 - \alpha_2) + \cos(\alpha_1 - \alpha_3) + \cos(\alpha_2 - \alpha_3)\} \\ &\times (\sin\alpha_1 + \sin\alpha_2 + \sin\alpha_3)^2 \end{aligned} \quad (8.26)$$

where $\alpha_1 + \alpha_2 + \alpha_3 = \pi$.

The use of the computer algebra system Mathematica is recommended. In the Euclidean limit, $s \to \infty$, the gyrotriangle defect δ tends to zero, so that the left-hand side of (8.23) tends to $\|-I + H\|^2$ and the right-hand side of (8.23) tends to a known result, giving rise to the following corollary of Theorem 8.14:

Corollary 8.15. *Let α_k, $k = 1, 2, 3$, R, I and H be the angles, circumradius, incenter and orthocenter of a triangle $A_1A_2A_3$ in a Euclidean space $\mathbb{R}^n$. Then, the squared distance between I and H is given by the equation*

$$\|-I+H\|^2 = 4R^2\{(1-\cos\alpha_1)(1-\cos\alpha_2)(1-\cos\alpha_3) - \cos\alpha_1\cos\alpha_2\cos\alpha_3\} \quad (8.27)$$

where $\alpha_1 + \alpha_2 + \alpha_3 = \pi$.

The result (8.27) of the Corollary is well-known [60, Eq. (18), p. 200].

Chapter 9

Ptolemy's Theorem: Euclidean and Hyperbolic

In this chapter we revisit gyrotrigonometry, paying attention to the trigonometry–gyrotrigonometry duality. This duality plays an important role in the translation of Ptolemy's theorem from Euclidean geometry into its counterpart in two models of hyperbolic geometry. The two models of hyperbolic geometry are (i) the relativistic ball model, also known as the Klein model; and (ii) the Poincaré ball model.

The (i) relativistic model of hyperbolic geometry is regulated algebraically by Einstein gyrovector spaces; and the (ii) Poincaré model of hyperbolic geometry is regulated algebraically by Möbius gyrovector spaces, just as (iii) Euclidean geometry is regulated algebraically by the standard model of vector spaces. The resulting gyrovector space approach to hyperbolic geometry enables hyperbolic geometry to be studied analytically, just as the standard vector space approach to Euclidean geometry enables Euclidean geometry to be studied analytically.

Vector spaces admit the introduction of trigonometry and, in full analogy, gyrovector spaces admit the introduction of gyrotrigonometry. Trigonometry and gyrotrigonometry, in turn, enjoy the duality property according to which any identity that is valid if viewed in trigonometry, remains valid if and only if viewed in gyrotrigonometry. Thus, the trigonometry–gyrotrigonometry duality is the principle that an identity is valid *trigonometrically* if and only if it is valid *gyrotrigonometrically*.

Accordingly, viewing *Ptolemy's trigonometric-gyrotrigonometric identity* (9.5) trigonometrically in (i) trigonometry; and gyrotrigonometrically in (ii) Einstein gyrotrigonometry; and in (iii) Möbius gyrotrigonometry, the trigonometry–gyrotrigonometry duality enables Ptolemy's Theorem to be uncovered in (i) Euclidean geometry; in the (ii) relativistic ball model of hyperbolic geometry; and in the (iii) Poincaré ball model of hyperbolic geometry.

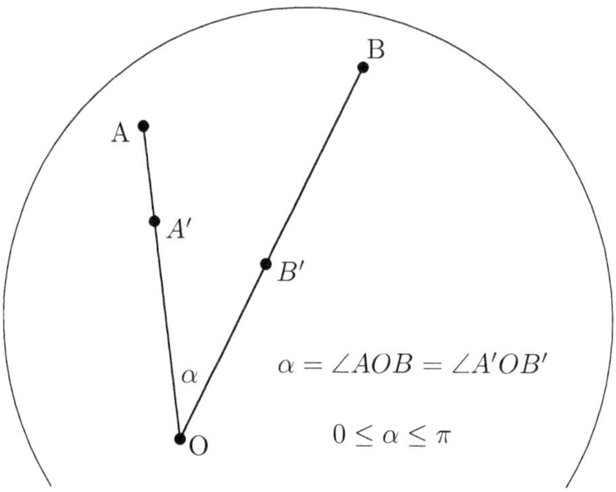

$$\cos\alpha := \frac{\ominus O\oplus A}{\|\ominus O\oplus A\|}\cdot\frac{\ominus O\oplus B}{\|\ominus O\oplus B\|} = \frac{\ominus O\oplus A'}{\|\ominus O\oplus A'\|}\cdot\frac{\ominus O\oplus B'}{\|\ominus O\oplus B'\|}$$

Fig. 9.1 Gyroangles share remarkable analogies with angles, allowing the use of the elementary trigonometric functions cos, sin, etc., in gyrotrigonometry as well. Let A' and B' be points different from O, lying arbitrarily on the gyrosegments OA and OB, respectively, that emanate from a common point O in an Einstein gyrovector space $(\mathbb{R}^n_s, \oplus, \otimes)$ as shown here for $n = 2$. The measure of the gyroangle α formed by the two gyrosegments OA and OB or, equivalently, formed by the two gyrosegments OA' and OB', is given by $cos\alpha$, as shown here. In full analogy with angles, the measure of gyroangle α is independent of the choice of A' and B'.

9.1 Gyroangles – The Hyperbolic Angles

Viewed in the Euclidean plane, the angle α in Fig. 9.1 is given by the equation

$$\cos\alpha = \frac{-O+A}{\|-O+A\|}\cdot\frac{-O+B}{\|-O+B\|}\,. \tag{9.1}$$

We wish to find the analogous counterpart of (9.1) in the hyperbolic plane, in order to translate the common trigonometry in the Euclidean plane into gyrotrigonometry in the hyperbolic plane.

The analogies between lines and gyrolines boil down to the translation of $+$ and $-$ into $\oplus$ and $\ominus$, shown in Fig. 2.1, p. 92, and in Fig. 2.12, p. 133. These analogies, in turn, suggest corresponding analogies between angles and gyroangles. Indeed, in full analogy with the notions of distance and angle, the notion of the gyroangle is deduced from the notion of the gyrodistance. Let O, A and B be any three distinct points in an Einstein gyrovector space $(\mathbb{R}^n_s, \oplus, \otimes)$. The resulting gyrosegments OA and OB that emanate from the point O include a gyroangle $\alpha = \angle AOB$ with gyrovertex O, as shown in Fig. 9.1 for $n = 2$.

Following the analogies between gyrolines and lines, the radian measure of gyroangle α in Fig. 9.1 is, suggestively, given by the equation

$$\cos\alpha = \frac{\ominus O \oplus A}{\|\ominus O \oplus A\|} \cdot \frac{\ominus O \oplus B}{\|\ominus O \oplus B\|}. \tag{9.2}$$

Here, $(\ominus O \oplus A)/\|\ominus O \oplus A\|$ and $(\ominus O \oplus B)/\|\ominus O \oplus B\|$ are unit *gyrovectors*, and cos is the common cosine function of trigonometry, which we apply to the inner product between unit gyrovectors rather than unit vectors. Accordingly, in the context of gyrovector spaces rather than vector spaces, we refer the function "cosine" of trigonometry to as the function "gyrocosine" of gyrotrigonometry. Similarly, all the other elementary trigonometric functions and their interrelationships survive unimpaired in their transition from the common trigonometry in Euclidean spaces $\mathbb{R}^n$ to a corresponding gyrotrigonometry in Einstein gyrovector spaces $\mathbb{R}^n_s$, as demonstrated in Sec. 2.19 and in [19, Chap. 7].

The center $\mathbf{0} = (0, \ldots, 0) \in \mathbb{R}^n_s$ of the ball $\mathbb{R}^n_s = (\mathbb{R}^n_s, \oplus, \otimes)$ is conformal (to Euclidean geometry) in the sense that the measure of any gyroangle with gyrovertex $\mathbf{0}$ is equal to the measure of its Euclidean counterpart. Indeed, if $O = \mathbf{0}$ then (9.2) descends to

$$\cos\alpha = \frac{A}{\|A\|} \cdot \frac{B}{\|B\|} \tag{9.3}$$

which is indistinguishable from its Euclidean counterpart.

We thus encounter the *cosine–gyrocosine duality*, according to which $\cos\alpha$ can be viewed simultaneously as (i) the cosine of an angle α; and as (ii) the gyrocosine of a gyroangle α. More about the trigonometry–gyrotrigonometry duality is presented in Sec. 9.2. It is this duality that enables Ptolemy's Theorem to be translated from Euclidean geometry into hyperbolic geometry, as we will see in Secs. 9.4 – 9.5.

Unlike Euclidean geometry, where the differences $-O + A$ and $A - O$ are equal, in general, the gyrodifferences $\ominus O \oplus A$ and $A \ominus O$ are distinct. The presence of the gyrodifferences $\ominus O \oplus A$ and $\ominus O \oplus B$ in the gyroangle definition (9.2), rather then $A \ominus O$ and $B \ominus O$, is dictated by the demand that gyroangles must be *gyroinvariant*, that is, invariant under gyromotions. Being invariant under the gyromotions of $(\mathbb{R}^n_s, \oplus, \otimes)$, which are left gyrotranslations and rotations about the origin of $\mathbb{R}^n_s$, gyroangles gain geometric significance, so that they are geometric objects of the hyperbolic geometry of the Einstein gyrovector space $(\mathbb{R}_s^n, \oplus, \otimes)$. In detail, the study of motions and gyromotions is found in [19, Secs. 3.10, 3.12].

9.2 Trigonometry – Gyrotrigonometry Duality

The *special positions of a gyroangle* are positions where the gyroangle gyrovertex coincides with the origin O of its gyrovector space. The measure of a gyroangle in a special position in $\mathbb{R}_s^n$ equals its measure when viewed as a corresponding angle in the Euclidean geometry of $\mathbb{R}^n$, as demonstrated in (9.3). Hence, the origin of an Einstein gyrovector space $\mathbb{R}_s^n$ is said to be *conformal.*

The result that every gyroangle can be left gyrotranslated without distortion to special positions where it can be viewed, without distorting its measure, as a Euclidean angle is *crucially important* in the gyrotrigonometry of Einstein gyrovector spaces. It implies that every trigonometric identity of trigonometric functions remains valid in gyrotrigonometry, giving rise to a corresponding gyrotrigonometric identity of gyrotrigonometric functions. Accordingly, we view these identities simultaneously as trigonometric identities and as gyrotrigonometric identities.

Thus, for instance, the familiar trigonometric identities

$$
\begin{aligned}
\cos^2\alpha + \sin^2\alpha &= 1 \\
\cot\frac{\alpha}{2} &= \frac{1+\cos\alpha}{\sin\alpha} \\
\sin^2\frac{\alpha}{2} &= \frac{1-\cos\alpha}{2}\,,
\end{aligned}
\tag{9.4}
$$

$\alpha \in \mathbb{R}$, are valid trigonometrically, where α is given by (9.1). Remarkably, they remain valid gyrotrigonometrically as well, where α is given by (9.2).

To see that trigonometric identities like (9.4) remain valid in gyrotrigonometry, we left gyrotranslate the gyroangle α to a special position, where it can be viewed as an angle satisfying (9.4). This angle α can be left gyrotranslated back to its original position, where it can no longer be viewed as an angle. But, since gyroangles are invariant under left gyrotranslations, gyroangle α still obeys (9.4) regardless of whether it is located in a special position.

Accordingly, one can use a computer algebra system for symbolic manipulation, like Mathematica or Maple, to manipulate gyrotrigonometric expressions in gyrotrigonometry. Thus, while a computer algebra system like Mathematica or Maple is designed to deal with symbolic manipulation in trigonometry, it can be used to deal with symbolic manipulation in gyrotrigonometry as well.

As a remarkable consequence of this intimate relationship between trigonometric and gyrotrigonometric functions, we observed in Chap. 4 the following result: Gyrobarycentric coordinates of gyrotriangle gyrocenters that are determined in terms of gyrotriangle gyroangles survive unimpaired, in form, in the transition from hyperbolic to Euclidean geometry.

A vivid example of the use of the trigonometry–gyrotrigonometry duality, according to which every trigonometric identity can simultaneously be viewed both trigonometrically and gyrotrigonometrically, is provided by the trigonometric identity

$$\sin\frac{\alpha}{2}\sin\frac{\gamma}{2}+\sin\frac{\beta}{2}\sin\frac{\delta}{2}=\sin\frac{\alpha+\beta}{2}\sin\frac{\beta+\gamma}{2} \tag{9.5}$$

for any $\alpha, \beta, \gamma \in \mathbb{R}$, where $\delta = 2\pi - (\alpha + \beta + \gamma)$. Like any trigonometric identity, Identity (9.5) can be realized both trigonometrically and gyrotrigonometrically.

As we will see in the sequel, Identity (9.5) is of particular interest since it made up of (i) only sine functions of half angles in trigonometry; and, owing to the trigonometry–gyrotrigonometry duality, (ii) only gyrosine functions of half gyroangles in gyrotrigonometry.

Furthermore, realizing Identity (9.5) trigonometrically yields in Sec. 9.4 the famous Ptolemy's Theorem in the Euclidean plane. As expected, it turns out to be invariant under the Euclidean motions, which are translations and rotations.

In full analogy, realizing identity (9.5) gyrotrigonometrically yields in Sec. 9.5 and in Sec. 9.8 the novel Ptolemy's Theorem in hyperbolic planes. As expected, it turns out to be invariant under the hyperbolic motions, which are left gyrotranslations and rotations, following Equations (2.131) and (2.132), p. 99.

We, therefore, call Identity (9.5) *Ptolemy's trigonometric (or, gyrotrigonometric) identity*, and will view it (i) in Euclidean geometry, (ii) in the relativistic ball model of hyperbolic geometry and (iii) in the Poincaré ball model of hyperbolic geometry.

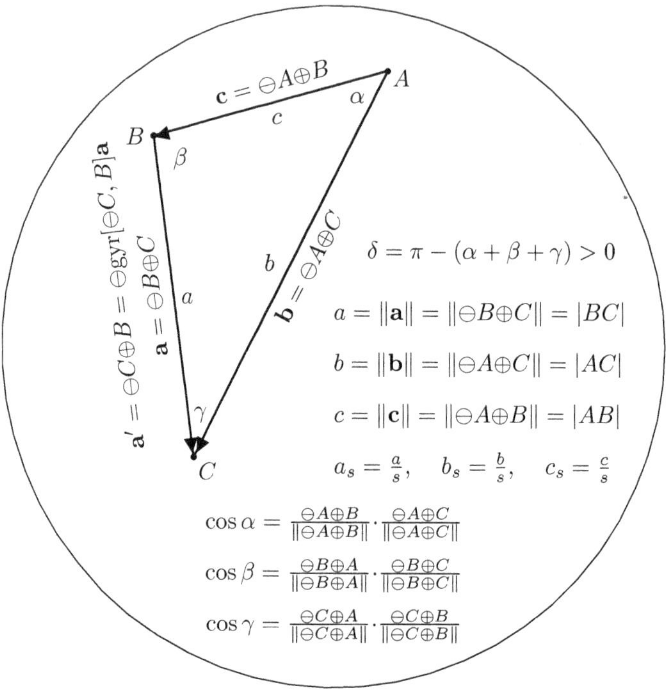

Fig. 9.2 Gyrotriangle ABC, along with its standard notation, in an Einstein gyrovector space. The notation that we use with a gyrotriangle ABC, its gyrosides, and its gyroangles in an Einstein gyrovector space $(\mathbb{R}^n_s, \oplus, \otimes)$ is shown here for the Einstein gyrovector plane $(\mathbb{R}^2_s, \oplus, \otimes)$.

9.3 The Law of Gyrocosines in Einstein Gyrovector Spaces

Let ABC be a gyrotriangle in an Einstein gyrovector space $(\mathbb{R}^n_s, \otimes, \oplus)$ along with its standard notation shown in Fig. 9.2. According to (2.142), p. 103, the gyrotriangle ABC obeys the following three identities, each of which represents its *law of gyrocosines*,

$$\begin{aligned} \gamma_a &= \gamma_b \gamma_c (1 - b_s c_s \cos\alpha) \\ \gamma_b &= \gamma_a \gamma_c (1 - a_s c_s \cos\beta) \\ \gamma_c &= \gamma_a \gamma_b (1 - a_s b_s \cos\gamma)\,. \end{aligned} \tag{9.6}$$

The elements a, b, c and a_s, b_s, c_s and the gyroangles α, β, γ in the law of gyrocosines (9.6) are defined in Fig. 9.2. The gamma factors γ_a, γ_b and γ_c in (9.6) are defined in (2.5).

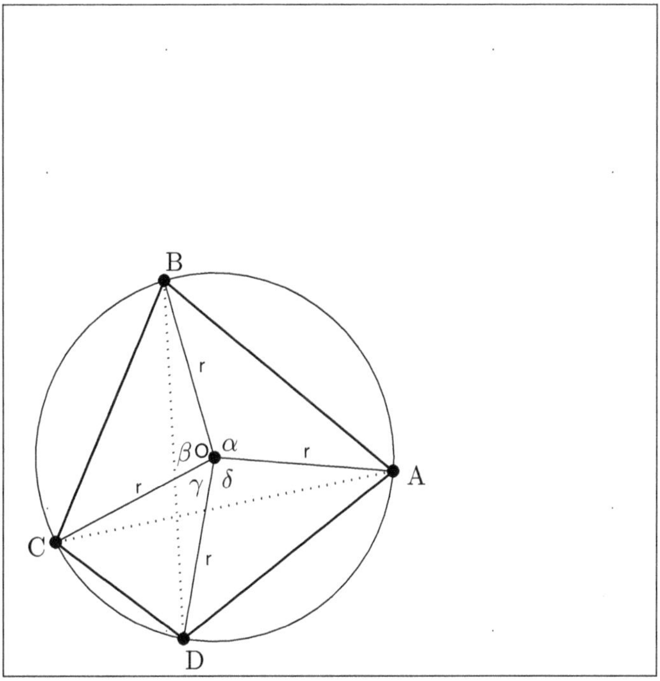

Fig. 9.3 Illustrating Ptolemy's Theorem in the Euclidean plane. $ABCD$ is a cyclic quadrilateral. It is inscribed in its circumcircle centered at its circumcenter O, with radius $r = \|-O+A\| = \|-O+B\| = \|-O+C\| = \|-O+D\|$. The O-vertex angles α, β, γ and δ satisfy the equation $\alpha+\beta+\gamma+\delta = 2\pi$. Ptolemy's Theorem asserts that $|AB||CD| + |AD||BC| = |AC||BD|$, where $|AB| = \|-A+B\|$, *etc.* The hyperbolic counterpart of this figure is shown in Fig. 9.4 (Einstein) and in Fig. 9.7 (Möbius).

In Sec. 9.4 we will apply the common law of cosines in trigonometry in order to establish the famous Ptolemy's Theorem, illustrated in Fig. 9.3.

In full analogy, in Sec. 9.5 we will apply the law of gyrocosines (9.6) in gyrotrigonometry in order to establish the novel hyperbolic Ptolemy's Theorem in the relativistic model of hyperbolic geometry.

9.4 Ptolemy's Theorem in the Euclidean Plane

The discovery of the Euclidean and hyperbolic Ptolemy's theorem is a vivid example of the use of the trigonometry–gyrotrigonometry duality.

Let us consider Ptolemy's trigonometric identity (9.5), viewed in Euclidean geometry,

$$\sin\frac{\alpha}{2}\sin\frac{\gamma}{2} + \sin\frac{\beta}{2}\sin\frac{\delta}{2} = \sin\frac{\alpha+\beta}{2}\sin\frac{\beta+\gamma}{2} \tag{9.7}$$

for any angles $\alpha, \beta, \gamma, \delta > 0$ such that $\alpha + \beta + \gamma + \delta = 2\pi$. We note that owing to the condition $\alpha + \beta + \gamma + \delta = 2\pi$ the angle $(\alpha + \beta)/2$ can be replaced with the angle $(\gamma + \delta)/2$ in (9.7).

In this section we show that the trigonometric identity (9.7) is equivalent to Ptolemy's Theorem in the Euclidean plane, described in Fig. 9.3.

Being a trigonometric identity, (9.7) can be viewed as a gyrotrigonometric identity as well, as explained in Sec. 9.2.

In Sec. 9.5 we will show that Identity (9.7), viewed gyrotrigonometrically, gives rise to the hyperbolic Ptolemy's Theorem in the hyperbolic plane.

In the context of Euclidean geometry we use the notation

$$|AB| := \|-A + B\| \tag{9.8}$$

for any $A, B \in \mathbb{R}^2$. Let $A, B, C, D \in \mathbb{R}^2$ be four points such that $ABCD$ is a cyclic quadrilateral inscribed in a circle centered at O with radius $r = |OA| = |OB| = |OC| = |OD|$ in a Euclidean plane, as shown in Fig. 9.3.

As indicated in Fig. 9.3, A, B, C, D are points lying on a circle and so arranged that the order of the points as one traverses the circle either anticlockwise or clockwise is as shown. Accordingly, the sides AB and CD do not intersect inside the circle.

Then, by the law of cosines, applied to triangle ABO in Fig. 9.3, we have

$$\begin{aligned} |AB|^2 &= 2r^2 - 2r^2 \cos\alpha \\ &= 2r^2(1 - cos\alpha) \\ &= 4r^2 \sin^2 \frac{\alpha}{2} \,. \end{aligned} \tag{9.9}$$

Accordingly, in the notation of Fig. 9.3, we have

$$\begin{aligned} |AB| &= 2r \sin \frac{\alpha}{2} \\ |BC| &= 2r \sin \frac{\beta}{2} \\ |CD| &= 2r \sin \frac{\gamma}{2} \\ |AD| &= 2r \sin \frac{\delta}{2} \\ |AC| &= 2r \sin \frac{\alpha + \beta}{2} \\ |BD| &= 2r \sin \frac{\beta + \gamma}{2} \,. \end{aligned} \tag{9.10}$$

The proof of the first equation in (9.10) is given in (9.9). The proof of each of the remaining equations in (9.10) is similar.

Hence,

$$\begin{aligned}|AB||CD| &= 4r^2 \sin\frac{\alpha}{2}\sin\frac{\gamma}{2}\\ |BC||AD| &= 4r^2 \sin\frac{\beta}{2}\sin\frac{\delta}{2}\\ |AC||BD| &= 4r^2 \sin\frac{\alpha+\beta}{2}\sin\frac{\beta+\gamma}{2}\,.\end{aligned} \tag{9.11}$$

By means of (9.11) and (9.7) we obtain Identity (9.12) of the following well known Ptolemy's Theorem in the Euclidean plane.

Theorem 9.1 (Ptolemy's Theorem in the Eucledean Plane). *Let ABCD be a cyclic quadrilateral, shown in Fig. 9.3. Then, the product of the diagonals equals the sum of the products of the opposite sides, that is,*

$$|AB||CD| + |AD||BC| = |AC||BD|\,. \tag{9.12}$$

In the special case when the two diagonals AC and BD of the cyclic quadrilateral $ABCD$ in Fig. 9.3 intersect at the circumcenter O, we have $\alpha = \gamma$, $\beta = \delta$ and $\alpha + \beta = \beta + \gamma$, implying $|AB| = |CD|$, $|BC| = |AD|$ and $|AC| = |BD|$. Consequently, in this special case triangle ABD is right-angled, and Ptolemy's Identity (9.12) descends to the Pythagorean identity

$$|AB|^2 + |AD|^2 = |BD|^2\,. \tag{9.13}$$

9.5 Ptolemy's Theorem in the Relativistic Model of Hyperbolic Geometry

Ptolemy's identity is viewed in Sec. 9.4 trigonometrically. Contrastingly, in this section we view it gyrotrigonometrically in the gyrotrigonometry of the relativistic model of hyperbolic geometry.

Accordingly, let us consider Ptolemy's gyrotrigonometric identity (9.5), viewed in the relativistic model of hyperbolic geometry,

$$\sin\frac{\alpha}{2}\sin\frac{\gamma}{2} + \sin\frac{\beta}{2}\sin\frac{\delta}{2} = \sin\frac{\alpha+\beta}{2}\sin\frac{\beta+\gamma}{2} \tag{9.14}$$

for any gyroangles $\alpha, \beta, \gamma \in \mathbb{R}$, where $\delta = 2\pi - (\alpha + \beta + \gamma)$. We note that owing to the condition $\alpha + \beta + \gamma + \delta = 2\pi$ the gyroangle $(\alpha + \beta)/2$ can be replaced with the gyroangle $(\gamma + \delta)/2$ in (9.14).

In this section we show that the gyrotrigonometric identity (9.14) gives rise to Ptolemy's Theorem in the relativistic model of the hyperbolic plane.

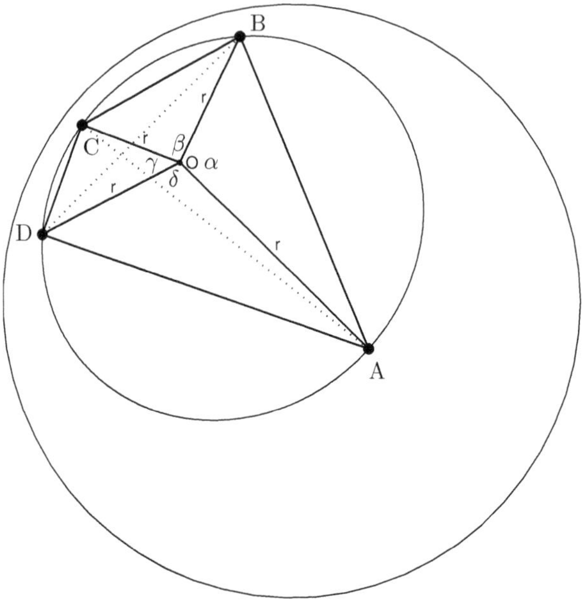

Fig. 9.4 Illustrating Ptolemy's Theorem in the hyperbolic plane regulated by Einstein gyrovector plane $(\mathbb{R}^2_c, \oplus, \otimes)$. $ABCD$ is a gyrocyclic gyroquadrilateral. It is inscribed in its circumgyrocircle gyrocentered at its circumgyrocenter O, with gyroradius $r = \|\ominus O \oplus A\| = \|\ominus O \oplus B\| = \|\ominus O \oplus C\| = \|\ominus O \oplus D\|$. The O-gyrovertex gyroangles α, β, γ and δ satisfy the equation $\alpha + \beta + \gamma + \delta = 2\pi$. The Hyperbolic Ptolemy's Theorem is fully analogous to its Euclidean counterpart, asserting that $|AB|_h |CD|_h + |AD|_h |BC|_h = |AC|_h |BD|_h$, where $|AB|_h$, *etc.*, is defined by (9.26) along with (9.15).

In full analogy with (9.8), in the context of the relativistic model of hyperbolic geometry we use the notation

$$|AB| := \|\ominus A \oplus B\| \tag{9.15}$$

for any $A, B \in \mathbb{R}^2_s$, where $\oplus$ denotes Einstein addition in $\mathbb{R}^2_s$.

It should be noted that in the Euclidean limit, $s \to \infty$, the hyperbolic $|AB|$, given by (9.15), descends to the Euclidean $|AB|$, given by (9.8), since

$$\lim_{s \to \infty} \|\ominus A \oplus B\| = \| - A + B\| \,. \tag{9.16}$$

Let $A, B, C, D \in \mathbb{R}^2_s$ be four points such that $ABCD$ is a gyrocyclic gyroquadrilateral inscribed in a gyrocircle gyrocentered at O, with gyroradius $r = |OA| = |OB| = |OC| = |OD|$ in the hyperbolic plane regulated by the Einstein gyrovector plane $(\mathbb{R}^2_s, \oplus, \otimes)$, as shown in Fig. 9.4.

Then, by the law of gyrocosines (9.6) applied to gyrotriangle ABO in Fig. 9.4, we have

$$\gamma_{|AB|} = \gamma_r^2(1 - r_s^2 \cos\alpha) \tag{9.17}$$

where we use the usual notation $\gamma_{|AB|} = (1 - |AB|^2/s^2)^{-1/2}$ and $r_s = r/s$, $\gamma_r = (1 - r_s^2)^{-1/2}$.

Solving (9.17) for $\cos\alpha$, noting (2.11), p. 66, yields

$$\cos\alpha = \frac{\gamma_r^2 - \gamma_{|AB|}}{\gamma_r^2 - 1} \tag{9.18}$$

so that

$$\sin^2\frac{\alpha}{2} = \frac{1 - \cos\alpha}{2} = \frac{1}{2}\frac{\gamma_{|AB|} - 1}{\gamma_r^2 r_s^2} \tag{9.19}$$

and, hence,

$$\sin\frac{\alpha}{2} = \frac{1}{\sqrt{2}}\frac{\sqrt{\gamma_{|AB|} - 1}}{\gamma_r\, r_s}\,. \tag{9.20}$$

Repeating the result in (9.20) to the O-gyrovertex gyroangles $\alpha, \beta, \gamma, \delta, \alpha+\beta$ and $\beta+\gamma$ in Fig. 9.4, where $\alpha+\beta+\gamma+\delta = 2\pi$, yields

$$\begin{aligned}
\sin\frac{\alpha}{2} &= \frac{1}{\sqrt{2}\gamma_r\, r_s}\sqrt{\gamma_{|AB|} - 1}\\
\sin\frac{\beta}{2} &= \frac{1}{\sqrt{2}\gamma_r\, r_s}\sqrt{\gamma_{|BC|} - 1}\\
\sin\frac{\gamma}{2} &= \frac{1}{\sqrt{2}\gamma_r\, r_s}\sqrt{\gamma_{|CD|} - 1}\\
\sin\frac{\delta}{2} &= \frac{1}{\sqrt{2}\gamma_r\, r_s}\sqrt{\gamma_{|AD|} - 1}\\
\sin\frac{\alpha+\beta}{2} &= \frac{1}{\sqrt{2}\gamma_r\, r_s}\sqrt{\gamma_{|AC|} - 1}\\
\sin\frac{\beta+\gamma}{2} &= \frac{1}{\sqrt{2}\gamma_r\, r_s}\sqrt{\gamma_{|BD|} - 1}\,.
\end{aligned} \tag{9.21}$$

By means of (2.11), p. 66, we have

$$\gamma_{|AB|}^2 - 1 = \frac{1}{s^2}\gamma_{|AB|}^2|AB|^2 \tag{9.22}$$

so that

$$\gamma_{|AB|} - 1 = \frac{1}{s^2}\frac{\gamma_{|AB|}^2}{1 + \gamma_{|AB|}}|AB|^2 \tag{9.23}$$

and, hence,

$$\sqrt{\gamma_{|AB|} - 1} = \frac{1}{s}\frac{\gamma_{|AB|}}{\sqrt{1+\gamma_{|AB|}}}|AB|\,. \tag{9.24}$$

Repeating the result in (9.24) to the gyrocircle gyrochords AB, BC, CD, AD, AC and BD in Fig. 9.4 yields

$$\begin{aligned}
\sqrt{\gamma_{|AB|} - 1} &= \frac{1}{s}\frac{\gamma_{|AB|}}{\sqrt{(1+\gamma_{|AB|})}}|AB| =: \frac{1}{s}|AB|_h \\
\sqrt{\gamma_{|BC|} - 1} &= \frac{1}{s}\frac{\gamma_{|BC|}}{\sqrt{(1+\gamma_{|BC|})}}|BC| =: \frac{1}{s}|BC|_h \\
\sqrt{\gamma_{|CD|} - 1} &= \frac{1}{s}\frac{\gamma_{|CD|}}{\sqrt{(1+\gamma_{|CD|})}}|CD| =: \frac{1}{s}|CD|_h \\
\sqrt{\gamma_{|AD|} - 1} &= \frac{1}{s}\frac{\gamma_{|AD|}}{\sqrt{(1+\gamma_{|AD|})}}|AD| =: \frac{1}{s}|AD|_h \\
\sqrt{\gamma_{|AC|} - 1} &= \frac{1}{s}\frac{\gamma_{|AC|}}{\sqrt{(1+\gamma_{|AC|})}}|AC| =: \frac{1}{s}|AC|_h \\
\sqrt{\gamma_{|BD|} - 1} &= \frac{1}{s}\frac{\gamma_{|BD|}}{\sqrt{(1+\gamma_{|BD|})}}|BD| =: \frac{1}{s}|BD|_h
\end{aligned} \tag{9.25}$$

where we define

$$\begin{aligned}
|AB|_h &:= \frac{\gamma_{|AB|}}{\sqrt{1+\gamma_{|AB|}}}|AB| \\
|AB| &= \|\ominus A \oplus B\|
\end{aligned} \tag{9.26}$$

for any $A, B \in \mathbb{R}_s^2$, where $\oplus$ denotes Einstein addition in $\mathbb{R}_s^2$. We call $|AB|_h$ the h-modified $|AB|$ in the relativistic model of hyperbolic geometry.

The equations in (9.21) and (9.25) yield

$$
\begin{aligned}
\sin\frac{\alpha}{2} &= \frac{1}{\sqrt{2}\gamma_r\, r}|AB|_h \\
\sin\frac{\beta}{2} &= \frac{1}{\sqrt{2}\gamma_r\, r}|BC|_h \\
\sin\frac{\gamma}{2} &= \frac{1}{\sqrt{2}\gamma_r\, r}|CD|_h \\
\sin\frac{\delta}{2} &= \frac{1}{\sqrt{2}\gamma_r\, r}|AD|_h \\
\sin\frac{\alpha+\beta}{2} &= \frac{1}{\sqrt{2}\gamma_r\, r}|AC|_h \\
\sin\frac{\beta+\gamma}{2} &= \frac{1}{\sqrt{2}\gamma_r\, r}|BD|_h\,.
\end{aligned}
\tag{9.27}
$$

Finally, Ptolemy's gyrotrigonometric identity (9.14) along with (9.27) yields Identity (9.28) of the following Hyperbolic Ptolemy's Theorem.

Theorem 9.2 (Ptolemy's Theorem in the relativistic model of the Hyperbolic Plane). *Let $ABCD$ be a gyrocyclic gyroquadrilateral in the relativistic model of a hyperbolic plane $(\mathbb{R}_s^2, \oplus, \otimes)$, shown in Fig. 9.4. Then, the product of the h-modified gyrodiagonals equals the sum of the products of the h-modified opposite gyrosides, that is,*

$$|AB|_h|CD|_h + |AD|_h|BC|_h = |AC|_h|BD|_h \tag{9.28}$$

where $|AB|_h$ is given by (9.26) for any $A, B \in \mathbb{R}_s^2$.

Clearly, in the Euclidean limit, $s \to \infty$, the hyperbolic Ptolemy's Identity (9.28) descends to its Euclidean counterpart (9.12).

9.6 Gyrodiametric Gyrotriangles

Definition 9.3. (Gyrodiametric Gyrotriangles). A triangle is diametric if one of its sides coincides with a diameter of its circumcircle. In full analogy, a gyrotriangle is gyrodiametric if one of its gyrosides coincides with a gyrodiameter of its circumgyrocircle.

A diametric triangle is right angled, the angle opposite to the diametric side being $\pi/2$. In contrast, gyrodiametric gyrotriangles are not right gyroangled. But, they obey the three equations in (9.29) and (9.30), which are shown in Fig. 9.5.

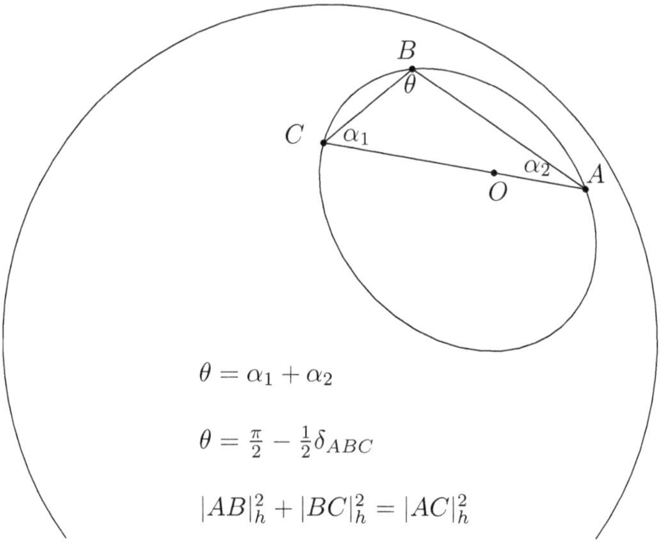

Fig. 9.5 A Gyrodiametric Gyrotriangle in an Einstein gyrovector plane. The circumgyrocenter O of gyrotriangle ABC lies on gyroside AC of the gyrotriangle. Hence, the gyrotriangle is gyrodiametric, obeying the three equations that are shown in the Figure, where δ_{ABC} is the defect of gyrotriangle ABC. $|AB|_h$ is given by (9.26), etc.

The first two properties of gyrodiametric gyrotriangles in Fig. 9.5 are

$$\begin{aligned} \theta &= \alpha_1 + \alpha_2 \\ \theta &= \frac{\pi}{2} - \frac{1}{2}\delta_{ABC} \end{aligned} \tag{9.29}$$

where δ_{ABC} is the defect of gyrotriangle ABC. These properties are established in [19, Sec. 8.11]. The third property of gyrodiametric gyrotriangles in Fig. 9.5 is a hyperbolic Pythagorean identity. It is established in (9.30) as a special case of Ptolemy's Theorem in the hyperbolic plane.

In the special case when the two gyrodiagonals AC and BD of the gyrocyclic gyroquadrilateral $ABCD$ in Fig. 9.4 intersect at the circumgyrocenter O, we have the gyroangle equalities $\alpha = \gamma$, $\beta = \delta$ and $\alpha + \beta = \beta + \gamma$, implying by (9.27) the equations $|AB|_h = |CD|_h$, $|BC|_h = |AD|_h$ and $|AC|_h = |BD|_h$. Consequently, in this special case the hyperbolic Ptolemy's Identity (9.28) descends to the hyperbolic Pythagorean identity

$$|AB|_h^2 + |BC|_h^2 = |AC|_h^2 \tag{9.30}$$

for the gyrodiametric gyrotriangle ABC in the relativistic model of hyperbolic geometry, shown in Fig. 9.5, where $|AB|_h$ is given by (9.26).

Clearly, in the Euclidean limit, $s \to \infty$, the hyperbolic Pythagorean identity (9.30) descends to the common Pythagorean identity (9.13) in Euclidean geometry.

By means of (9.25), the hyperbolic Pythagorean identity (9.30) can be written as

$$(\gamma_{|AB|} - 1) + (\gamma_{|BC|} - 1) = \gamma_{|AC|} - 1)\,. \tag{9.31}$$

However, in the Euclidean limit, $s \to \infty$, (9.31) tends to the trivial identity $0 = 0$.

9.7 Gyrotrigonometry in Möbius Gyrovector Spaces

The *gyrosegment* AB that links the two distinct points $A, B \in \mathbb{R}^n_s$ in a Möbius gyrovector space $(\mathbb{R}^n_s, \oplus, \otimes)$ is given by $A\oplus(\ominus A\oplus B)\otimes t$ with $0 \le t \le 1$, where $\oplus$ and $\otimes$ denote Möbius addition and multiplication. The *gyrolength* of a gyrosegment AB is

$$|AB| := \|\ominus A\oplus B\|\,, \tag{9.32}$$

$A, B \in \mathbb{R}^n_s$, just as the length of a segment AB is

$$|AB|_e := \|-A + B\|\,, \tag{9.33}$$

$A, B \in \mathbb{R}^n$. Contrasting (9.15), here $|AB|$ is the gyrolength of gyroside AB of gyrotriangle ABC in Fig. 9.6 in the Poincaré ball model of hyperbolic geometry.

A gyrotriangle ABC whose gyrosides are the gyrosegments AB, BC and AC in $\mathbb{R}^n_s$, along with its gyroangles, is depicted in Fig. 9.6. The gyrolengths of the gyrosides of gyrotriangle ABC are

$$\begin{aligned} a &:= |BC| = \|\ominus B\oplus C\| \\ b &:= |AC| = \|\ominus A\oplus C\| \\ c &:= |AB| = \|\ominus A\oplus B\|\,, \end{aligned} \tag{9.34}$$

as shown in Fig. 9.6.

Gyrotrigonometry in Möbius gyrovector spaces is studied, for instance, in [3, Sec. 8.5]. In gyrotrigonometry the gyrocosine of a gyroangle $\alpha = \angle BAC$, shown in Fig. 9.6, is given by

$$\cos\alpha = \frac{\ominus A\oplus B}{\|\ominus A\oplus B\|}\cdot\frac{\ominus A\oplus C}{\|\ominus A\oplus C\|}\,, \tag{9.35}$$

just as in trigonometry the cosine of an angle α, in the Euclidean counterpart of Fig. 9.6, is given by

$$\cos\alpha = \frac{-A + B}{\|-A + B\|}\cdot\frac{-A + C}{\|-A + C\|}\,. \tag{9.36}$$

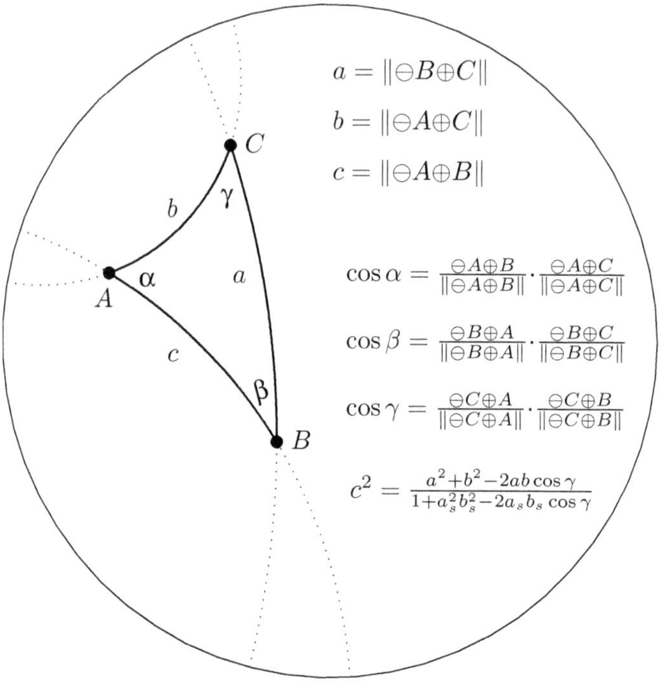

Fig. 9.6 Möbius Law of Gyrocosines. Möbius gyrovector spaces regulate algebraically the Poincaré ball model of hyperbolic geometry. A gyrotriangle ABC in a Möbius gyrovector plane $(\mathbb{R}_s^2, \oplus, \otimes)$ is shown, along with its (i) gyrovertices A, B, C, (ii) gyroside gyrolengths a, b, c, and (iii) gyroangles α, β, γ and their gyrocosines. The Möbius law of gyrocosines is presented in this figure and in (9.38).

Accordingly, viewing $\cos\alpha$ trigonometrically (that is, within the framework of trigonometry) yields (9.36), while viewing $\cos\alpha$ gyrotrigonometrically (that is, within the framework of gyrotrigonometry) yields (9.35).

The trigonometry–gyrotrigonometry duality enables us to uncover Ptolemy's Theorem in both Euclidean and hyperbolic geometry.

The trigonometric identity that we will now view gyrotrigonometrically in the Poincaré ball model of hyperbolic geometry is the Ptolemy's gyrotrigonometric identity (9.5),

$$\sin\frac{\alpha}{2}\sin\frac{\gamma}{2} + \sin\frac{\beta}{2}\sin\frac{\delta}{2} = \sin\frac{\alpha+\beta}{2}\sin\frac{\beta+\gamma}{2} \tag{9.37}$$

for any $\alpha, \beta, \gamma \in \mathbb{R}$, where $\delta = 2\pi - (\alpha + \beta + \gamma)$.

Ptolemy's identity (9.37) is realized trigonometrically (i) in the Euclidean plane, and gyrotrigonometrically (ii) in the Einstein gyrovector plane, and (iii) in the Möbius gyrovector plane. These three distinct realizations of (9.37) are valid owing to the trigonometry–gyrotrigonometry duality.

Indeed, (i) realizing (9.37) in trigonometry gives rise to Ptolemy's Theorem in Euclidean geometry, as shown in Sec. 9.4; (ii) realizing (9.37) in Einstein gyrotrigonometry gives rise to Ptolemy's Theorem in the relativistic model of hyperbolic geometry in Sec. 9.5; and (iii) realizing (9.37) in Möbius gyrotrigonometry gives rise to Ptolemy's Theorem in the Poincaré ball model of hyperbolic geometry, as we will see in Sec. 9.8.

Having the gyrotriangle notation in Fig. 9.6, we are now in the position to state the *law of gyrocosines* in a Möbius gyrovector plane, established in [2, Sec. 8.4] and [3, Sec. 8.5].

Theorem 9.4. (Law of Gyrocosines in Möbius Gyrovector Spaces). *Let ABC be a gyrotriangle in a Möbius gyrovector plane $(\mathbb{R}^2_s, \oplus, \otimes)$ with gyrovertices $A, B, C \in \mathbb{R}^2_s$, gyroside gyrolengths $0 < a, b, c < s$, and gyroangles α, β, γ, as shown in Fig. 9.6. Then, the gyroside gyrolengths and gyroangles of the gyrotriangle satisfy the law of gyrocosines*

$$c^2 = \frac{a^2 + b^2 - 2ab\cos\gamma}{1 + a_s^2 b_s^2 - 2a_s b_s \cos\gamma} \tag{9.38}$$

and, hence equivalently,

$$\cos\gamma = \frac{a^2 + b^2 - c^2 - a_s^2 b_s^2 c^2}{2ab(1 - c_s^2)} \tag{9.39}$$

where $a_s = a/s$, $b_s = b/s$, $c_s = c/s$, with similar relations involving the other gyrosides and gyroangles of the gyrotriangle.

Note that in the Euclidean limit, $s \to \infty$, the law of gyrocosines (9.38) descends to the law of cosines,

$$c^2 = a^2 + b^2 - 2ab\cos\gamma \tag{9.40}$$

as expected. Also, note that in the context of Euclidean geometry, $\cos\gamma$ in (9.40) is viewed trigonometrically, as in (9.36), while in the context of hyperbolic geometry, $\cos\gamma$ in (9.38)–(9.39) is viewed gyrotrigonometrically, as in (9.35) and as shown in Fig. 9.6.

9.8 Ptolemy's Theorem in the Poincaré Ball Model of Hyperbolic Geometry

Ptolemy's Theorem in Euclidean geometry is established in Sec. 9.4 by employing Ptolemy's trigonometric identity (9.5) along with the law of cosines in trigonometry.

In full analogy, Ptolemy's Theorem in the relativistic model of hyperbolic geometry is established in Sec. 9.5 by employing Ptolemy's gyrotrigonometric identity (9.5) along with the law of gyrocosines (9.6) in the gyrotrigonometry of the relativistic model of hyperbolic geometry.

Again, in full analogy, in this section we study Ptolemy's Theorem in the Poincaré ball model of hyperbolic geometry by employing Ptolemy's gyrotrigonometric identity (9.5) along with the law of gyrocosines in the gyrotrigonometry of the Poincaré ball model of hyperbolic geometry.

Applying the law of gyrocosines (9.39) to the O-gyrovertex gyroangle $\alpha = \angle AOB$ of gyrotriangle ABO, shown in Fig. 9.7, yields

$$\cos\alpha = \frac{2r^2 - |AB|^2 - \frac{r^4}{s^4}|AB|^2}{2r^2(1 - \frac{|AB|^2}{s^2})}, \tag{9.41}$$

noting that a, b, c and γ in (9.39) are, respectively, realized in (9.41) as $r, r, |AB|$ and α.

Here $|AB|$ is the gyrolength of gyroside AB of gyrotriangle ABC in Fig. 9.6 in the Möbius gyrovector plane $(\mathbb{R}_s^2, \oplus, \ominus)$, given by (9.32).

We now employ the trigonometric and, hence, gyrotrigonometric identity $\sin^2(\alpha/2) = (1 - \cos\alpha)/2$, obtaining from (9.41) the equation

$$\sin^2\frac{\alpha}{2} = \frac{1-\cos\alpha}{2} = \frac{(s^2-r^2)^2}{4s^2r^2(s^2-|AB|^2)}|AB|^2 \tag{9.42}$$

so that, finally,

$$\sin\frac{\alpha}{2} = \frac{s^2-r^2}{2sr\sqrt{s^2-|AB|^2}}|AB| = \frac{1-r_s^2}{2r}\gamma_{|AB|}|AB| \tag{9.43}$$

where $r_s = r/s$, and where $\gamma_{|AB|}$ is the Lorentz gamma factor of $|AB|$, given by

$$\gamma_{|AB|} = \frac{1}{\sqrt{1-\frac{|AB|^2}{s^2}}} \tag{9.44}$$

Repeating the result in (9.43) for the six O-gyrovertex gyrotriangle gyroangles α, β, γ, δ, $\alpha+\beta$, and $\beta+\gamma$ in Fig. 9.7 yields the following six

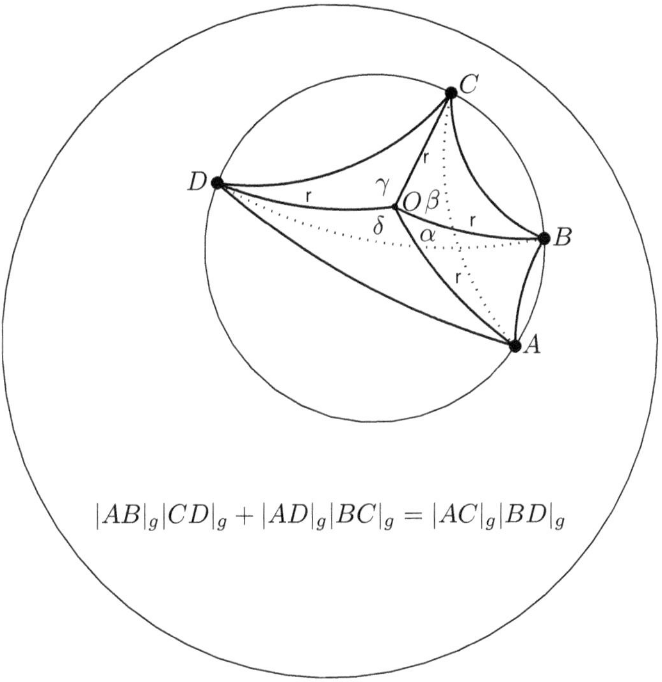

Fig. 9.7 Illustrating Ptolemy's Theorem in the hyperbolic plane regulated by the Möbius gyrovector plane $(\mathbb{R}^2_s, \oplus, \otimes)$. A gyrocyclic gyroquadrilateral $ABCD$ inscribed in its circumgyrocircle gyrocentered at its circumgyrocenter O, with gyroradius $r = \|\ominus O \oplus A\| = \|\ominus O \oplus B\| = \|\ominus O \oplus C\| = \|\ominus O \oplus D\|$ is shown. The O-gyrovertex gyroangles α, β, γ and δ satisfy the equation $\alpha + \beta + \gamma + \delta = 2\pi$. The Hyperbolic Ptolemy's Theorem is fully analogous to its Euclidean counterpart, asserting that $|AB|_g|CD|_g + |AD|_g|BC|_g = |AC|_g|BD|_g$, where $|AB|_g$, *etc.*, is defined by (9.46).

equations

$$
\begin{aligned}
\sin\frac{\alpha}{2} &= \frac{1-r_s^2}{2r}\gamma_{|AB|}|AB| =: \frac{1-r_s^2}{2r}|AB|_g \\
\sin\frac{\beta}{2} &= \frac{1-r_s^2}{2r}\gamma_{|BC|}|BC| =: \frac{1-r_s^2}{2r}|BC|_g \\
\sin\frac{\gamma}{2} &= \frac{1-r_s^2}{2r}\gamma_{|CD|}|CD| =: \frac{1-r_s^2}{2r}|CD|_g \\
\sin\frac{\delta}{2} &= \frac{1-r_s^2}{2r}\gamma_{|AD|}|AD| =: \frac{1-r_s^2}{2r}|AD|_g \\
\sin\frac{\alpha+\beta}{2} &= \frac{1-r_s^2}{2r}\gamma_{|AC|}|AC| =: \frac{1-r_s^2}{2r}|AC|_g \\
\sin\frac{\beta+\gamma}{2} &= \frac{1-r_s^2}{2r}\gamma_{|BD|}|BD| =: \frac{1-r_s^2}{2r}|BD|_g
\end{aligned}
\tag{9.45}
$$

where

$$|AB|_g = \gamma_{|AB|}|AB|$$
$$|AB| = \|\ominus A \oplus B\| \tag{9.46}$$

for any $A, B \in \mathbb{R}_s^2$, where $\oplus$ denotes Möbius addition in $\mathbb{R}_s^2$. We call $|AB|_g$ the g-modified gyrolength of gyroside AB in the Poincaré model of hyperbolic geometry.

The proof of the first equation in (9.45) is given in (9.43). The proof of each of the remaining equations in (9.45) is similar.

Substituting (9.45) into the gyrotrigonometric identity (9.37) yields the result (9.47) of Ptolemy's Theorem in the Poincaré ball model of hyperbolic geometry.

Theorem 9.5. (Ptolemy's Theorem in the Poincaré Ball Model of Hyperbolic Geometry). *Let $ABCD$ be a gyrocyclic gyroquadrilateral in a Möbius gyrovector plane $(\mathbb{R}_s^2, \oplus, \otimes)$, shown in Fig. 9.7. Then, the product of the g-modified gyrodiagonals equals the sum of the products of the g-modified opposite gyrosides, that is*

$$|AB|_g|CD|_g + |AD|_g|BC|_g = |AC|_g|BD|_g \tag{9.47}$$

where $|AB|_g$ is given by (9.46) for any $A, B \in \mathbb{R}_s^2$.

Clearly, in the Euclidean limit, $s \to \infty$, the hyperbolic Ptolemy's Theorem (9.47) descends to its Euclidean counterpart, studied in Sec. 9.4.

An interesting special case of Theorem 9.5 is depicted in Fig. 9.8.

9.9 Gyrodiametric Gyrotriangles

By definition, a gyrotriangle is gyrodiametric if one of its gyrosides coincides with a gyrodiameter of its circumgyrocircle. In Euclidean geometry, a diametric triangle is right angled, the angle opposite to the diametric side being $\pi/2$. In contrast, gyrodiametric gyrotriangles are not right gyroangled. But, they obey a hyperbolic Pythagorean identity, as Theorem 9.6 asserts.

By definition, a gyrocyclic gyroquadrilateral is gyrodiametric if its gyrodiagonals intersect at its circumgyrocenter, as shown in Fig. 9.8. A gyrodiametric gyroquadrilateral $ABCD$, shown in Fig. 9.8, gives rise to the two gyrodiametric gyrotriangles ABC and ADC. In the notation shown in Fig. 9.8 we clearly have

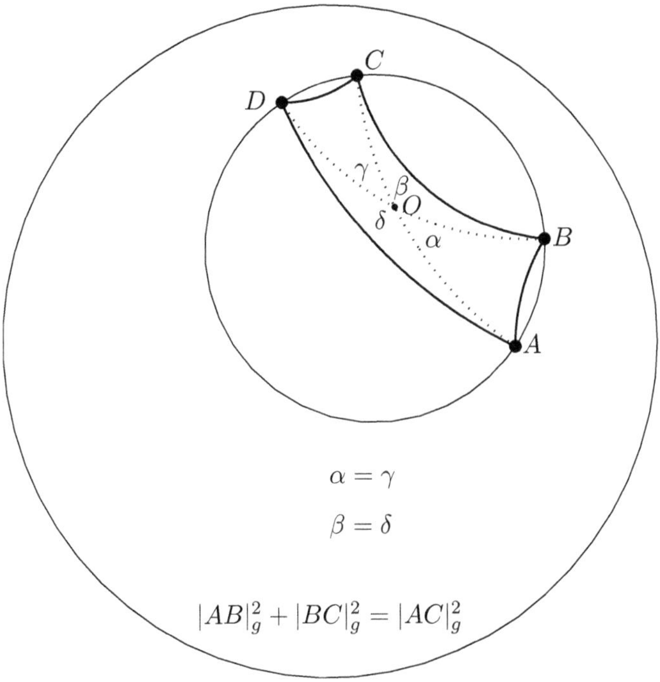

Fig. 9.8 A Gyrodiametric Gyroquadrilateral in a Möbius gyrovector plane. The gyrocyclic gyroquadrilateral $ABCD$ of Fig. 9.7 is depicted here in the special position where the two gyrodiagonals AC and BD intersect at the gyroquadrilateral circumgyrocenter O, resulting in the two gyrodiametric gyrotriangles ABC and ADC. The common gyroside, AC, of these gyrotriangles is a gyrodiameter of the gyroquadrilateral circumgyrocircle. The hyperbolic Pythagorean identity that the g-modified gyroside gyrolengths of a gyrodiametric gyrotriangle obey in the Poincaré ball model of hyperbolic geometry is shown.

$$\begin{aligned}\alpha &= \gamma \\ \beta &= \delta \\ \alpha + \beta &= \beta + \gamma\,.\end{aligned} \tag{9.48}$$

Hence, by (9.45), the equations in (9.48) yield, respectively, the equations

$$\begin{aligned}|AB|_g &= |CD|_g \\ |BC|_g &= |AD|_g \\ |AC|_g &= |BD|_g\,.\end{aligned} \tag{9.49}$$

Hence, following (9.47) we have the hyperbolic Pythagorean identity

$$|AB|_g^2 + |BC|_g^2 = |AC|_g^2 \tag{9.50}$$

where $|AB|_g$, $|BC|_g$ and $|AC|_g$ are the g-modified gyrolengths of the gyrosides of the gyrodiametric gyrotriangle ABC, as shown in Fig. 9.8.

Formalizing the result in (9.50), we have proved the following theorem, where we use the notation in Fig. 9.8.

Theorem 9.6. (Hyperbolic Pythagorean Identity). *Let ABC be a gyrodiametric gyrotriangle in a Möbius gyrovector plane $(\mathbb{R}_s^2, \oplus, \otimes)$, where AC is the gyrodiametric gyroside. Then, the gyrotriangle gyrosides obey the hyperbolic Pythagorean identity*

$$|AB|_g^2 + |BC|_g^2 = |AC|_g^2 \tag{9.51}$$

in the Poincaré model of hyperbolic geometry, where

$$\begin{aligned} |AB|_g &= \gamma_{|AB|}|AB| \\ |AB| &= \|\ominus A \oplus B\| \end{aligned} \tag{9.52}$$

for any $A, B \in \mathbb{R}_s^2$, where $\oplus$ denotes Möbius addition in $\mathbb{R}_s^2$.

Chapter 10

Cyclic Antipodal Segments

A cyclic antipodal segment AA' in a Euclidean plane is a segment between two points A and A' on a circle and passing through its center. In full analogy, a gyrocyclic antipodal gyrosegment AA' in a hyperbolic plane is a gyrosegment between two gyropoints A and A' on a gyrocircle and passing through its gyrocenter.

The study of cyclic antipodal segments in both Euclidean and hyperbolic geometry in this chapter is motivated by the study of Ptolemy's theorem in Chap. 9 in the following sense. The mathematical tools developed in Chap. 9 in order to discover Ptolemy's theorem in both Euclidean and hyperbolic geometry are well suited for the study of cyclic antipodal segments in both Euclidean and hyperbolic geometry, leading to interesting novel results in both Euclidean and hyperbolic geometry.

10.1 Introduction

A trigonometric identity made up of sines of half angles is said to be *half-angled* in trigonometry, if viewed in Euclidean geometry, or *half-gyroangled* in gyrotrigonometry, if viewed in hyperbolic geometry. Thus, for instance, the trigonometric identities (10.1), (10.4), (10.44) and (10.55)–(10.56) are half-angled when viewed in Euclidean geometry, and are half-gyroangled gyrotrigonometric identities when viewed in hyperbolic geometry. The significance of half-angled trigonometric identities rests on the result that they admit geometric realizations in both Euclidean and hyperbolic geometry. They, accordingly, lead to novel, interesting results, as demonstrated by the study of Ptolemy's Theorem in Chap. 9.

The study of Ptolemy's Theorem

(1) in Euclidean geometry in Sec. 9.4; and
(2) in the relativistic model of hyperbolic geometry in Sec. 9.5; and
(3) in the Poincaré ball model of hyperbolic geometry in Sec. 9.8

stems from the geometric realization of the half-angled trigonometric identity (or, simultaneously, the half-gyroangled gyrotrigonometric identity)

$$\sin\frac{\alpha}{2}\sin\frac{\gamma}{2}+\sin\frac{\beta}{2}\sin\frac{\delta}{2}=\sin\frac{\alpha+\beta}{2}\sin\frac{\beta+\gamma}{2} \qquad (10.1)$$
$$\alpha+\beta+\gamma+\delta=2\pi$$

for any $\alpha, \beta, \gamma, \delta > 0$ such that $\alpha+\beta+\gamma+\delta=2\pi$.

Following the trigonometry–gyrotrigonometry duality, Ptolemy's identity (10.1) is viewed in Chap. 9 both trigonometrically (that is, within the framework of trigonometry) and gyrotrigonometrically (that is, within the framework of gyrotrigonometry).

The success of the process of geometric realization of a half-angled trigonometric identity simultaneously in both Euclidean and hyperbolic geometry in Chap. 9 motivates to explore the geometric realization of various half-angled trigonometric identities simultaneously in both Euclidean and hyperbolic geometry. To stimulate the exploration of geometric realization of half-angled trigonometric identities, in this chapter we explore the geometric realization of the half-angled trigonometric identities (10.4), (10.44) and (10.55)–(10.56).

Accordingly, the study of the first cyclic antipodal segments in this chapter is based on the geometric realization of the half-angled trigonometric identity (10.4). Following the trigonometry–gyrotrigonometry duality, Identity (10.4), p. 356, is simultaneously viewed

(1) within the framework of trigonometry, as a half-angled trigonometric identity in Euclidean geometry, in Sec. 10.2, giving rise to identity (10.3) in Euclidean geometry; and
(2) within the framework of gyrotrigonometry, as a half-gyroangled gyrotrigonometric identity in the relativistic model of hyperbolic geometry, in Sec. 10.4, giving rise to identity (10.37), p. 362, in hyperbolic geometry.

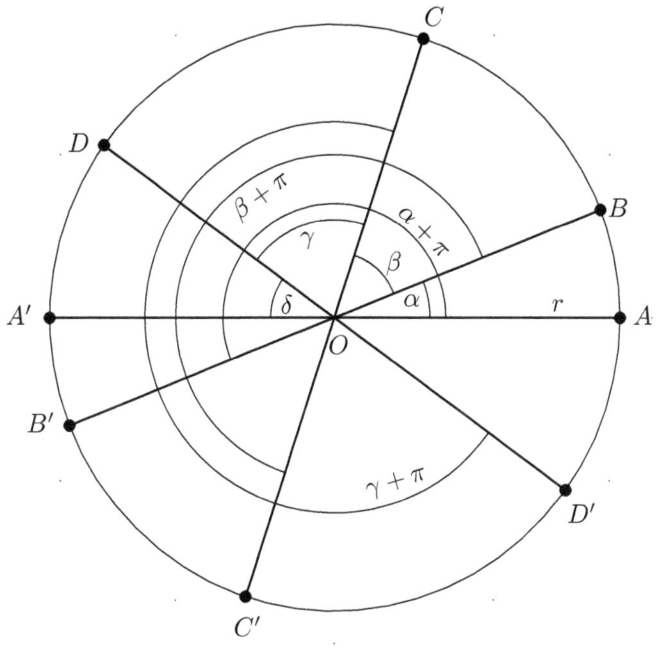

$$|AB'||BC'||CD'| - |AB'||BC||CD| - |AB||BC'||CD| - |AB||BC||CD'|$$
$$= 4r^2|A'D|$$

Fig. 10.1 Four cyclic antipodal segments, AA', BB', CC', DD', on a circle $\Sigma(O,r)$ centered at the origin $O = (0,0) \in \mathbb{R}^2$ of the Euclidean plane, with radius r, and their corresponding O-vertex angles α, β, γ, δ. The points $A, B, C, D, A', B', C', D'$ are arbitrarily ordered counterclockwise, implying $\alpha + \beta + \gamma + \delta = \pi$. The identity of Theorem 10.1 is shown, where $|AB| = \|{-A}+B\|$, etc. Since the point O is conformal, this figure can be viewed in the hyperbolic plane $(\mathbb{R}_s^2, \oplus, \otimes)$ as well, where $|AB| = \|\ominus A \oplus B\|$, etc. and where $\Sigma(O,r)$ is a gyrocircle of gyroradius r. The Euclidean distance $|AB|$ and the Euclidean circle $\Sigma(O,r)$ are invariant under the Euclidean motions, while the hyperbolic gyrodistance $|AB|$ and the hyperbolic gyrocircle $\Sigma(O,r)$ are invariant under the hyperbolic motions.

10.2 Four Cyclic Antipodal Segments

A cyclic antipodal segment of a circle in a Euclidean plane $\mathbb{R}^2$ is a segment AA' between points $A, A' \in \mathbb{R}^2$ that are the intersection of the circle with a diameter of the circle. Illustrating Theorem 10.1, a circle with four cyclic antipodal segments is depicted in Fig. 10.1.

The Euclidean distance $|AB|$ between two points $A, B \in \mathbb{R}^2$ is given by

$$|AB| := \| -A + B\| \,. \tag{10.2}$$

Equivalently, $|AB|$ is the Euclidean length of the segment AB that joins the points A and B. Contrastingly, in the context of hyperbolic geometry, $|AB|$ is a corresponding hyperbolic length, called *gyrolength*, as in (10.28).

Theorem 10.1. (Four Cyclic Antipodal Segments). *Let $\Sigma(O, r)$ be a circle in the Euclidean plane $\mathbb{R}^2$ with radius r, centered at the origin $O = (0, 0) \in \mathbb{R}^2$, with four cyclic antipodal segments AA', BB', CC' and DD', such that the eight points $A, B, C, D, A', B', C', D' \in \mathbb{R}^2$ are arbitrarily ordered counterclockwise (or clockwise), as shown in Fig. 10.1.*

Then, the four cyclic antipodal segments satisfy the identity

$$\begin{aligned} |AB'||BC'||CD'| - |AB'||BC||CD| - |AB||BC'||CD| \\ - |AB||BC||CD'| = 4r^2|A'D|\,. \end{aligned} \tag{10.3}$$

Proof. Clearly, $r = \frac{1}{2}|AA'| = \frac{1}{2}|BB'| = \frac{1}{2}|CC'| = \frac{1}{2}|DD'|$.

The proof of the theorem is based on the elegant half-angled trigonometric identity

$$\begin{aligned} \sin\frac{\alpha+\pi}{2}\sin\frac{\beta+\pi}{2}\sin\frac{\gamma+\pi}{2} - \sin\frac{\alpha+\pi}{2}\sin\frac{\beta}{2}\sin\frac{\gamma}{2} \\ - \sin\frac{\alpha}{2}\sin\frac{\beta+\pi}{2}\sin\frac{\gamma}{2} \\ - \sin\frac{\alpha}{2}\sin\frac{\beta}{2}\sin\frac{\gamma+\pi}{2} = \sin\frac{\delta}{2}\,, \end{aligned} \tag{10.4}$$

which holds for any $\alpha, \beta, \gamma, \delta \in \mathbb{R}^2$ that satisfy the condition

$$\alpha + \beta + \gamma + \delta = \pi\,. \tag{10.5}$$

Noting that $\sin\frac{\alpha+\pi}{2} = \cos\frac{\alpha}{2}$, the trigonometric identity in (10.4)–(10.5) can readily be verified. When realized geometrically in Fig. 10.1, this trigonometric identity reveals its unexpected grace and elegance.

In order to realize (10.4)–(10.5) geometrically by a circle with four cyclic antipodal segments in the Euclidean plane $\mathbb{R}^2$, shown in Fig. 10.1, we define the four O-vertex angles in Fig. 10.1 as follows:

$$\begin{aligned} \alpha &= \angle AOB \\ \beta &= \angle BOC \\ \gamma &= \angle COD \\ \delta &= \angle DOA'\,. \end{aligned} \tag{10.6}$$

We note that $\alpha + \beta + \gamma + \delta = \pi$, as required by Condition (10.5), since the points $A, B, C, D, A', B', C', D'$ are ordered counterclockwise.

Then, consequently, the remaining three O-vertex angles in Fig. 10.1 are

$$\begin{aligned}\angle AOB' &= \alpha + \pi \\ \angle BOC' &= \beta + \pi \\ \angle COD' &= \gamma + \pi \,.\end{aligned} \tag{10.7}$$

Applying the law of cosines to triangle AOB in Fig. 10.1 yields

$$\begin{aligned}|AB|^2 &= 2r^2 - 2r^2 \cos\alpha \\ &= 2r^2(1 - \cos\alpha) \\ &= 4r^2 \sin^2 \frac{\alpha}{2} \,,\end{aligned} \tag{10.8}$$

so that

$$|AB| = 2r \sin \frac{\alpha}{2} \,. \tag{10.9}$$

Similarly to (10.9), by means of (10.6) – (10.7), we obtain the following seven equations for angles shown in Fig. 10.1:

$$\begin{aligned}\sin \frac{\alpha}{2} &= \frac{1}{2r}|AB| \\ \sin \frac{\beta}{2} &= \frac{1}{2r}|BC| \\ \sin \frac{\gamma}{2} &= \frac{1}{2r}|CD| \\ \sin \frac{\delta}{2} &= \frac{1}{2r}|A'D| \\ \sin \frac{\alpha+\pi}{2} &= \frac{1}{2r}|AB'| \\ \sin \frac{\beta+\pi}{2} &= \frac{1}{2r}|BC'| \\ \sin \frac{\gamma+\pi}{2} &= \frac{1}{2r}|CD'| \,.\end{aligned} \tag{10.10}$$

The proof of the first equation in (10.10) is given by Equation (10.9) for triangle AOB in Fig. 10.1. The proof of each of the remaining equations in (10.10) is similar.

Substituting the sines in (10.10) into the half-angled trigonometric identity (10.4) yields (10.3), as desired. □

We say that (10.3) realizes the half-angled trigonometric identity (10.4) geometrically in the Euclidean plane.

The proof of Theorem 10.1 is motivated by the proof of Ptolemy's theorem in Chap. 9, where it is extended to hyperbolic geometry. Similarly, Theorem 10.1 can be extended to hyperbolic geometry as well, as we will see in Section 10.5.

10.3 Special Cases

In the special case when $A = B$ and, hence, $A' = B'$, the result (10.3) of Theorem 10.1 descends to

$$\begin{aligned}|BB'||BC'||CD'| - |BB'||BC||CD| - |BB||BC'||CD| \\ - |BB||BC||CD'| = 4r^2|B'D|\,.\end{aligned} \tag{10.11}$$

Noting that $|BB'| = 2r$ and $|BB| = 0$, (10.11) yields

$$|BC'||CD'| - |BC||CD| = 2r|B'D|\,. \tag{10.12}$$

Formalizing the result in (10.12) we obtain the result (10.17) of Corollary 10.1.

In a second special case, when $D = A'$ and, hence, $D' = A$, the result (10.3) of Theorem 10.1 descends to

$$\begin{aligned}|AB'||BC'||AC| - |AB'||BC||A'C| - |AB||BC'||A'C| \\ - |AB||BC||AC| = 4r^2|A'A'| = 0\,.\end{aligned} \tag{10.13}$$

Equation (10.13) gives rise to the elegant equation

$$\frac{|AB'|}{|AB|}\frac{|BC'|}{|BC|} - \frac{|AB'|}{|AB|}\frac{|A'C|}{|AC|} - \frac{|BC'|}{|BC|}\frac{|A'C|}{|AC|} = 1\,. \tag{10.14}$$

Equation (10.14) gives rise to the inequalities

$$\frac{|AB'|}{|AB|} > \frac{|A'C|}{|AC|} \tag{10.15}$$

and

$$\frac{|BC'|}{|BC|} > \frac{|A'C|}{|AC|}\,. \tag{10.16}$$

Formalizing, we have the following Corollary.

Corollary 10.1. (Three Cyclic Antipodal Segments). *Let $\Sigma(O, r)$ be a circle in the Euclidean plane $\mathbb{R}^2$ with radius r, centered at $O \in \mathbb{R}^2$, with three cyclic antipodal segments AA', BB' and CC'. The six points $A, B, C, A', B', C' \in \mathbb{R}^2$ are arbitrarily ordered counterclockwise (or clockwise), as shown in Fig. 10.2.*

Then,

$$|AB'||BC'| - |AB||BC| = 2r|A'C| \tag{10.17}$$

and

$$\frac{|AB'|}{|AB|}\frac{|BC'|}{|BC|} - \frac{|AB'|}{|AB|}\frac{|A'C|}{|AC|} - \frac{|BC'|}{|BC|}\frac{|A'C|}{|AC|} = 1 \tag{10.18}$$

and

$$|AB||BC'| + |AB'||BC| = 2r|AC|\,. \tag{10.19}$$

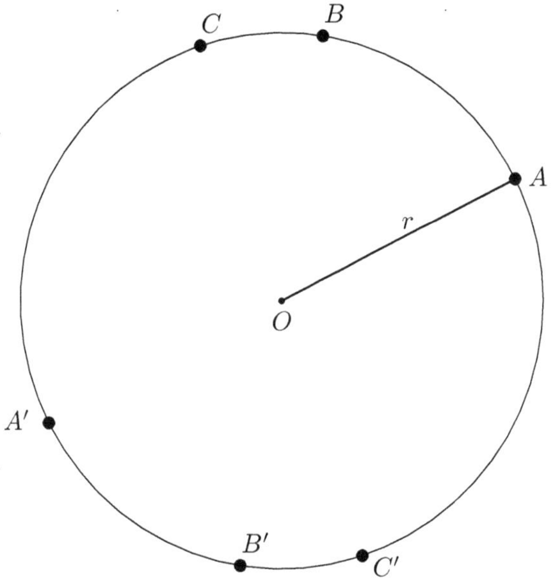

$$|AB'||BC'| - |AB||BC| = 2r|A'C|$$
$$|AB||BC'| + |AB'||BC| = 2r|AC|$$
$$\frac{|AB'|}{|AB|}\frac{|BC'|}{|BC|} - \frac{|AB'|}{|AB|}\frac{|A'C|}{|AC|} - \frac{|BC'|}{|BC|}\frac{|A'C|}{|AC|} = 1$$

Fig. 10.2 Three cyclic antipodal segments, AA', BB', CC', of a circle $\Sigma(O, r)$ centered at the origin $O = (0, 0) \in \mathbb{R}^2$ of the Euclidean plane, with radius r. The points A, B, C, A', B', C' are arbitrarily ordered counterclockwise. The identities of Corollary 10.1 are shown, where $|AB| = \| -A + B\|$, etc.

Proof. Equations (10.17) and (10.18) are proved, respectively, in (10.12) and (10.14). Equation (10.19) follows from (10.17) and (10.18) by eliminating the term

$$(|AB'|/|AB|)(|BC'|/|BC|)$$

between these two equations, □

In order to recover the Pythagorean identity we consider (10.17) and (10.19) in the special case when $C = A'$ and, hence, $C' = A$. In this special case (10.17) descends to

$$|AB'||AB| - |AB||A'B| = 0\,, \tag{10.20}$$

implying

$$|AB'| = |A'B| \tag{10.21}$$

and (10.19) descends to

$$|AB|^2 + |AB'||A'B| = 2r|AA'| = 4r^2 \,. \tag{10.22}$$

Equations (10.21) and (10.22) recover the Pythagorean identity

$$|AB|^2 + |AB'|^2 = 4r^2 = |BB'|^2 \tag{10.23}$$

for right angled triangles ABB', where AA' and BB' are any distinct cyclic antipodal segments of a circle.

Formalizing the result in (10.23) we have the following well-known theorem, expressed in terms of two cyclic antipodal segments.

Theorem 10.2. (Two Cyclic Antipodal Segments – Pythagorean Identity). *Let AA' and BB' be two distinct cyclic antipodal segments of a circle in a Euclidean plane $\mathbb{R}^2$. Then, triangle ABB' is right-angled satisfying the Pythagorean identity*

$$|AB|^2 + |AB'|^2 = |BB'|^2 \,. \tag{10.24}$$

10.4 The Law of Gyrocosines in the Relativistic Model of Hyperbolic Geometry

Let $\mathbf{v} \in \mathbb{R}^n_s$ be a gyrovector in an Einstein gyrovector space $(\mathbb{R}^n_s, \oplus, \otimes)$. It follows from (2.11), p. 66, that

$$\frac{\|\mathbf{v}\|^2}{s^2} = \frac{\gamma_{\mathbf{v}}^2 - 1}{\gamma_{\mathbf{v}}^2} \tag{10.25}$$

so that

$$\gamma_{\mathbf{v}}^2 - 1 = \frac{1}{s^2}\gamma_{\mathbf{v}}^2\|\mathbf{v}\|^2 \tag{10.26}$$

and, hence,

$$\sqrt{\gamma_{\mathbf{v}} - 1} = \frac{1}{s}\frac{\gamma_{\mathbf{v}}}{\sqrt{\gamma_{\mathbf{v}} + 1}}\|\mathbf{v}\| \,. \tag{10.27}$$

Equation (10.27) will prove useful in calculating (10.33).

The proof of Theorem 10.1 is based on the law of cosines of trigonometry. Naturally, for translating Theorem 10.1 from Euclidean geometry into the relativistic model of hyperbolic geometry, we consider the law of gyrocosines of gyrotrigonometry in the relativistic model of hyperbolic geometry, studied in Sec. 9.3.

In full analogy with (10.2), in the context of the relativistic model of hyperbolic geometry we view Fig. 10.1 using the notation

$$|AB| := \|\ominus A \oplus B\| \tag{10.28}$$

for any $A, B \in \mathbb{R}_s^2$, where $\oplus$ denotes Einstein addition (2.4), p. 65, in $\mathbb{R}_s^2$, and where $\ominus A = -A$. Here $|AB|$ represents the gyrodistance between A and B, that is, the hyperbolic distance between A and B in the relativistic model. Equivalently, $|AB|$ is the gyrolength of the gyrosegment AB that joins the points A and B.

It should be noted that in the Euclidean limit, $s \to \infty$, the hyperbolic $|AB|$, given by (10.28), descends to the Euclidean $|AB|$, given by (10.2), since

$$\lim_{s\to\infty} \|\ominus A \oplus B\| = \| - A + B\| \,. \tag{10.29}$$

Let $A, B \in \mathbb{R}_s^2$ be two gyropoints of a gyrocircle with gyroradius r gyrocentered at $O \in \mathbb{R}_s^2$, as shown in Fig. 10.1. The gyroradius of the gyrocircle in Fig. 10.1 is given by $r = |OA| = |OB|$ in the hyperbolic plane regulated by the Einstein gyrovector plane $(\mathbb{R}_s^2, \oplus, \otimes)$.

Then, by the law of gyrocosines (2.142), p. 103, applied to gyrotriangle ABO in Fig. 10.1, we have

$$\gamma_{|AB|} = \gamma_r^2(1 - r_s^2 \cos\alpha) \,, \tag{10.30}$$

where we use the usual notation $\gamma_{|AB|} = (1 - |AB|^2/s^2)^{-1/2}$ and $r_s = r/s$, $\gamma_r = (1 - r_s^2)^{-1/2}$.

Solving (10.30) for $\cos\alpha$, noting (10.25), yields

$$\cos\alpha = \frac{\gamma_r^2 - \gamma_{|AB|}}{\gamma_r^2 - 1} \,, \tag{10.31}$$

so that

$$\sin^2\frac{\alpha}{2} = \frac{1-\cos\alpha}{2} = \frac{1}{2}\frac{\gamma_{|AB|} - 1}{\gamma_r^2 r_s^2} \tag{10.32}$$

and, hence,

$$\sin\frac{\alpha}{2} = \frac{1}{\sqrt{2}}\frac{\sqrt{\gamma_{|AB|} - 1}}{\gamma_r\, r_s} \,. \tag{10.33}$$

Hence, finally, by means of (10.27) we have

$$\sin\frac{\alpha}{2} = \frac{1}{\sqrt{2}\gamma_r\, r}|AB|_h \,, \tag{10.34}$$

where we define

$$|AB|_h := \frac{\gamma_{|AB|}}{\sqrt{1 + \gamma_{|AB|}}}|AB| \tag{10.35}$$

for any $A, B \in \mathbb{R}_s^2$.

Note that here *sin* and *cos* represent the *gyrosine* and the *gyrocosine* functions in the gyrotrigonometry of the relativistic model of hyperbolic geometry. Accordingly, here $|AB|$ is the gyrolength of gyrosegment AB in the relativistic model of hyperbolic geometry.

The result in (10.34) is obtained by applying the law of gyrocosines to the O-gyrovertex gyrotriangle OAB with the O-gyrovertex gyroangle α. Expressing (10.34) to all the O-gyrovertex gyroangles in Fig. 10.1, which are $\alpha, \beta, \gamma, \delta, \alpha+\pi, \beta+\pi$ and $\gamma+\pi$, where $\alpha+\beta+\gamma+\delta=\pi$, yields the following seven equations:

$$\begin{aligned}
\sin\frac{\alpha}{2} &= \frac{1}{\sqrt{2}\gamma_r\, r}|AB|_h \\
\sin\frac{\beta}{2} &= \frac{1}{\sqrt{2}\gamma_r\, r}|BC|_h \\
\sin\frac{\gamma}{2} &= \frac{1}{\sqrt{2}\gamma_r\, r}|CD|_h \\
\sin\frac{\delta}{2} &= \frac{1}{\sqrt{2}\gamma_r\, r}|A'D|_h \\
\sin\frac{\alpha+\pi}{2} &= \frac{1}{\sqrt{2}\gamma_r\, r}|AB'|_h \\
\sin\frac{\beta+\pi}{2} &= \frac{1}{\sqrt{2}\gamma_r\, r}|BC'|_h \\
\sin\frac{\gamma+\pi}{2} &= \frac{1}{\sqrt{2}\gamma_r\, r}|CD'|_h\,.
\end{aligned} \tag{10.36}$$

The proof of the first equation in (10.36) follows from (10.34). The proof of each of the remaining equations in (10.36) is similar.

Substituting the sines in (10.36) into the half-gyroangled gyrotrigonometric identity (10.4) yields the equation

$$\begin{aligned}
|AB'|_h|BC'|_h|CD'|_h - |AB'|_h|BC|_h|CD|_h - |AB|_h|BC'|_h|CD|_h \\
- |AB|_h|BC|_h|CD'|_h = 2\gamma_r^2 r^2 |A'D|_h\,.
\end{aligned} \tag{10.37}$$

We say that (10.37) realizes the half-gyroangled gyrotrigonometric identity (10.4) geometrically in the relativistic model of the hyperbolic plane.

10.5 Four Gyrocyclic Antipodal Segments

Formalizing the result in (10.37), we obtain the following theorem.

Theorem 10.3. (Four Gyrocyclic Antipodal Gyrosegments). *Let $\Sigma(O, r)$ be a gyrocircle in the relativistic model of the hyperbolic plane $(\mathbb{R}^2_s, \oplus, \otimes)$ with gyroradius r, gyrocentered at $O \in \mathbb{R}^2_s$, with four gyrocyclic antipodal gyrosegments AA', BB', CC' and DD', such that the eight gyropoints $A, B, C, D, A', B', C', D' \in \mathbb{R}^2_s$ are arbitrarily ordered counterclockwise (or clockwise), as shown in Fig. 10.1 (viewed in the relativistic model of the hyperbolic plane).*

Then, the four gyrocyclic antipodal gyrosegments satisfy the identity

$$\begin{aligned}|AB'|_h|BC'|_h|CD'|_h - |AB'|_h|BC|_h|CD|_h - |AB|_h|BC'|_h|CD|_h \\ - |AB|_h|BC|_h|CD'|_h = 2\gamma_r^2 r^2 |A'D|_h \,,\end{aligned} \tag{10.38}$$

where

$$|AB|_h := \frac{\gamma_{|AB|}}{\sqrt{1+\gamma_{|AB|}}}|AB| \,. \tag{10.39}$$

for any $A, B \in \mathbb{R}^2_s$, where $|AB|$ is the gyrolength given by (10.28).

10.6 Three Gyrocyclic Antipodal Segments

In the special case when $D = A'$ and, hence, $D' = A$ the result (10.38) of Theorem 10.3 descends to

$$\begin{aligned}|AB'|_h|BC'|_h|AC|_h - |AB'|_h|BC|_h|A'C|_h - |AB|_h|BC'|_h|A'C|_h \\ - |AB|_h|BC|_h|AC|_h = 2\gamma_r^2 r^2 |A'A'| = 0 \,.\end{aligned} \tag{10.40}$$

Equation (10.40) gives rise to the elegant equation (10.41) in Corollary 10.2

Formalizing, we have the following Corollary.

Corollary 10.2. (Three Gyrocyclic Antipodal Gyrosegments). *Let $\Sigma(O, r)$ be a gyrocircle in the relativistic model of the hyperbolic plane $\mathbb{R}^2_s$ with gyroradius r, gyrocentered at $O \in \mathbb{R}^2_s$, with three gyrocyclic antipodal gyrosegments AA', BB' and CC'. The six gyropoints $A, B, C, A', B', C' \in \mathbb{R}^2_s$ are arbitrarily ordered counterclockwise (or clockwise), as shown in Fig. 10.2 (viewed in the relativistic model of the hyperbolic plane).*

Then,

$$\frac{|AB'|_h}{|AB|_h}\frac{|BC'|_h}{|BC|_h} - \frac{|AB'|_h}{|AB|_h}\frac{|A'C|_h}{|AC|_h} - \frac{|BC'|_h}{|BC|_h}\frac{|A'C|_h}{|AC|_h} = 1 \,. \tag{10.41}$$

Proof. Equation (10.41) follows immediately from (10.40). □

Equation (10.41) is the relativistic hyperbolic counterpart of (10.14) in Euclidean geometry.

10.7 The Pythagorean Identity is Recovered

The study of half-angled trigonometric identities is rewarding since they admit geometric realizations associated with cyclic antipodal segments of a circle, in both Euclidean and hyperbolic geometry. Accordingly, in this section we explore a simple half-angled trigonometric identity, which we realize in both Euclidean geometry (in this section) and the relativistic model of hyperbolic geometry (in Section 10.8).

Let us consider the trigonometric identity

$$\cos^2\frac{\alpha}{2}+\sin^2\frac{\alpha}{2}=1\,. \tag{10.42}$$

Noting that

$$\cos\frac{\alpha}{2}=\sin\frac{\alpha+\pi}{2} \tag{10.43}$$

identity (10.42) can be written as a half-angled trigonometric identity,

$$\sin^2\frac{\alpha+\pi}{2}+\sin^2\frac{\alpha}{2}=1\,. \tag{10.44}$$

In order to realize (10.44) geometrically in the Euclidean plane we consider Fig. 10.1 in a Euclidean plane with only the three points $A, B, B' \in \mathbb{R}^2$ on the circle $\Sigma(O,r)$, and the associated O-vertex angles α and $\alpha+\pi$. Then, by (10.10),

$$\begin{aligned}\sin\frac{\alpha}{2}&=\frac{1}{2r}|AB|\\ \sin\frac{\alpha+\pi}{2}&=\frac{1}{2r}|AB'|\,.\end{aligned} \tag{10.45}$$

Substituting the sines in (10.45) into the half-angled trigonometric identity (10.44) yields

$$|AB|^2+|AB'|^2=4r^2=|BB'|^2\,, \tag{10.46}$$

thus recovering the Pythagorean identity for the right-angled triangle ABB' in Fig. 10.1 when viewed in a Euclidean plane.

It seems that exploring the half-angled trigonometric identity (10.44) is pointless since it yields an expected, well-known result. But, we are not finished. In Section 10.8 we will find that identity (10.44), when viewed gyrotrigonometrically as a half-gyroangled gyrotrigonometric identity, does yield an interesting novel result in hyperbolic geometry, that is, a hyperbolic Pythagorean identity.

10.8 A Hyperbolic Pythagorean Identity

In Section 10.7 we consider Identity (10.44) trigonometrically as a half-angled trigonometric identity and, accordingly, we realize it geometrically in Euclidean geometry. In contrast, in this section we consider Identity (10.44) gyrotrigonometrically as a half-gyroangled gyrotrigonometric identity and, accordingly, we realize it geometrically in the relativistic model of hyperbolic geometry.

Consequently, in the same way that in Section 10.7 we recover from Identity (10.44) the well known Pythagorean identity in Euclidean geometry, in this section we discover from Identity (10.44) a novel hyperbolic Pythagorean identity. As such, the study in Sections 10.7 – 10.8 illustrates the trigonometry–gyrotrigonometry duality.

In order to realize (10.44) geometrically in the relativistic model of the hyperbolic plane we consider Fig. 10.1 in the relativistic model of the hyperbolic plane, with only the three gyropoints $A, B, B' \in \mathbb{R}_s^2$ on the gyrocircle $\Sigma(O, r)$, and the associated O-vertex gyroangles α and $\alpha + \pi$. Then, by (10.36),

$$\begin{aligned} \sin\frac{\alpha}{2} &= \frac{1}{\sqrt{2}\gamma_r\, r}|AB|_h \\ \sin\frac{\alpha+\pi}{2} &= \frac{1}{\sqrt{2}\gamma_r\, r}|AB'|_h\,, \end{aligned} \tag{10.47}$$

where

$$\begin{aligned} |AB|_h &:= \frac{\gamma_{|AB|}}{\sqrt{1+\gamma_{|AB|}}}|AB| \\ |AB| &= \|\ominus A \oplus B\| \end{aligned} \tag{10.48}$$

for any $A, B \in \mathbb{R}_s^2$, where $\oplus$ denotes Einstein addition in $\mathbb{R}_s^2$.

Substituting the sines in (10.47) into the half-gyroangled gyrotrigonometric identity (10.44) yields the equation

$$|AB|_h^2 + |AB'|_h^2 = 2\gamma_r^2 r^2\,. \tag{10.49}$$

It can be shown by means of (10.48) and $BB' = r \oplus r = 2r/(1 + r_s^2)$ that $|BB'|_h^2 = 2\gamma_r^2 r^2$. Hence, (10.49) can be written as

$$|AB|_h^2 + |AB'|_h^2 = |BB'|_h^2 \tag{10.50}$$

for gyrotriangle ABB', shown in Fig. 10.1 when viewed in the relativistic model of the hyperbolic plane (as in (9.30), p. 344).

Identity (10.50) is a novel hyperbolic Pythagorean identity in the relativistic model. Note that gyrotriangle ABB', shown in Fig. 10.1, associated with (10.50), is not right-gyroangled. However, as in the Euclidean case, gyrotriangle ABB' is *gyrodiametric*, that is, one of its gyrosides coincides with a gyrodiameter of its circumgyrocircle.

Formalizing the result in (10.50), we have the following theorem.

Theorem 10.4. (A Hyperbolic Pythagorean Identity). *Let ABB' be a gyrotriangle of which gyroside BB' coincides with a gyrodiameter of its circumgyrocircle in the relativistic model of the hyperbolic plane (Fig. 10.1). Then*

$$|AB|_h^2 + |AB'|_h^2 = |BB'|_h^2 \,. \tag{10.51}$$

10.9 Cyclic Antipodal Segments – Example I

The trigonometric identities

$$\sin\frac{\alpha+\beta}{2} = \sin\frac{\alpha}{2}\cos\frac{\beta}{2} + \cos\frac{\alpha}{2}\sin\frac{\beta}{2} \tag{10.52}$$

and

$$\cos\frac{\alpha+\beta}{2} = \cos\frac{\alpha}{2}\cos\frac{\beta}{2} - \sin\frac{\alpha}{2}\sin\frac{\beta}{2} \tag{10.53}$$

along with the trigonometric identity

$$\cos\frac{\alpha}{2} = \sin\frac{\alpha+\pi}{2} \tag{10.54}$$

yield the half-angled trigonometric identities

$$\sin\frac{\alpha+\beta}{2} = \sin\frac{\alpha}{2}\sin\frac{\beta+\pi}{2} + \sin\frac{\alpha+\pi}{2}\sin\frac{\beta}{2} \tag{10.55}$$

and

$$\sin\frac{\alpha+\beta+\pi}{2} = \sin\frac{\alpha+\pi}{2}\sin\frac{\beta+\pi}{2} - \sin\frac{\alpha}{2}\sin\frac{\beta}{2} \tag{10.56}$$

for any $\alpha, \beta \in \mathbb{R}$.

Let us view in a Euclidean plane $\mathbb{R}^2$ the circle in Fig. 10.1 with its three cyclic antipodal segments AA', BB', CC', where the points $A, B, C, A'B'C'$ are arbitrarily ordered counterclockwise, along with the notation in the figure.

Then, by means of the law of cosines in trigonometry, as in (10.10) and as in (9.10), in the notation of Fig. 10.1, we have the following six equations:

$$\begin{aligned} \sin\frac{\alpha}{2} &= \frac{1}{2r}|AB| \\ \sin\frac{\beta}{2} &= \frac{1}{2r}|BC| \\ \sin\frac{\alpha+\pi}{2} &= \frac{1}{2r}|AB'| \\ \sin\frac{\beta+\pi}{2} &= \frac{1}{2r}|BC'| \\ \sin\frac{\alpha+\beta}{2} &= \frac{1}{2r}|AC| \\ \sin\frac{\alpha+\beta+\pi}{2} &= \frac{1}{2r}|AC'| \end{aligned} \tag{10.57}$$

where, as in (10.2),

$$|AB| := \|-A+B\| \tag{10.58}$$

is the Euclidean length of segment AB, for any $A, B \in \mathbb{R}^2$.

The first equation in (10.57) is verified in (10.9), p. 357. The proof of each of the remaining equations in (10.57) is similar.

Finally, substituting (10.57) into (10.55)–(10.56) yields the following remarkable identities that the three cyclic antipodal segments AA', BB', CC' in Fig. 10.1 obey in Euclidean geometry:

$$|AC| = \frac{1}{2r}(|AB||BC'| + |AB'||BC|) \tag{10.59}$$

and

$$|AC'| = \frac{1}{2r}(|AB'||BC'| - |AB||BC|)\,. \tag{10.60}$$

The result in (10.59)–(10.60) in a Euclidean plane $\mathbb{R}^2$ are invariant under the Euclidean motions of $\mathbb{R}^2$, which are translations of $\mathbb{R}^2$ and rotations of $\mathbb{R}^2$ about its origin. Hence, the result has geometric significance in Euclidean geometry.

10.10 Gyrocyclic Antipodal Gyrosegments – Example II

In Sec. 10.9 we consider identities within the framework of trigonometry in Euclidean geometry. Accordingly, the law of cosines is employed in the derivation of the equations in (10.57). In contrast, in this section we consider the same identities, but within the framework of gyrotrigonometry in the relativistic model of hyperbolic geometry. Accordingly, the law of gyrocosines (9.6), p. 336, in the relativistic model of hyperbolic geometry is employed in the derivation of the equations in (10.66) below.

By the trigonometry–gyrotrigonometry duality, any trigonometric identity remains valid in gyrotrigonometry as well. Accordingly, the now gyrotrigonometric identities

$$\sin\frac{\alpha+\beta}{2} = \sin\frac{\alpha}{2}\cos\frac{\beta}{2} + \cos\frac{\alpha}{2}\sin\frac{\beta}{2} \tag{10.61}$$

and

$$\cos\frac{\alpha+\beta}{2} = \cos\frac{\alpha}{2}\cos\frac{\beta}{2} - \sin\frac{\alpha}{2}\sin\frac{\beta}{2} \tag{10.62}$$

along with the gyrotrigonometric identity

$$\cos\frac{\alpha}{2} = \sin\frac{\alpha+\pi}{2} \tag{10.63}$$

yield the half-gyroangled gyrotrigonometric identities

$$\sin\frac{\alpha+\beta}{2} = \sin\frac{\alpha}{2}\sin\frac{\beta+\pi}{2} + \sin\frac{\alpha+\pi}{2}\sin\frac{\beta}{2} \tag{10.64}$$

and

$$\sin\frac{\alpha+\beta+\pi}{2} = \sin\frac{\alpha+\pi}{2}\sin\frac{\beta+\pi}{2} - \sin\frac{\alpha}{2}\sin\frac{\beta}{2} \tag{10.65}$$

for any $\alpha, \beta \in \mathbb{R}$.

The center of the circle in Fig. 10.1 is the origin O of its plane, which is conformal. Hence, the circle in a Euclidean plane can also be viewed as a gyrocircle in a hyperbolic plane.

Accordingly, let us view in a hyperbolic plane $(\mathbb{R}^2_s, \oplus, \otimes)$, relativistic model, the gyrocircle in Fig. 10.1 with its three gyrocyclic antipodal gyrosegments AA', BB', CC', where the gyropoints $A, B, C, A', B', C' \in \mathbb{R}^2_s$ are arbitrarily ordered counterclockwise, along with the notation in the figure.

Then, by means of the law of gyrocosines (9.6), p. 336, in gyrotrigonometry, as in (9.27), p. 343, in the notation of Fig. 10.1, we have the following six equations:

$$\begin{aligned}
\sin\frac{\alpha}{2} &= \frac{1}{\sqrt{2}\gamma_r\, r}|AB|_h \\
\sin\frac{\beta}{2} &= \frac{1}{\sqrt{2}\gamma_r\, r}|BC|_h \\
\sin\frac{\alpha+\pi}{2} &= \frac{1}{\sqrt{2}\gamma_r\, r}|AB'|_h \\
\sin\frac{\beta+\pi}{2} &= \frac{1}{\sqrt{2}\gamma_r\, r}|BC'|_h \\
\sin\frac{\alpha+\beta}{2} &= \frac{1}{\sqrt{2}\gamma_r\, r}|AC|_h \\
\sin\frac{\alpha+\beta+\pi}{2} &= \frac{1}{\sqrt{2}\gamma_r\, r}|AC'|_h
\end{aligned} \tag{10.66}$$

where, as in (9.26), p. 342,

$$\begin{aligned}
|AB|_h :&= \frac{\gamma_{|AB|}}{\sqrt{1+\gamma_{|AB|}}}|AB| \\
|AB| &= \|\ominus A \oplus B\|
\end{aligned} \tag{10.67}$$

for any $A, B \in \mathbb{R}^2_s$, where $\oplus$ is Einstein addition in $\mathbb{R}^2_s$. We call $|AB|_h$ the h-modified $|AB|$ in the relativistic model of hyperbolic geometry.

The first equation in (10.66) is verified in (9.27), p. 343. The proof of each of the remaining equations in (10.66) is similar.

Finally, substituting (10.66) into (10.64)–(10.65) yields the following remarkable identities that the three gyrocyclic antipodal gyrosegments AA', BB', CC' in Fig. 10.1 obey in the relativistic model of hyperbolic geometry:

$$|AC|_h = \frac{1}{\sqrt{2}\gamma_r\, r}(|AB|_h|BC'|_h + |AB'|_h|BC|_h) \tag{10.68}$$

and

$$|AC'|_h = \frac{1}{\sqrt{2}\gamma_r\, r}(|AB'|_h|BC'|_h - |AB|_h|BC|_h)\,. \tag{10.69}$$

As expected, under the Euclidean limit, $s \to \infty$, the hyperbolic result in (10.68)–(10.69) tends to the Euclidean result in (10.59)–(10.60).

The result in (10.68)–(10.69) in a hyperbolic plane $\mathbb{R}^2_s$, relativistic model, is invariant under the hyperbolic motions of $\mathbb{R}^2_s$, which are left gyrotranslations of $(\mathbb{R}^2_s, \oplus, \otimes)$ and rotations of $\mathbb{R}^2_s$ about its origin. Hence, the result has geometric significance in hyperbolic geometry.

10.11 Gyrocyclic Antipodal Gyrosegments – Example III

In Sec. 10.9 we consider identities within the framework of trigonometry in Euclidean geometry. Accordingly, the law of cosines is employed in the derivation of the equations in (10.57). In Sec. 10.10 we consider the same identities, but within the framework of gyrotrigonometry in the relativistic model of hyperbolic geometry. Accordingly, the law of gyrocosines (9.6), p. 336, in the relativistic model of hyperbolic geometry is employed in Sec. 10.10 in the derivation of the equations in (10.66).

Contrasting Secs. 10.9 and 10.10, in this section we consider the same identities, but within the framework of gyrotrigonometry in the Poincaré model of hyperbolic geometry. Accordingly, the law of gyrocosines 9.38, p. 347, in the Poincaré model of hyperbolic geometry is employed in the derivation of the equations in (10.75) below.

By the trigonometry–gyrotrigonometry duality, any trigonometric identity remains valid in gyrotrigonometry as well. Accordingly, the now gyrotrigonometric identities

$$\sin\frac{\alpha+\beta}{2} = \sin\frac{\alpha}{2}\cos\frac{\beta}{2} + \cos\frac{\alpha}{2}\sin\frac{\beta}{2} \tag{10.70}$$

and

$$\cos\frac{\alpha+\beta}{2} = \cos\frac{\alpha}{2}\cos\frac{\beta}{2} - \sin\frac{\alpha}{2}\sin\frac{\beta}{2} \tag{10.71}$$

along with the gyrotrigonometric identity

$$\cos\frac{\alpha}{2} = \sin\frac{\alpha+\pi}{2} \tag{10.72}$$

yield the half-gyroangled gyrotrigonometric identities

$$\sin\frac{\alpha+\beta}{2} = \sin\frac{\alpha}{2}\sin\frac{\beta+\pi}{2} + \sin\frac{\alpha+\pi}{2}\sin\frac{\beta}{2} \tag{10.73}$$

and

$$\sin\frac{\alpha+\beta+\pi}{2} = \sin\frac{\alpha+\pi}{2}\sin\frac{\beta+\pi}{2} - \sin\frac{\alpha}{2}\sin\frac{\beta}{2} \tag{10.74}$$

for any $\alpha, \beta \in \mathbb{R}$.

The center of the circle in Fig. 10.1 is the origin O of its plane, which is conformal. Hence, the circle can also be viewed as a gyrocircle in the Poincaré model of a hyperbolic plane.

Accordingly, let us view in a hyperbolic plane $(\mathbb{R}_s^2, \oplus, \otimes)$, Poincaré model, the gyrocircle in Fig. 10.1 with its three gyrocyclic antipodal gyrosegments AA', BB', CC', where the gyropoints A, B, C, A', B', C' are arbitrarily ordered counterclockwise, along with the notation in the figure.

Then, by means of the law of gyrocosines (9.38), p. 347, in gyrotrigonometry, as in (9.45), p. 349, in the notation of Fig. 10.1, we have the following six equations:

$$\begin{aligned} sin\frac{\alpha}{2} &= \frac{1-r_s^2}{2r}|AB|_g \\ sin\frac{\beta}{2} &= \frac{1-r_s^2}{2r}|BC|_g \\ sin\frac{\alpha+\pi}{2} &= \frac{1-r_s^2}{2r}|AB'|_g \\ sin\frac{\beta+\pi}{2} &= \frac{1-r_s^2}{2r}|BC'|_g \\ sin\frac{\alpha+\beta}{2} &= \frac{1-r_s^2}{2r}|AC|_g \\ sin\frac{\alpha+\beta+\pi}{2} &= \frac{1-r_s^2}{2r}|AC'|_g \end{aligned} \tag{10.75}$$

where, as in (9.46), p. 350,

$$\begin{aligned} |AB|_g &= \gamma_{|AB|}|AB| \\ |AB| &= \|\ominus A \oplus B\| \end{aligned} \tag{10.76}$$

for any $A, B \in \mathbb{R}_s^2$, where $\oplus$ is Möbius addition in $\mathbb{R}_s^2$. We call $|AB|_g$ the g-modified $|AB|$ in the Poincaré model of hyperbolic geometry.

The first equation in (10.75) is verified in (9.45), p. 349. The proof of each of the remaining equations in (10.75) is similar.

Finally, substituting (10.75) into (10.73)–(10.74) yields the following remarkable identities that the three gyrocyclic antipodal gyrosegments AA', BB', CC' in Fig. 10.1 obey in the Poincaré model of hyperbolic geometry:

$$|AC|_g = \frac{1-r_s^2}{2r}(|AB|_g|BC'|_g + |AB'|_h|BC|_g) \tag{10.77}$$

and

$$|AC'|_g = \frac{1-r_s^2}{2r}(|AB'|_g|BC'|_g - |AB|_g|BC|_g)\,. \tag{10.78}$$

As expected, under the Euclidean limit, $s \to \infty$, the hyperbolic result in (10.77)–(10.78) tends to the Euclidean result in (10.59)–(10.60).

The result in (10.77)–(10.78) in a hyperbolic plane $\mathbb{R}_s^2$, Poincaré model, is invariant under the hyperbolic motions of $\mathbb{R}_s^2$, which are left gyrotranslations of $(\mathbb{R}^2, \oplus, \otimes)$ and rotations of $\mathbb{R}_s^2$ about its origin. Hence, the result has geometric significance in hyperbolic geometry.

Notation and Special Symbols

$\oplus$ Gyroaddition, Gyrogroup operation.
$\ominus$ Gyrosubtraction, Inverse gyrogroup operation.
$\boxplus$ Cogyroaddition, Gyrogroup cooperation.
$\boxminus$ Cogyrosubtraction, Inverse gyrogroup cooperation.
$\oplus_{\mathrm{E}}$ Einstein addition (of relativistically admissible coordinate velocities, and generalizations).
$\ominus_{\mathrm{E}}$ Einstein subtraction.
$\boxplus_{\mathrm{E}}$ Einstein coaddition.
$\boxminus_{\mathrm{E}}$ Einstein cosubtraction.
$\oplus_{\mathrm{M}}$ Möbius addition.
$\ominus_{\mathrm{M}}$ Möbius subtraction.
$\boxplus_{\mathrm{M}}$ Möbius coaddition.
$\boxminus_{\mathrm{M}}$ Möbius cosubtraction.
$\otimes$ Scalar multiplication (scalar gyromultiplication) in a gyrovector space.
$\otimes_{\mathrm{E}}$ Einstein scalar multiplication.
$\otimes_{\mathrm{M}}$ Möbius scalar multiplication.
$:=$ Equality, where the lhs is defined by the rhs.
$=:$ Equality, where the rhs is defined by the lhs.
m^t Transpose of matrix m.
AB A gyrosegment with distinct endpoints A and B of a gyroline. A gyroline containing the distinct points A and B.
ABC A gyrotriangle with vertices A, B and C.
Aut An automorphism group.
Aut_0 A subgroup of an automorphism group.
CM Center of Momentum.
c The vacuum speed of light.

gyr Gyrator. gyr$[\mathbf{a}, \mathbf{b}]$ the gyration (gyroautomorphism) generated by $\mathbf{a}$ and $\mathbf{b}$.

s Gyrovector space analogue of the vacuum speed of light c. It is the radius of the ball in ball models of hyperbolic geometry and their gyrogroups and gyrovector spaces.

$\gamma_{\mathbf{v}}$ The gamma factor, $\gamma_{\mathbf{v}} = (1 - \|\mathbf{v}\|^2/s^2)^{-1/2}$ in the ball $\mathbb{V}_s$.

$\mathbb{R}$ The real line.

$\mathbb{R}^n$ The Euclidean n-space.

$\mathbb{R}^n_s$ The s-ball of the Euclidean n-space, $\mathbb{R}^n_s = \{\mathbf{x} \in \mathbb{R}^n : \|\mathbf{x}\| < s\}$.

$SO(n)$ The group of all rotations of $\mathbb{R}^n$ about its origin.

$(S, +)$ A groupoid, a set S with a binary operation $+$.

$\mathbb{V}$ A real inner product space $\mathbb{V} = (\mathbb{V}, +, \cdot)$ with a binary operation $+$ and an inner product $\cdot$.

$\mathbb{V}_s$ The s-ball of the real inner product space $\mathbb{V}$.

Bibliography

[1] A. A. Ungar, *Beyond the Einstein addition law and its gyroscopic Thomas precession: The theory of gyrogroups and gyrovector spaces*, *Fundamental Theories of Physics*, Vol. 117. Kluwer Academic Publishers Group, Dordrecht (2001), ISBN 1-4020-0353-6.

[2] A. A. Ungar, *Analytic hyperbolic geometry: Mathematical foundations and applications.* World Scientific Publishing Co. Pte. Ltd., Hackensack, NJ (2005), ISBN 981-256-457-8.

[3] A. A. Ungar, *Analytic hyperbolic geometry and Albert Einstein's special theory of relativity.* World Scientific Publishing Co. Pte. Ltd., Hackensack, NJ (2008, 2nd ed. 2022), ISBN 978-981-277-229-9; 981-277-229-4.

[4] A. A. Ungar, *Analytic hyperbolic geometry and Albert Einstein's special theory of relativity, 2nd ed.* World Scientific Publishing Co. Pte. Ltd., Hackensack, NJ (2022).

[5] M. J. Crowe, *A history of vector analysis.* Dover Publications Inc., New York (1994), ISBN 0-486-67910-1, the evolution of the idea of a vectorial system, Corrected reprint of the 1985 edition.

[6] V. Varičak, *Darstellung der Relativitätstheorie im dreidimensionalen Lobatchefskijschen Raume [Presentation of the theory of relativity in the three-dimensional Lobachevskian space].* Zaklada, Zagreb (1924).

[7] S. Walter, The non-Euclidean style of Minkowskian relativity, in J. J. Gray (ed.), *The symbolic universe: Geometry and physics 1890–1930.* Oxford Univ. Press, New York, pp. 91–127 (1999).

[8] C. Kimberling, Clark Kimberling's Encyclopedia of Triangle Centers - ETC, (2012), `http://faculty.evansville.edu/ck6/encyclopedia/ETC.html`.

[9] C. Kimberling, *Triangle Centers and central triangles*, Congressus numerantium, 129. Utilitas Mathematica Pub., Winnipeg (1998).

[10] J. E. Marsden, *Elementary classical analysis.* W. H. Freeman and Co., San Francisco (1974), with the assistance of Michael Buchner, Amy Erickson, Adam Hausknecht, Dennis Heifetz, Janet Macrae and William Wilson, and with contributions by Paul Chernoff, István Fáry and Robert Gulliver.

[11] A. A. Ungar, The intrinsic beauty, harmony and interdisciplinarity in Einstein velocity addition law: gyrogroups and gyrovector spaces, *Math. Interdisc. Res.* **1**, 1, pp. 5–51 (2016).

[12] D. Mumford, C. Series and D. Wright, *Indra's pearls: The vision of Felix Klein.* Cambridge University Press, New York (2002), ISBN 0-521-35253-3.
[13] M. J. Greenberg, *Euclidean and non-Euclidean geometries*, 3rd edn., W. H. Freeman and Company, New York (1993), ISBN 0-7167-2446-4, development and history.
[14] J. Gray, *Möbius's geometrical Mechanics* (1993), ISBN 0-19-853969-X, In John Fauvel, Raymond Flood, and Robin Wilson, eds. *Möbius and his band, Mathematics and astronomy in nineteenth-century Germany*, The Clarendon Press Oxford University Press, New York, 1993, 78–103.
[15] J. G. Hocking and G. S. Young, *Topology*, 2nd edn., Dover Publications Inc., New York (1988), ISBN 0-486-65676-4.
[16] D. Veljan, The 2500-year-old Pythagorean theorem, *Math. Mag.* **73**, 4, pp. 259–272 (2000).
[17] H. S. M. Coxeter, *Introduction to geometry.* John Wiley & Sons Inc., New York (1961).
[18] A. A. Ungar, *Hyperbolic triangle centers: The special relativistic approach.* Springer-Verlag, New York (2010).
[19] A. A. Ungar, *Analytic hyperbolic geometry in n dimensions: An introduction.* CRC Press, Boca Raton, FL (2015).
[20] A. A. Ungar, The proper-time Lorentz group demystified, *J. Geom. Symmetry Phys.* **4**, pp. 69–95 (2005).
[21] V. Fock, *The theory of space, time and gravitation.* The Macmillan Co., New York (1964), second revised edition. Translated from the Russian by N. Kemmer. A Pergamon Press Book.
[22] C. Møller, *The theory of relativity.* Oxford, at the Clarendon Press (1952).
[23] R. U. Sexl and H. K. Urbantke, *Relativity, groups, particles*, Springer Physics. Springer-Verlag, Vienna (2001), ISBN 3-211-83443-5, special relativity and relativistic symmetry in field and particle physics, Revised and translated from the third German (1992) edition by Urbantke.
[24] A. Sommerfeld, Über die Zusammensetzung der Geschwindigkeiten in der Relativtheorie, *Physikalische Zeitschrift* **10**, pp. 826–829 (1909).
[25] V. Varičak, Anwendung der Lobatschefskjschen Geometrie in der Relativtheorie, *Physikalische Zeitschrift* **11**, pp. 93–96 (1910).
[26] A. A. Ungar, Einstein's special relativity: Unleashing the power of its hyperbolic geometry, *Comput. Math. Appl.* **49**, pp. 187–221 (2005).
[27] A. Einstein, Zur Elektrodynamik Bewegter Körper [on the electrodynamics of moving bodies] (We use the English translation in [28] or in [62], or in `http://www.fourmilab.ch/etexts/einstein/specrel/www/`), *Ann. Physik (Leipzig)* **17**, pp. 891–921 (1905).
[28] A. Einstein, *Einstein's Miraculous Years: Five Papers that Changed the Face of Physics.* Princeton, Princeton, NJ (1998), edited and introduced by John Stachel. Includes bibliographical references. Einstein's dissertation on the determination of molecular dimensions – Einstein on Brownian motion – Einstein on the theory of relativity – Einstein's early work on the quantum hypothesis. A new English translation of Einstein's 1905 paper on pp. 123–160.

[29] A. A. Ungar, Thomas rotation and the parametrization of the Lorentz transformation group, *Found. Phys. Lett.* **1**, 1, pp. 57–89 (1988).

[30] A. A. Ungar, Gyrations: the missing link between classical mechanics with its underlying Euclidean geometry and relativistic mechanics with its underlying hyperbolic geometry, in *Panos M. Pardalos and Themistocles M. Rassias (eds.): Essays in mathematics and its applications in honor of Stephen Smale's 80th birthday.* Springer, Heidelberg, pp. 463–504 (2012), doi:10.1007/978-3-642-28821-0_18, `http://dx.doi.org/10.1007/978-3-642-28821-0_18`.

[31] A. A. Ungar, Thomas precession: its underlying gyrogroup axioms and their use in hyperbolic geometry and relativistic physics, *Found. Phys.* **27**, 6, pp. 881–951 (1997).

[32] J. Vermeer, A geometric interpretation of Ungar's addition and of gyration in the hyperbolic plane, *Topology Appl.* **152**, 3, pp. 226–242 (2005).

[33] A. A. Ungar, *A gyrovector space approach to hyperbolic geometry.* Morgan & Claypool Pub., San Rafael, California (2009).

[34] T. Foguel and A. A. Ungar, Involutory decomposition of groups into twisted subgroups and subgroups, *J. Group Theory* **3**, 1, pp. 27–46 (2000).

[35] T. Foguel and A. A. Ungar, Gyrogroups and the decomposition of groups into twisted subgroups and subgroups, *Pac. J. Math* **197**, 1, pp. 1–11 (2001).

[36] S. Mahdavi, A. R. Ashrafi, M. A. Salahshour and A. A. Ungar, Construction of 2-gyrogroups in which every proper subgyrogroup is either a cyclic or a dihedral group, *Symmetry* **13**, 2 (2021), doi:10.3390/sym13020316, `https://www.mdpi.com/2073-8994/13/2/316`.

[37] T. Suksumran, The algebra of gyrogroups: Cayley's theorem, Lagrange's theorem and isomorphism theorems, in *Essays in Mathematics and its Applications: in Honor of Vladimir Arnold*, Springer. Springer, New York, pp. 369–437 (2016), eds: Themistocles M. Rassias and Panos M. Pardalos.

[38] A. A. Ungar, The relativistic hyperbolic parallelogram law, in *Geometry, integrability and quantization.* Softex, Sofia, pp. 249–264 (2006).

[39] A. A. Ungar, *Beyond Pseudo-Rotations in Pseudo-Euclidean Spaces: An introduction to the theory of bi-gyrogroups and bi-gyrovector spaces*, Mathematical Analysis and its Applications. Elsevier/Academic Press (2018), ISBN 978-0-12-811773-6.

[40] S. Walter, Book Review: *Beyond the Einstein Addition Law and its Gyroscopic Thomas Precession: The Theory of Gyrogroups and Gyrovector Spaces*, by Abraham A. Ungar, *Found. Phys.* **32**, 2, pp. 327–330 (2002).

[41] A. A. Ungar, Möbius transformation and Einstein velocity addition in the hyperbolic geometry of bolyai and lobachevsky, in *Nonlinear analysis.* Springer, pp. 721–770 (2012).

[42] J. McCleary, *Geometry from a differentiable viewpoint.* Cambridge University Press, Cambridge (1994), ISBN 0-521-41430-X; 0-521-42480-1.

[43] E. Kreyszig, *Differential geometry.* Dover Publications Inc., New York (1991), ISBN 0-486-66721-9, reprint of the 1963 edition.

[44] A. A. Ungar, Gyrovector spaces and their differential geometry, *Nonlinear Funct. Anal. Appl.* **10**, 5, pp. 791–834 (2005).

[45] J. A. Rhodes and M. D. Semon, Relativistic velocity space, Wigner rotation, and thomas precession, *Amer. J. Phys.* **72**, 7, pp. 943–960 (2004).

[46] A. A. Ungar, Hyperbolic trigonometry in the Einstein relativistic velocity model of hyperbolic geometry, *Comput. Math. Appl.* **40**, 2–3, pp. 313–332 (2000).

[47] A. A. Ungar, Hyperbolic trigonometry and its application in the Poincaré ball model of hyperbolic geometry, *Comput. Math. Appl.* **41**, 1/2, pp. 135–147 (2001).

[48] A. A. Ungar, From Möbius to gyrogroups, *Amer. Math. Monthly* **115**, 2, pp. 138–144 (2008).

[49] L. V. Ahlfors, *Conformal invariants: topics in geometric function theory.* McGraw-Hill Book Co., New York (1973), mcGraw-Hill Series in Higher Mathematics.

[50] S. G. Krantz, *Complex analysis: the geometric viewpoint.* Mathematical Association of America, Washington, DC (1990), ISBN 0-88385-026-5.

[51] T. Needham, *Visual complex analysis.* The Clarendon Press Oxford University Press, New York (1997), ISBN 0-19-853447-7.

[52] A. A. Ungar, Seeing the möbius disc-transformation like never before, *Comput. Math. Appl.* **45**, pp. 805–822 (2003).

[53] J. G. Ratcliffe, *Foundations of hyperbolic manifolds, Graduate Texts in Mathematics*, Vol. 149. Springer-Verlag, New York (1994), ISBN 0-387-94348-X.

[54] A. A. Ungar, Hyperbolic barycentric coordinates, *Aust. J. Math. Anal. Appl.* **6**, 1, pp. 1–35 (2009).

[55] A. A. Ungar, On the study of hyperbolic triangles and circles by hyperbolic barycentric coordinates in relativistic hyperbolic geometry, in *Mathematics Without Boundaries: Surveys in Pure and Applied Mathematics.* Springer, New York, pp. 569–649 (2014), eds: Themistocles M. Rassias and Panos Pardalos.

[56] A. A. Ungar, On the origin of the dark matter/energy in the universe and the Pioneer anomaly, *Prog. Phys.* **3**, pp. 24–29 (2008).

[57] A. A. Ungar, When relativistic mass meets hyperbolic geometry, *Commun. Math. Anal.* **10**, 1, pp. 30–56 (2011).

[58] N. Altshiller-Court, *Modern pure solid geometry*, Second edition. Chelsea Publishing Co., New York (1964).

[59] A. A. Ungar, The hyperbolic triangle defect, in *Geometry, integrability and quantization.* Softex, Sofia, pp. 225–236 (2004).

[60] E. W. Hobson, *A treatise on plane and advanced trigonometry.* Dover Publications Inc., New York (1957), 7th ed.

[61] H. S. M. Coxeter and S. L. Greitzer, *Geometry revisited.* Math. Assoc. Amer., New York (1967).

[62] H. A. Lorentz, A. Einstein, H. Minkowski and H. Weyl, *The principle of relativity.* Dover Publications Inc., New York, N. Y. (undated), with notes by A. Sommerfeld, Translated by W. Perrett and G. B. Jeffery, A collection of original memoirs on the special and general theory of relativity.

Index

www.ingramcontent.com/pod-product-compliance
Ingram Content Group UK Ltd.
Pitfield, Milton Keynes, MK11 3LW, UK
UKHW020401150926
8624IPUK00004B/279